D1554290

TACTICS OF ORGANIC SYNTHESIS

TACTICS OF ORGANIC SYNTHESIS

TSE-LOK HO
National Chiao Tung University
Taiwan, ROC

A Wiley-Interscience Publication
JOHN WILEY & SONS, INC.
New York • Chichester • Brisbane • Toronto • Singapore

This text is printed on acid-free paper.

Library of Congress Cataloging in Publication Data:

Ho, Tse-Lok.
Tactics of organic synthesis / Tse-Lok Ho.
p. cm.
"A Wiley–Interscience publication."
Includes index.
ISBN 0-471-59896-8 (acid-free)
1. Organic compounds—Synthesis. I. Title.
QD262.H618 1994
547.2—dc20 93-38710
CIP

Printed in the United States of America

10 9 8 7 6 5 4 3 2

FOREWORD

Tactics of Organic Synthesis is the latest in an excellent series of monographs on modern synthetic organic chemistry by Dr. Tse-Lok Ho. It is a valuable companion to Ho's *Carbocycle Construction in Terpene Synthesis* (1988), *Polarity Control For Synthesis* (1991), *Enantioselective Synthesis* (1992), and *Tandem Organic Reaction* (1992). The subject of synthesis is approached from a very contemporary point of view with emphasis on both the general aspects of molecular assembly ("tactics") and the important new reactions and reaction combinations of the last several years.

The science of organic synthesis has grown very rapidly in size over the past four decades to the point where only serious participants can hope to make major contributions. In intellectual depth and intrinsic complexity, there are not many other areas of science as formidable and demanding. It is fortunate, therefore, that Dr. Ho has undertaken to produce a series of important texts on synthesis, including the present work which is certain to be useful to synthetic chemists at all levels.

E. J. Corey

PREFACE

The general who does not understand the variation of tactics will be unable to use his troops effectively.

Sun Tze: *The Art of War*

Chemical synthesis is frequently likened to war in which strategy is laid down and tactics are implemented in deploying troops (applying reactions) for gaining access to the target. Sun Tze's statement which underscores the utmost importance of the tactical aspects in warfare applies equally well to a chemist's struggles toward a set goal. In pursuit of a synthesis plan, a chemist, however thoroughly proficient in strategy formulation (e.g., retrosynthetic analysis) and acquainted with numerous reactions, still needs tactical coordination to smooth the progression, otherwise his/her success will be arduous and unspectacular. Unfortunately, despite its importance and the recognition of the term by many synthetic chemists, "tactics" are rarely emphasized in synthesis courses taught in a university.

Perhaps the neglect of tactics is much less consequential to the execution of a synthesis in comparison with the lack of an overall plan, yet a chemist can hardly afford to spend effort in obtaining inferior results. An ordinary student not previously exposed to explicit discussion on tactics of chemical synthesis cannot be expected to appreciate its profound role in the dispersed, concealed and varied forms. To this author a course of organic synthesis is incomplete without introducing tactical maneuverability as a unit, thus I am motivated to contribute the present volume with the hope that both teachers and students of organic synthesis will be encouraged to pay due attention to synthetic tactics.

This book should be used in conjunction with other texts that deal with design (e.g., retrosynthetic analysis [Corey, 1989a]) and the reader is supposed to possess a fundamental understanding of synthesis principles and many reactions. Again, because of space consideration, references are selected from recent literature reports and those of the more basic aspects or previous work are largely omitted. It is possible for the more assiduous and acquisitive readers to track them down from those listed. Of course, it is impossible to include even a fraction of the relevant literature, and I can only state that the more interesting or important examples have been included, while being limited by my knowledge and memory.

The demarcation between certain tactics and strategies is difficult to make. For example, convergency and reiterative processes may be more properly considered as strategies, but for the sake of completeness such topics and some others are also discussed in this book. Since the exact partition point of strategy and tactic is always arguable, controversy on the definition is unavoidable. In fact, some maneuvers I consider as tactical have frequently been stated in the chemical literature as strategies. However, I earnestly hope my judgments of the inclusion or exclusion of topics do not seriously detract from the importance of the whole. Furthermore, I must make excuse for placing closely related tactics under two different subject headings, but most of the time such placements are not arbitrary.

I am most grateful to Professor E.J. Corey for his kind statement in the Foreword and to the Li-Ching Foundation for a grant to support my writing endeavor. The unflagging sustenance from my family has made this effort possible.

Tse-Lok Ho

Taiwan, ROC

CONTENTS

TACTICS OF ORGANIC SYNTHESIS

1

CONVERGENCY AND REITERATIVE PROCESSES

1.1 CONVERGENT SYNTHESIS

In complex synthesis two lines of pursuit are practiced. These are the linear strategy and convergent strategy. In a linear strategy the molecular skeleton is built up more or less piece by piece, on the other hand, in a convergent synthesis [Velluz, 1967; Hendrickson, 1977] small molecular fragments are joined together (affixations) and the necessary assemblages are then made from these larger pieces. Ideally the fragments are of comparable sizes and complexity.

One of the advantages of convergent synthesis over the linear strategy becomes clear on considering a seven-step process leading to a molecule having eight subunits. By assuming a yield of 90% in each step the overall yield from a convergent method is 73% ($=0.9^3 \times 100\%$) as compared with 48% ($=0.9^7 \times 100\%$) from a linear approach. As the number of steps increases the difference in yields widens dramatically. However, one must bear in mind that these estimations are for reference only, and the overall yield is misleading since it is calculated on the basis of one starting material whereas in fact several are used. When dealing with convergent sequences the matter is even more complicated and therefore it will be less meaningful.

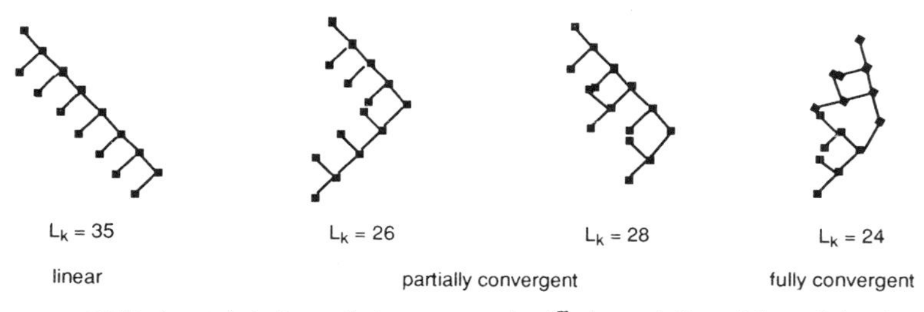

smaller L_k ranks better; perfect convergency: $k = 2^m$ where m is the mainline path length

The qualitative basis for convergent economy is related to the fact that a reaction of an intermediate does not involve all synthons in the desired fashion, thus uninvolved synthons comprising the intermediate are subjected to needless waste from yield loss. A significant relationship of convergency to synthesis design [Hendrickson, 1977] is that it defines whole bondsets and their construction order; only a comparatively small number of bondsets satisfy full convergency. Bondset is a set of bonds in the target skeleton to be constructed in a synthesis; the bondset defines the synthons.

Direct comparison of the two approaches is difficult, but enough evidence has been accumulated in the domain of polypeptide synthesis. However, it must be emphasized that in the event of nearly quantitative yield for each operation the difference in yields also becomes much smaller. The development of solid phase peptide synthesis [Merrifield, 1964] has changed the perspective to a certain degree because many technical problems associated with the linear synthesis have been resolved, and protection, deprotection, activation, and coupling steps all proceed in high yields. The most significant aspect of the solid phase synthesis is the anchoring and hence insolubilization of the product which permits convenient purification by washing away all side-products and excess reagents. This simplification also enables the operator to use excess reagents to drive each reaction to completion without worrying about product purification. Furthermore, the solid phase synthesis is amenable to automation, a powerful technique demonstrated in a synthesis of ribonuclease [Gutte, 1969] which is an enzyme containing 124 amino acid residues. There are 369 chemical reaction and 11,931 operation. The method is certainly not infallible due to some miscouplings, but the product is bioactive (13–24% activity).

Dendritic molecules are particularly amenable to elaboration by reiterative and convergent approaches [Mekelburger, 1992]. Thus, a convergent method involves construction of large reactive fragments and their linkage to a core species. The former task is accomplished by union of several copies of a reactive species (A) with another species (B) containing several coupling sites and one protected functionality to give a product from which a new reactive species (A′) is unveiled through deprotection. Coupling of the latter with (B) to arrive at the second generation product [Hawker, 1990]. The process may be repeated several times. The divergent synthesis that builds up the molecule from the center outwards also relies on reiterative application of reaction sequence in each level.

A few synthetic examples of more conventional organic molecules which clearly show advantages of the convergent approach are mentioned here. For straight chain compounds such as lipoxin-A_4 the dissection into three building blocks [Nicolaou, 1985] allows, in addition to labor division, the use of chiral precursors in its elaboration. Thus, a Wittig reaction furnished a chiral dienyne which was coupled to a vinylic bromide under Pd(0)/Cu(I) catalysis. The product already contained a full skeleton of the lipoxin so that only semihydrogenation and deprotection of the functional groups remained to be carried out.

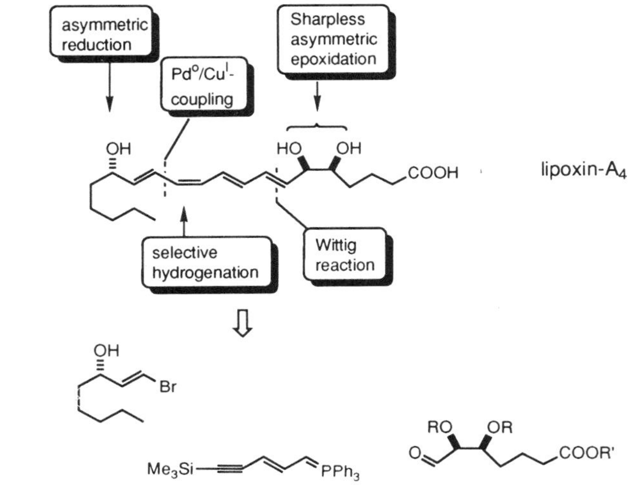

asymmetric reduction
Pd^0/Cu^I-coupling
Sharpless asymmetric epoxidation
selective hydrogenation
Wittig reaction
lipoxin-A_4

The linear triquinane sesquiterpene hirsutene is amenable to convergent assembly, and a particularly interesting route is one involving twofold Michael addition to assemble a B-seco intermediate [Ramig, 1992].

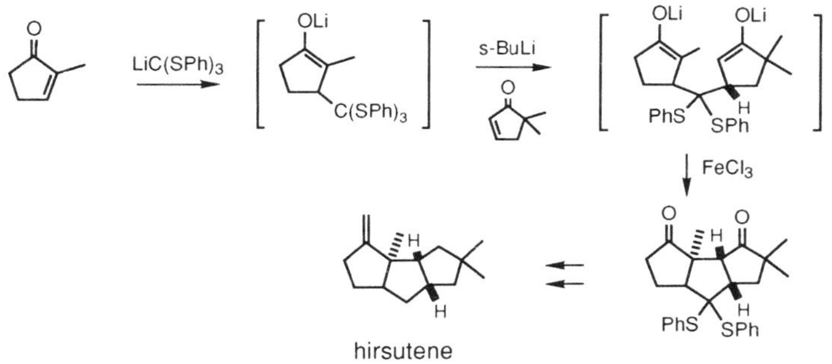

hirsutene

The long-standing interest in the efficient synthesis of steroids has prompted several different approaches, among which the Torgov route [Ananchenko, 1963] can be regarded as a classic.

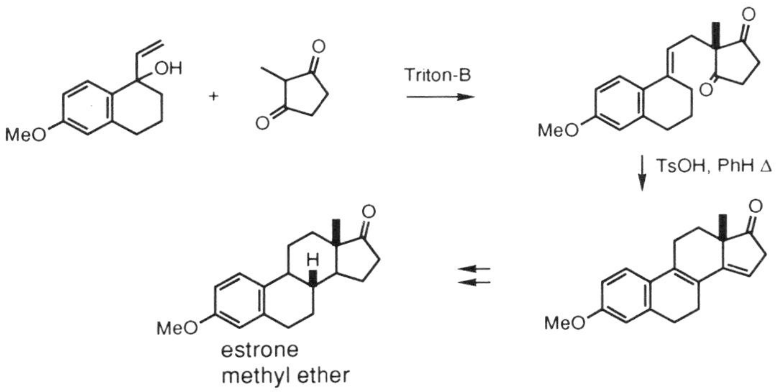

estrone methyl ether

The beauty of establishing a tetracyclic precursor of steroids by the remarkable biomimetic cyclization of a polyene is further enhanced by a convergent elaboration of the substrate [W.S. Johnson, 1970a].

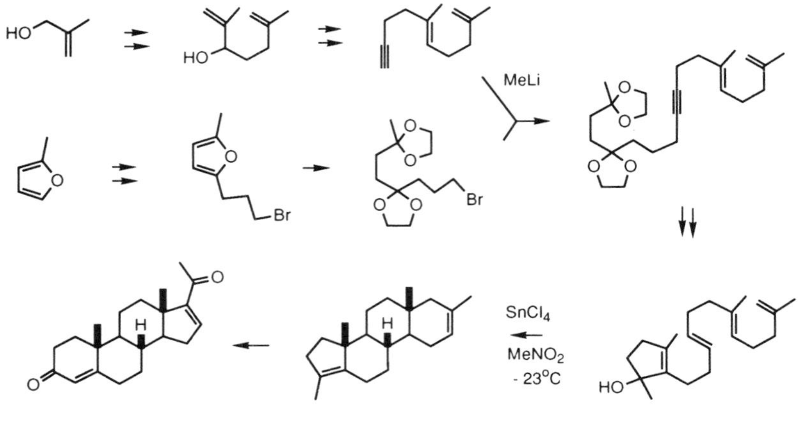

16,17-dehydroprogesterone

An economic route to steroids consists of condensation of a CD-ring synthon with 1,7-octadien-3-one [Tsuji, 1979]. The dienone is a readily obtainable telomer of butadiene, providing C-1 to C-7 and C-10 of 19-norsteroids. Serving in the same capacity are 2-methyl-6-vinylpyridine [Danishefsky, 1975] and substituted isoxazoles [Stork, 1967; J.W. Scott, 1972a,b]. Emphasis should be made that many previous steroid syntheses focused on annulation one ring at a time, for example by Robinson annulation.

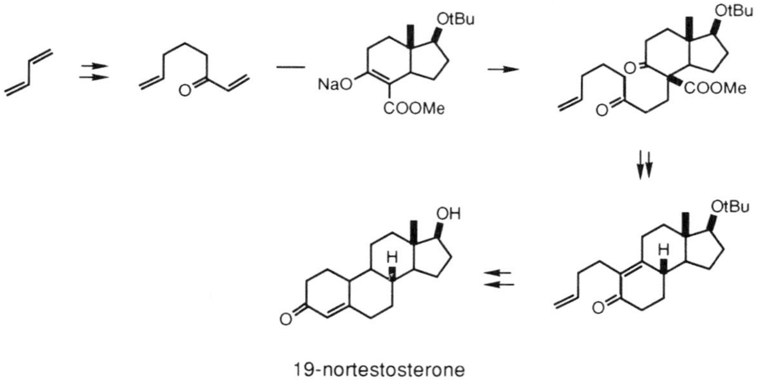

19-nortestosterone

Triply convergent routes to several intriguing natural products have been reported. Thus two eight-carbon units, in the roles of Michael donor and acceptor, respectively, were attached to a five-membered ring by means of *trans*-selective conjugate addition/enolate trapping tandem. These units constituted the sidechain, and large portions of the six- and seven-membered rings of gascardic acid [Boeckman, 1979]. A similar type of assemblage is seen in the elaboration of equilenin [Posner, 1978], estrone [Posner, 1981] and 1-aryltetralin lignans such as galactin [Mpango, 1980].

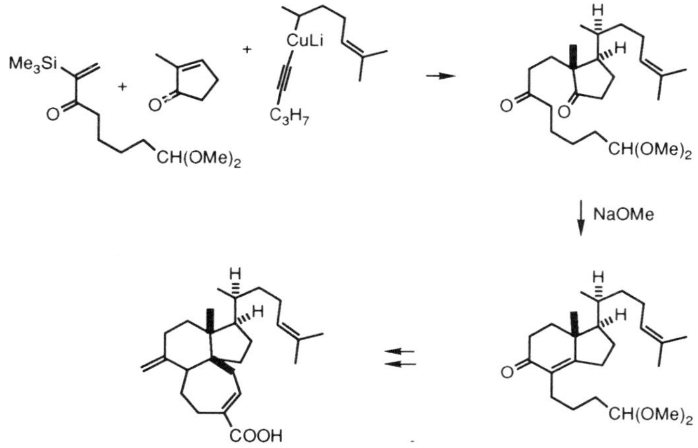

Me_3Si + O + CuLi C_3H_7 → H H O O $CH(OMe)_2$

$CH(OMe)_2$

NaOMe

H H COOH ⇇ H H O $CH(OMe)_2$

gascardic acid

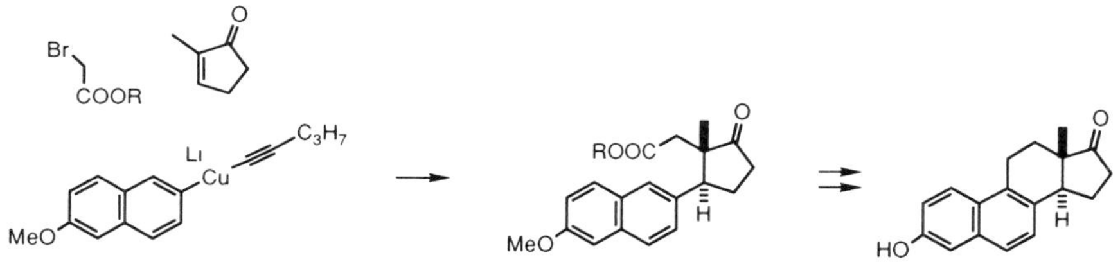

Br COOR O

Li Cu C_3H_7 MeO → ROOC O H MeO ⇉ O H HO

equilenin

S S Li O O $CONMe_2$ OHC OMe OMe → S S HO $CONMe_2$ OMe OMe

⇊

H OMe OMe

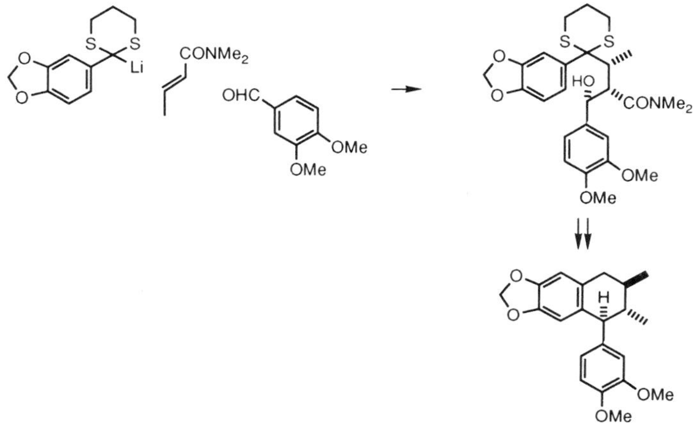

galactin

Owing to profound physiological activities and scarcity of the prostaglandins, the development of efficient synthetic routes for such substances has commanded much attention from chemists. An early success which led to PGE_1 [Corey, 1968b] was based on a convergent assemblage plan comprising Michael and Wittig reaction for affixing the skeletal elements, which was followed by an intramolecular aldol reaction and refunctionalization steps.

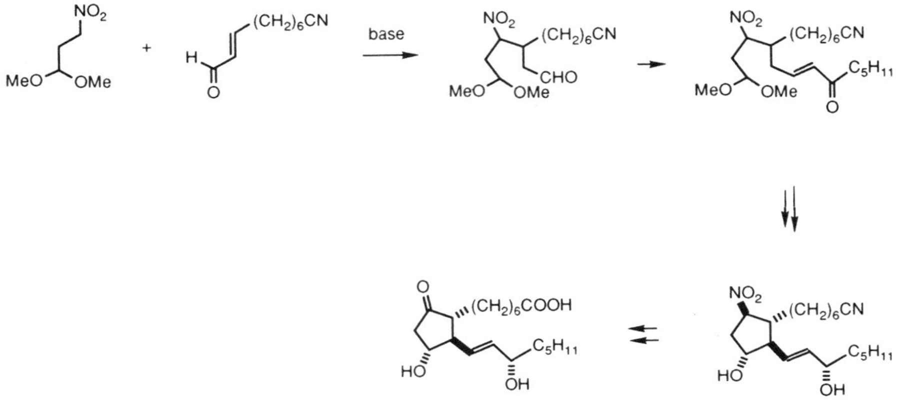

prostaglandin-$E_{1\alpha}$

The one-pot synthesis of prostaglandin precursors from three components: a protected 4-hydroxy-2-cyclopentenone, a cuprate reagent containing the ω-chain, and an alkylating agent which provides the α-chain [Noyori, 1990], is of significant implications. Only the removal of protecting groups remains to be performed after this operation.

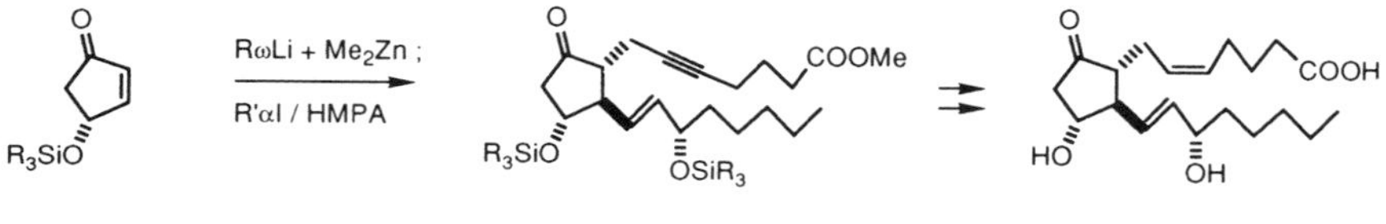

prostaglandin-$E_{2\alpha}$

Other variations have also been explored. These methods include the use of an unsaturated nitro compound as trapping agent (α-chain) [T. Tanaka, 1983] and an oxime ether instead of the ketone group of the cyclopentenone derivative [Corey, 1986].

A pentakisnor intermediate for prostaglandin-E_2 has been acquired by a Pd-promoted three-component, one-step coupling [Larock, 1991].

OH
R_3SiO
$Pd(OAc)_2$
OEt
OEt
PdX
O
R_3SiO
OEt
O
PdX
R_3SiO
COC_5H_{11}
NaOAc
(NaI)
OEt
O
PdX
C_5H_{11}
O
R_3SiO
OEt
O
C_5H_{11}
O
R_3SiO

An α,β-unsaturated sulfone can act in the same manner as the enone [R.E. Donaldson, 1983] in the elaboration of prostaglandins although a different set of transformations is required in the wake of the conjugate addition/alkylations tandem.

NMe_2
SO_2Ph
R_3SiO
Li
C_5H_{11}
OR
I
COOMe
NMe_2
COOMe
SO_2Ph
C_5H_{11}
R_3SiO
$OSiR_3$
O
COOH
C_5H_{11}
HO
OH

Prostaglandin-$E_{2\alpha}$

Extension of the latter theme successfully achieved the assemblage of the two carbocycles and all the other skeletal carbon atoms of cephalotaxine [Burkholder, 1990].

O
O
O
O
O
Li
Br
SO_2Ph
O
O
O
O
O
O
O
SO_2Ph
H
O
O
O
O
H
N
HO
OMe

cephalotaxine

A cyclization initiated by intramolecular addition of an aryllithium to an α,β-unsaturated sulfone and terminated by alkylation constituted the key step to a morphine synthesis [Toth, 1988]. The precursor was prepared from a phenol and a cyclohexenol, the two building blocks which contain all the skeletal carbon atoms of the alkaloid. Recent syntheses of dihydroisocodeine (i.e. formal synthesis of morphine) [K.A. Parker, 1992a] and lycoramine [K.A. Parker, 1992b] are similarly patterned in terms of convergency, although the bond formation steps are of free radical nature.

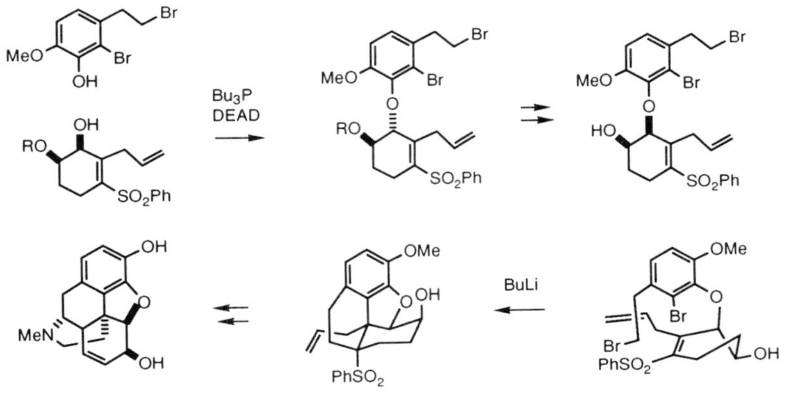

Bu_3P
DEAD

BuLi

morphine

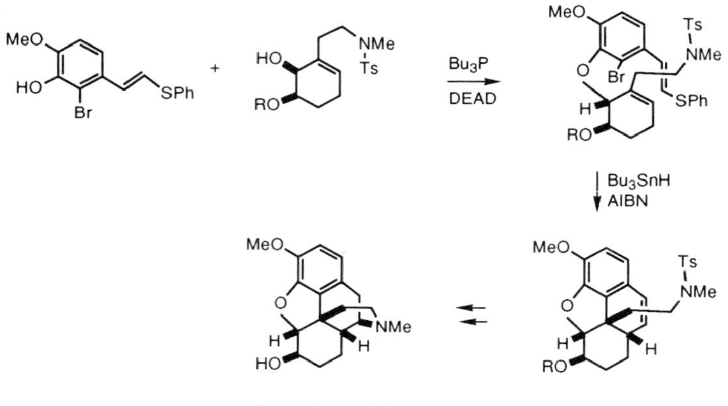

Bu_3P
DEAD

Bu_3SnH
AIBN

dihydroisocodeine

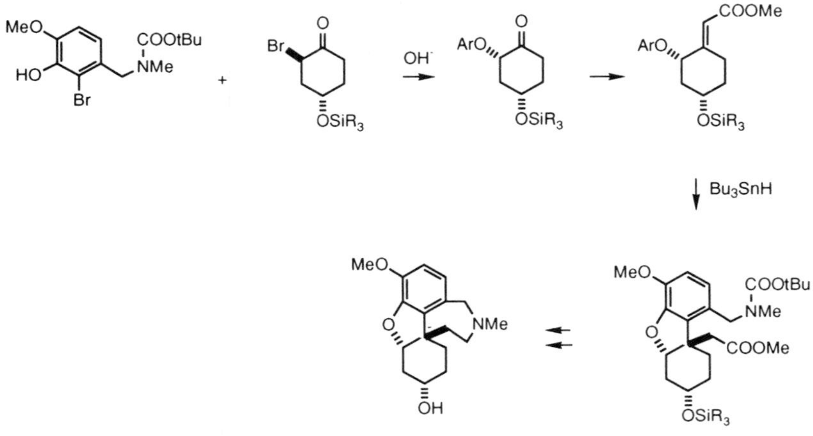

OH^-

Bu_3SnH

lycoramine

The indole alkaloid aspidofractinine is a truly challenging target for synthesis. A concise convergent elaboration [Cartier, 1989] was based on the union of two building blocks in a Mannich reaction which was followed by lactamization to form a tetracyclic product in one operation. The hexacyclic skeleton was formed in another tandem process involving aldol and Mannich condensations.

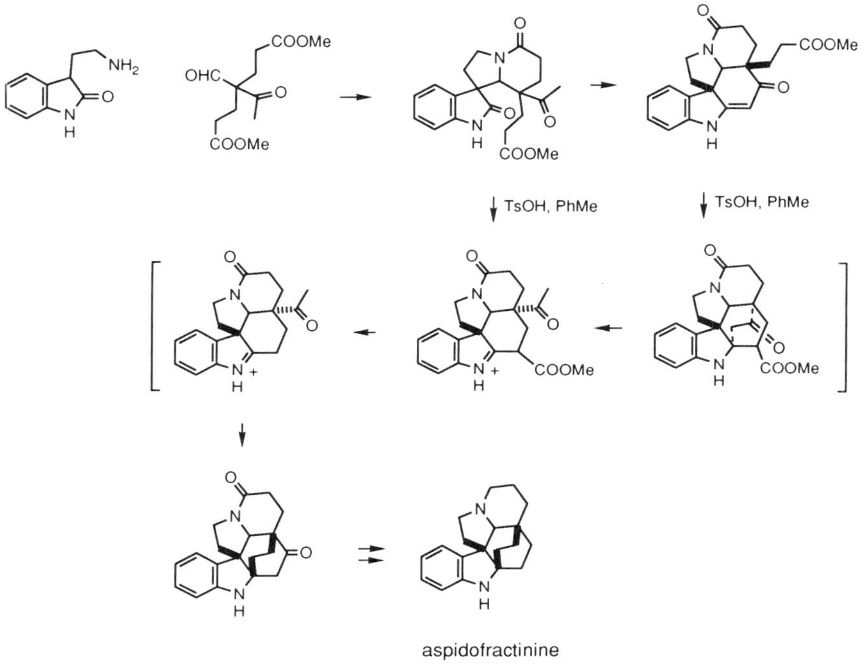

TsOH, PhMe

TsOH, PhMe

aspidofractinine

(−)-Homaline has been synthesized from putrescine and β-phenyl-β-alanine [Wasserman, 1983a]. The formation of the eight-membered heterocyclic units was accomplished by intramolecular transamidation.

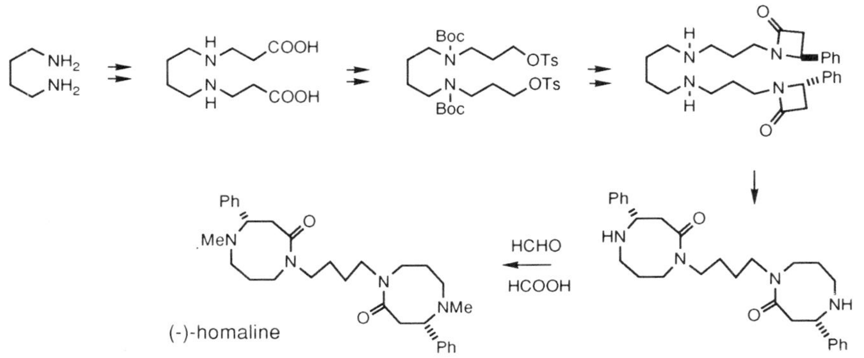

HCHO

HCOOH

(-)-homaline

Comparison of the convergent routes to several alkaloids such as thebaine [Schwartz, 1975], (+)-galanthamine. [Tomioka, 1977], colchicine [A.I. Scott, 1965; Kotani, 1974; Evans, 1981b; Wenkert, 1989], and a potential precursor of lysergic acid [Julia, 1969], with the other methods is instructive. In general, the convergent syntheses are shorter and higher yielding.

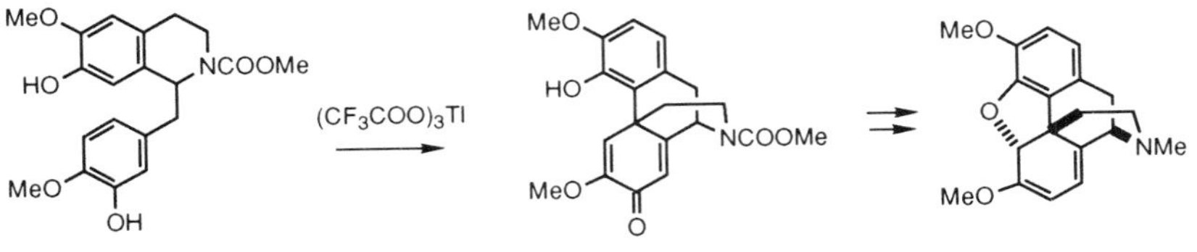

thebaine

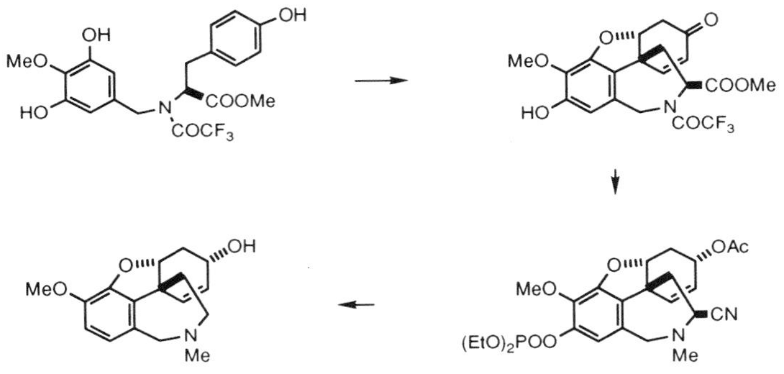

(+)-galanthamine

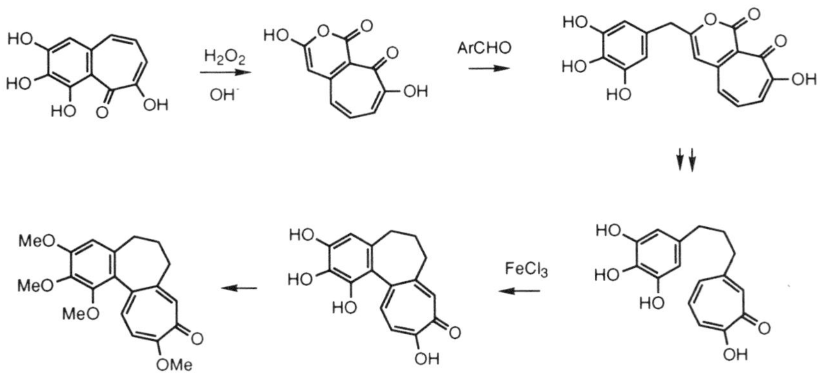

desacetamidocolchicine

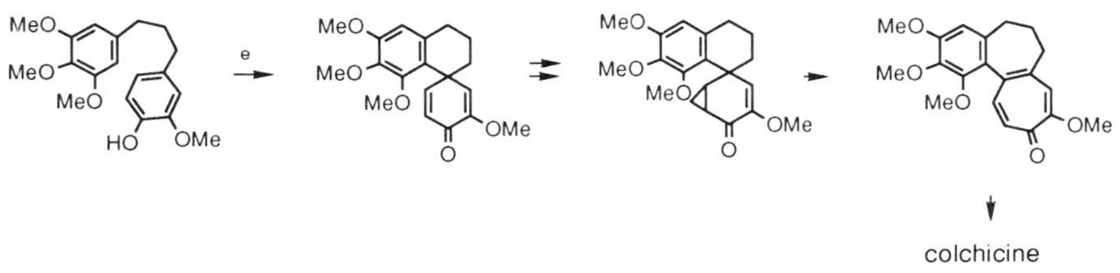

e

colchicine

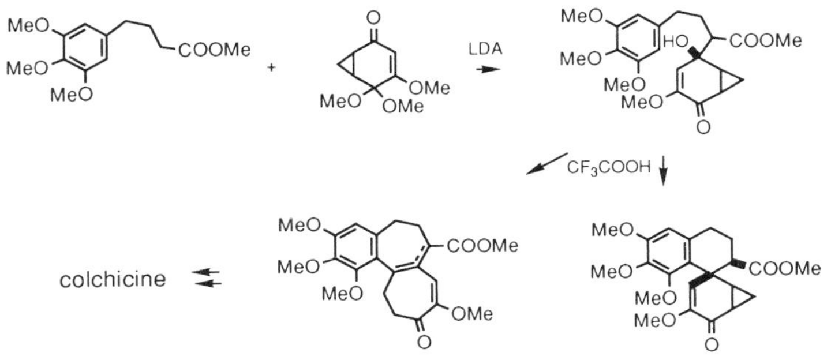

LDA

CF_3COOH

colchicine

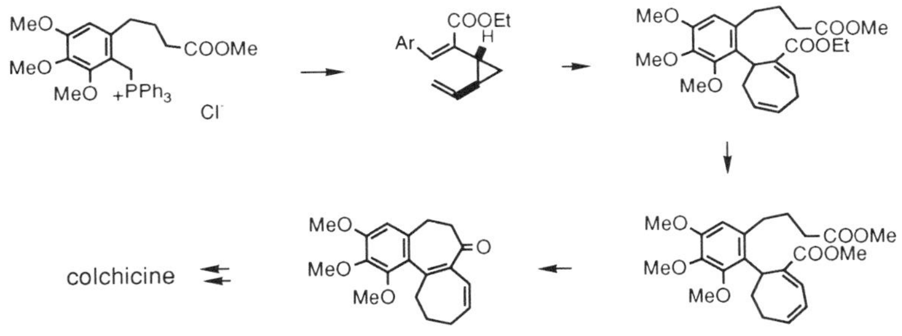

colchicine

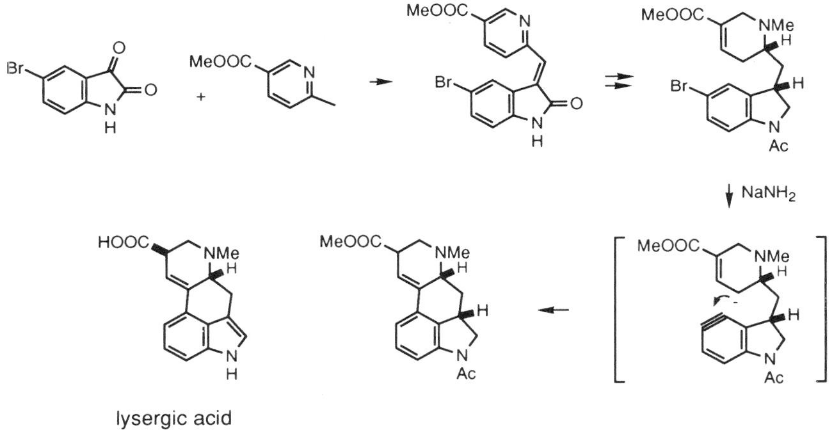

$NaNH_2$

lysergic acid

Molecules such as asteltoxin are clearly divisible into several structural units. Consequently their syntheses can be advantageously undertaken in the convergent sense [S.L.Schreiber, 1984b].

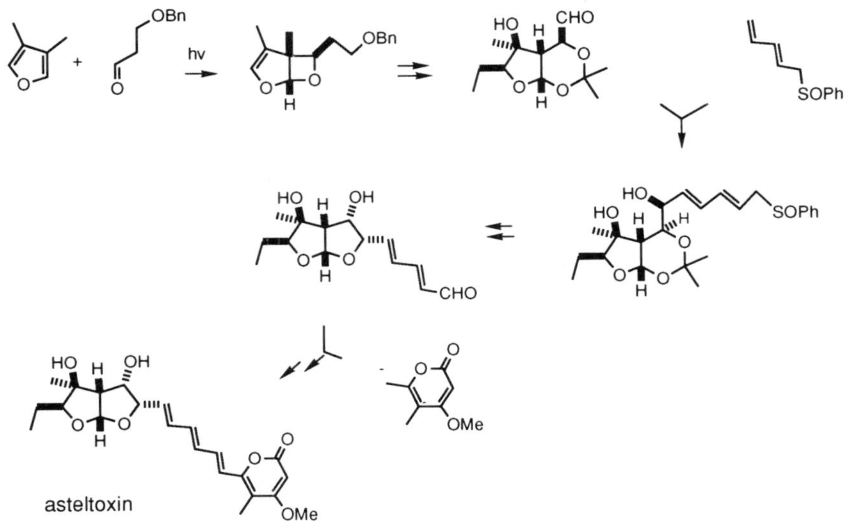

asteltoxin

Many of the major contributions of R.B. Woodward in the latter part of his career adopted convergent approaches: reserpine [Woodward, 1958], chlorophyll-*a* [Woodward 1960, 1961], vitamin B_{12} [Woodward 1973a], marasmic acid [Greenlee, 1976].

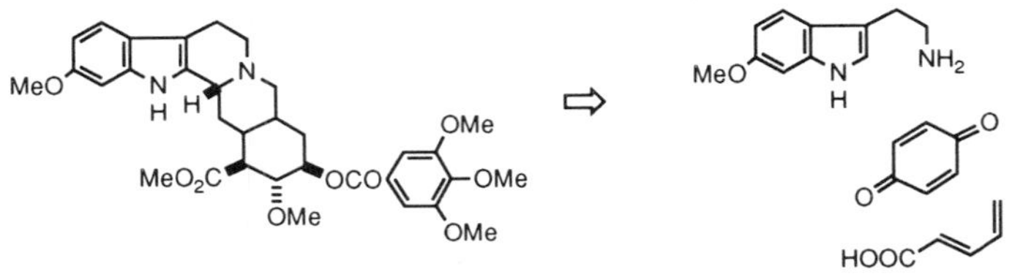

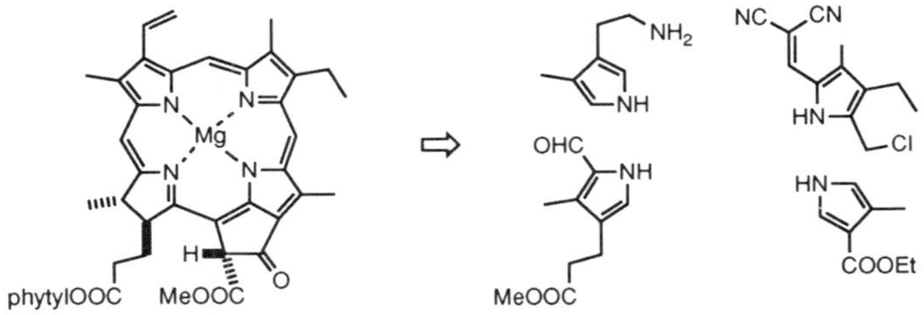

reserpine

chlorophyll-*a*

marasmic acid

Practically all syntheses of complex carbogenic molecules in recent times adopt convergent approaches. It would be prohibitively difficult to prepare compounds such as palytoxin and halichondrin-B by linear assemblage. A scheme for the elaboration of halichondrin-B [Aicher, 1992] is delineated below.

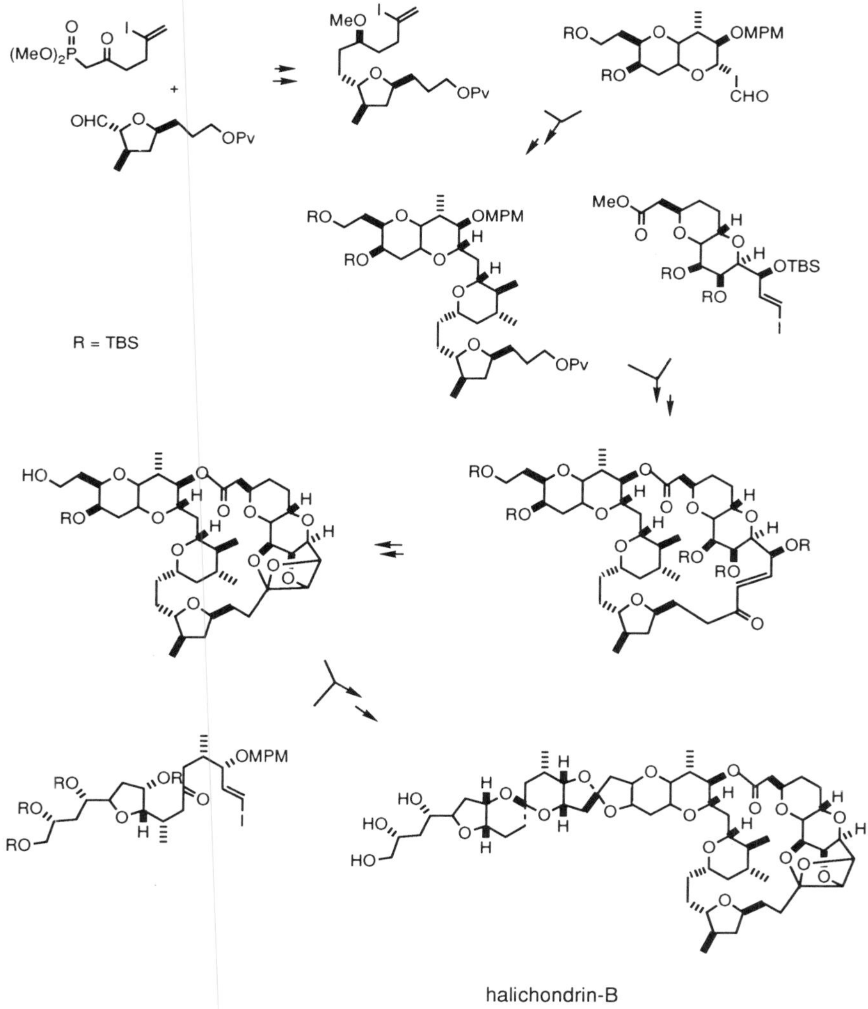

R = TBS

halichondrin-B

For a further examination of some other molecules synthesized in the convergent manner the following reports may be consulted: 2,3-oxidosqualene [Corey, 1993a], phyllanthocin [S.F. Martin, 1987], sordaricin methyl ester [N. Kato, 1993], bertyadionol [A.B. Smith, 1986], retigeranic acid-A [Wright, 1988; Hudlicky, 1989], recifeiolide [Trost, 1980], compactin [Hirama, 1982], miroestrol [Corey, 1993b], zearalenone [Kalivretenos, 1991; Solladie, 1991], erythronolide-A precursor [Stork, 1982], jasplakinolide [Grieco, 1988], lasalocid-A [Ireland, 1980], monensin [Ireland, 1993], indanomycin [Edwards, 1983, 1984], maysine [Meyers, 1983], hygromycin-A [Chida, 1991], aklavinone [Boeckman, 1983], resistomycin [Keay, 1982], aspidospermidine [Laronze, 1974], and ancistrocladine [Bringmann, 1986].

Finally, the availability of enantiomeric and/or diastereomeric compounds from a single precursor by biotransformations has greatly facilitated the synthesis of complex chiral molecules. However, this situation also presents a challenge for the chemist to devise the utility of the "unwanted" isomer(s). In this respect it is instructive to examine an approach to the prostaglandin $PGF_{2\alpha}$ [J. Davies, 1981] from bicyclo[3.2.0]hept-2-en-6-one through reduction with baker's yeast. The two diasteromeric alcohols were processed in different ways to converge into the target compound.

yeast

(enantiomeric)

prostaglandin-$F_{2\alpha}$

1.2. REITERATIVE PROCESSES

The use of the same reaction or reaction sequence more than once in a synthesis is highly regarded on the ground of economics (same reagents/solvents used) and technical effectiveness (or experience). Although for synthesis of molecules such as polypeptides or oligonucleotides which contain repeating subunits, the approach is unexceptional, when the target molecules are devoid of apparently uniform or congeneric fragments, the design and application of reiterative methodologies in their synthesis would convey a sense of surprise and elegance.

In recent years much attention has been paid to the controlled union of monosaccharide moieties to form oligosaccharides. While in this book the focus is placed on synthesis of conventional organic molecules, one reiterative approach to β-linked oligosaccharides [Halcomb, 1989] is mentioned here. In this route a glycal containing nonparticipating protecting groups is epoxidized with dimethyldioxirane, and then treated with another glycal in the presence of $ZnCl_2$. After proper protection the epoxidation–glycosylation sequence is performed anew.

The Kolbe electrolysis of carboxylic acids is a useful procedure for carbon chain building. When a dicarboxylic acid monoester is employed as one component the coupling product can be reactivated by saponification. The reiterative application of such a procedure was the basis of a tuberculostearic acid synthesis [Linstead, 1950, 1951].

$Me(CH_2)_2COOH$ + $HOOC(CH_2)_4COOMe$ → (- e [Pt/MeOH] ; NaOH) → $Me(CH_2)_6COOH$

- e, HOOC–CH₂–CH(CH₃)–CH₂–COOMe ↓

$Me(CH_2)_7$–CH(CH₃)–CH₂–COOH

- e, $HOOC(CH_2)_7COOMe$ ↓

$Me(CH_2)_7$–CH(CH₃)–CH₂–$(CH_2)_7COOH$

tuberculostearic acid

Many disubstituted benzenes including biphenyl and terphenyl derivatives are not readily available. Thus, the synthesis from a dichlorobenzene [Tiecco, 1982] by displacement with an isopropylthio group, conversion of the remaining chlorine and then the thio group to carbon substituents by nickel-catalyzed cross-coupling reactions with two different Grignard reagents, is a valuable technique.

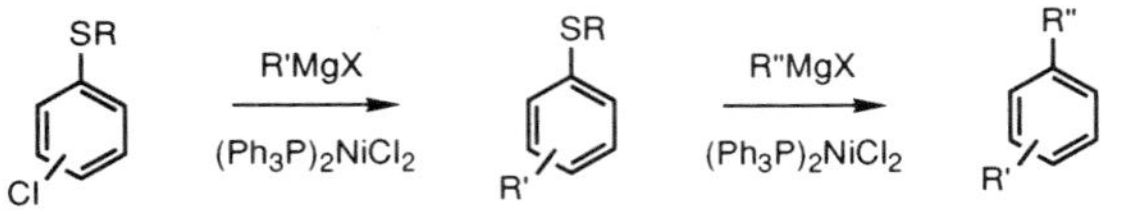

A method for stereoselective synthesis of di- and trisubstituted olefins by using an organonickel reagent to replace a vinylic oxygen functionality [Wenkert, 1984a] is amenable to assembly of insect pheromones with several (*Z*)-double bonds [Ducoux, 1990] by judiciously choosing dihydropyran and dihydrofuran as substrates in a reiterative fashion. Similarly the fragments of C-8 to C-20 of premonensin-B has also been acquired [Kocienski, 1988].

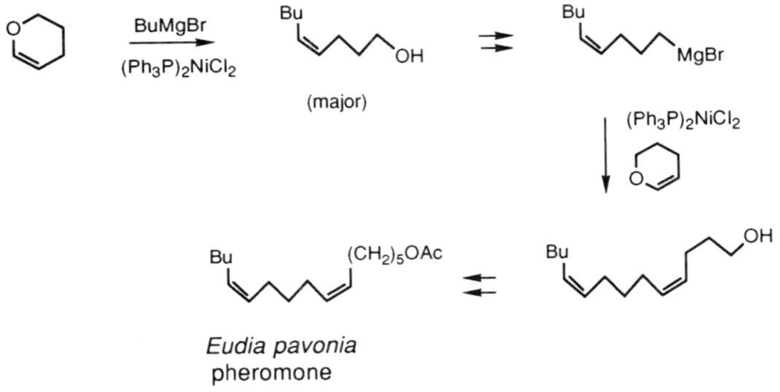

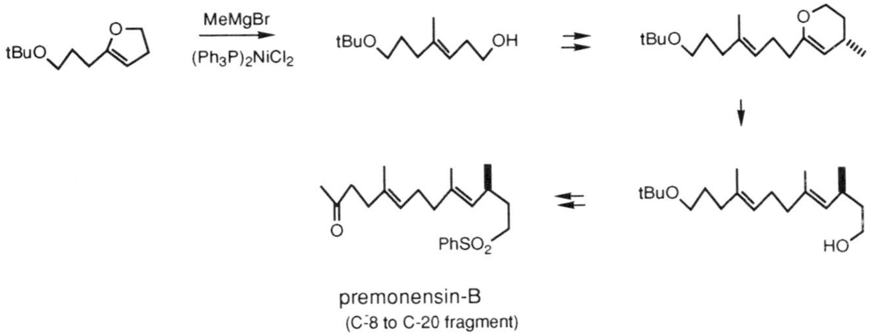

Eudia pavonia pheromone

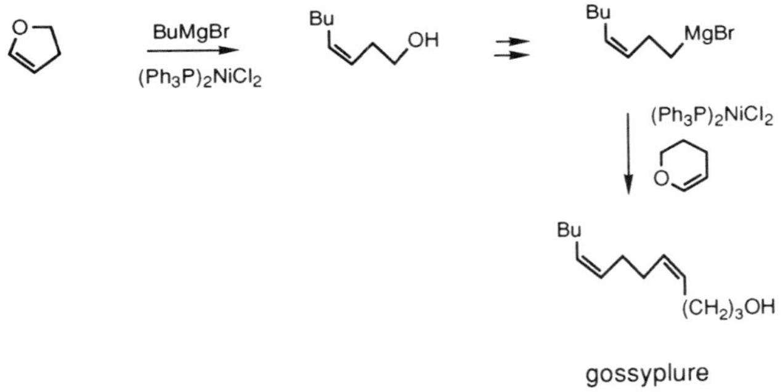

gossyplure

premonensin-B
(C-8 to C-20 fragment)

Diketones separated by polyene units can be prepared by reiterative desilylative acylation [Babudri, 1991] of 1,ω-polyenes using two different acyl chlorides, when the first reaction is carried out at 0°C.

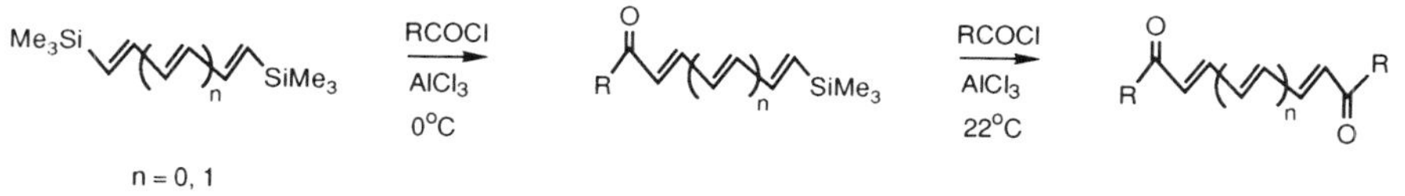

Me_3Si … $SiMe_3$ (n = 0, 1) → RCOCl, $AlCl_3$, 0°C → … → RCOCl, $AlCl_3$, 22°C → …

Of practical importance is a process of nerolidol/farnesol synthesis [Nazarov, 1958] using a combination of Carroll reaction, acetylide addition, and semihydrogenation. From 2-methyl-3-buten-2-ol (itself obtainable from acetone by acetylide addition and semihydrogenation) the first cycle gives linalool.

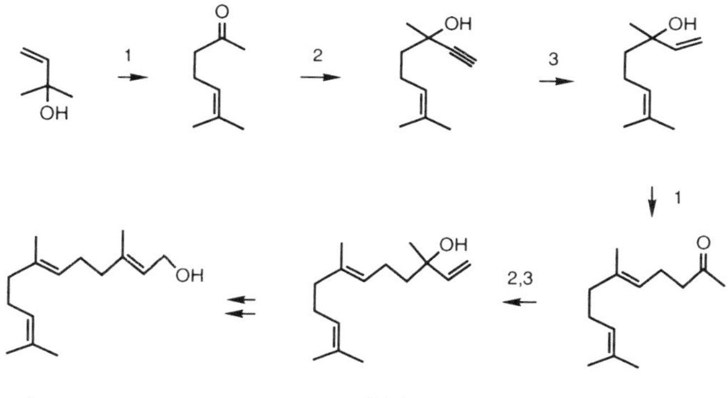

farnesol nerolidol

(1) $MeCOCH_2COOEt$, $(iPrO)_3Al$; (2) HC≡CH, KOH; (3) H_2, Pd-$SrCO_3$

An efficient route to the *cecropia* juvenile hormone JH-I [W.S. Johnson, 1970] consists of two series of reduction (to acquire an allylic alcohol) and a Claisen rearrangement.

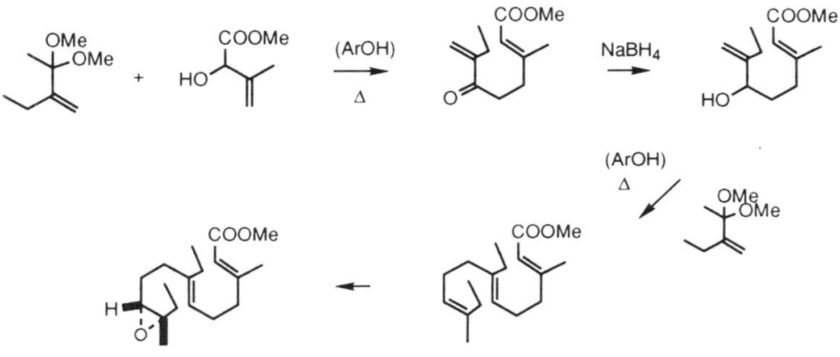

(ArOH) Δ NaBH$_4$ (ArOH) Δ

The Ireland version of the Claisen rearrangement is capable of reiterative manipulation [McKew, 1993]. Excellent enantiomeric and diasteromeric excesses for the synthesis of dienoic acids have been shown.

An excellent method for the elaboration of a trisubstituted olefin involves reductive iodination of a propargyl alcohol and replacement of the iodine atom by an alkyl group based on organocopper chemistry. Reiterative application of this procedure allowed rapid construction of the *cecropia* juvenile hormone JH-I [Corey, 1968a].

OH LiAlH$_4$ / NaOMe; I$_2$ I OH Et$_2$CuLi; EtI OH OH Me$_2$CuLi I OH OH O COOMe JH-I

Another reiterative method is that which consists of conjugate addition of benzenethiol to an ynoic ester and stereoretentive substitution of the benzenethio group by the Kharasch reaction. Accordingly, the precursors of JH-I and JH-II were acquired by repetitive employment of the reaction sequence [Kobayashi, 1974].

OH COOMe PhSNa PhS COOMe EtMgBr CuI COOMe OH COOMe SPh COOMe COOMe JH-I

A synthesis of the once-claimed sex pheromone of codling moth employed iterative construction of the trisubstituted alkene units [Marfat, 1979].

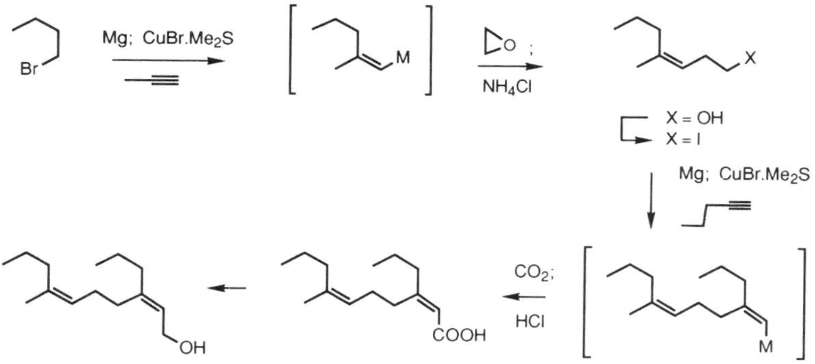

Mg; CuBr.Me$_2$S

NH$_4$Cl

X = OH

X = I

Mg; CuBr.Me$_2$S

CO$_2$;

HCl

COOH

OH

The homoallylic bromide preparation by the Julia method has found use in the synthesis of terpenoid compounds. However, its inherent weakness regarding the lack of configurational control about the double bond becomes apparent when applied to JH-I synthesis [Cochrane, 1972]. This is an extension of the original method for a nerolidol synthesis [Julia, 1960].

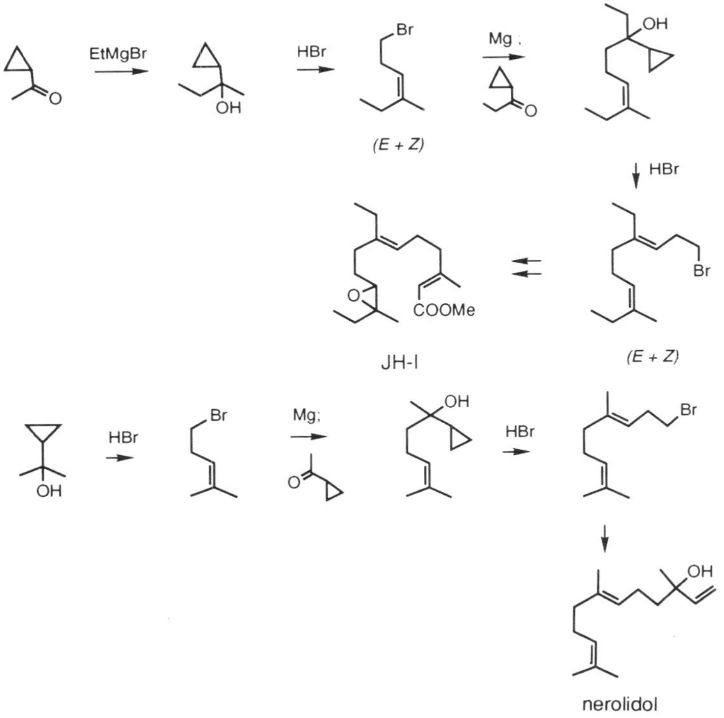

EtMgBr

HBr

Mg ;

(E + Z)

HBr

COOMe

JH-I

(E + Z)

HBr

Mg;

HBr

nerolidol

Talaromycin-B has been constructed from a ketopentaol [S.L. Schreiber, 1983] which relied on spiroketalization to control remote stereochemical relationships. In other words, a more stable isomer emerged from the cyclization which enlisted one of the diastereotopic hydroxymethyl groups. Formation of a 1,3-dioxane unit from two hydroxyl groups of the exposed triol allowed deoxygenative homologation on the remaining one. The precursor of the ketopentaol was readily assembled by two dithiane alkylations using the same allylic chloride; because hydrogenation was required in between these operations, they could not be done in one step.

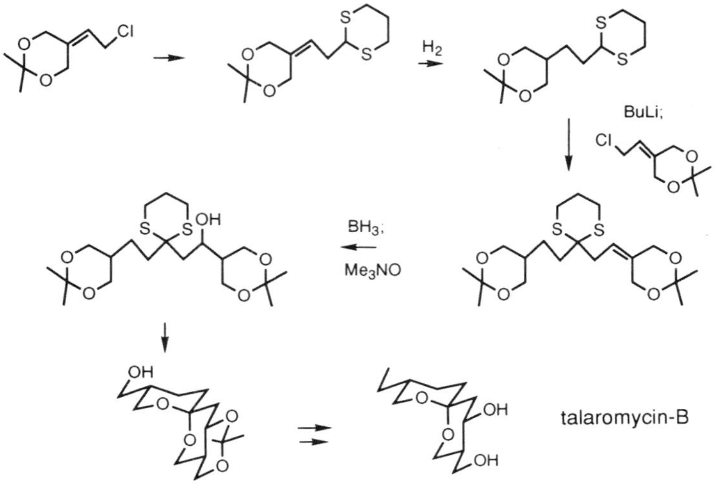

H_2

BuLi;

BH_3;

Me_3NO

talaromycin-B

Reiterative application of a cyclobutanone synthesis by reaction of a carbonyl compound with 1-phenylthiocyclopropyllithium and pinacol-type rearrangement of the product has led to a spirocyclic substance for elaboration of grandisol [Trost, 1977].

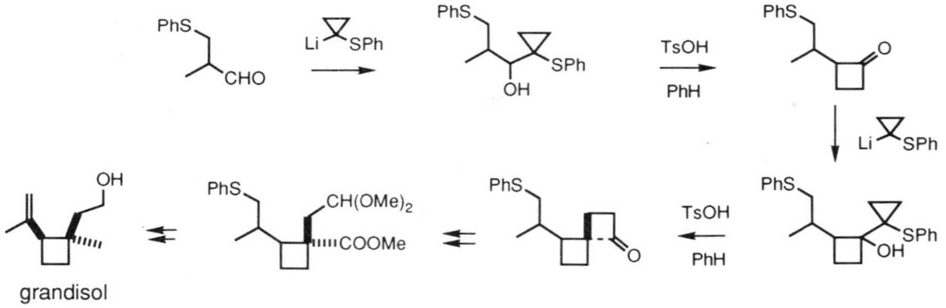

TsOH

PhH

grandisol

A sequence of $LiAlH_4$ reduction of malonate ester, tosylation, and alkylation can be used to construct spirocyclic systems composed of cyclobutane units [Buchta, 1966].

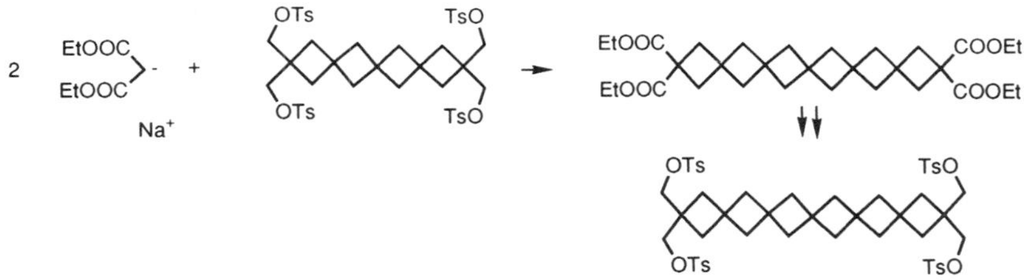

A regenerative annulation procedure that gives rise to a linear dihydroacene skeleton consists of Diels–Alder reaction between a 1,2-dimethylenecyclohexane and allenyl chloromethyl sulfone, and a Ramberg–Bäcklund reaction [Block, 1990].

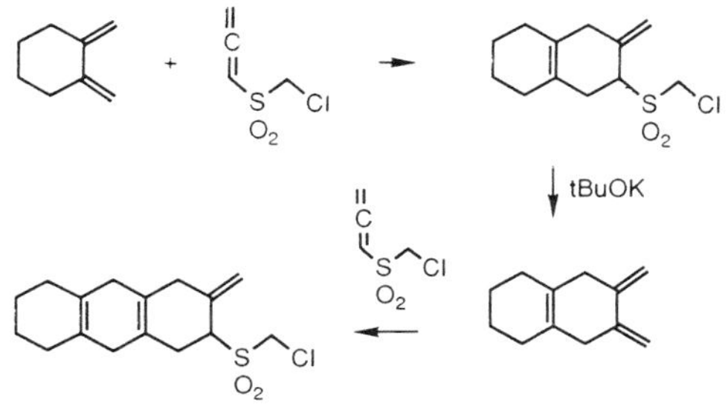

Fascinating molecules have been constructed by stereoregular Diels–Alder oligomerizations [Ashton, 1992]. One example is the [12]cyclacene derivative formed from two molecules each of a bisdiene and a bisdienophile by the same reaction. Three intermolecular Diels–Alder reactions were followed by an intramolecular cycloaddition.

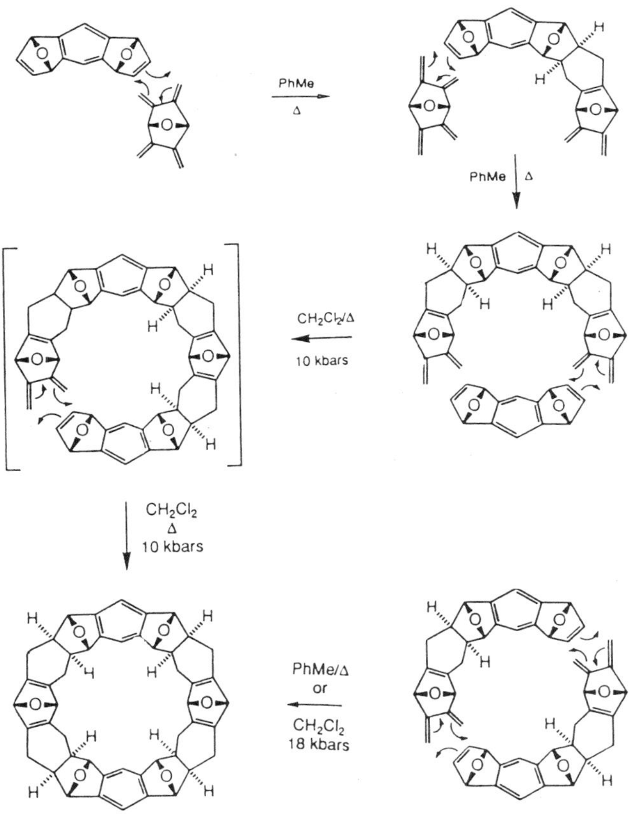

The potential of the Lewis acid-catalyzed diene/aldehyde cyclocondensation is amply demonstrated in its application to a synthesis of 6-deoxyerythronolide-B [Myles, 1990]. Reiterative employment of this methodology is a remarkable feature.

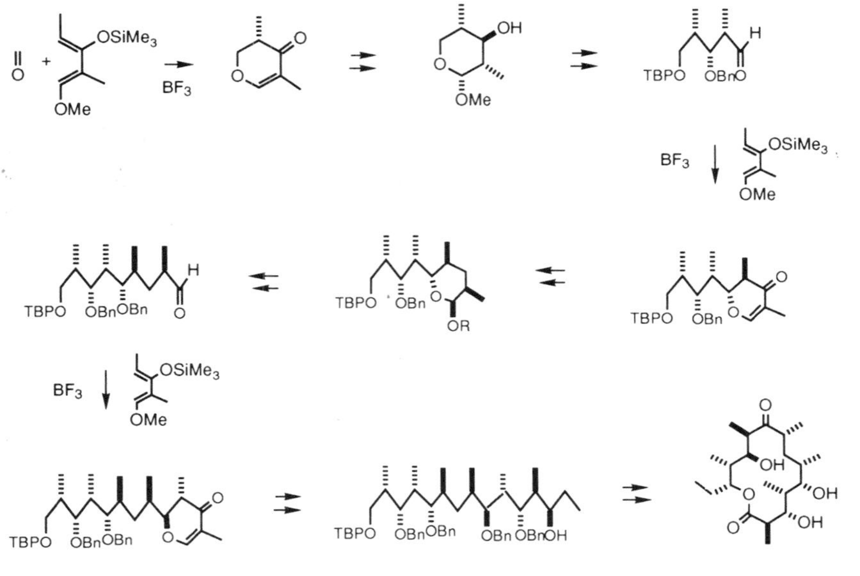

6-deoxyerythronolide

Certain marine toxins contain condensed cyclic ethers of 6-, 7-, and 8-membered rings. Oxepane units bearing vinyl and hydroxyl groups with *trans* relationships at C-2 and C-3, respectively, can be created by intramolecular reaction of γ-stannyl-(Z)-enol ether with aldehyde. Reiterative application of this protocol has been demonstrated [Y. Yamamoto, 1991].

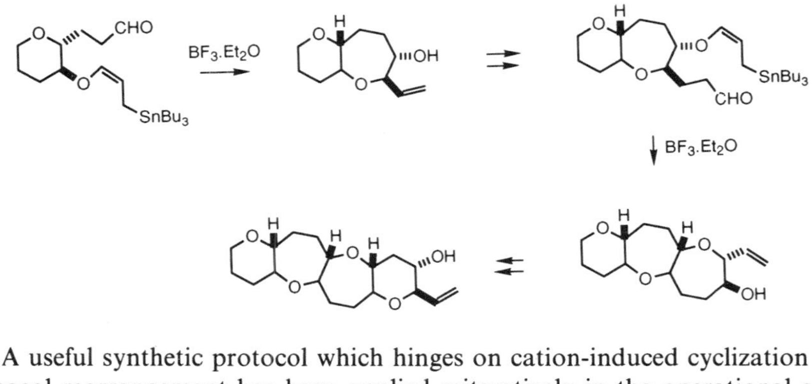

A useful synthetic protocol which hinges on cation-induced cyclization and pinacol rearrangement has been applied reiteratively in the operational sense [G.C. Hirst, 1989]. Formation of two five-membered rings in a fused manner and expansion of the original cycloalkanone by two carbon units (one in each operation) are readily accomplished. Thus this rapid assemblage of the dicyclopenta[*a*,*d*]cyclooctane system may be gainfully employed in an access to the fusicoccins and ophiobolins.

$(MeO)_2CH$; Me_3SiCl imidazole ; R = $SiMe_3$; $SnCl_4$, -78° to -23°; Et_3N ; MeOH ; MeO ; $CH(OMe)_2$; $SnCl_4$, -78° to -23° ; RO ; OMe

Duplicate umpolung at C-4 of 2-cyclohexenone has been achieved through electrophilic transition metal π-complexes. When such a π-complex is regenerated from the first adduct, a second nucleophile can enter. To be synthetically useful the reiterative process must be subjected to regiocontrol. In a synthesis of *O*-methyljoubertiamine [Stephenson, 1993] the first complexed cation was formed by hydride abstraction from tricarbonyl[(1,2,3,4-η-1,4-dimethoxy)-1,3-cyclohexadiene]iron(O) and the cation for the second coupling was made by ionization of the tertiary methoxy group of the adduct.

$(OC)_3Fe$; OMe ; MeO ; $(OC)_3Fe^+$; $Fe(CO)_3$; NC ; ROOC ; Me_2N

O-methyljoubertiamine

The cyclic ether formation method by intramolecular displacement of an alkylthio group of a dithioketal in the presence of silver(I) perchlorate has been applied to the construction of the seven- and eight-membered ring moieties of brevitoxin-A [Nicolaou, 1991b]. Removal of the residual alkylthio unit was accomplished by treatment with a tin hydride reagent, and if an angular methyl group is desired, the transformation can be carried out via oxidation (to the sulfone) and displacement with trimethylaluminum.

A ring expansion technique consisting of [2.3]-sigmatropic rearrangement of sulfonium ylides derived from 2-vinylthiacycloalkanes has the attributes of efficiency and sterocontrol on the cyclic periphery. Moreover, its versatility is increased by reiterative applicability [Vedejs, 1984a]. Thus, a synthesis of methynolide [Vedejs, 1987] based on two such ring expansion processes using slightly different *S*-alkylating agents deserves mention here.

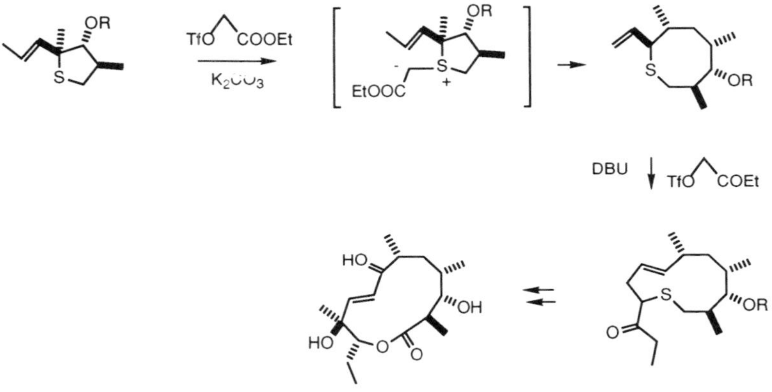

methynolide

Many alkaloids have been prepared via 1,3-dipolar cycloadditions of nitrones with alkenes. Reiterative use of this method has been witnessed in approaches to anatoxin-*a* [Tufariello, 1985] and α-isosparteine [Oinuma, 1983]. In the latter work, reiteration of the cycloaddition was automatic. Hydrogenation of the 2:1 cycloadduct of Δ^1-piperidine-1-oxide and 4*H*-pyran directly furnished the alkaloid.

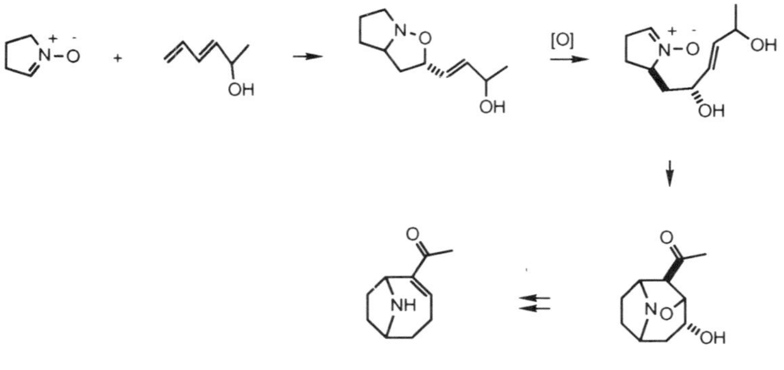

anatoxin-*a*

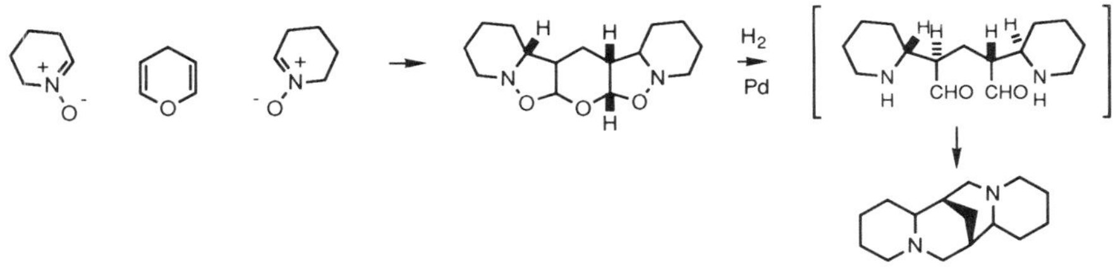

α-isosparteine

Reiterative nitrile oxide cycloaddition to unsaturated hydrocarbons to form isoxazoles and thence *β*-tetraones is illustrated in the following preparation of phenolic compounds [Auricchio, 1974].

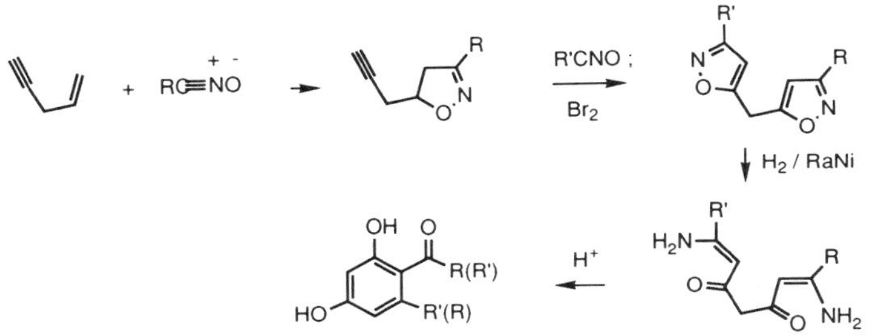

By far the most elegant concept of utilizing nitrile oxide cycloaddition in synthesis is that dealing with the assembly of a seco precursor of cobyric acid [Stevens, 1986]. Both the "northern" and the "southern" halves of the molecule were linked together by formation of 3,5-disubstituted isoxazoles, which in turn were joined by another such cycloaddition.

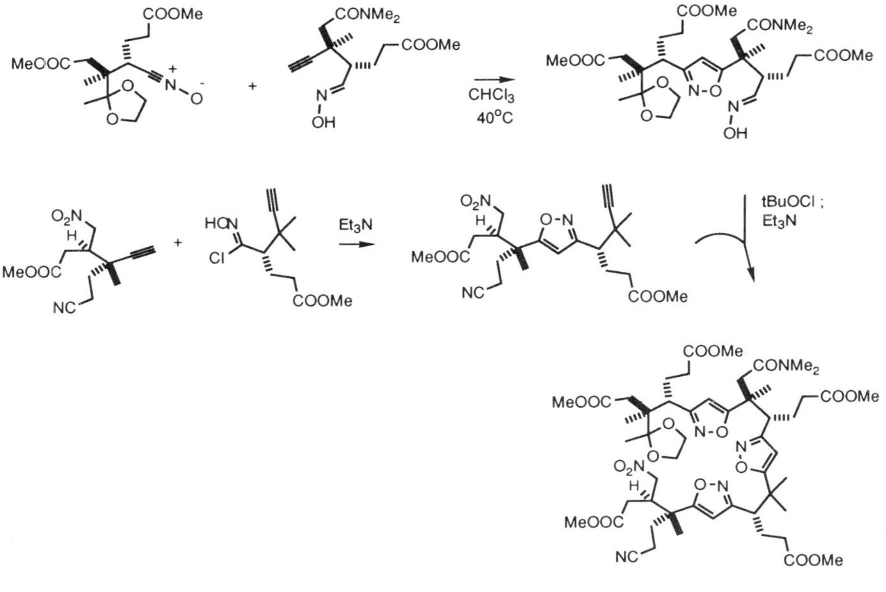

An efficient synthesis of hirsutene [A.E. Greene, 1980] is based on a three-carbon annulation, in duplicate via cycloaddition of a chloro ketene to a cyclopentene and subsequent ring expansion and generation of the double bond from the resulting α-chloroketone. The advantage of a chloro ketene in the cycloaddition step is threefold: its better stability (as compared with the nonchlorinated species), control over the regiochemistry of the ring expansion step, and facilitation of the olefin regeneration.

Cl
C=O
Cl
CH_2N_2
Cl
O
$NaBH_4$;
$Cr(ClO)_4$
Cl_2C=C=O
Cl Cl
O
H
CH_2N_2
Cl Cl
O
H
$NaBH_4$;
$Cr(ClO)_4$
Cl
H
H

hirsutene

One-carbon homologation of lactones via chlorocarbene addition to the derived enol silyl ketals and thermolytic elimination of chlorosilane (with cyclopropane cleavage) can be repeated [Fouque, 1990] after hydrogenation. Such series of reactions are valuable for the preparation of mesocyclic lactones.

The classic Robinson annulation was originally developed for the synthesis of steroids. Two successive Robinson annulations were employed to establish the AB ring system of aldosterone [W.S. Johnson, 1963] in the hydrochrysene approach.

OMe
O
O
O
OMe
O
OH
OH
O
O
H
O

aldosterone

In the acquisition of all-*syn* 1,6;8,4n + 1;10,4n − 1;..-polykismethano[4n + 2]-annulenes the reiterative process involving the Wittig–Horner–Emmons–Wadsworth reaction on dialdehyde intermediates proved advantageous [Wagemann, 1978; Vogel, 1980].

CHO
CHO
+
O
$(EtO)_2P$ COOEt
$(EtO)_2P$ COOEt
O
COOEt
COOEt
CHO
CHO
repeat
COOEt
COOEt

Pyrolysis of ethynyl ketones leads to the formation of vinylcarbenes by 1,2-hydrogen migration. Such a vinylcarbene would undergo intramolecular insertion in the $C^{\beta'}$-H bond to afford the cyclopentenone. An excellent exploitation of this process in a reiterative manner is demonstrated in a synthesis of $\Delta^{9(12)}$-capnellene [Huguet, 1982].

620°C

620°C

$\Delta^{9,12}$-capnellene

The intramolecular metallo-ene reactions for organic synthesis have been exploited in recent years. The magnesium version appeared twice in a synthesis of $\Delta^{9(12)}$-capnellene [Oppolzer, 1982a] as ring-forming steps that also achieved active site transfer. As noted there is variation of the trapping agent.

Mg; 60°C

MgCl

CHO ; $SOCl_2$

Cl

Mg

MgCl

$\Delta^{9(12)}$-capnellene

A reaction sequence with capability to recurring operation consists of conjugate addition of a vinyl group to a 2-carbalkoxy-2-cycloalkenone, α-alkylation of the product with phenyl vinyl sulfoxide, and thermolysis [Bruhn, 1979]. The last reaction involves elimination of phenylsulfenic acid and Cope rearrangement. The ring expansion regenerates the original functional group (and a double bond at a remote site which may be independently manipulated).

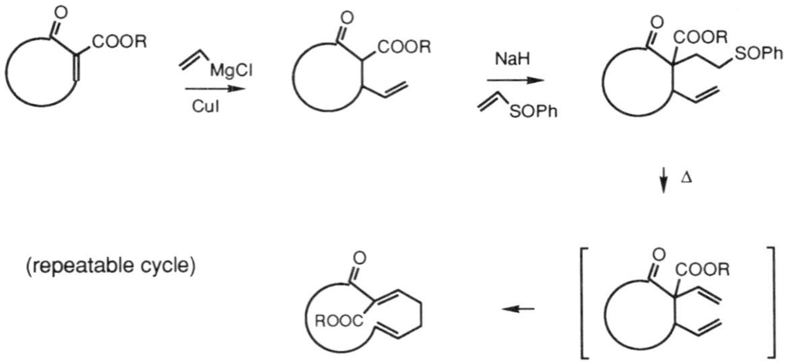

(repeatable cycle)

Reiterative annulation via a three-step sequence of reduction (or RM reaction), ester Claisen rearrangement, and cyclodehydration gives rise to a drill-shaped product in which the cyclopentane rings appear as blades projecting from the molecular axis [Dorsch, 1984]. On the other hand, homologation (+2C) of the Claisen rearrangement product to a terminal alkyne and exposing the latter, which is a 1,6-enyne, to tin hydride results in a stannylmethylenecyclopentane. The reaction sequence of destannylation, oxidative cleavage, introduction of a conjugated double bond, and reduction of the carbonyl group reconstitutes an allylic alcohol unit for reiterative cyclopentannulation [Clive, 1993].

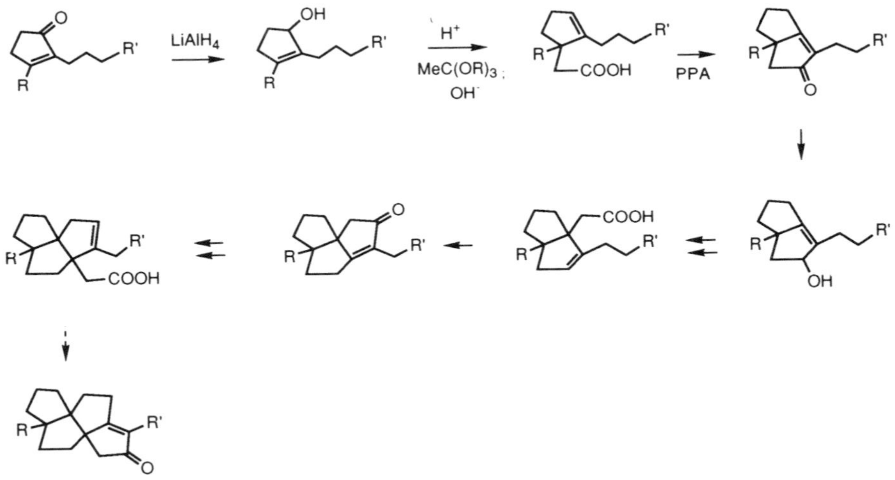

Based on the finding that adducts of α-lithio-α-methoxyallene and ketones undergo cyclization to give 2,2-disubstituted dihydrofuran-3-ones on exposure to potassium *t*-butoxide and then aqueous acid, helical molecules containing spiroannulated tetrahydrofuran units have been synthesized by repetitive application of the reaction sequence [Gange, 1978].

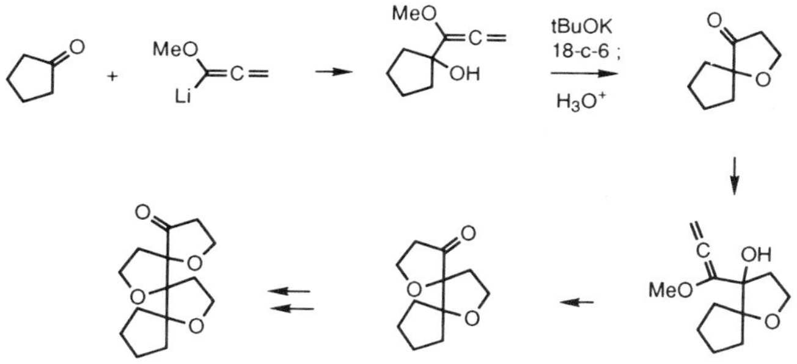

Pyrolidinone-based peptidomimetics possessing a conformation of the β-strand has been prepared by a reiterative procedure [A.B. Smith, 1992].

From an aldehyde it is possible to extend the chain to yield 1,4-dicarbonyl compound in a protected form, and this process is amenable to repetition [T. Sato, 1988]. Thus, the aldehyde is reacted with methoxy(phenylthio)methyllithium and treated with mesyl chloride-triethylamine to the homologous α-phenylthio aldehyde, which is subjected to Wittig or Horner–Emmons–Wadsworth condensation and deconjugation. The vinyl sulfide can be kept while the functional terminus is manipulated until proper time for carbonyl group generation.

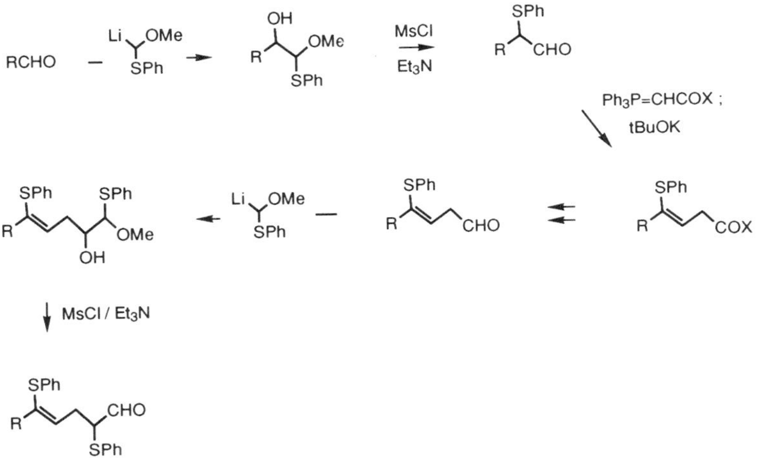

Acyclic stereocontrol of adjacent centers has become a prominent problem in synthesis as related to polypropionate antibiotics. Chelation control in aldol and related condensation reactions is an enormously successful feature that has been exploited, but the development of a general method for the construction of segments incorporating hydroxyl and methyl substituents of defined relative configurations in alternative carbon atoms is highly desirable.

To this end a solution emerged from elaboration of butenolide templates [Stork, 1987], which allows preparation of any of the four diastereomers at will, and its applicability to complex situations has been demonstrated by a synthesis of (+)(9*S*)-dihydroerythronolide-A.

Me_3SiCl; $BnOCH_2COOEt$ base; K_2CO_3

$(PhO)_2POCl$; Me_2Zn / $Ni(acac)_2$

Me_3SiCl; $BnOCH_2COOEt$ base; K_2CO_3

(9*S*)-dihydro-erythronolide-A

$(PhS)_3CLi$; MoOPh; RaNi EtOH

Ph_3P, DEAD, PhCOOH; K_2CO_3

H_2 / Rh-Al_2O_3

Ph_3P, DEAD, PhCOOH; K_2CO_3

A similar idea of repeatedly utilizing the butanolide and butenolide which are initially derived from glutamic acid has allowed stereocontrolled construction of chain segments differing in substituents [Hanessian, 1985a,b,c].

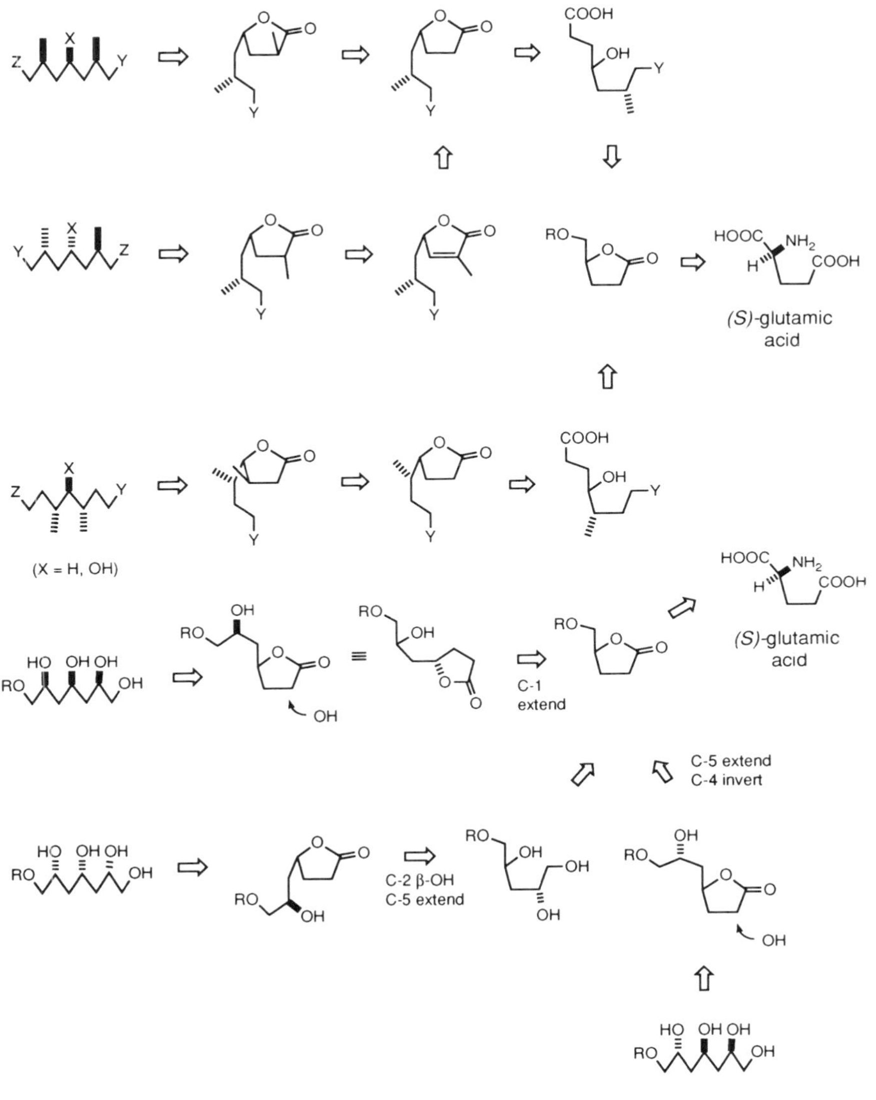

The powerful method of asymmetric epoxidation of allylic alcohols using chiral tartaric esters as catalysts constitutes a pivot to the synthesis of various sugar molecules. Horner–Emmons reaction of aldehyde followed by reduction results in an allylic alcohol which can then be epoxidized. With transformation of the epoxide into *vic*-diol and the primary alcohol into a phenyl sulfide, and subsequent acetonide formation and oxidative desulfurization, the sequence can be repeated as often as desired. All eight hexoses have thus been acquired [Ko, 1983]. A similar reiterative approach was employed in a synthesis of swainsonine [Adams, 1985].

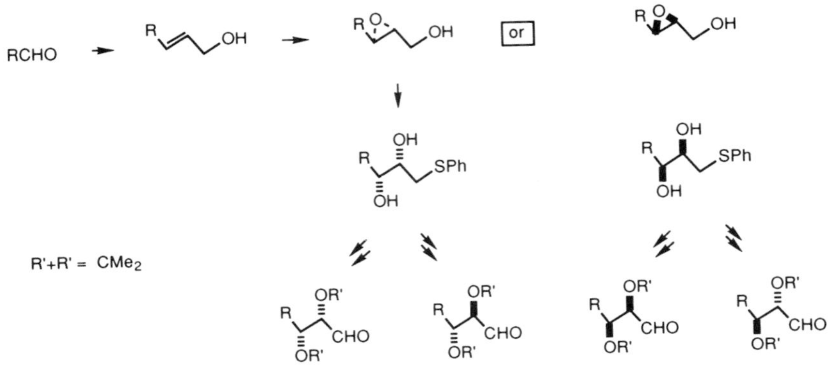

Most sugar molecules contain a polyhydroxylated carbon chain, and individual molecules have characteristic hydroxylation patterns and configurations. In principle, reiterative methodologies are well suited for the stepwise construction of sugar molecules, but it is imperative to achieve stereocontrol during establishment of each new sterocenter. (Note the classic Kiliani–Fischer synthesis of sugars via cyanohydrin formation, hydrolysis, lactonization, and reduction is amenable to reiterative treatment to homologate aldose sequentially, but the general lack of stereocontrol is a major defect.) The use of 2-trimethylsilylthiazole as a formyl anion equivalent favors formation of the *anti*-adduct, therefore a carbon chain adorned with *anti*-1,2-polyol branches is easily assembled [Dondoni, 1989]. Transformation of the thiazole ring into a formyl group is achieved by a reaction sequence of methylation, reduction, and mercury(II) ion-assisted hydrolysis.

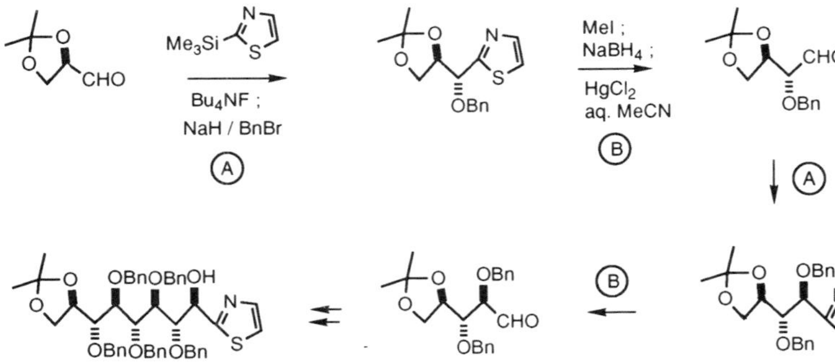

1,2-Polyol systems with either *syn* or *anti* arrangement have been synthesized using a α-alkoxy silanes as intermediates [Yoshida, 1992]. Electrolysis of these silanes in methanol gives dimethyl acetals, and by chain extension of the latter compounds into alkenylsilanes, asymmetric epoxidation, methanolytic ring opening, a new unit of α-alkoxy silane is introduced. Repetitive employment of the reaction sequence serves to build up the functionalized chain rapidly. Moreover, it is possible to invert the configuration of the hydroxyl group (Mitsunobu reaction) of a hydroxy epoxysilane to enter the *syn* series.

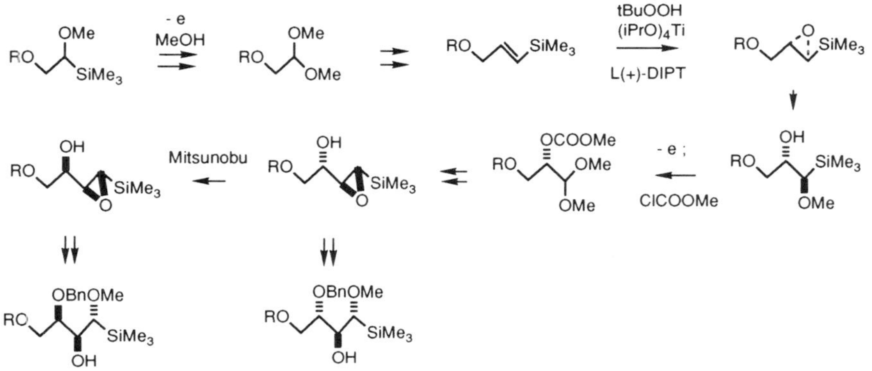

1,3-Polyols can be acquired by repetitive employment of the Sharpless method of asymmetric epoxidation of allylic alcohols, chain elongation of the resulting glycidols via Wittig–Horner reaction, reductive cleavage of the epoxide ring, and proper protection of the hydroxyl group [Nicolaou, 1982b; Finan, 1982].

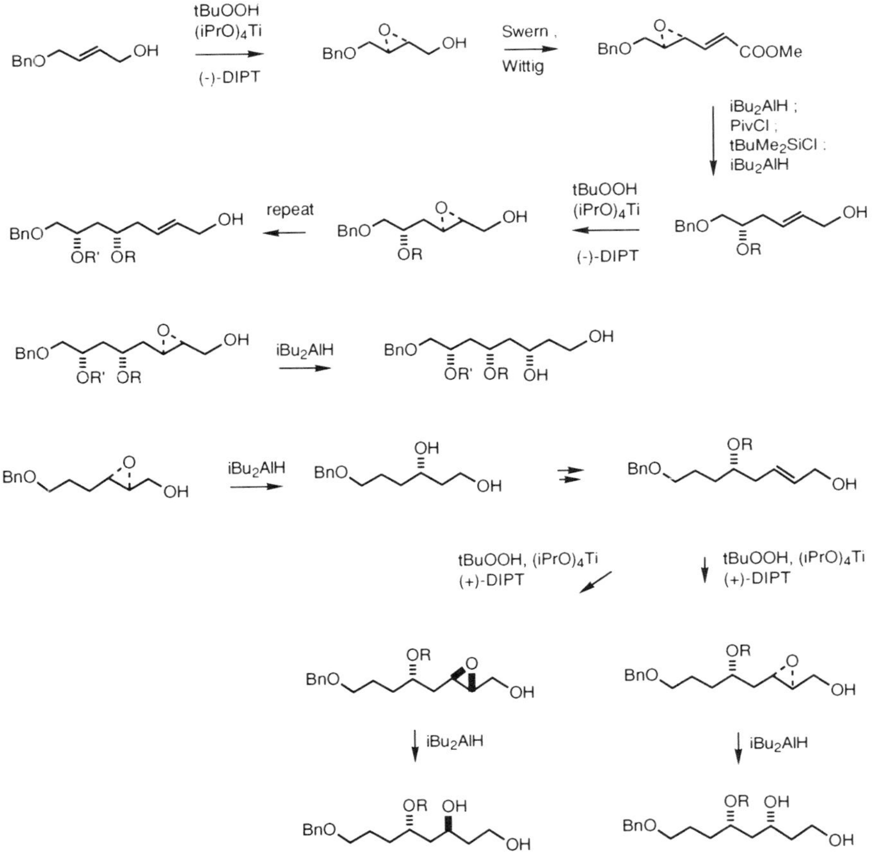

Another convenient route to chiral all-*syn* 1,3-polyols from benzyl ether of glycidol which is available in both enantiomers consists of reaction with a vinylcuprate reagent, *O*-carboxylation, *in situ* iodocarbonation, and mild hydrolysis. The hydrolytic step serves to cleave the carbonate ring and induce epoxide formation. The new epoxide can initiate a new reaction cycle [Lipshutz, 1984].

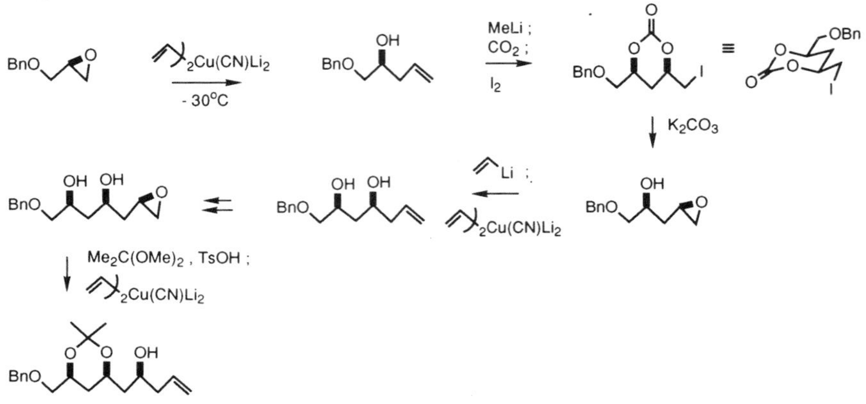

An efficient carbon chain synthesis by adding *syn*-1,3-diol blocks is based on the activating properties of the cyano group and its ready removal by reductive cleavage. Accordingly, 4-cyano-2,2-dimethyl-1,3-dioxane undergoes alkylation with 2,2-dimethyl-4-iodomethyl-1,3-dioxane to give a product in which the small cyano group is an axial orientation. Replacement of the cyano group by a hydrogen atom with retention of configuration then furnishes a protected all-*syn* skipped tetraol. Convergent and reiterative maneuver based on this technique has permitted a relatively concise synthesis of a *syn*-polyol permethyl ether [Rychnovsky, 1992a].

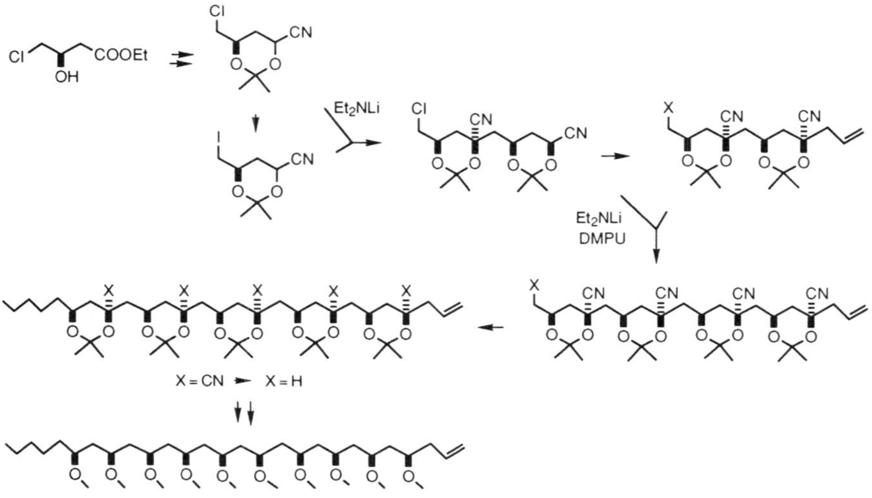

The alkylation of 2-(2,2-dimethyl-1,3-dioxolan-4-yl)methyl-1,3-dithiane with a 4-alkoxybutene oxide followed by recovery of the carbonyl group from the dithiane moiety and stereoselective hydride reduction constitutes a method for synthesis of 1,3-polyols [Y. Mori, 1989a, 1990]. After proper protection of the hydroxyl groups, the dioxolane can be converted into an epoxide and the reaction sequence repeated to increase the chain length by a four-carbon unit. It is possible to form either all-*syn* or alternating *syn-anti* 1,3-polyols. Furthermore, proper manipulation of the dioxolane group can give rise to either epimeric epoxide, and the change of the dithiane from one enantiomer to the other furnishes another variation such that the acquisition of different series of 1,3-polyols is readily accommodated [Y. Mori, 1989b].

Bidirectional chain homologation is most suitable for the preparation of chain segments with C_2 symmetry. Applying the same set of reactions on a symmetrical precursor causes double growth in chain units, however, the methodology is useful only if the reactions are highly stereoselective and the two termini of the chain can be differentiated at the proper stage. In this regard the protocols for creating skipped all-*syn* polyols from *meso*-1,4-pentadiene dioxide [S.L. Schreiber, 1987c] are versatile indeed. After the epoxides are opened by reaction with allyloxide ion, and with the resulting diol protected as an acetonide, [1.2]-Wittig rearrangement from both ends gives two allylic alcohol units which may be resolved kinetically by asymmetric epoxidation. When vinylmagnesium bromide is used (in the presence of CuI) in the initial epoxide opening, and the product is processed via acetonide formation, ozonolysis, another vinylmagnesium bromide reaction, the location of molecular symmetry plane is altered. Note the formation of an equilibratable dialdehyde to ensure the all-*syn* stereochemistry. After proper transformations a new substrate for the reaction sequence involving the [1.2]-Wittig rearrangement is obtained. The two series of reactions can be repeated as many times as desired.

ONa ; Me_2CO $CuSO_4$ Amberlyst-15

BuLi

MgBr ; CuI Me_2CO $CuSO_4$ Amberlyst-15

tBuOOH $(iPrO)_4Ti$, D(-)-DIPT

tBuOOH $(iPrO)_4Ti$, L(+)-DIPT

O_3; Ph_3P MgBr ; $TfOSiMe_3$, Me_2CO

O_3; Me_2S; $NaBH_4$

NaH, Br

BuLi

tBuOOH $(iPrO)_4Ti$, L(+)-DIPT

In connection with a synthesis of (-)-hikizimycin [Ikemoto, 1992] the bidirectional technique was exploited. The C-2 to C-11 segment of the antiparasitic agent, when rendered in an aldose form, is shown to possess a repeating *syn-anti* arrangement of vicinal hydroxyl groups if the 4-amino group is replaced by OH with inversion of configuration. As such, a synthetic approach based on bidirectional chain extension from a building block containing the C-6, C-7 stereocenters, employing two sets of olefination/osmylation, is appealing. Osmylation is highly stereoselective: (*E*)-olefins yield *syn*-diols and (*Z*)-olefins yield *anti*-diols, furthermore, allylic ether group can determine facial stereoselectivity. The combined effect is of practical value in stereocontrolled homologation.

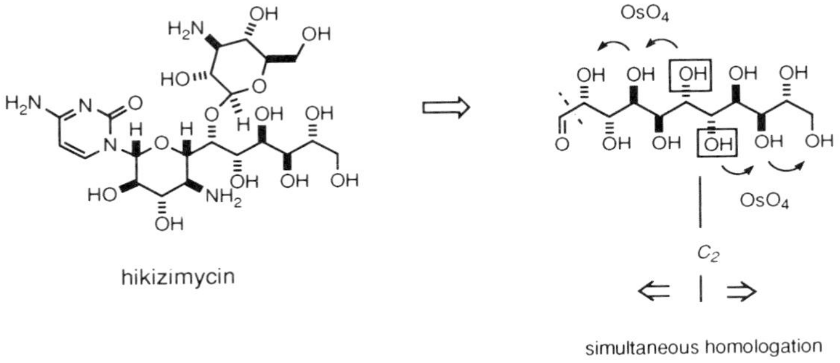

hikizimycin

OsO_4

OsO_4

C_2

simultaneous homologation

Another method for the construction of skip polyols relies on formation of pyran-4-one intermediates which are subjected to stereocontrolled reduction with K-selectride [Oishi, 1984].

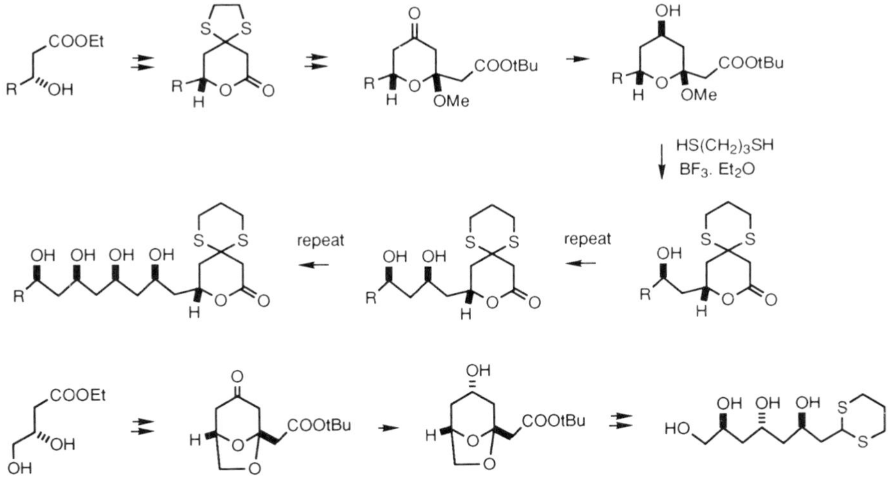

$HS(CH_2)_3SH$
$BF_3.\ Et_2O$

repeat

repeat

The azetidinone framework can be used as chiral template for elaboration of *syn*-1,3-polyols and 2-amino-1,3-polyols [Palomo, 1993]. Reiterative processing is feasible.

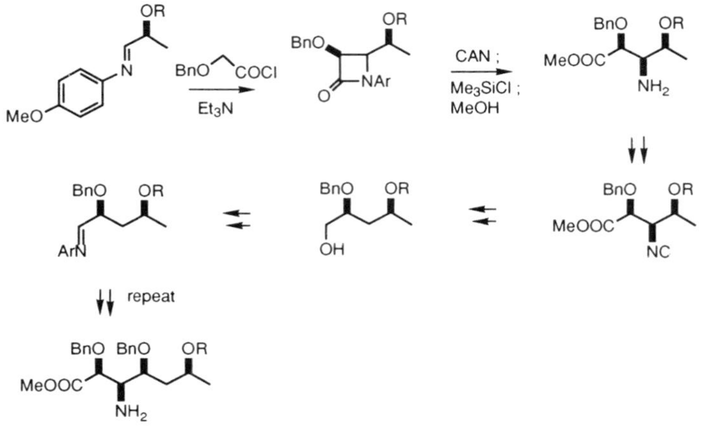

Allylsilane reactions [Reetz, 1983a, 1984b, 1988] are potentially amenable to construction of skipped *syn*- or *anti*-1,3-polyols by reiterative application. The Ti(IV)-catalyzed reaction of a β-alkoxy aldehyde with allyltrimethylsilane gives predominantly an *anti*-1,3-diol monoether. Etherification followed by ozonolysis would regenerate the β-alkoxy aldehyde functionality. Intramolecular $Si \rightarrow C$ transfer of an allyl group from a β-allyldimethylsiloxy aldehyde can give rise to the *syn*-1,3-diol or *anti*-1,3-diol predominantly, depending on whether the catalyst is $TiCl_4$ or $SnCl_4$, respectively.

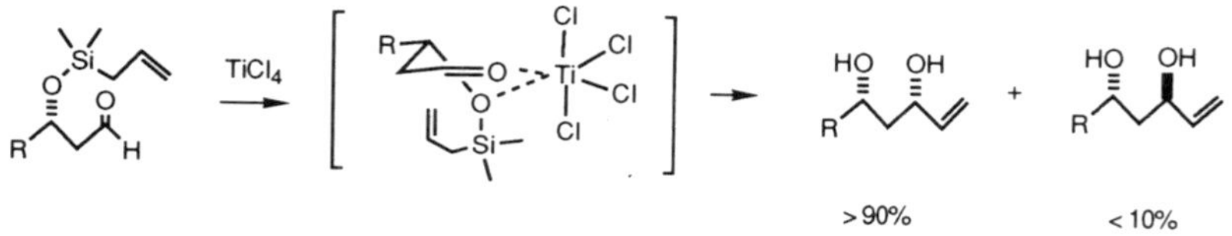

>90% <10%

The allyl transfer from chiral allylisopinocampheylborane to a β-alkoxy aldehyde, coupled with *O*-protection, and ozonolysis, constitutes another reiterative sequence for adding a chiral hydroxyethylene unit to a carbon chain [Nicolaou, 1989].

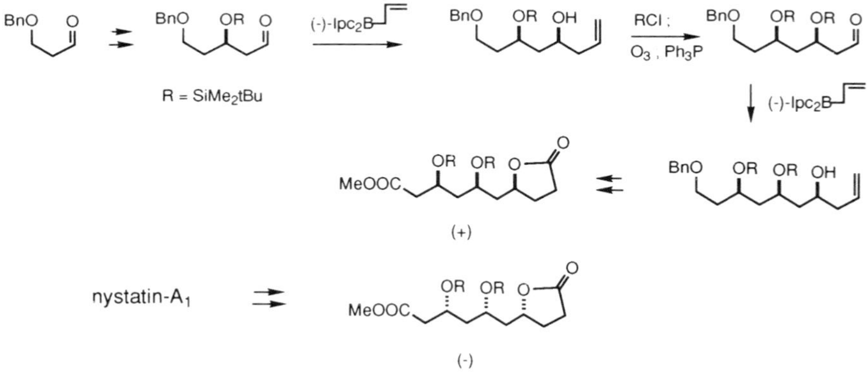

R = SiMe$_2$tBu

(+)

nystatin-A$_1$

(-)

Tricylco[4.4.1.1^{2,5}]decane was found to be a useful material for the preparation of diasterane (tricylco[3.1.1.1^{2,4}]octane) [Otterbach, 1987]. Since the simultaneous ring contraction processing at both ethano bridges was unsatisfactory, reiterative application of a reaction sequence of photobromination, dehydrobromination, hydroboration, oxidation, formation of the α-diazoketone, Wolff rearrangement, saponification, and Barton decarboxylation was implemented.

diasterane

The reiterative functionalization of C_δ-H by photolysis of α-peracetoxynitriles has been demonstrated [Watt, 1976]. Thus it is possible to prepare 1,4,7,...-alkanepolyones from nitriles via reaction of the α-anions with oxygen and acetylation of the resulting hydroperoxides.

An anecdotal publication ["Dinsburg", 1982] described an approach to the hydrocarbons israelane and helvetane as formulated by A. Eschenmoser. Purportedly a ladderlane precursor would undergo cyclization and sulfur extrusion to give israelane, and such a ladderlane would be derived from a reiterative process.

israelane

helvetane

2

ACTIVITY MODULATION, GROUP PROTECTION, AND LATENT FUNCTIONALITIES

2.1. PREEMPTIVE ACTIVITY MODULATION

In synthesis of a complex carbogen, reaction selectivity depends on functional group reactivities and manipulations. If reaction conditions can be found to differentiate two or more functional groups in a synthetic operation, a smoother progress is expected. A useful tactic for achieving such differentiation is by selective activation of the functional group designed to undergo a reaction. While many techniques have been developed to increase group reactivity, those of special merit pertain to preemptive activity modulation.

Preemptive activity modulation is modification of reactivity of a molecule such that an existing functionality can no longer exert its dominant (including overactivating and suppressive) effect during a reaction. The most celebrated representative of suppressed reactivity is that shown by the enediyne antibiotics such as the calicheamicins. Nature inserts a bridge containing a double bond orthogonal to the enediyne system in the 10-membered ring to enforce spatial separation of the terminal carbon atoms of the enediyne, thereby preventing cyclization [M.D. Lee, 1992]. Bioactivation is achieved by cleavage of the trisulfide group to expose a Michael donor which then adds intramolecularly to the enone. The bridgehead tetrahedralization brings the enediyne termini within bonding distance so that cyclization to give a benzene-1,4-diyl is inevitable. It can be said that the latent diradical is divested through release of the thiol function.

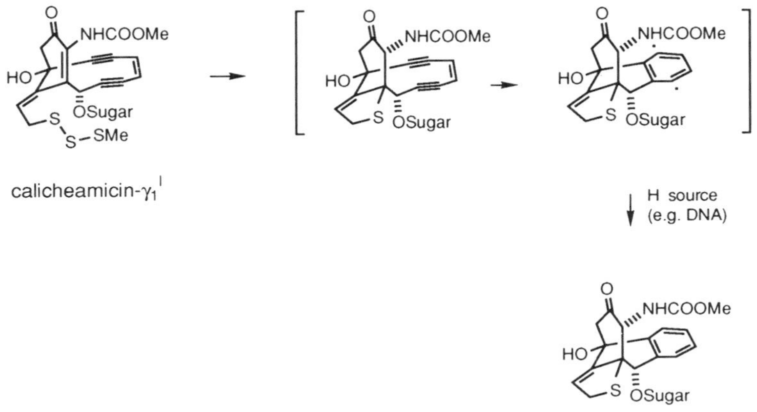

In many reactions the consequence of preemptive activity modulation of substrates is a change of regioselectivity. For illustration of this mode of activation we refer to the following examples.

The achievement of selectively oxidizing a dihydroxyl compound depends on the nature of each of the hydroxyl groups. *A priori*, it is not very easy to operate on one to the exclusion of an allylic alcohol. However, if the coexisting functionality is a primary alcohol, advantage can be taken to its selective tritylation, with modification of the reagent by incorporating an *o*-bromo substituent which allows the ether to generate a free radical and initiate oxidative cleavage via intramolecular hydrogen abstraction [Curran, 1992b].

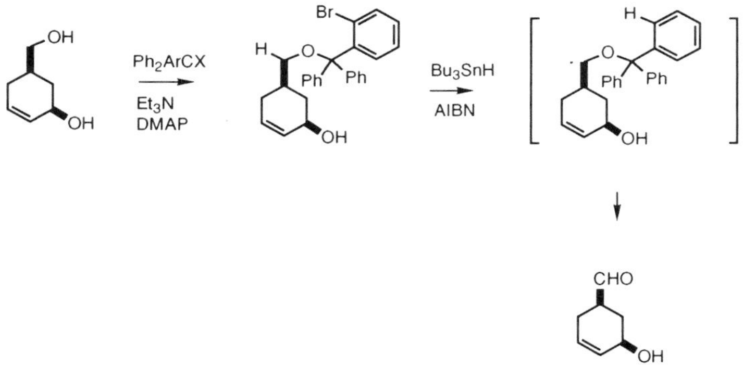

The alkylation of dimethyl malonate with 3-acetoxy-8-bromo-1-octene gives a product in which the long chain contains the allylic acetate; however, in the presence of Pd(0) catalyst the reaction product is the ω-bromoalkenylmalonate ester [Trost, 1980b]. Activation of the allyloxy function by palladium (with formation of a π-allylpalladium intermediate) supersedes the normal reactivity of the bromoalkane.

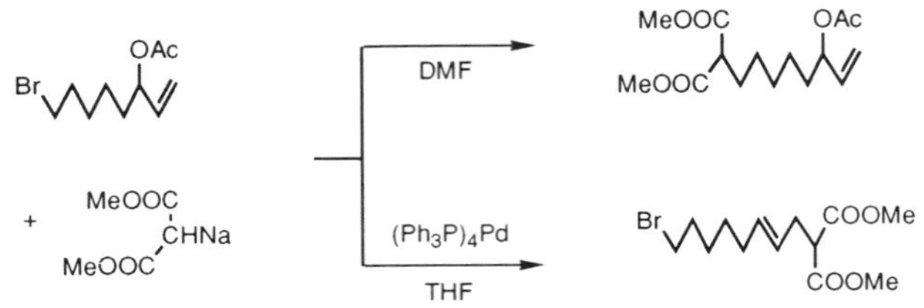

By taking advantage of the preferred coordination of low-valent ruthenium species to the cyano group the Cα-H bond of an alkanonitrile can be activated by metal insertion in the presence of an active carbon acid of similar pK_a value [Naota, 1989]. For example, a compound combining both malonate and cyanoacetate moieties would undergo Michael reaction at the α-carbon atom of the cyanoacetate.

The Nazarov cyclization generally produces more highly substituted cyclopentenones. The cyclization of 1-trimethylsilyl-1,4-alkadien-3-ones leads to products in which the conjugate double bond occupies C-1 and C-2 of the original dienones [T.K. Jones, 1983], due to the effect of the silyl substituent.

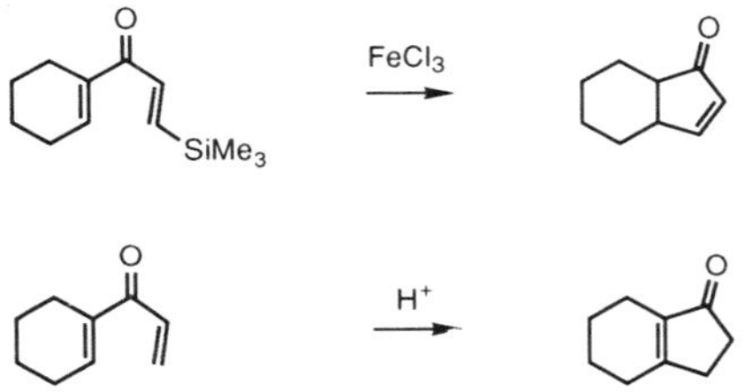

In fact, a silyl group can be used to direct electrophilation of alkenes irrespective of the Markovnikov rule [Fleming, 1981]. The desilylative cyclization of iminium ion as illustrated in a synthesis of deplancheine [Overman, 1982] also provides a solution to the stereochemical problem concerning establishment of a trisubstituted double bond.

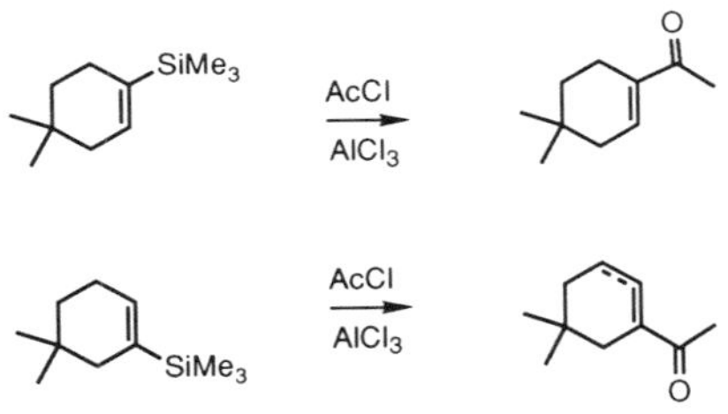

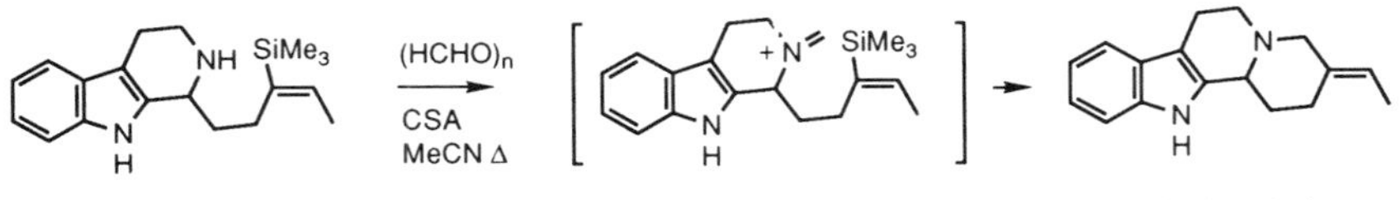

deplancheine

In the application of the cationic aza-Cope rearrangement/Mannich cyclization tandem to the synthesis of 6a-epitazettine and related alkaloids the hydroxyl group in the pyrrolidine ring is incompatible because of the difficulty in its maintenance in the precursor (as an enol). Consequently, a silyl substituent was used instead [Overman, 1989].

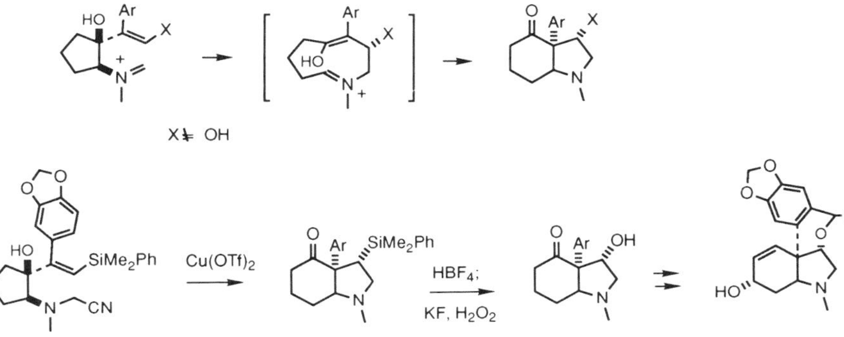

6a-epipretazettine

A bicyclic lactone has been identified as a precursor for the DE-ring portion of reserpine [Stork, 1989a]. This lactone was acquired from a reflexive Michael reaction in which the required methyl β-methoxyacrylate component was replaced with a β-silyl surrogate based on reactivity considerations. The silyl group of the adduct was subsequently transformed into the oxygen substituent with retention of configuration.

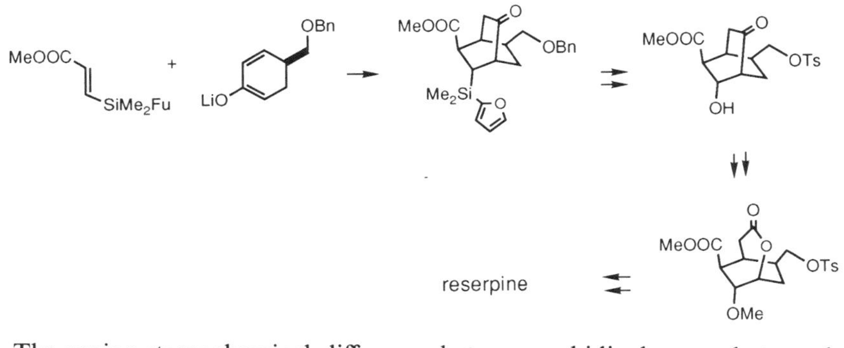

The major stereochemical difference between aphidicolane and stemodane diterpenes is the epimerism at C-9 and C-12, that is the bridged ring systems. In terms of synthesis both ring systems could be derived by intramolecular alkylation of a spirocyclic precursor. The key is the regioselectivity of the alkylation step, especially when the objective is in the stemodane series. Consequently, in the synthesis of a stemodane intermediate a Δ^{11}-enone would be the required substrate [Corey, 1980b]. In the absence of the double bond to block the enolization toward C-12 the course of alkylation leading to the aphidicolane-type structure [Corey, 1980a] can be achieved.

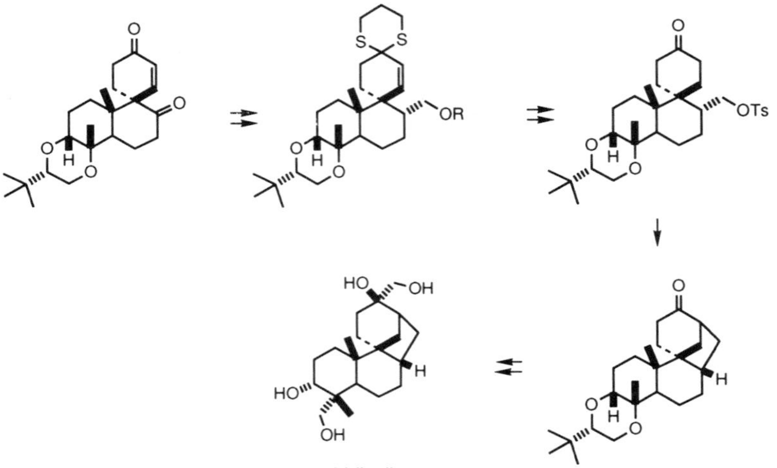

aphidicolin

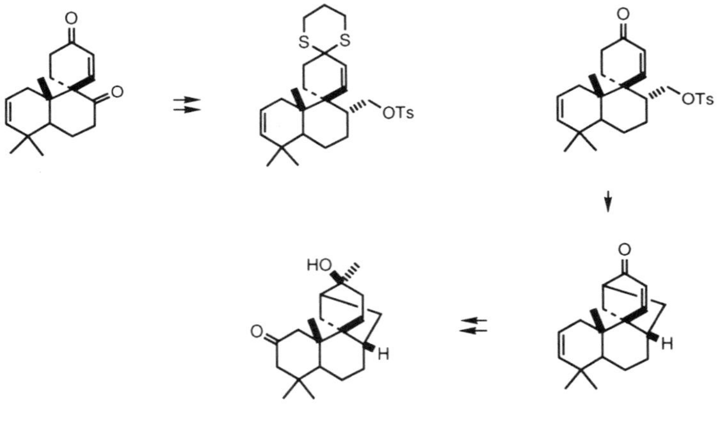

stemodinone

Generally, a ketone is more reactive than an ester toward Grignard reagents. However, the Grignard reaction of 8-levulinoxylquinoline gives predominantly 1,4-diketones (selectivity *ca* 3:1) [Sakan, 1973]. The chemoselectivity is further increased on addition of silicon tetrachloride or phosphorus tribromide. Cocomplexation of the ester and the nitrogen atom to a Lewis acid center elevates the reactivity of the ester over that of the ketone group

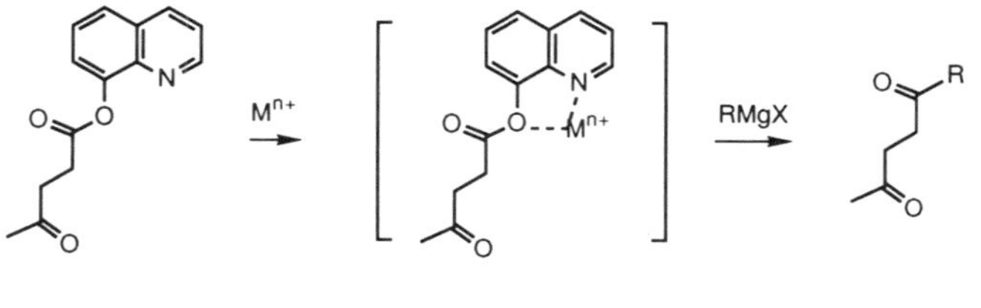

Note that enollactones derived from certain ketoacids serve the same purpose of chemoselection in reaction with organometallic reagents and complex hydride reductions. An example of the application is in a synthesis of olean-11,12;13,18-diene [Corey, 1963].

RMgX

olean-11,12;13,18-diene

Introduction of detachable activating groups to inactive reaction sites is another preemptive activation tactic. Thus *C*-alkylation of a 1,4-naphthoquinone is impossible under normal conditions, however, the enolizability of the ketone groups is restored in a Diels–Alder adduct. 2-Prenyljuglone can be obtained in this manner [Matsutomo, 1985].

OH O OMe Δ

juglone

Another example pertains to alkylation at C-3 of 2-cyclohexenone which is rendered feasible by using the conjugate hydrocyanation adduct. Subsequent dehydrocyanation achieves the objective.

Under normal conditions acrylic esters do not undergo alkylation because enolate formation is not possible. A way to circumvent this difficulty is by adding a Michael donor catalyst to generate zwitterionic intermediates *in situ*. Thus admixture of an acrylic ester, an aldehyde and catalytic amount of triphenylphosphine affords the α-(1-hydroxyalkyl)acrylic ester via a tandem sequence of Michael addition, aldol-type condensation, and elimination to regenerate the triphenylphosphine.

The same tactic is applicable to α-allylation of acrylic acid [Hanamoto, 1993]. When the acid is derivatized into an allylic ester and treated with a trialkylphosphine, a chlorotrialkylsilane and a non-nucleophilic base, a tandem Michael addition/Claisen rearrangement ensues. *In situ* elimination of the phosphine completes a reaction cycle.

R''_3P, R'_3SiCl, DBU; $OSiR'_3$, PR''_3^+ Cl^-

Occasions may demand the use of masked reagents/substrates in reactions in which the active forms are liberated in situ. It is also possible that higher selectivities may be attainable in such modified reactions. The formation of β-cyanohydrins from *gem*-disubstituted epoxides is conveniently achieved by using acetone cyanohydrin with catalytic amount of triethylamine [Mitchell, 1992].

A Michael adduct rendered the acrylic ester sidechain capable of forming a ketene silyl acetal and undergoing a Claisen rearrangement, which was a crucial step for annexing the α-methylene-γ-lactone ring to the dicarbocyclic precursor of frullanolide [Still, 1977].

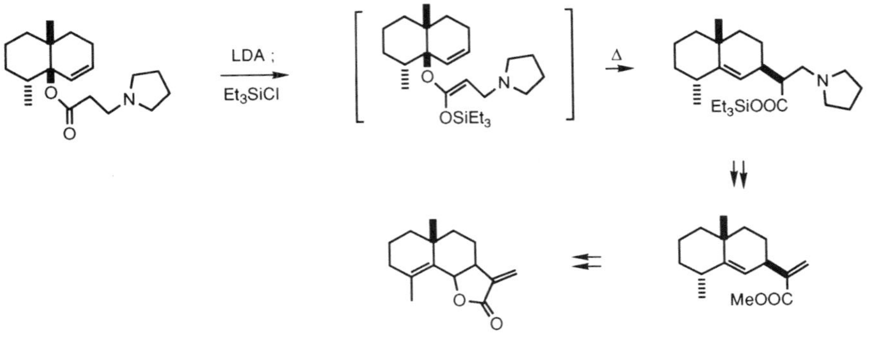

LDA ; Et_3SiCl → [$OSiEt_3$] Δ → Et_3SiOOC ⇊ MeOOC ⇇

frullanolide

Alkylation at the β-position of an enone must be accomplished indirectly. For example, activation at that site is possible via addition of an arenesulfonyl group to the substrate with ketalization to create the necessary reactivity profile [Yoshida, 1982]. Regeneration of the ketone after the alkylation prompts the arenesulfinic acid elimination.

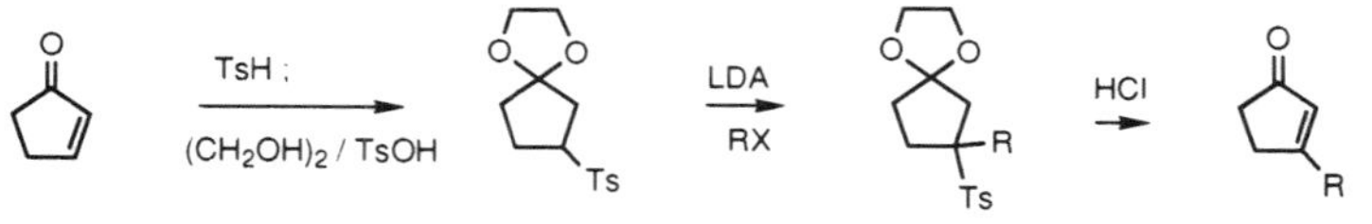

TsH ; $(CH_2OH)_2$ / TsOH → Ts; LDA, RX → R, Ts; HCl → R

Trialkylphosphines are good catalysts for the conjugate addition of an alcohol to propynoic esters [Inanaga, 1993]. The mode of action involves adduct formation of the catalyst with the acceptor molecule, the adduct being apparently more susceptible to attack by the alcohol.

The normal Diels–Alder adduct of 2-cyclohexenone and 1,3-pentadiene is an octalone in which the methyl group is peri to the carbonyl. It is possible to reverse the regiochemistry by inserting a nitro group into C-3 of the cyclohexenone [Ono, 1982]. After exerting a dominant effect the nitro group can be removed reductively.

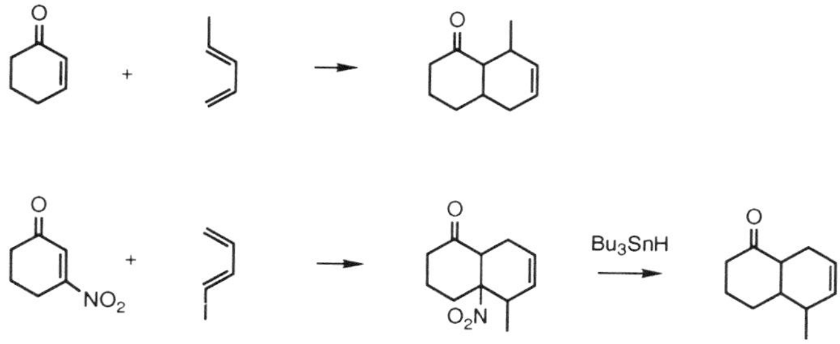

Although metallation of the pyridine ring at C-2 is a well-known process, further activation [Taylor, 1983] *in situ* by forming a complex with hexafluoroacetone at $-107°C$ is noteworthy. The zwitterionic species now has a more acidic hydrogen at C-2, and the oxide anion of the species scavenges the lithium amide reagent, bringing it close to the deprotonation site,and stabilizes the resulting 2-lithiopyridine by chelation. Electrophilation is accomplished readily, and thereafter hexafluoroacetone dissociates from the adduct on warming the solution.

Pyridine undergoes electrophilic substitutions with difficulty because the nitrogen atom is electron-withdrawing. Its nitration occurs at C-3, as expected from the polarity alternation rule [T.L. Ho, 1991]. The reaction site can be changed to C-4 by converting it into pyridine *N*-oxide. The oxygen atom in this compound is a donor and it takes over the direction.

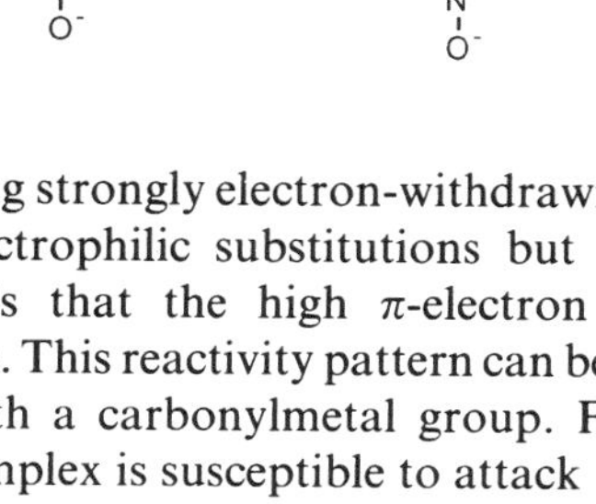

Excepting those bearing strongly electron-withdrawing substituents aromatic compounds undergo electrophilic substitutions but rarely nucleophilic substitutions. The reason is that the high π-electron density tends to repel approaching nucleophiles. This reactivity pattern can be modified by complexing the aromatic nucleus with a carbonylmetal group. For example, the anisole tricarbonylchromium complex is susceptible to attack by certain organolithium reagents, and the CC bond formation process which takes place at the *m*-position is complementary to *o*/*p*-alkylations and acylations of the uncomplexed substrate. A synthesis of acorenone [Semmelhack, 1980] demonstrates this versatility.

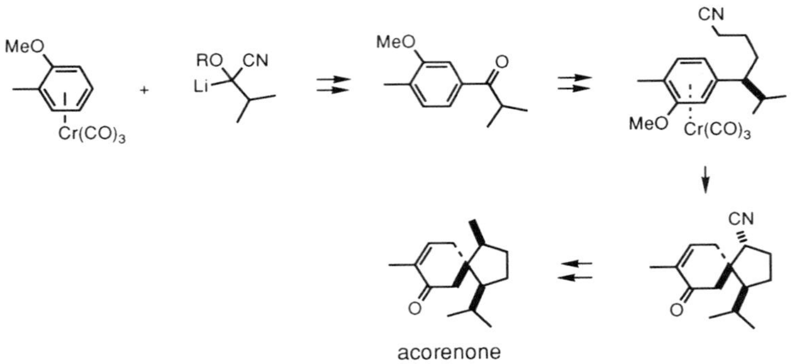

MeO, Cr(CO)3, RO, CN, Li, MeO, O, CN, MeO, Cr(CO)3, CN, O, O

acorenone

Normally it is very difficult if not impossible to displace the amino group of unactivated anilines yet the aryl group can be activated on converting these anilines into pyridinium salts on reaction with pyrylium ions. Arylation of an arylamine has been achieved on further facilitation by intramolecularization [Katritzky, 1983].

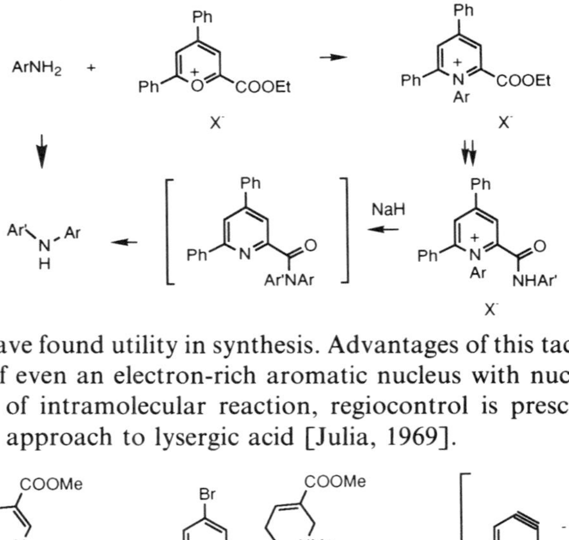

ArNH2, Ph, Ph, COOEt, X-, Ph, Ph, N, Ar, COOEt, X-, Ar', N, Ar, H, Ph, Ph, N, O, Ar'NAr, NaH, Ph, Ph, N, Ar, O, NHAr', X-

Arynes have found utility in synthesis. Advantages of this tactic include bond formation of even an electron-rich aromatic nucleus with nucleophilic agents. In the case of intramolecular reaction, regiocontrol is prescribed, as shown in a concise approach to lysergic acid [Julia, 1969].

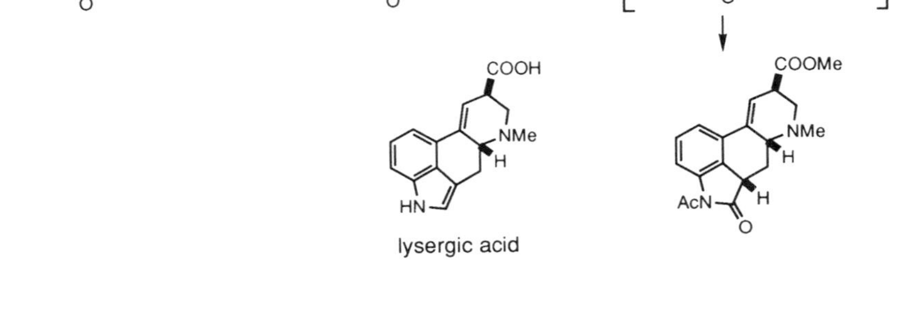

Br, COOMe, N, HN, O, Br, COOMe, NMe, H, AcN, H, O, COOMe, NMe, H, AcN, H, O, COOH, NMe, H, HN, COOMe, NMe, H, AcN, H, O

lysergic acid

The common sense of changing reagents (and conditions) to achieve the desired results cannot be overly asserted in chemical research, and the above is but an extension of this extremely important tactic. Inspired by Nature's synthetic prowess which involves sulfur-based leaving groups (thioesters in acyl transfer, adenosylmethionine in methylation, etc.), chemists have developed several efficient methods directly or indirectly mediated by sulfur-containing reagents. For example, the activation of a carboxylic acid by a combination of a disulfide and a phosphine [Mukaiyama, 1976] is due to successive formation of a thiophosphonium species and carboxylphosphonium ion.

Ph_3P + (2-pyridyl disulfide) → $Ph_3\overset{+}{P}$-S-(2-pyridyl) + (2-pyridinethiolate, S^-)

↓ RCOOH

$RCONH_2$ ← ($R'NH_2$; $-Ph_3P{=}O$) $RCOO\text{-}\overset{+}{P}Ph_3$ + (pyridine-2(1H)-thione)

The carbonyl group of amides/lactams does not behave as an acceptor in aldol-type condensations. The reactivity problem particularly concerning the union of amides with enamides has been circumvented by changing them into the thioamides and oxidizing them (to the disulfides) in the presence of the enamides. Extrusion of the sulfur atom from the products to give the vinylogous amidines can then be achieved in the presence of a phosphine or phosphite.

2 (thiolactam, COOMe) → $(PhCOO)_2$; HCl → (disulfide salt, COOMe, COOMe, Cl^-) → + (enamide, COOMe) → (sulfide bridged product, COOMe) ⇌ (episulfide, COOMe) → $(EtO)_3P$, 120°C → AD-component of vitamin-B_{12} (COOMe)

In the context of vitamin-B_{12} synthesis [Woodward, 1968, 1971, 1973a; Eschenmoser, 1970] this sulfur-extrusion method was employed in the B + C, BC + DA fragments and the cyclization of the linear ADCB precursor.

Finally, it must be emphasized that reactivity modulation does not only involve activation, sometimes it is necessary to moderate the reactivity of a reaction partner. A case in point is the alkanethiol preparation from an alkyl halide, in which the reagent of choice is thiourea instead of sodium sulfide, as the *S*-alkylisothiouronium salt does not react further. In the same sense the Gabriel synthesis of primary amine is superior to displacement with ammonia in terms of product purity. Such a tactic can be viewed as reagent protection; the more prevalent concern of substrate protection is outlined in the following sections.

2.2. PROTECTION AND LATENT FUNCTIONALITIES

Although an ideal synthesis [Hendrickson, 1976] consists of only skeletal construction steps, that is it begins with available structural units requiring no initial functional preparation for construction, proceeds further without intervening functional alteration, and arrives at a fully constructed target, it is rarely possible because of various functional constraints. The presence of functional groups is not always compatible with reactions at other sites of the molecule and their protection is almost unavoidable. Such protection necessarily increases the number of steps and decreases the efficiency of a synthesis, therefore the choice is critical to avoid further complication of the situation. In this regard it is noteworthy that in a synthesis of the anthelmintic macrolide avermectin-B1a [Ley, 1989] oxidation of an intermediate containing five hydroxyl groups of allylic, cyclic secondary (two), tertiary, and primary nature could be achieved selectively to give the desired acid without affecting the other four alcohol functions. Such is an exceptional case. The evolvement of chemoenzymatic methods may also help avoid functionality protection in certain steps, but the number of such methods is still quite limited as to the applicability to substrates of diverse structures.

Introduction of a functional group in a latent form alleviates the situation somewhat, but the tactic is not always feasible. There are excellent works in which a latent functionality was incorporated in an earlier step of a synthesis, serving the protective role until the time when that functionality is required. It must be emphasized that this tactic is related to synthetic convergency when its implementation is well conceived.

Before delineating conventional protection methods we shall sample several syntheses which demonstrate the elegance and utility of latent functionalities. The first and foremost of these examples concerns a synthesis of strychnine [Woodward, 1963] which started from the preparation of 2-veratrylindole by the Fischer indolization method. After spiroannulating a pyrrolidine moiety to the β-position of the indoline system, the veratryl residue was cleaved by ozonolysis at the CC bond of the highest electron density. One of the two carbon chains thus generated was immediately used to form a pyridone ring with the indolic nitrogen, while the other chain would participate in construction of the innermost ring of strychnine via a Dieckmann condensation. Consequently, the service of the veratryl group included blocking C-2 of the indole nucleus to smooth elaboration of the pyrrolidine ring, as well as its own mutation into parts of two other cyclic substructures.

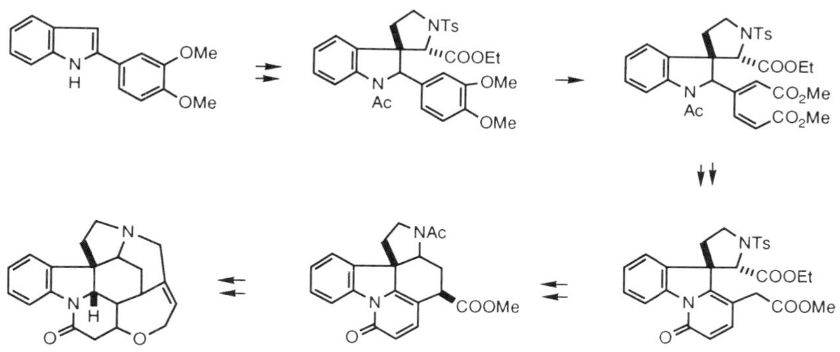

strychnine

The reader is urged to consult a report on the synthesis of lycopodine [Stork, 1968] for a similar tactic. In this case an anisole ring was submitted to Birch reduction and isomerization prior to its cleavage. The long chain was then involved in lactamization.

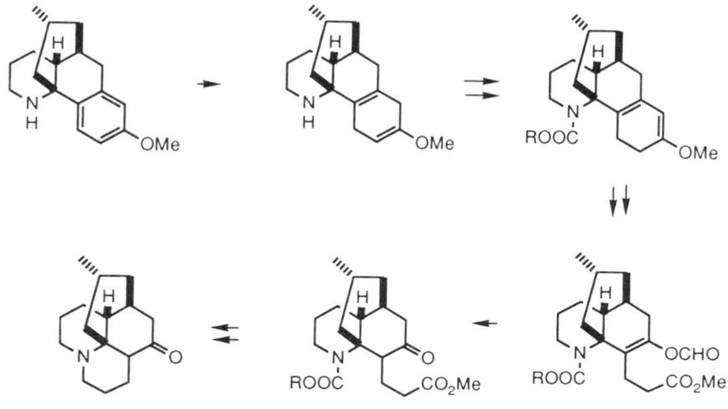

lycopodine

3,5-Dialkylisoxazole is a latent β-diketone unit. Its synthetic use has been developed by Stork. For example, 4-chloromethyl-3,5-dimethylisoxazole is a synthetic equivalent of methyl vinyl ketone by virtue of its high alkylating reactivity and ready regeneration of a butanone chain via cleavage of the heterocycle and hydrolvsis. Its application in a ferruginol synthesis [Ohashi, 1968] is most interesting because the substituted 1,3-pentanedione chain was fully utilized, as part of the aromatic ring and part of the isopropyl group.

COOEt + NaH → COOEt

Et_3OBF_3; NaOH

ferruginol

The furan ring of 2-substituted furans is a latent function of a carboxyl group and protocols for the conversion by oxidative cleavage (e.g. using ozone) of the heterocycle are well established. Accordingly, when one faces availability and incompatibility problems concerning an exposed and conventional carboxyl unit in a substrate, a furan derivative should be considered as surrogate. Furfural is a useful substitute for glyoxylic acid in the construction of the chiral tetrahydropyran moiety of indanomycin [Danishefsky, 1987].

Et_3SiO R*O Eu(hfc)$_3$ Et_3SiO R*O COOMe

COOMe BnO COOH HN

indanomycin

The ethyleneketal of 2-cyclopentenone is a latent 2-hydroxy-1,3-cyclopentadiene in Diels–Alder reactions [Ohkita, 1991]. In terms of stability and availability in large quantities this compound appears to be superior to 2-trimethylsiloxy-1,3-cyclopentadiene. Moreover, the Diels-Alder cycloadducts are more stable because they contain an ethyleneketal group instead of the highly strained and hydrolytically sensitive enol silyl ether. (It should be noted that *p*-hydroquinone has been used in rare occasions to react as a diene in the Diels-Alder reaction. The situation is different because *p*-hydroquinone is not a latent diene, and its stability greatly affects its reactivity in such reactions).

Concerning the issue of standard functional group protection the most commonly encountered problems involve masking hydroxyl, amino, carbonyl, carboxyl groups [T.W. Greene, 1991]. As there is no such device as a universal protective group, each case must be considered separately. Fortunately, the general properties of various derivatives are quite well known to chemists, therefore the selection of proper protective groups is a relatively minor matter in synthetic operations, although the noninterference with other functionalities is an obligatory condition. Of course the protection of more than one type of functional group in a molecule requires closer attention.

2.2.1. Hydroxyl Protection

Methods for protecting a hydroxyl group include acylation and etherification. Accordingly, acetylation and benzoylation to remove temporarily the active hydrogen of alcohols became a common practice a long time ago. Regeneration of an alcohol from its esters can be effected by treatment with alkali.

Esters are commonly prepared from an alcohol by Fischer esterification, by condensation with carboxylic acids in the presence of a dehydration agent such as dicyclohexylcarbodimide, by acylation with acid chlorides or anhydrides using a tertiary amine (pyridine, triethylamine, etc.) as proton scavenger. A very effective catalyst for acylation is 4-dimethylaminopyridine (DMAP) which enables acylation of tertiary alcohols.

Selective acylation of primary and secondary alcohols is still not well developed. One useful procedure appears to be that involving *N*-acylthiazolidine-2-thiones in the presence of sodium hydride [S. Yamada, 1992].

Many special esters have been employed in the protective schemes by taking advantage of other functionalities present in the acyl moieties which permit deblocking by ways other than saponification, aminolysis/hydrazinolysis, or lithium aluminium hydride reduction. For example, a carbonate may undergo selective cleavage by special manipulation of the second alkyl group, as in the case of an alkyl 2,2,2-trichloroethyl carbonate which liberates the alcohol by treatment of zinc dust, and in the case of an alkyl benzyl carbonate, whose cleavage can be effected by hydrogenolysis.

Benzyloxycarbonyl group removal from an alkyl benzyl carbonate by hydrogenolysis provided solution to a synthetic problem pertaining to a longifolene [Oppolzer, 1978] which was based on an intramolecular deMayo reaction to construct the bridged tricyclic nucleus of the sesquiterpene. The initially employed enol acetate was undesirable due to *in situ* intramolecular aldolization of the diketone released on saponification of the photocycloadduct. Avoidance of alkaline conditions was essential to the preservation of the tricyclic skeleton.

OBn OCOBn hv H_2 Pd-C HOAc longifolene

Ethers constitute another class of alcohol derivatives commonly used for the purpose of hydroxyl group protection. Methyl ethers, usually made by Williamson reaction (ROH + NaH/Me_2SO_4) are very stable but there are not many methods available for their selective deblocking. Perhaps the mildest conditions are those involving treatment with iodotrimethylsilane. It is possible to oxidize a methyl ether to furnish a formate ester, and the alcohol is recovered upon hydrolysis.

Phenols are frequently masked as methyl ethers. In a model study for a free radical cyclization route to fredericamycin-A [Clive, 1991] the naphthoquinone precursor in which all the oxygen atoms are methylated was found to be unsuitable because the nascent vinyl radical pursued hydrogen abstraction in both inter- and intramolecular manners. When the relevant *peri*-methoxy substituent was replaced with the trideuterio analog, the offending intramolecular abstraction pathway was greatly suppressed. Exploitation of isotope effect (stronger C-D bond) in synthesis such as this is rare.

X = SePh

R = H	1.15-1.4	:	1
R = D	9.7	:	1

Besides the Williamson reaction, benzyl ethers may be prepared by admixture of alcohols with benzyl trichloroacetimidate under essentially neutral conditions. The latter compound is in turn obtained from benzyl alcohol and trichloroacetonitrile. Removal of the benzyl group is achievable by hydrogenolysis, either catalytically or using Li/NH_3.

Allyl ethers are used frequently in the carbohydrate field. Older procedures for hydrodeallylation involve isomerization with a strong base system (*t*BuOK-Me_2SO) to give the propenyl ethers which are then subjected to acid or Hg(II)-catalyzed hydrolysis or oxidative cleavage. Milder conditions for the isomerization include Rh(I) catalysis and ene reaction with a dialkyl azodicarboxylate. The cleavage can also be effected via *in situ* isomerization with Pd-C.

t-Butyl ethers are made from alcohols and isobutene in the presence of a strong protic acid (usually H_2SO_4). They are labile to aqueous acids and iodotrimethylsilane.

A classical method for the selective protection of primary alcohols (e.g., of carbohydrate molecules) is by tritylation. Actually, such ethers may provide effective stereocontrol to reactions at a proximal site.

A labile yet favorite choice of hydroxyl protection is in the form of a tetrahydropyranyl ether, which is actually a mixed acetal. Its popularity is due to the ease of formation and cleavage. However, another drawback besides lability toward aqueous acids is the presence of an additional stereocenter. Because of this structural defect, mixed ketals derived from 4-methoxy-Δ^3-dihydropyran have been advocated to replace dihydropyran. Unfortunately, the much more difficult availability of this methoxy enol ether negates its attractiveness. A better candidate is the acyclic mixed acetal derived from the alcohol and methyl 2-propenyl ether.

Gaining popularity as protective devices for alcohols are the alkoxymethyl and methylthiomethyl (MTM) ethers. The Williamson method is often used to generate these ethers, but in the preparation of methoxymethyl (MOM) ethers the exchange reaction of alcohols with dimethoxymethane in the presence of a catalytic amount of phosphorus pentoxide is preferred, as it avoids the cacinogenic chloromethyl methyl ether.

The cleavage of MOM ethers can be carried out by exposure to aqueous acids, to a mixture of benzenethiol and boron trifluoride etherate, or trityl tetrafluoroborate. MTM ethers survive in the mildly acidic media for hydrolysis of THP ethers and acetonides, but they can be cleaved in the presence of heavy metal (Hg and Ag) salts or via *S*-methylation.

2-Methoxyethoxymethyl (MEM) ethers can withstand mild acids but are cleavable in fluoroboric acid at 0°C. They were designed on the basis of cleavage in the presence of a metal cation (e.g. Zn(II), Ti(IV)) under aprotic conditions.

RO–O O– (Zn, Br Br) → $RO^+{=}CH_2$ Br^- + O O– (Zn, Br)

$\downarrow H_2O$

ROH

The silyl group provides a convenient device for hydroxyl protection. The versatility of this method lies in the possibility of variation of substituents on silicon in steric bulk and electronic nature. The simplest of the silyl groups is the trimethylsilyl (TMS), numerous reagents (e.g. Me_3SiCl, $(Me_3Si)_2NH$, Me_3Si

imidazole, etc) are useful for trimethylsilylation of alcohols. The second most popular protecting group of this family is the *t*-butyldimethylsilyl (TBDMS) group whose derivatives are less acid-sensitive than those of a trimethylsilyl or isopropyldimethylsilyl, and slightly more stable than THP ethers, but they can be selectively deblocked by fluoride ion. They are not affected under reduction ($Zn/MeOH$, H_2/Pd-C, Na/NH_3, iBu_2AlH), oxidation (CrO_3/py; H_2O_2/OH^-), and many other reaction conditions (e.g. MeI/Ag_2O). Even more acid-stable are the *t*-butyldiphenylsilyl ethers.

Protective group manipulation is highlighted in a synthesis of brefeldin-A [Corey, 1977]. At one stage there were four hydroxyl groups which appeared as ethers with a TBDMS, a MTM, a MEM and a THP derivative. The progress of the synthesis continued by removal of the MTM group so that the exposed primary alcohol could be converted into a carboxylic acid. The secondary alcohol on the other sidechain was then released from the TBDMS ether on treatment with tetrabutylammonium fluoride to participate in macrolactonization. Next, the THP ether was hydrolyzed, and the epimeric alcohols were subjected to an oxidoreduction sequence to afford predominantly the desired alcohol. The final step of the synthesis consisted of MEM ether cleavage.

$HgCl_2$; H_2O, MeCN, $CaCO_3$

Bu_4NF

HOAc; H_2O, THF

$TiCl_4$; CH_2Cl_2

brefeldin-A

Besides individual protection, 1,2- and 1,3-diols can form acid sensitive cyclic acetals/ketals and orthoesters, and base sensitive cyclic esters (carbonates and boronates).

Among acetals, 2-benzyl-1,3-dioxolanes and 1,3-dioxanes are the most popular. They are stable to conditions for alkylation and acylation, certain reductions and oxidations. When exposed to ozone [Deslongchamps, 1975] or

N-bromosuccinimide [Hanessian, 1966] they are transformed into the monobenzoates and bromoalkyl benzoates, respectively. Acetonides are by far the most frequently protected form of cyclic ketals.

It should be noted that a 1,n-diol mono-*p*-methoxybenzyl ether can be converted into the 2-*p*-anisyl-1,3-dioxa heterocycle, and such a tactic of exploiting proximity is particularly valuable for differentiation of several hydroxyl groups in a molecule, as shown in the elaboration of a rifamycin-S segment [Lautens, 1992]. Remarkable is the cleavage of the oxabicycle with the *endo*-alcohol, apparently due to enhanced electrophilicity of the double bond by coordination with the lithium atom of the alkoxide. Both the corresponding silyl ether or the *exo*-alcohol failed to undergo the analogous reaction.

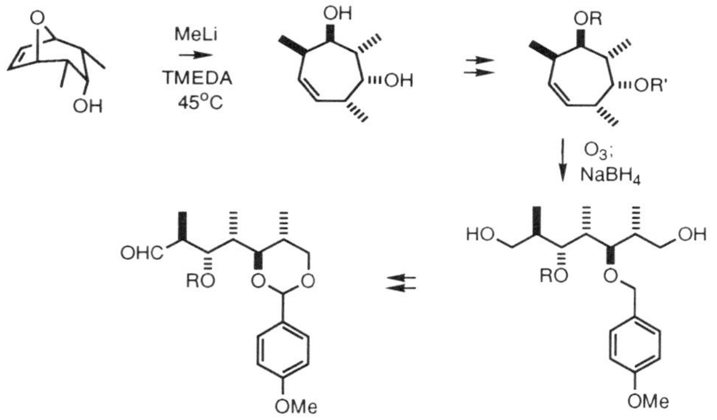

rifamycin-S segment

Selective participation of one hydroxyl group of a symmetrical diol in ketalization is a very important desymmetrization technique. A diastereotopic selection below is significant, because deketalization at a later stage would allow preparation of two epimers [S.L. Schreiber, 1985].

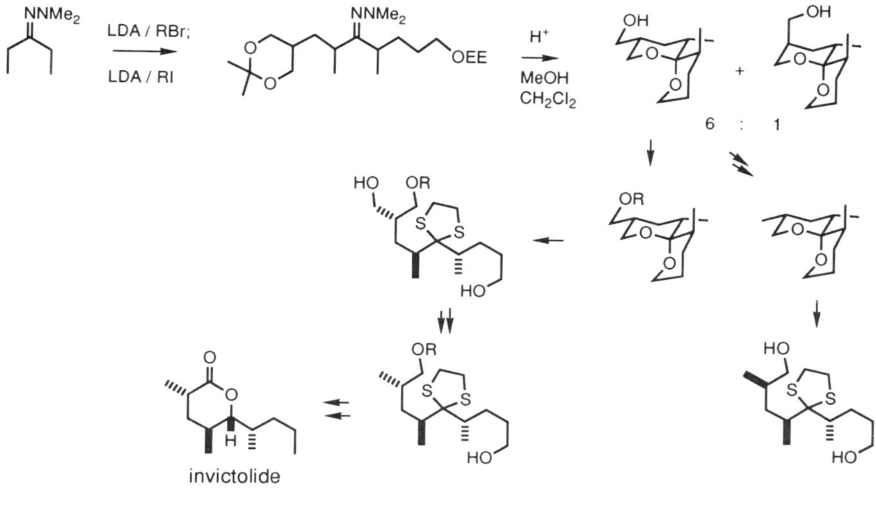

invictolide

Stannoxane derivatives are prepared from diols by reaction with dibutyltin oxide. They are converted into diol monobenzoates on treatment with one equivalent of benzoyl chloride [Shanzer, 1980].

Simultaneous protection of a *vic*-diol as trimethylsilyl ether on the primary hydroxyl and a pivalic ester at the secondary hydroxyl is possible by reaction of the dioxastannacyclopentanes with pivalyl chloride and with chlorotrimethylsilane [Reginato, 1990].

Protection of *vic*-diol systems in the preserence to 1,3-diols is feasible by forming dispiroketals using 3,3′,4,4′-tetrahydro-6,6′-bi-*2H*-pyran [Ley, 1992b]. The method achieves selective protection of diequatorial *vic*-diols [Ley, 1992a] and is particularly valuable in manipulation of carbohydrates.

2.2.2. Amino Protection

In many respects the primary and secondary amino groups are similar to the hydroxyl. The occurrence of the amino group in alkaloids, amino acids and in some of the nitrogen bases of nucleosides, means that their protection is required during synthesis and/or utility of such compounds. Two general types of protection are carbamate and amide derivatives, whereas numerous special NH protective groups have also been developed. Formation of carbamates usually involves reaction of amines with an alkyl chloroformate, azidoformate, or a carbonate. Ordinary acid chlorides are used in amidation.

Carbamates enjoy a favorite position in amino protection during peptide synthesis owing to the fact that very little racemization occurs in the coupling step when the amino group is changed into a carbamate. The best representatives are benzyl and *t*-butyl carbamates which undergo deblocking by hydrogenolysis and acid treatment, respectively. Other special types of carbamates include those containing a 2,2,2-trichloroethyl group which can be removed with zinc, and an allyl residue which is detachable on reaction with tributyltin hydride in the presence of a Pd(II) catalyst.

It is appropriate to emphasize here the value of allosteric trigger typified by the 2,2,2-trichloroethoxycarbonyl group. The deprotection, involving dechlorinative fragmentation to generate 1,1-dichloroethene and carbon dioxide, is sufficiently mild not to interfere with most other functionalities. A mechanistic kin is the 2-trimethylsilylethoxycarbonyl group in which the silicon atom is susceptible to attack by fluoride ion.

In view of the stability of amides toward hydrolysis, other methods are frequently employed to cleave them. For example, their transformation into imino ethers by Meerwein's reagent facilitates deacylation; sometimes hydrazinolyis is useful (particularly of phthalimides). On the other hand, amides possessing an additional substituent in the acyl residue are valuable as protected amino derivatives. The special substituent may be used to introduce an internal nucleophile by which the deacylation becomes favorable. α-Haloacetamides belong to this class of substrates, and in the case of a trifluoroacetamide, very mildly basic conditions suffice to dislodge the protecting group.

Protection of the amino group of a calicheamicinone predecessor was severely hampered by a vinylogous aldehyde attaching to an adjacent carbon atom [A.L. Smith, 1992]. this interference was removed by engaging both functionalities with a phthaloyl group to form a nine-membered heterocycle which was induced to rearrange to the conventional derivative and re-expose the aldehyde for its participation in CC bond formation.

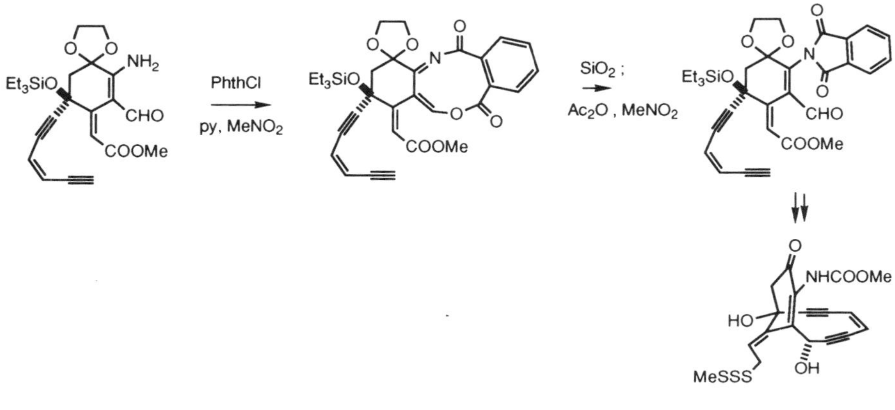

PhthCl; py, $MeNO_2$; SiO_2; Ac_2O, $MeNO_2$

(-)-calicheamicinone

2.2.3. Carbonyl Protection

The virtual ubiquity of the carbonyl group in the starting materials, intermediates, or target molecules of most syntheses means that protection of the functionality can hardly be avoided. Since the reactivity of a carbonyl group toward nucleophiles varies according to its environment, it may be possible to selectively protect one carbonyl among two or more such groups in a molecule. Thus, an isolated ketone may form ethyleneketal in the presence of a conjugate ketone, for example Wieland–Miescher ketone. (Numerous synthetic applications of this diketone in the monoprotected form have appeared in the literature; for terpene synthesis, see [T.L. Ho, 1988]).

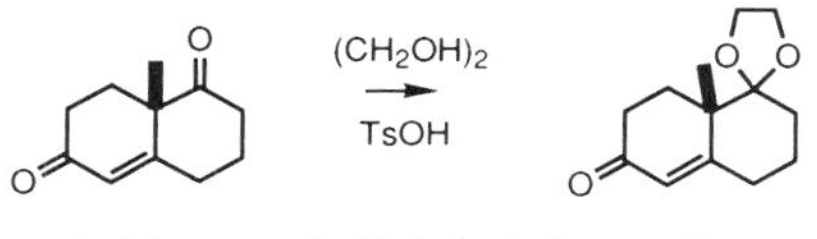

$(CH_2OH)_2$; TsOH

Acetals/ketals and thioacetals/thioketals are the most general protected forms of carbonyl compounds. Their formation generally involves admixture of carbonyl compounds with alcohols/diols or thiols/dithiols in the presence of acid, sometimes with continuous water removal during the reaction. Acetals/ketals and their thio analogs show complementary stability toward aqueous acids, oxidants, and hydrolytic conditions involving heavy metal ion catalysis.

Thioacetals/thioketals are quite stable in aqueous acids, while acetals/ketals are labile, and the acyclic members are more readily hydrolyzed than cyclic members. It is notable that 2,2-disubstituted 1,3-dioxanes are hydrolyzed faster than the corresponding 1,3-dioxolanes, and in the 2-monosubstituted analogs the relative selectivities are reversed. Special 1,3-dioxanes are 5-methylene and 5,5-dibromo derivatives which offer unusual opportunities for selective deblocking, for example via Rh(I)-catalyzed isomerization, oxidation of the double bond, or hydride transfer in the former, and reductive elimination from the latter derivatives.

Less commonly employed carbonyl protection methods include conversion into C=N derivatives. *N*,*N*-Dimethylhydrazones are synthetically useful on account of directed *C*-alkylations they mediate, and they are readily hyrolyzed oxidatively ($^1O_2$, O_3, $NaIO_4$), with Cu(II)-catalysis, or via *N*-methylation. A ketone intermediate in a sativene synthesis [McMurry, 1968] was most suitably protected as a 2,4-dinitrophenylhydrazone (2,4-DNP) while hydroborating a double bond. Oxime and ketal group could not be used due to their reducibility and double bond migration accompanying their respective formation. The 2,4-DNP group is removable by many oxidizing and reducing agents. The method involving reaction with titanium(III) chloride is mild.

ArNHN ArNHN OH O_3 OH TsCl, py; $MeSOCH_2Na$

sativene

The sensitivity to organolithiums requires a carbonyl group to be in a protected form when it is exposed to such reagents. Activation of the benzylic position of an aromatic aldehyde toward lithiation while preserving the carbonyl has been achieved by the addition of *N*,*N*,*N*′-trimethylethylenediamine [Comins, 1984] whereby an α-amino alkoxide species mediates the deprotonation. This method has been applied to a synthesis of schumanniophytine and isoschumanniophytine [Kelly, 1992].

CHO LiN(Me) NMe_2 OLi NMe NMe_2 BuLi OLi N Li N RX; H_3O^+ CHO R

In a dicarbonyl compound containing both an aldehyde and a ketone, it is possible to achieve selective protection of the aldehyde. However, the more interesting techniques are those which exploit the higher reactivity of the aldehyde group toward certain components of a mixed reagent, such as the exclusive formation of 6-hydroxy-6-methylheptanal from reaction of 6-oxoheptanal with the triphenylphosphine complex of methyltitanium trichloride [Kauffmann, 1988]. It has been postulated that the aldehyde group rapidly forms an α-titanoxyalkylphosphonium salt and is thereby immune to attack by the nucleophilic reagent. Very subtle chemoselectivities are exhibited by the ate complexes formed from titanium tetrakis(dialkylamides) and titanium tetralkoxides with Grignard reagents or organolithiums [Reetz, 1982c, 1983c].

Returning to the conventional protection protocols the selective masking of the ketone again relies on its temporary (*in situ*) transformation. For example, ketalization of such a difunctional substrate can be performed by treatment with 1,2-bistrimethylsiloxyethane and dimethyl sulfide in the presence of trimethylsilyl triflate [Kim, 1992]. The first reaction is the formation of an α- trimethylsiloxy dimethylsulfonium triflate in which only the ketone is exposed to ketalization. Desilylation triggers release of dimethyl sulfide without affecting the ketal.

CHO
Me_2S
Me_3SiOTf
-78°C
$\overset{+}{S}Me_2$
TfO^-
$OSiMe_3$
$OSiMe_3$
$OSiMe_3$
Me_3SiOTf
$\overset{+}{S}Me_2$
TfO^-
$OSiMe_3$
H_2O
CHO

A synthetic requirement of introducing two substituents to a *p*-quinone may be fulfilled by converting one of the ketone groups into a ketal. A route to deoxyfrenolicin from juglone [Semmelhack, 1985b] involved such a tactic. The more hindered carbonyl group was protected to expose unambiguously an enone moiety for effecting a conjugate addition/electrophilation tandem.

OH O
MeO
nPrLi
$Ni(CO)_4$
MeO
Pr
OLi
Br
OH O
COOH
deoxyfrenolicin
MeO

Neighboring group participation may be used to advantage in the selective monoprotection of a dialdehyde. For example, during synthesis of coriolin [T. Ito, 1984] the etheno bridge of a dicyclopentadiene derivative was cleaved, and with the lactolization to differentiate two carbonyl groups, degradation of the excessive pendant could be accomplished readily.

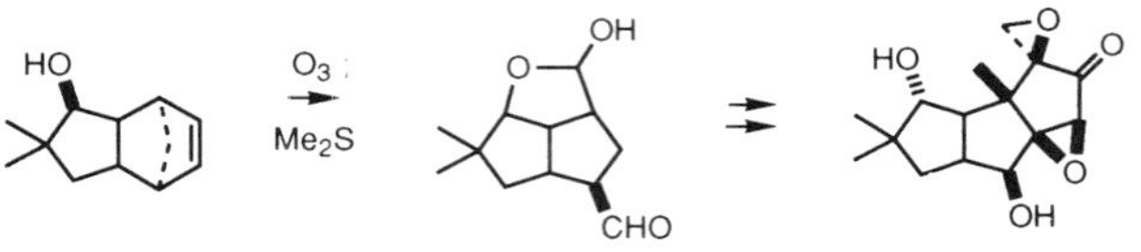

coriolin

A similar tactic was involved in the synthesis of methyl jasmonate [H. Tanaka, 1975] from a hydrindanone. Here, both chains were completely utilized.

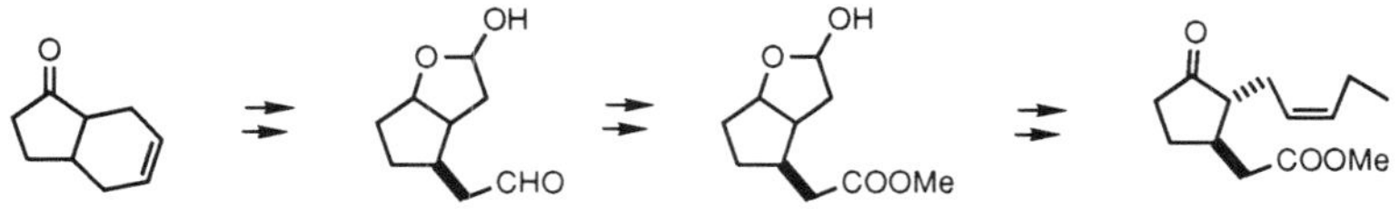

methyl jasmonate

2.2.4. Carboxyl Protection

Protection of the carboxyl group depends on whether it is necessary to mask only O-H or the whole. Esterification is the simplest way of removing the active hydrogen, whereas formation of orthoesters or oxazolines reduces the electrophilicity of the carboxyl group almost completely. However, for a long synthetic sequence it is often better to install the carboxyl group in a lower oxidation state, that is as a primary alcohol or aldehyde in their respective protected form.

For the protection of a carboxyl group of robust molecules as an ester, the Fischer esterification procedure is convenient, particularly on a large scale. If much milder esterification conditions are required a proper dehydration agent such as dicyclohexylcarbodiimide is added. Ester formation via *O*-alkylation is a complementary tactic.

Regarding regeneration of the carboxyl group from an ester, many alternative and often milder methods than saponification have been developed. These include S_N2 displacement with good nucleophiles (e.g. PrSLi, PhSeNa), and reaction with hard-soft reagents (Me_3SiI, AlX_3-RSH).

Esters which permit selective cleavage are represented by methoxymethyl esters (cleaved with R_3SiBr), methylthiomethyl and *t*-butyl esters (cleaved by CF_3COOH), benzyl, benzyloxymethyl and phenacyl esters (cleaved by hydrogenolysis), allyl esters (cleaved by Me_2CuLi), and 2,2,2-trichloroethyl esters (cleaved by Zn-HOAc, electroreduction). triorganosilyl esters are relatively labile, and they are used transitorily.

Regarding an approach to $\Delta^{9(12)}$capnellene [Sternbach, 1989] which involved devolution of a diquinane intermediate from a 3,4-cyclopentanobicyclo[2.2.1] hept-5-ene-2-carboxylic acid, itself derived from an intramolecular Diels–Alder reaction, lactonization of the ozonized product (from reductive workup) allowed differentiation of the two primary alcohols. Subsequently the angular hydroxymethyl group was deoxygenated.

COOH O_3; $NaBH_4$ H O O H OH H O CHO H H H

$\Delta^{9(12)}$-capnellene

Activated esters are necessarily very reactive toward nucleophiles and therefore unstable. Among them are the thioesters which have been used in macrolactonization, and *N*-hydroxysuccinimide esters which frequently mediate peptide synthesis. Methylthiomethyl esters which can be activated *in situ* by *S*-methylation should find some applications.

Complete masking of the carboxyl group in the form of a 4,4-dimethyloxazoline derivative enables its survival in the presence of Grignard and hydride reagents. A bicyclic orthoester is the best choice among such protecting groups because those deriving from simple alcohols undergo very ready hydrolysis. In a synthetic approach to vindorosine [Winkler, 1990] the use of this bulky group as a stereocontrol element in an intramolecular photocycloaddition was delineated.

O O O NH N R O — O O O N N R O — LDA; RX — O O O NR' N R H O

N OH OAc N Me H COOMe

vindorosine

The carbonyl group of lactones can be protected as dithioketal. The preparation requires aluminum reagents.

2.2.5. Miscellaneous Protective Devices

Besides those common functionalities indicated above, many other groups need protection during synthesis. Naturally, if two or more functionalities can be masked in one operation, all the better. Such a case is the bromoetherification for mutual protection of the isopropylidene group and the tertiary alcohol during a synthesis of picrotoxinin [Corey, 1979b].

OH ''OBz NBS O Br ''OBz ⇉ O O O O O Br Zn O O O O OH

picrotoxinin

The regioselective electrophilation of unsymmetrical ketones is a perennial problem in synthesis. Although solutions of varying success have been developed, the dianion tactic [Harris, 1969] is still a viable one. Accordingly, a ketone which contains a CH_2 group at the α-position is formylated and deprotonated to give the α,γ-dianion before addition of the electrophile. The subsequent reaction occurs at the γ-position. Since deformylation of the product is very facile the technique can be regarded as a transient protection method for an active methylene group adjacent to a carbonyl. For preexisting β-dicarbonyl compounds (diketones, keto esters, etc.), α-phosphonyl, α-sulfinyl and α-sulfonyl ketones, the dianions can be generated directly.

O CHOH NaH; BuLi O CHO⁻ Li⁺ Na⁺ RX R O CHO⁻ M⁺ mild base R O

During a synthesis of methyl homodaphniphyllate [Heathcock, 1992] rearrangement occurred in the transformation of an α-chloromethylidene ketone into the α-isopropyl ketone by organocuprate reaction because the intermediate underwent a retro-Michael/Michael reaction sequence. This unwanted process was avoided by replacing the chloromethylidene moiety by an ethylidene unit, and delaying addition of the second methyl group (to form the isopropyl group). The retro-Michael reaction (C-N bond cleavage) was blocked, and a normal intramolecular Michael reaction (C-C bond formation) could take place.

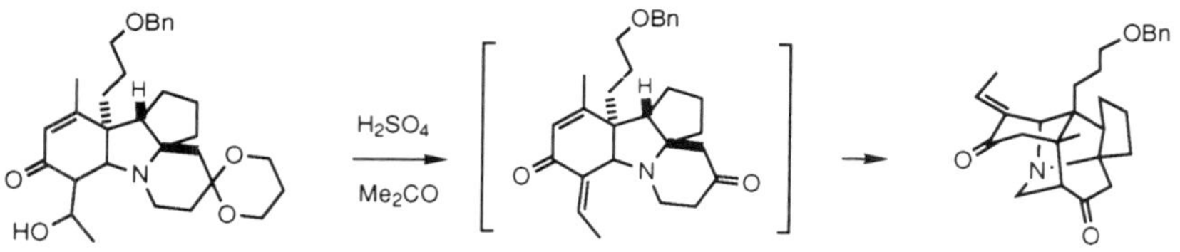

The preparation of 3,3-dimethyl-1-cyclohexenecarbaldehyde from cyclohexanone via exhaustive methylation of an α-heteromethylene derivative, hydride reduction, and hydrolysis shows expanded utility of a protective group in transpositional functionalization.

Protection of CC multiple bonds against destruction usually relies on addition and cycloaddition reactions, provided that such multiple bonds can be regenerated without affecting other functionalities. Bromination and epoxidation of an alkene are two frequently employed methods, in which under reductive conditions the addend atoms are removed from the adducts. Cycloaddition such as the Diels–Alder reaction also serves to protect a double bond, and the cycloadducts often dissociate on thermolysis. For example, a synthesis of *ar*-turmerone [T.-L. Ho, 1974a] from mesityl oxide required the masking of the existing double bond so that the *p*-cymyl unit could be achieved via aldol condensation and conjugate addition of the methyl group. A cyclopentadiene adduct of mesityl oxide was used in this work. Cycloadducts of fulvenes and 5-trimethylsilylcyclopentadiene will probably find more extensive utility in view of the mild conditions associated with their decomposition [Ichihara, 1979; Magnus, 1987].

A conceptually similar tactic is involved in an intramolecular Diels–Alder approach to lysergic acid [Oppolzer, 1981a]. Here a sensitive double bond which is part of the diene system was installed in a latent form.

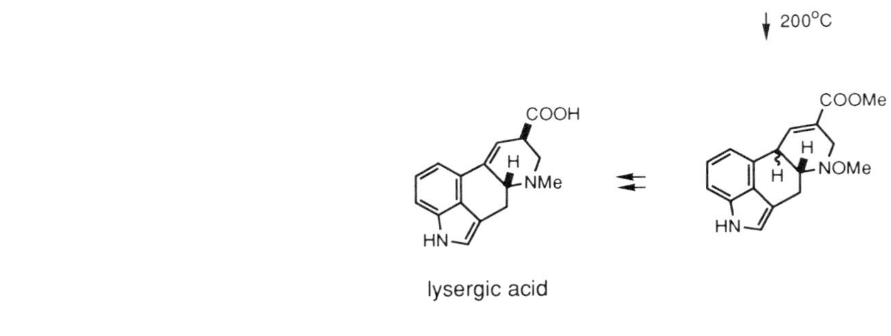

lysergic acid

Dicyclopentadiene may be considered as masked cyclopentadiene which permits synthetic operations at three consecutive carbon atoms. A route to *cis*-jasmone [Stork, 1971] was developed on the basis of this consideration.

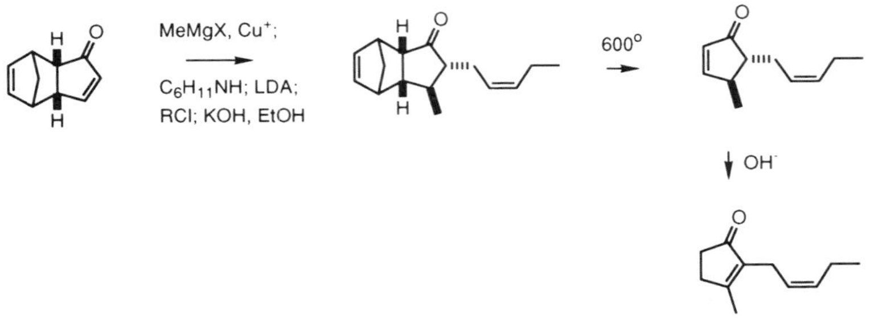

MeMgX, Cu^+; $C_6H_{11}NH$; LDA; RCl; KOH, EtOH

600°

OH^-

cis-jasmone

Removing one level of unsaturation of dimethyl acetylenedicarboxylate in the form of a Diels–Alder adduct, as described in the preparation of a potential intermediate of 2*β*-hydroxyjatrophone [Trost, 1986], is a necessity. The concern is not that of the survival of the unsaturation under various reaction conditions, but it is a matter of reactivity because electron-deficient alkynes do not undergo Pd(O)-mediated trimethylenemethane cycloaddition.

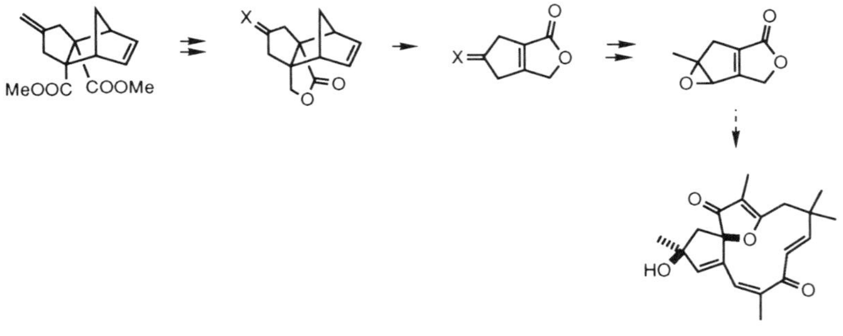

MeOOC COOMe

2β-hydroxyjatrophone

The formation of dicobalt hexacarbonyl complexes from propargyl alcohols has two favorable consequences. The complexation protects the triple bond, and it also activates the carbinol toward ionization. Even at −78°C, a primary hydroxyl in such a complex undergoes substitution with numerous nucleophiles [Nicholas, 1987]. Furthermore, the change in hybridization of the carbon atoms (sp to sp^3 state) permits intramolecular alkylation which, for steric reasons, cannot be accomplished. Thus a bicyclic enediyne system related to the calicheamicin antibiotics has been acquired [Magnus, 1988] by an intramolecular alkylation technique.

Conjugated dienes are more sensitive to various reagents and reaction conditions than isolated olefins, and their protection as 3-sulfolenes [Chou, 1989] by reaction with sulfur dioxide is adequate. The protection also renders the terminal positions of the original diene nucleophilic by sulfonyl group activation of the temporarily rehybridized carbon atoms.

3

UMPOLUNG

Heterolytic disconnection of a single bond leads to two synthons which guide the selection of starting materials or building blocks for the formation of that particular bond. As defined, synthons are cationic and anionic fragments of neutral molecules, and *a priori*, two sets of complementary synthons can be generated, viz, A^+/B^- and A^-/B^+ from disconnection of A-B.

In most cases, one set of synthons is logical, conforming to natural distribution of electric charges as favored by functional groups present in the synthons. For example, the disconnection of C—C bond linking a carbonyl group to its α-carbon atom to generate the $C^-/^+C{=}O$ pair is definitely a superior choice on account of the stability of acylium species, and consequently existence of numerous synthetic equivalents. On the other hand, the possibility of forming the C—C bond by the alternative mode of combining synthetic equivalents of $C^+/^-C{=}O$ is difficult but exciting because of its ramification of flexibility in synthetic operations. In certain situations this latter combination becomes more expedient. The unnatural or "illogically" derived synthons are the umpoled sythons [Hase, 1987].

The consequence of combining an umpoled synthon with another synthon is dependent on the nature of this synthon. Thus the umpoled and normal synthon combination would result in a disjoint molecular segment, whereas the union of two umpoled synthons gives rise to a conjoint species. Naturally, a conjoint segment always rises when two normal synthons combine.

Tactical guidelines for synthesis can be derived from such considerations. The synthesis of a disjoint molecular framework must involve reactivity umpolung. While two possibilities exist for the formation of a conjoint unit, it is usually far more inconvenient to employ two umpoled species in the process, unless special building blocks which are inherently disjoint are readily available.

Concerning the tactic with reference to demand of functionality distribution and starting material availability, it is instructive to examine a synthesis of vermiculine [Seebach, 1977]. It seems that the choice of a four-carbon bromo epoxide obtained from malic acid determined the approach to this macrodiolide which is characterized by the presence of a continuous segment of conjoint, disjoint, and disjoint oxygen functions. The bromo epoxide is conjoint between the bromine-bearing and the proximal epoxy carbon atoms. Its reaction with an umpoled acyl anion (at the brominated site) generates the required disjoint unit, and ring opening of the disjoint epoxide by means of attack with another umpoled species establishes the conjoint sidechain. Extension at the other chain terminus to furnish the enedicarbonyl segment by two reactions involves an initial umpolung. The one-carbon homologation results in a disjoint *vic*-functionality which requires coupling with a normal (conjoint) synthon.

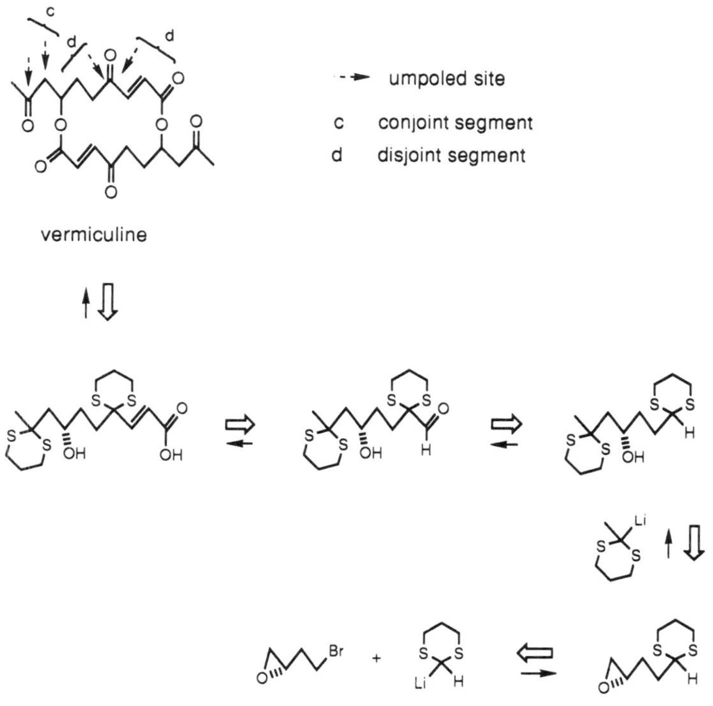

umpoled site
c conjoint segment
d disjoint segment

vermiculine

3.1. ACYL ANIONS

By far the most highly developed umpoled synthons are acyl anions. 2-Lithio-1,3-dithianes are readily generated because the hydrogen atom(s) at C-2 of the dithianes are quite acidic (pKa = 31.1) [Streitwieser, 1975]. They react with a wide range of electrophiles and therefore serve as acyl anion equivalents [Corey,

1965c; Page, 1989], thus unsubstituted 1,3-dithiane provides a formyl group to an alkyl chain in a homologation process and further reaction reaches a masked ketone. Several methods are available for converting 2-substituted 1,3-dithianes into carbonyl compounds. Cyclic ketone formation by intramolecular reaction has been demonstrated [Seebach, 1968; Grotjahn, 1981].

R = H, alkyl, ...

2-Lithio-1,3-dithianes attack the carbonyl group of cycloalkenones, but they behave as Michael donors when the reaction media contain 1–2 equivalents of HMPA [C.A. Brown, 1979].

1,3-Dithianes are also accessible from reduction of the dehydro derivatives, that is ketene dithioacetals [F.A. Carey, 1972]. Related acyl anion equivalents are lithium methylthioformaldine [Balanson, 1977] and the conjugate base of methyl methylthiomethyl sulfoxide [Ogura, 1971, 1974]. Also of expected usefulness are bisselenoketal carbanions [Burton, 1979] in serving the same purpose. Interestingly, the carbanion of (phenylselenyl)methyl-trimethyl-silane is a formyl equivalent [Sachdev, 1976]. An aldehyde is obtained from the alkylation product by oxidation with hydrogen peroxide.

Other *gem*-difunctional methanes and alkanes that have been employed as acyl anion equivalents are α-aminonitriles [Stork, 1978; Ahlbrecht, 1979], α-cyanohydrin ethers [Evans, 1974b; Hünig,1975]. These substrates can be deprotonated with amide bases. Quite remarkably, macrocyclic ketones can be prepared by intramolecular alkylation of ω-haloalkyl cyanohydrin ethers without resorting to high-dilution techniques [T. Takahashi, 1981a,b]. Enantioselective synthesis of 1,4-diketones via asymmetric Michael addition using metallated chiral α-aminonitriles has been described [Enders, 1992].

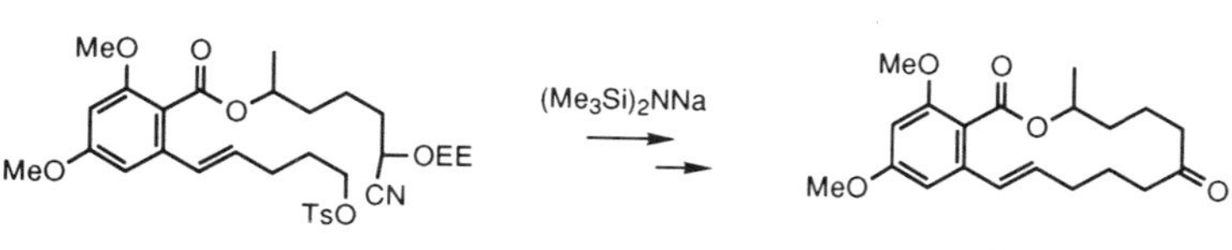

$(Me_3Si)_2NNa$

zearalenone

Such compounds as vinyl sulfides [Cookson, 1976; Harirchian, 1977; Oda, 1983] and vinyl ethers [J.E. Baldwin, 1974] by virtue of inductive effects of the heteroatoms in ethers, and the coordinative stabilization of the α-lithio derivatives, represent another class of valuable precursors of acyl anions. It should be noted that aryl vinyl ethers cannot be used because of preferential *o*-lithiation [Muthukrishnan, 1976]. α-Methoxyallenyllithium is a special acyl anion equivalent to acrolein [Gange, 1978], its reaction products with carbonyl compounds readily furnish dihydrofuran-3(2*H*)-ones.

RX → RX–Li ≡ O
X = O, S

PhS–SiMe3 —F⁻→ PhS⁻ —RCHO→ PhS R OH → O R OH

MeO C Li ≡ O

α-Trimethylsilylvinyllithium is a masked acetyl anion [Gröbel, 1974]. The electrophilation products are converted into ketones via the epoxides. The cuprate reagent obtained from the vinyllithium adds to enones [Boeckman, 1974], therefore 1,4-diketones can be prepared using such a reagent.

Me3Si Li —RX→ Me3Si R → Me3Si O R → O R

Me3Si CuLi (2) – O → Me3Si O → O O

Of particular significance is that acyl anions can be generated *in situ* from aldehydes [Stetter, 1976] in the presence of a thiazolium ylide. With aromatic aldehydes the cyanide anion is an equally effective catalyst. However, such masked acyl anions do not react with most of the common electrophiles and they partake in Michael addition mainly. (In the absence of a Michael acceptor, aldehydes undergo self-condensation to give α-ketols, the benzoin condensation being a special case for aromatic aldehydes).

R'' N+ R''' S + RCHO → R'' N+ O⁻ R''' S R ⇌ R'' N+ OH R''' S R ≡ O R

R' O

O R R' O ← R'' N+ O⁻ R''' S R R' O ⇌ R'' N+ OH R''' S R R' O⁻

Dianions derived from nitroalkanes can serve as acyl anion equivalents [Lehr, 1979]. After electrophilation the carbon atom bearing the nitro group can be converted into a carbonyl site by methods such as the Nef reaction, oxidation and reduction procedures.

Furans are masked γ-oxoacyl anions. Lithiation of these substances has been demonstrated [Gilman, 1934] a long time ago, and the application is illustrated in a synthesis of *cis*-jasmone [Büchi, 1966b].

H Li O O O O O

The α-oxy anion is at a lower oxidation level than an acyl carbanion. Three tactics to generate α-oxy anions or equivalents involve acceptor-stabilized carbanions [J.W. Wilson, 1980; Tamao, 1983], deprotonation of dipole-stabilized carbanions from hindered esters [Beak, 1981], and metal-exchange of α-stannyl carbinols [Still, 1980c]. As shown in the accompanying equations certain boryls and silyls are suitable acceptor groups for stabilizing α-carbanions and they themselves are transformable into a hydroxyl function.

R BMes$_2$ → R(Li) BMes$_2$ —R$^+$; HOO$^-$→ R(R') OH

ClMg Si(OiPr)$_2$ —RX, M→ R Si(OiPr)$_2$ —H_2O_2→ R OH

M= Pd, Ni R = Ar, viny, allyl

R OH → R O(C=O)Ar —sBuLi→ R(Li) O(C=O)Ar ⇉ R(R') OH

Bu$_3$Sn OH —BuLi→ [BuSn O]$^{2-}$ 2 Li$^+$ —R$^+$→ R OH

≡ $^-$ OH

3.2. α-ACYL CARBOCATIONS

α-Bromocarboxylic acids and esters represent classical equivalents of α-acyl carbocation synthons when they act as electrophiles, as seen in a preparation of glycine in which ammonia displaces the bromine atom. As acetonyl carbocation one can use 2,3-dihalopropene [Negishi, 1983]. The allylic halide is readily displaced and the acetonyl unit can be uncovered from vinylic halide group of the products by treatment with mercuric acetate in 88% formic acid.

In the role of a Michael acceptor vinyl sulfoxides form C—C bond with appropriate carbon nucleophiles [Seki, 1975; P.J. Brown, 1984]. The adducts are convertible into carbonyl compounds by Pummerer rearrangement.

Similarly, conjugated nitroalkenes undergo additions with 1,3-dicarbonyl compounds, carboxylic acid dianions or ester enolates, and silyl enol ethers and ketene silyl acetals, the last reaction in the presence of a Lewis acid [Yoshikoshi, 1985]. The adducts are precursors of 1,4-dicarbonyl compounds.

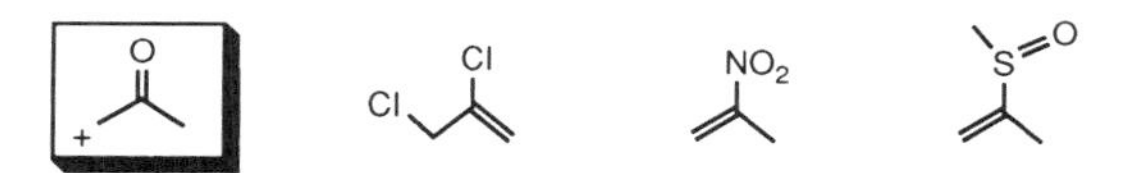

A useful synthetic equivalent of acetone α,α′-dication is 2-nitro-3-pivalyloxypropene [Seebach, 1984]. Usually this compound undergoes C—C bond formation with loss of the allylic ester group. The initial adducts can then participate in another alkylation in the Michael fashion.

3.3. HOMOENOLATES

These umpoled synthons find their synthetic equivalents in the forms of acetalized organometallic reagents such as Grignard reagents [Büchi, 1969, Bal, 1982] and carbanions such as those stabilized by a sulfonyl [Kondo, 1975], a phosphonyl [A. Bell, 1978], or a nitro group [Corey, 1969b]. The activating/stabilizing acceptor functionality is removed from the adducts at a later stage. Homoenolates are mostly involved in the synthesis of a disjoint difunctional substances, such as 1,4-diketones.

$X = SO_2Ph, NO_2, P(O)R_2$

2-Methoxycyclopropyllithium reagents react with carbonyl compounds, and the products undergo ring opening to give β,γ-unsaturated aldehydes [Corey, 1975b]. 3-Methoxypropylidenetriphenylphosphorane is another homologation reagent [S.F. Martin, 1977b]. 1-Ethoxy-1-trimethylsiloxycyclopropane is a source of ester β-enolate [Nakamura, 1977], in the presence of titanium(IV) chloride its reaction with aldehydes furnishes γ-lactones.

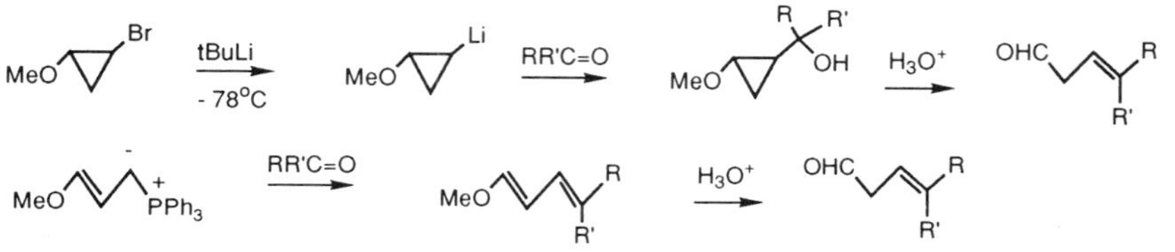

An indirect method of introducing a propanoic acid β-carbanion is via reaction of a carbonyl compound with the ylide derived from a cyclopropyl diphenylsulfonium salt, rearrangement of the products to give the cyclobutanones, and Baeyer–Villiger oxidation [Trost, 1973].

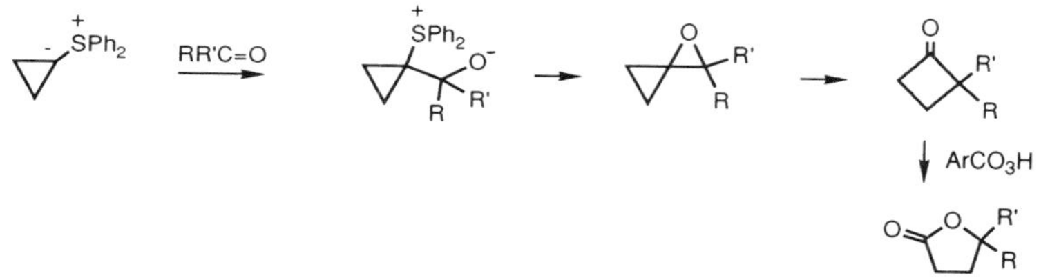

In certain circumstances α,β-enone β-anion equivalents may be derived from the enones via conjugate hydrocyanation and deprotonation at both α- and β-positions. The cyano-stabilized anions are more reactive than the enolates, therefore they undergo electrophilation more rapidly. Elimination of HCN from the products generates the β-substituted α,β-enones [Debal, 1977].

The same sacrificial act can be played by other strong acceptor groups. Thus β-nitropropanoic esters are useful for appending an acrylic acid chain to a carbon skeleton [Bakuzis, 1978]. The disjoint system of pyrenophorin can thereby be assembled.

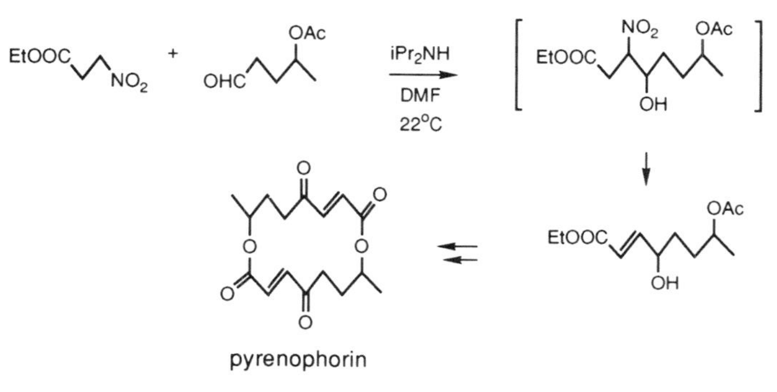

pyrenophorin

Heteroatom substituted allyl anions constitute another class of homoenolate equivalents. With two bulky groups, for example mesityl, an allylborane can be deprotonated and alkylated [Pelter, 1983]. Oxidation of the products yields aldehydes. Allylsilyl anions react with electrophiles, the γ-products (vinylsilanes) derived from their reactions with aldehydes and ketones have been converted into lactol ethers via epoxidation and BF_3 treatment [Ehlinger, 1980].The regiochemistry of the addition step can be controlled by metal ions (Li favors γ, Mg favors α) [Lau, 1978].

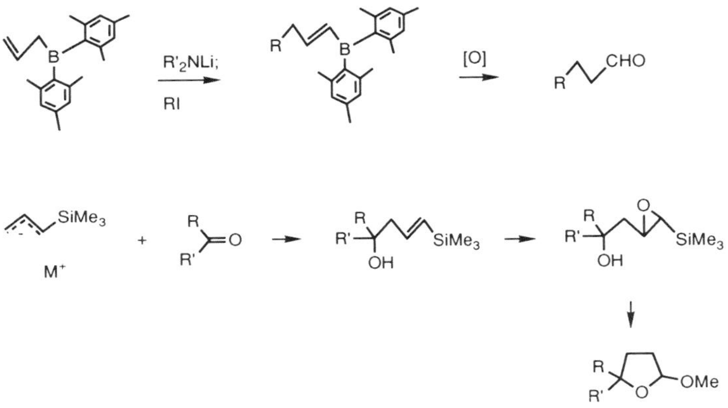

R'$_2$NLi; RI → [O] → CHO

α-Methoxyallyl boronates condense with aldehydes at the γ-position [Hoffmann, 1988] to give (Z)-enol ethers. There is a transfer of the boron residue from the carbon atom of the reagents to the oxygen atom of the substrates. It should be noted that the reagents contain a captodative center which is removed during the reaction.

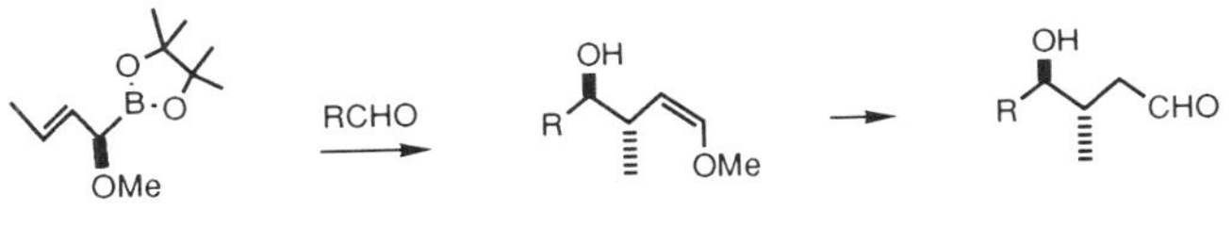

RCHO →

γ-Alkoxyallyl boronates [Hoffmann, 1982b] and aluminates [Yamaguchi, 1982] also react with aldehydes at their γ-position to furnish *vic*-dioxygenated products.

α-Lithiated 2-alkenyl N,N-diisopropylcarbamates undergo metal exchange for titanium in an interesting manner [Hoppe, 1990]. The configuration of the α-carbon is retained on treatment with $(i\mathrm{PrO})_4\mathrm{Ti}$, and it is inverted when reacted with chlorotris(diethylamino)titanium. Since lithiation of the carbamates in the presence of a chiral amine (e.g. (−)-sparteine) is remarkably stereoselective, enantioselectivity of homoaldols can be secured by the γ-selective condensation of these corresponding Ti-species which gives rise to enantiomeric products.

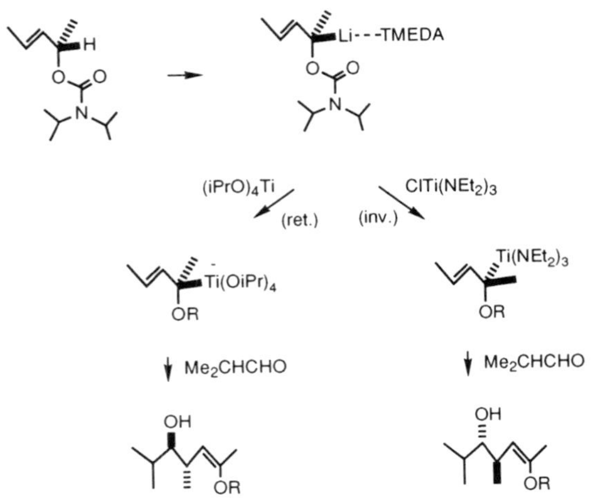

Allyl ethers are metallated with *sec*butyllithium in THF at $-65°C$ to give ambident anions [Evans, 1974a]. The regioselectivity of their reactions with electrophiles is dependent of the countercation. The allylzinc species undergo alkylation mainly at the α-position, and when the addition to α,β-unsaturated carbonyl compounds [Evans, 1978] is coupled with an oxy-Cope rearrangement of the adducts, the process should provide 1,6-dicarbonyl compounds.

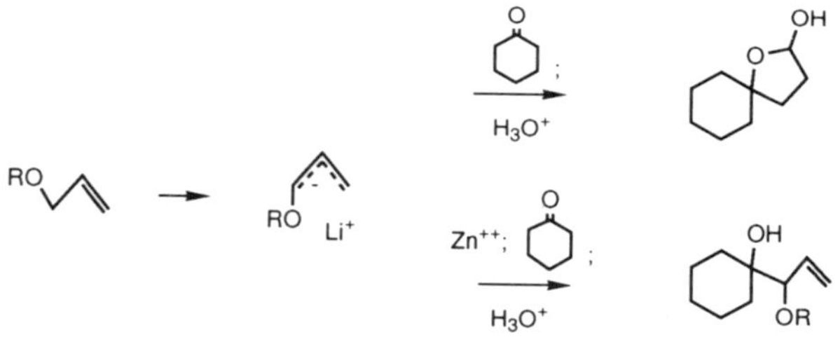

γ-Acylation of α-siloxyallysilanes with acid chlorides [Hosomi, 1978] followed by hydrolysis leads to γ-ketoaldehyes.

The reactivity of lithiated allyl sulfides has been examined [Biellmann, 1968; Altani, 1974]. The regiochemistry can be manipulated by the addition of complexing agents, for example, exclusive γ-alkylation with allylic bromides via an S_N2' mechanism has been observed when the allyllithium is treated with copper(I) iodide first [Oshima, 1973]. While thioallylic monoanions react preferentially at the α-position, thioacrolein dianion shows γ-selectivity in its electrophilation in the presence of TMEDA [Geiss, 1974].

1,3-Bis(methylthio)allyllithium, an equivalent of acrolein β-anion [Corey, 1971], presents no ambiguities in regiochemistry of its alkylation. The products afford unsaturated aldehydes on hydrolysis with mercuric chloride in aqueous acetonitrile.

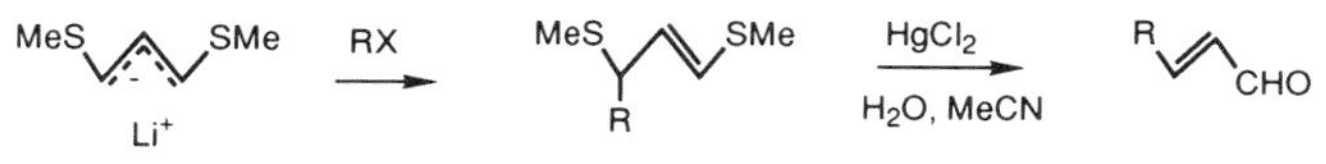

Metallated enamines are also suitable homoenolate equivalents [Ahlbrecht, 1977]. The two other substituents of the amino group can vary greatly, including pyrrolidine [S.F. Martin, 1977a] and carbazole derivatives [Julia, 1974]. A chiral allylamine has been deprotonated and alkylated [Ahlbrecht, 1980], and moderate asymmetric induction was observed.

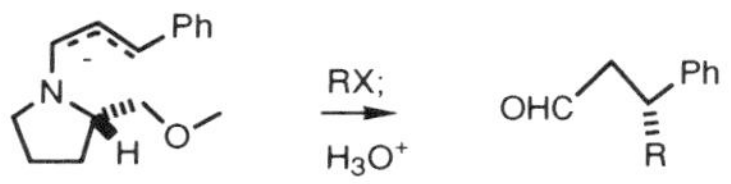

While γ-selectivity is an implicit requisite in the electrophilation of metallated enamines for acting as homoenolate equivalents, it should be indicated that reactions of 1-nitrosaminoallyllithiums kinetically favor α-reactivity with carbonyl substrates [Renger, 1977], but the α,γ-ratio dependence is related to the thermodynamic stability of the primary lithium alkoxide adducts which controls product formation.

3.4. MISCELLANEOUS UMPOLUNGS

This section describes some techniques and/or possibilities that reverse the polar characters of reactants in such cases the demand is not particularly strong. For example, rendering the α-carbon of an amine into a carbanion can be achieved via the *N*-nitrosamine. Deprotonation with an organolithium reagent affords the umpoled species which readily undergoes electrophilation [Seebach, 1975]. The α-substituted amine is recovered by reductive cleavage of the N—N bond.

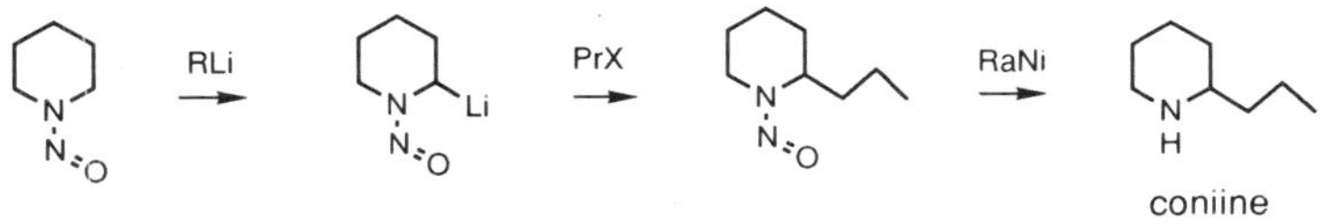

coniine

Alkylation of Reissert compounds from quinoline and isoquinoline followed by elimination of the cyano group from the products virtually performs an umpolung process.

γ-Acyl cations have their synthetic equivalents in acylcyclopropanes, which are usually doubly activated (i.e. with another acceptor group geminal to the acyl functionality). Their capability of undergoing homoconjugate addition [Danishefsky, 1977a] proved valuable to synthesis, for example as a basis for the development of a route to several necine bases.

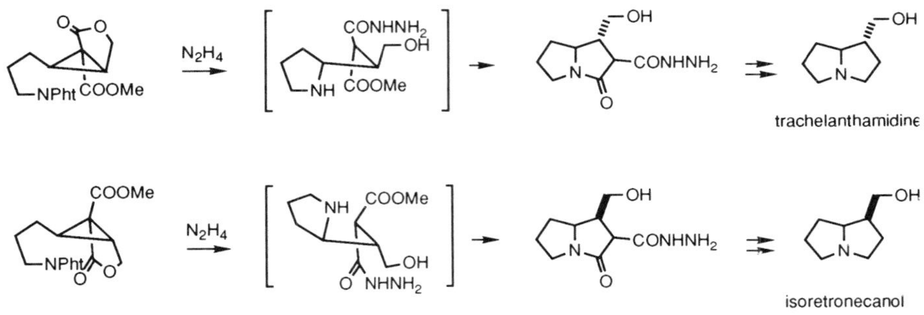

trachelanthamidine

isoretronecanol

To coax an allyl alcohol (and its acetate, mixed carbonate, etc.) into the role of an allyl anion equivalent, by *in situ* deoxystannylation using a Pd(O)-$SnCl_2$ system [Masuyama, 1988] is rather novel. The allylstannane species react with aldehydes in a γ- and *anti*-selective manner. Interestingly, a *syn* product is obtained predominantly from salicylaldehyde, owing to chelation of the hydroxyl group with the tin atom in the transition state.

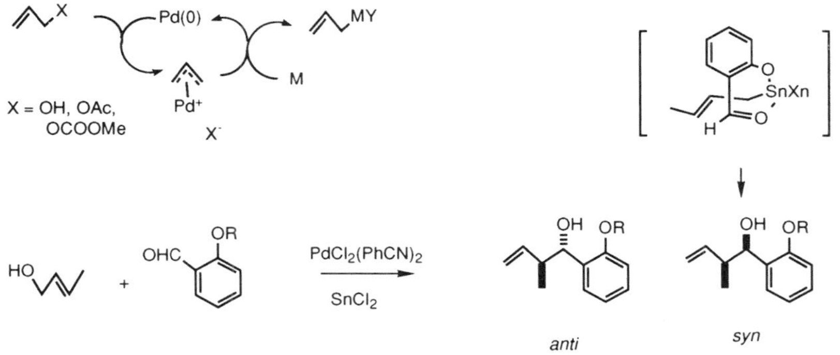

X = OH, OAc, OCOOMe

	anti		syn
R = Me	94	:	6
R = H	6	:	94

An even more interesting reaction is the formation of a 2-alken-3-one from allyl acetate and a primary alcohol in the presence of tris(triphenylphosphine)-ruthenium dichloride [T. Kondo, 1991]. Contrasting to π-allylpalladium complexe., the π-allylruthenium species are nucleophilic, coupled with their ability to dehydrogenate primary alcohols to generate aldehydes the said tranformation eventuates.

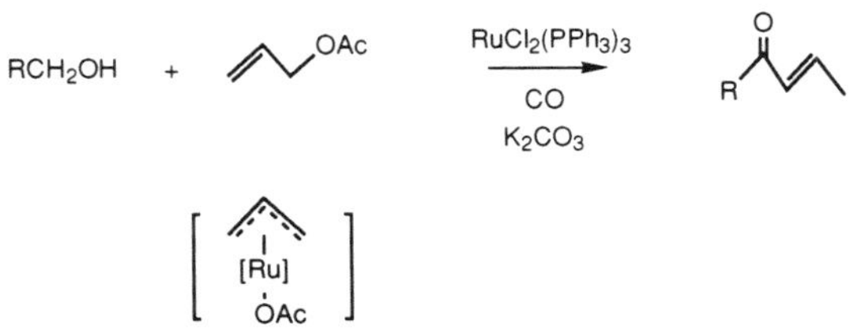

4

TANDEM REACTIONS

To attain high efficiency is one of the most important objectives of chemical synthesis. The employment of tandem reactions in a multistep synthesis usually helps the effort. While there are restrictions to effecting tandem reactions, many such processes are known [T.-L. Ho, 1992b].

Tandem reactions are those in which under the applied reaction conditions the structural features of the initial products are favorable to undergo further transformation(s). Necessarily the secondary reaction is not possible before the first reaction is accomplished. Clever combination of different reactions in timed order can generate very unusual compounds. In the following paragraphs some examples are delineated to demonstrate the diversity and richness of such reactions. Undoubtedly tandem reactions constitute a major area for rapid development.

4.1. ALDOL CONDENSATION

Tandem reactions ending in an aldol condensation have a long history. For example, the Robinson annulation [Gawley, 1976] consists of a base-catalyzed Michael reaction between a ketone enolate and an enone which is followed by aldolization to close a six-membered ring. In situ dehydration of the β-ketols often drives the annulation into completion. An extension of the method enables the formation of two rings [Danishefsky, 1971].

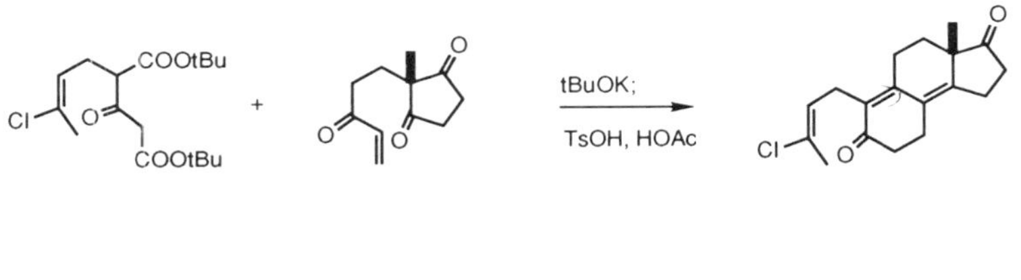

Bridged ring systems can be created from an intramolecular Michael/aldol tandem, and such a reaction sequence is the basis for an approach to patchouli alcohol and seychellene [K. Yamada, 1979].

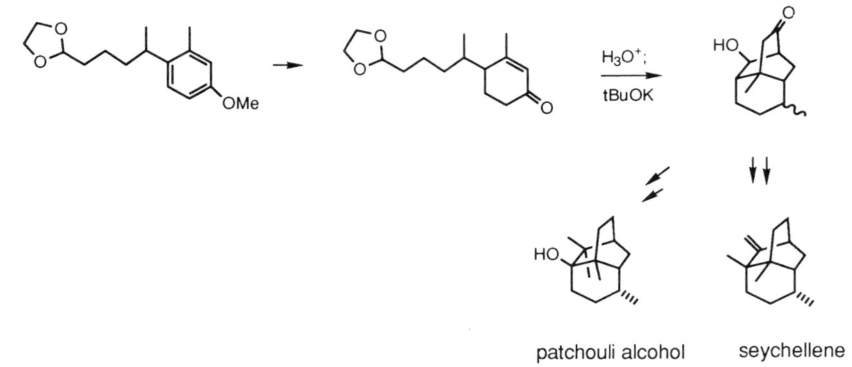

H_3O^+; tBuOK

patchouli alcohol seychellene

The Claisen/aldol condensation tandem has analogy to the ring formation step(s) for the genesis of many polycyclic polyketide molecules in nature. Biomimetic synthesis of them has received much attention, and records show successful elaboration of numerous such compounds, for example emodin, chrysophanol, and eleutherin [Harris, 1976].

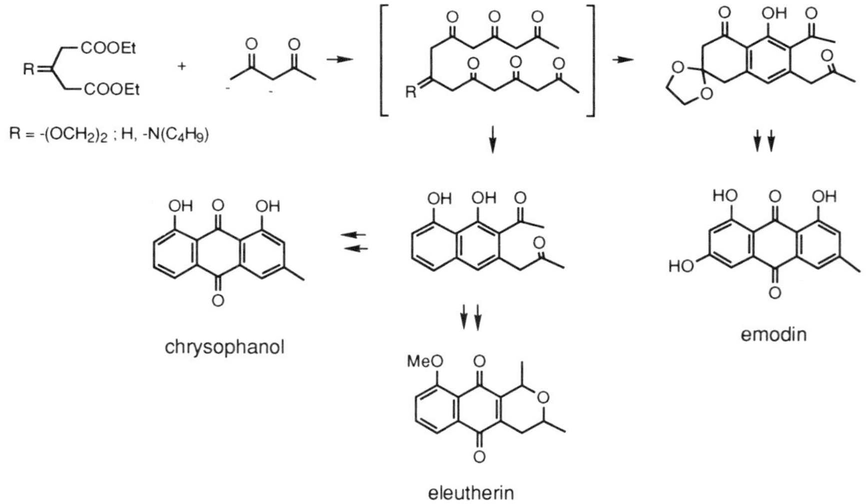

R = -$(OCH_2)_2$; H, -$N(C_4H_9)$

chrysophanol

emodin

eleutherin

The retroaldol/aldol reaction tandem occurs quite often. The sequence can be planned into synthetic schemes, as illustrated in a route to *cis*-jasmone [Wenkert, 1970; McMurry, 1971] and a synthesis of atisine [Guthrie, 1966]. Although not a prerequisite, in both cases the retroaldol fission was facilitated by strain relief on cleavage of a small ring.

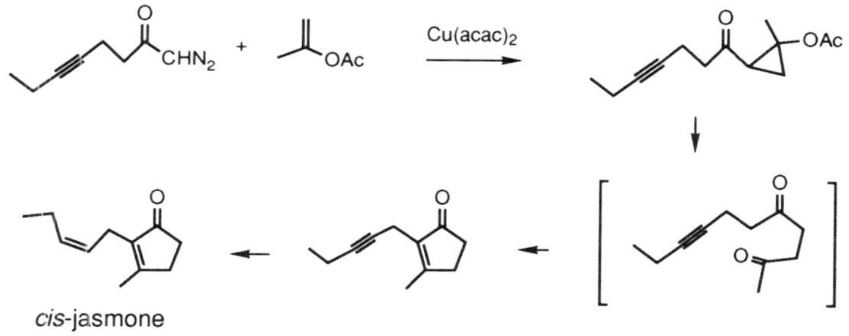
CHN2
OAc
Cu(acac)2
OAc
cis-jasmone

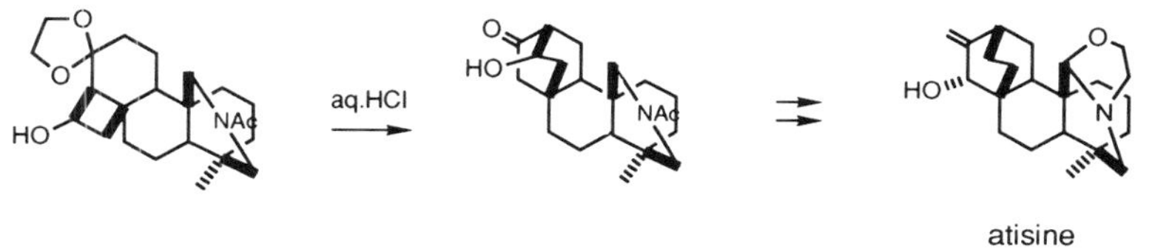
HO
NAc
aq.HCl
HO
NAc
HO
N
atisine

The Oppenaur oxidation proceeds in basic solutions, therefore a dicarbonyl compound generated by the method may undergo aldol reaction. Because of this tandem transformation a synthetic roue of lycopodine [Heathcock, 1982] was shortened by one step.

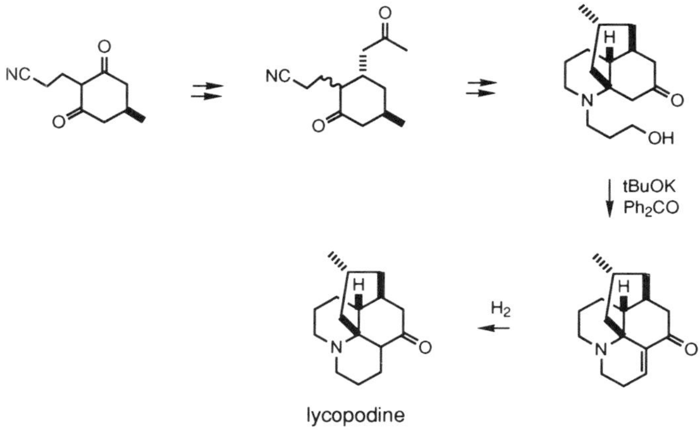
NC
NC
H
N
OH
tBuOK
Ph2CO
H
N
H2
H
N
lycopodine

Perhaps it is appropriate to consider the Horner–Emmons reaction as a variant of aldol condensation. Of the many uses of this method a tandem process intervened by a retroaldol fission is noteworthy. The result is a skeletal rearrangement, as shown in its application to a synthesis of α-acoradiene [Y. Yamamoto, 1990].

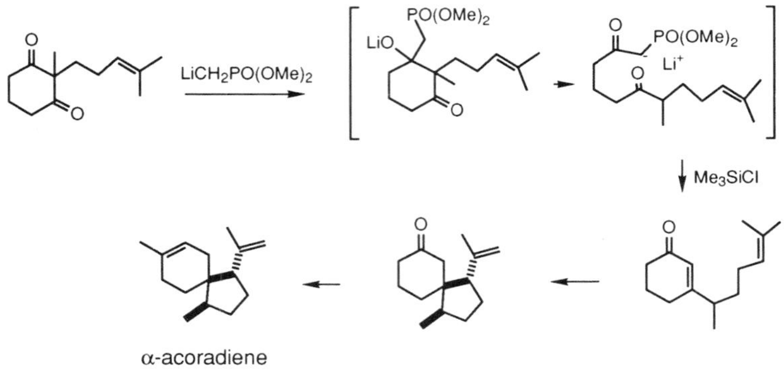

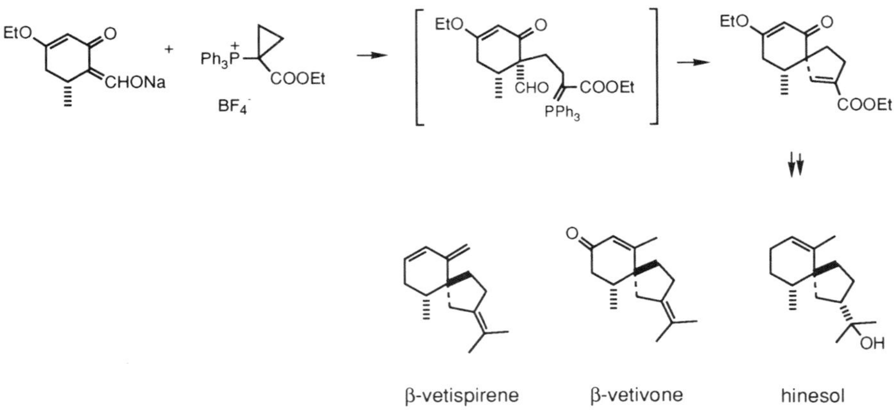

α-acoradiene

β-vetispirene β-vetivone hinesol

Aldoı condensation is also effected by acid catalysis. A useful modification is the Wichterle–Lansbury cyclization in which the enol/enolate is surrogated by a chloroalkene. An alkyne linkage can act the same role, and such is involved in the annulation [Sisko, 1992] represented by the following equation.

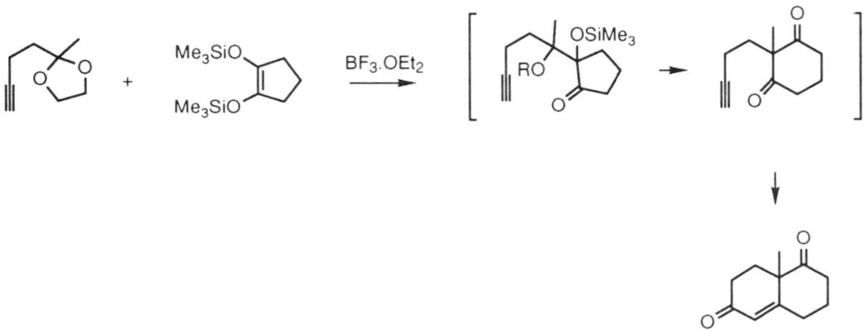

The Wittig reaction with a carbonyl-stabilized ylide is related to the aldol reaction. Thus the three-component synthesis of α,β-unsaturated esters from ketenylidenetriphenylphosphorane, an alcohol and an aldehyde is a tandem process apparently consisting of acylation and intramolecular Wittig reaction [Bestmann, 1985]. This reaction sequence has been extended to the obtention of certain macrocyclic α,β-unsaturated lactones.

$n = 8, 10$

4.2. MICHAEL, DIECKMANN, AND CLAISEN REACTIONS

All steps of a Michael reaction involving stabilized donor species are reversible, therefore the structures of substrates and adducts, and reaction conditions are of paramount importance to the success of reactions. Tandem reactions terminated with a Michael reaction are rather common, but only a few examples can be mentioned here.

A widely used ring-forming method is that comprising a double Michael reaction. For example, the cyclohexenone moiety of griseofulvin was rapidly generated by such a process employing a cross-conjugated enynone as a Michael acceptor [Stork, 1964b]. Interestingly, this seems to be the result of a kinetically controlled reaction, in view of the higher thermodynamic stability of epigriseofulvin.

griseofulvin

An $\alpha,\beta;\gamma,\delta$-dienone may participate in a double Michael reaction as the acceptor, with bond formation occurring at the β and δ carbon atoms. This efficient annulation has been applied to a synthesis of occidentalol [Irie, 1978; Mizuno, 1980].

occidentalol

Two Michael reactions may be interposed by another reaction. Thus, in the Weiss cyclization [Gupta, 1991] which is a convenient method for access to bicyclo[3.3.0]octane-3,7-diones the closure of the second ring must await the emergence of the enones by dehydration.

Reflexive Michael reaction represents a special kind of double Michael reaction between a kinetic enolate of an enone (cross-conjugated dienolate ion) and a Michael acceptor, wherein each reactant acts as both acceptor and donor in appropriate stages. A six-membered ring is formed in the second stage, and the overall result is the creation of a bicyclo[2.2.2]octanone system.

The complete tetracyclic skeleton of (+)-atisirene has been generated by an intramolecular version of the reflexive Michael reaction from a substrate containing two isolated six-membered rings [Ihara, 1986].

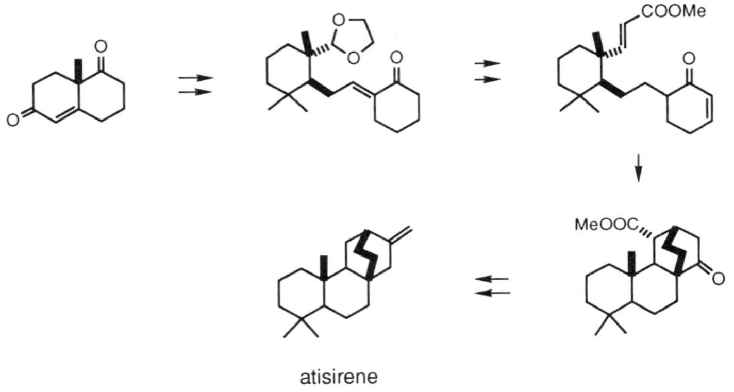

atisirene

Many other reactions may precede a Michael reaction. For example, in a synthesis of ibogamine [Büchi, 1966a] the molecular skeleton of an intermediate required modification, and after reductive cleavage of a C—N bond an intramolecular Michael reaction, with the nitrogen atom joining the β-carbon atom of the enone system, established the correct framework.

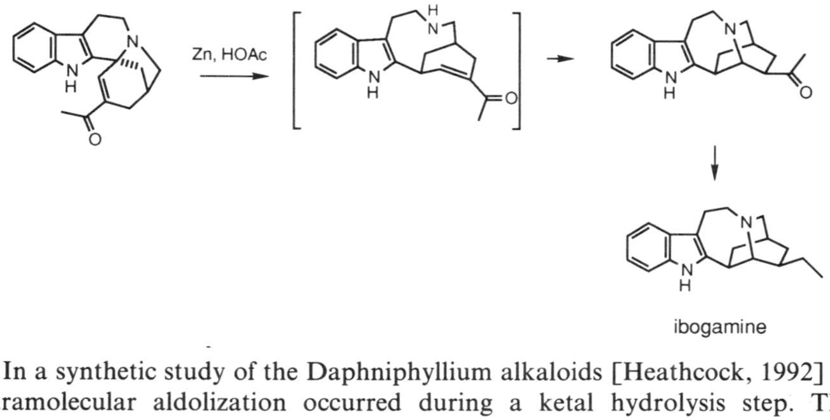

ibogamine

In a synthetic study of the Daphniphyllium alkaloids [Heathcock, 1992] an intramolecular aldolization occurred during a ketal hydrolysis step. This inadvertent reaction was reversed in a subsequent base treatment which also induced the desired Michael addition.

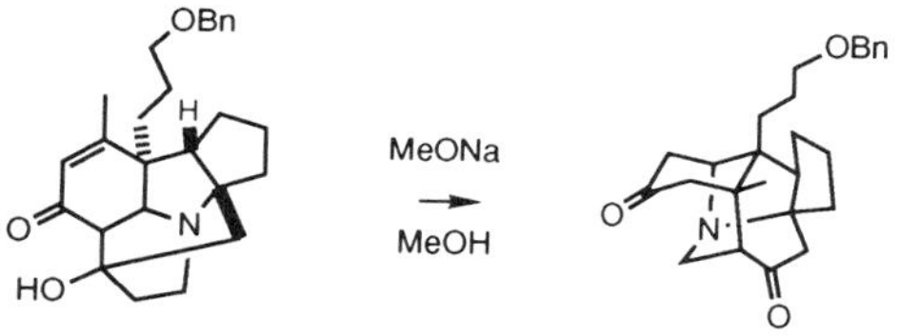

MeONa / MeOH

In the presence of a strong base system bromobenzene reacts with the kinetic enolates of enones to give α-tetralones (and sometimes indanones) [Essiz, 1976]. Apparently the reaction proceeds via benzyne and benzocyclobutenoxides and the latter species undergo ring opening and finally an intramolecular Michael reaction.

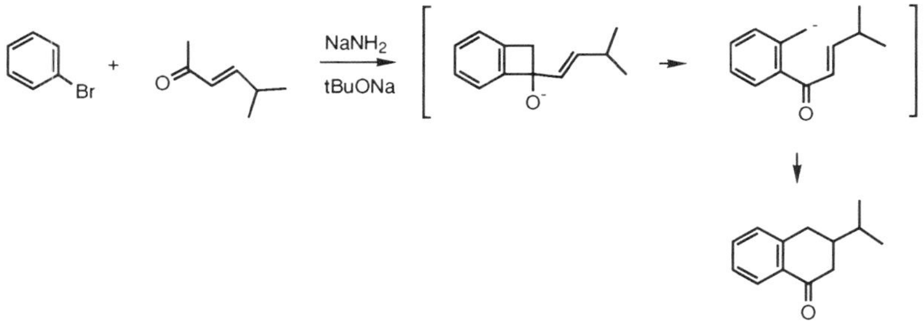

$NaNH_2$ / tBuONa

The carbanion generated from cleavage of a nascent 2-hydroxycyclopropanecarboxylic ester has been trapped by an unsaturated sulfone [Marino, 1988], and the product is an apparent intermediate of compactin.

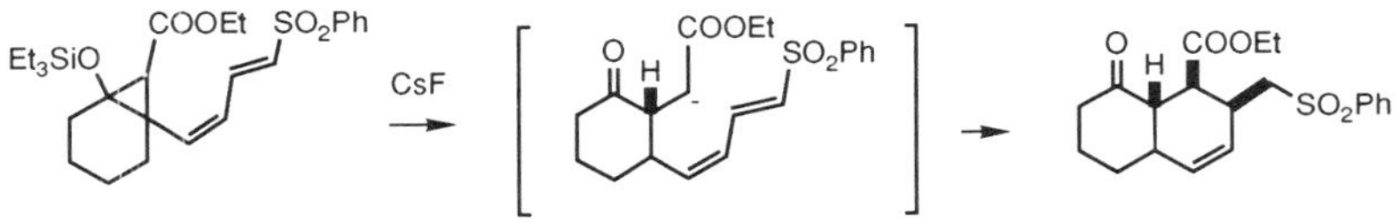

CsF

A convenient entry into the 2-substituted benzofuran series involves a sigmatropic rearrangement–Michael reaction sequence [Gray, 1992].

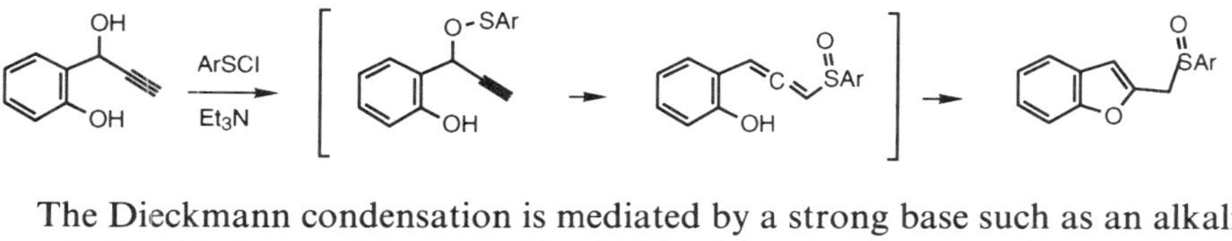

ArSCl / Et_3N

The Dieckmann condensation is mediated by a strong base such as an alkali metal alkoxide. Its combination with a Michael reaction is expected to constitute a useful synthetic process. An expedient preparation of 2,6-adamantanedione-1,3,5,7-tetracarboxylic ester [Stetter, 1974] is based on such a reaction tandem.

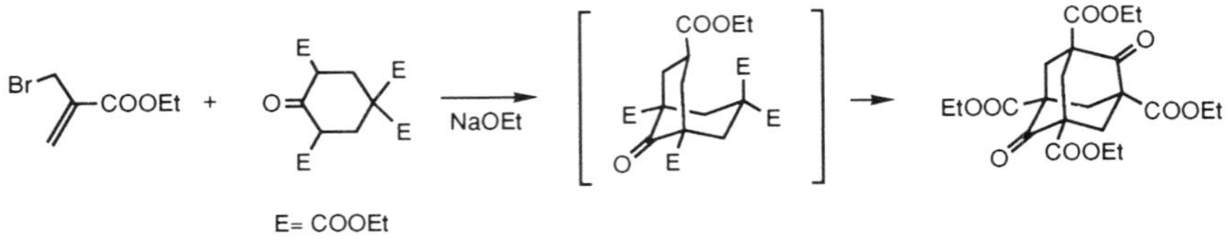

E= COOEt

A more recent application of this tandem in the formation of a ketoester precursor of (*R*)-(−)-muscone [Ogawa, 1991] demonstrated its usefulness in macrocyclic synthesis.

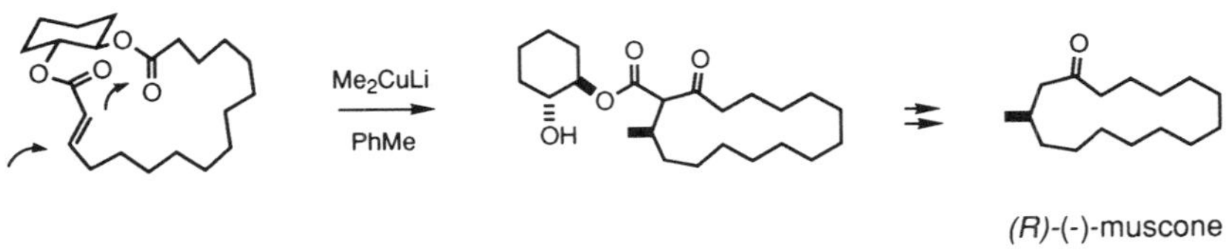

(*R*)-(-)-muscone

The homo-Michael ring opening of cyclopropane-1,1-dicarboxylic esters with malonate carbanions furnishes cyclopentanone derivatives directly, due to the occurrence of a subsequent Dieckmann condensation. Among many applications of this method a synthesis of methyl jasmonate [Quinkert, 1982] may be mentioned.

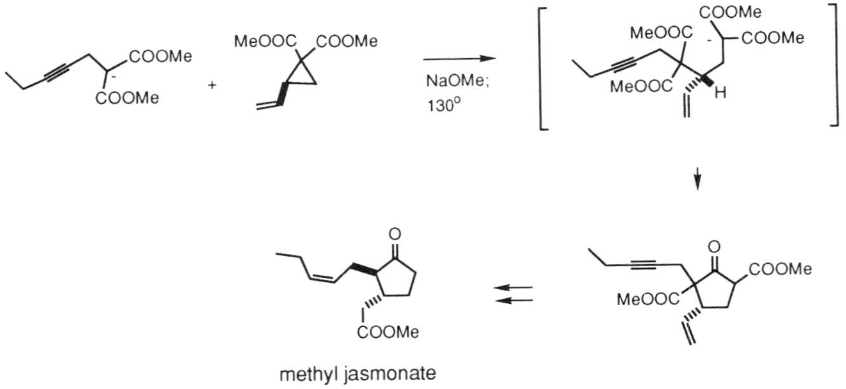

methyl jasmonate

The Michael–Claisen tandem has been known for a long time. Thus, the preparation of 5,5-dimethyl-1,3-cyclohexanedione from mesityl oxide and diethyl malonate [Vorländer, 1897] is initiated by this reaction sequence.

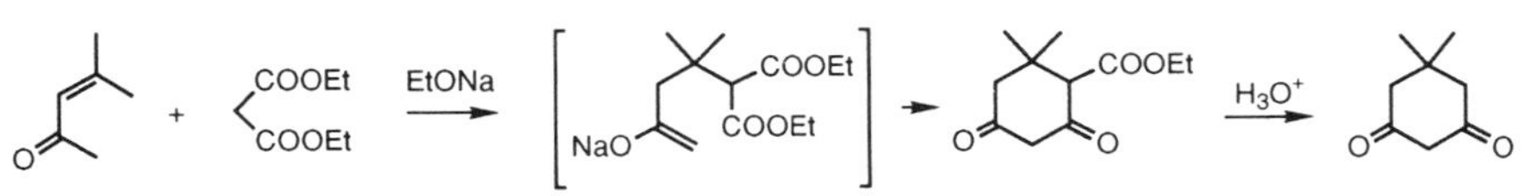

4.3. MANNICH REACTION

The Mannich reaction itself is a tandem process consisting of Schiff reaction and then a condensation similar to the aldol reaction. To highlight its utility we examine two variants of a route to lycopodine [Heathcock, 1982], both involving an intra-molecular Mannich cyclization to form three of the four rings. It should be noted that there are two ketones to be released during the initial stage of the reaction, but the Schiff reaction is regioselective due to the preferential formation of a six-membered ring to an eight-membered ring. On the other hand, the symmetrical nature of the Mannich reaction precursor for the synthesis of trachelanthamidine [Takano, 1981a] renders a concern for regioselectivity totally unnecessary.

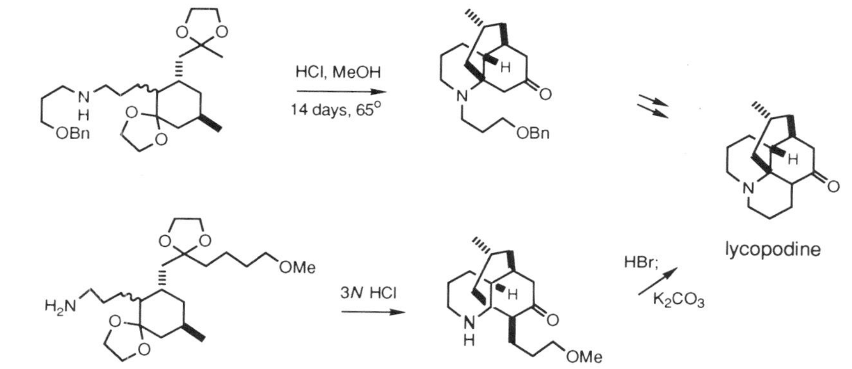

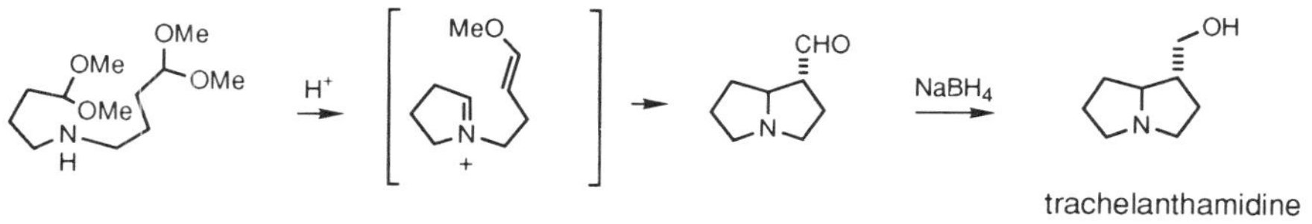

HCl, MeOH

14 days, 65°

3*N* HCl

HBr; K_2CO_3

lycopodine

H^+

$NaBH_4$

trachelanthamidine

Manifold Mannich reactions are very versatile synthetic processes. The famous tropinone synthesis [Robinson, 1917; Schöpf, 1937] consists of essentially the admixture of methylamine, succindialdehyde, and acetone (or its activated surrogates). In recent years the manifold Mannich reaction has attracted renewed interest including its application in the elaboration of ladybug alkaloids such as precocinelline [Stevens, 1979].

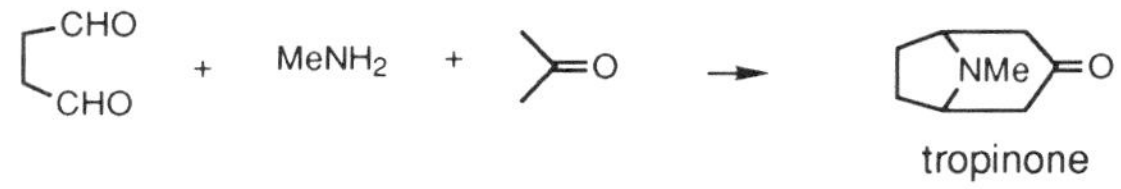

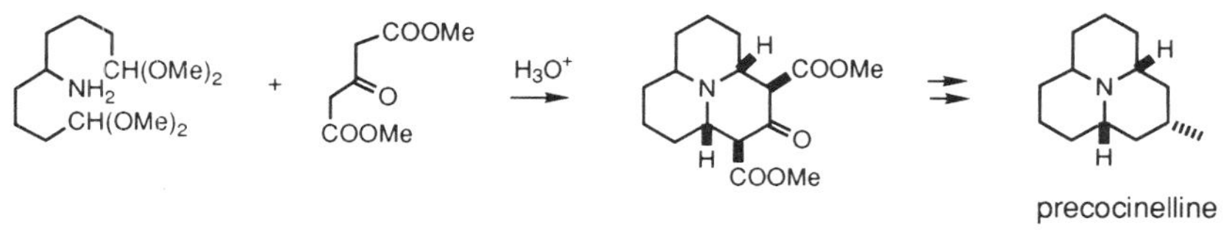

tropinone

H_3O^+

precocinelline

The Stork annulation is a combination of Michael and Mannich reactions. Using endocyclic enamines as the Michael donors, routes to many alkaloids have been developed. A relatively simple example is mesembrine [Curphey, 1968; Stevens, 1968].

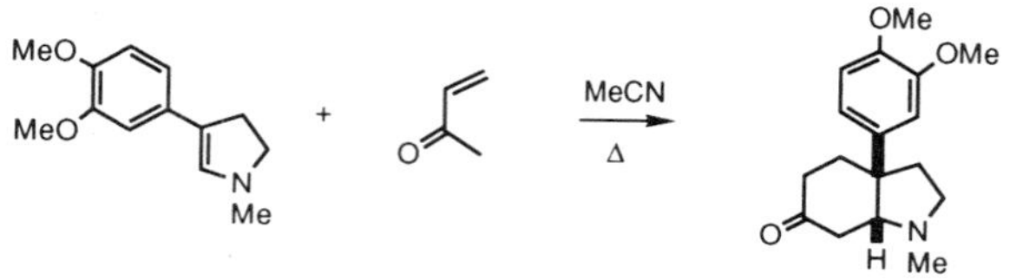

MeCN, Δ

mesembrine

A one-step synthesis of karachine [Stevens, 1983] from berberine involves formation of three C—C bonds in a sequence of Mannich, Michael, and Mannich reactions. Although the final step requires a boat transition state, such is not prohibitive because the precursor has already invested in such a conformation.

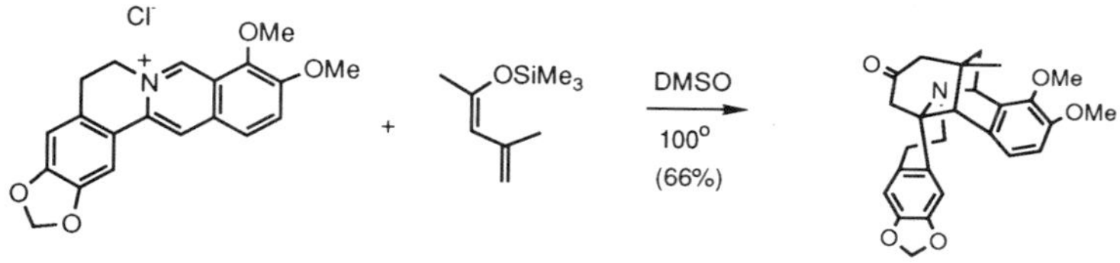

DMSO, 100°, (66%)

karachine

Acyliminium species are excellent electrophiles. In a case relevant to the Mannich reaction is the formation of a tricyclic lactam for an elaboration of methyl homodaphniphyllate [Heathcock, 1986].

A tandem reaction which starts from a cationic aza-Cope rearrangement and proceeds through a Mannich cyclization is the core of a series of elegant achievements in the synthesis of complex alkaloids. To illustrate the value of this method an outline of the relevant steps in a route to 16-methoxytabersonine [Overman, 1983] is shown below. Note that the preexisting five-membered ring is expanded while a new ring is formed.

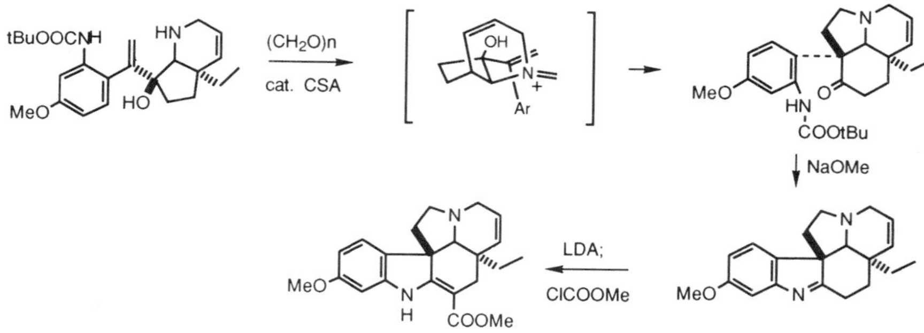

$(CH_2O)_n$, cat. CSA; NaOMe; LDA; ClCOOMe

16-methoxytabersonine

A very expedient route to the pentacyclic skeleton of the Aspidosperma alkaloids [A.H. Jackson, 1987] consists of a tandem Bischler–Napieralski and Mannich reactions.

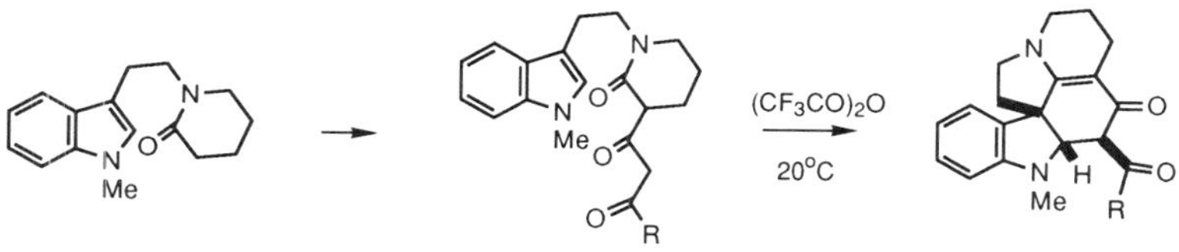

It should be stated that the Pictet–Spengler cyclization is an extension of the Mannich reaction. The consecutive formation of the quinolizidine moiety of deplancheine [Rosenmund, 1992] by a tandem process involving Michael reaction and Pictet–Spengler cyclization rendered the synthesis of the tetracyclic indole alkaloid very efficient.

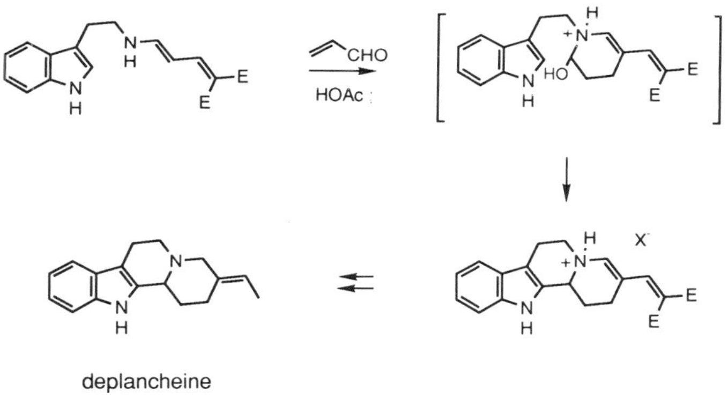

deplancheine

4.4. *vic*-DIALYLATIONS

While 1,2-additions to multiple bonds are very common reactions, the controlled formation of two C—C bonds in one step was not a simple process until more recently. Previously, the best known methods for achieving such goals were free radical polymerizations.

The development of organocopper chemistry, particularly the Michael-type addition of organocopper species to conjugated carbonyl compounds, has enabled the facile attachment of a carbon residue to the β-position of these substrates. The equally important discovery that these adducts can be directly alkylated has enabled the design of highly convergent synthetic routes to many organic molecules including the highly valued prostaglandins [Noyori, 1990].

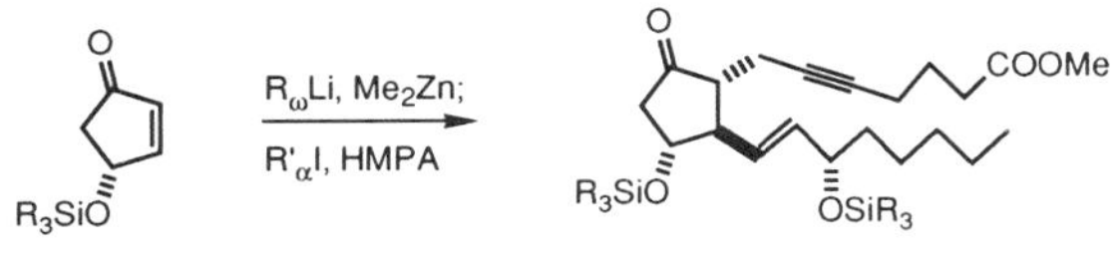

Metal-halide exchange, Michael reaction, and intramolecular alkylation in tandem constitute a very succinct step in a synthesis of morphine [Toth, 1988].

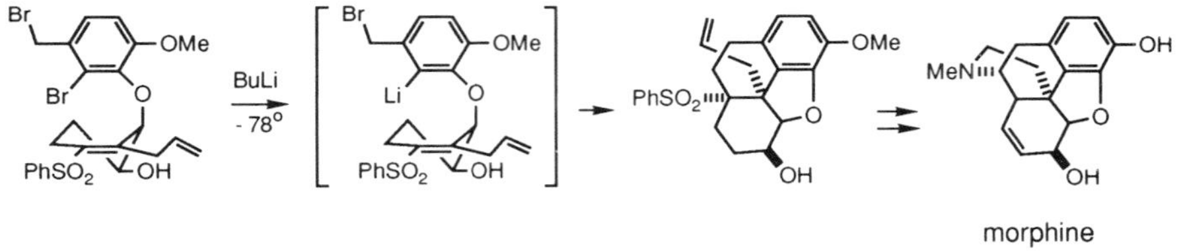

morphine

A process involving alkylative trapping of a Michael adduct provided a cyclopentanecarboxylic ester containing all the framework carbon atoms for elaboration of methyl homosecodaphniphyllate [Heathcock, 1992].

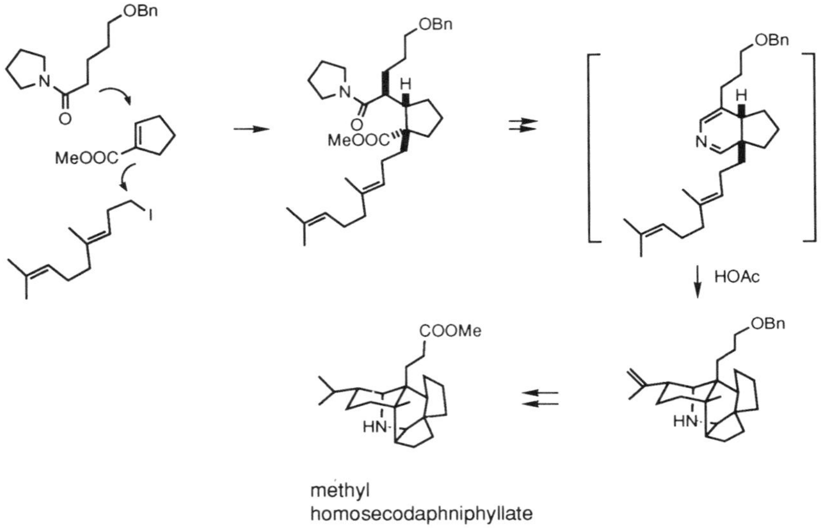

methyl homosecodaphniphyllate

The tandem process involving a reflexive Michael reaction and intramolecular alkylation to form three C—C bonds in one step is extremely desirable. Such is evident from a direct elaboration of the ishwarane skeleton [Hagiwara, 1980] from an octalone.

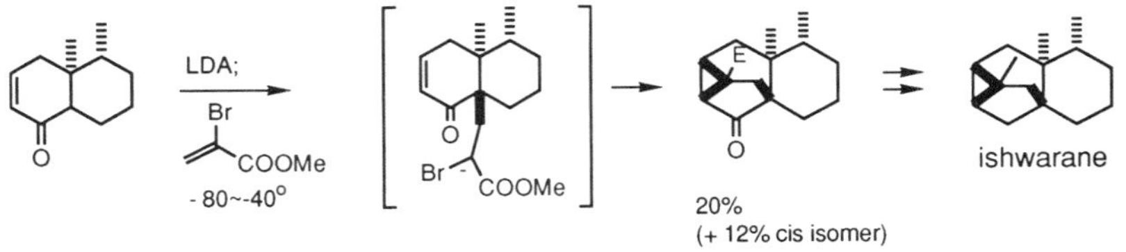

20% (+ 12% cis isomer)

ishwarane

A splendid achievement in organic synthesis of the past three decades is the stereocontrolled construction of polycyclic compounds by cationic polyene cyclizations [W.S. Johnson, 1991]. This biogenetically inspired work has culminated in the elaboration of steroids in several variants by changing the initiating and terminating groups. A brief outline of the key steps for a synthesis of 16,17-dehydroprogesterone [W.S. Johnson, 1968] is shown below.

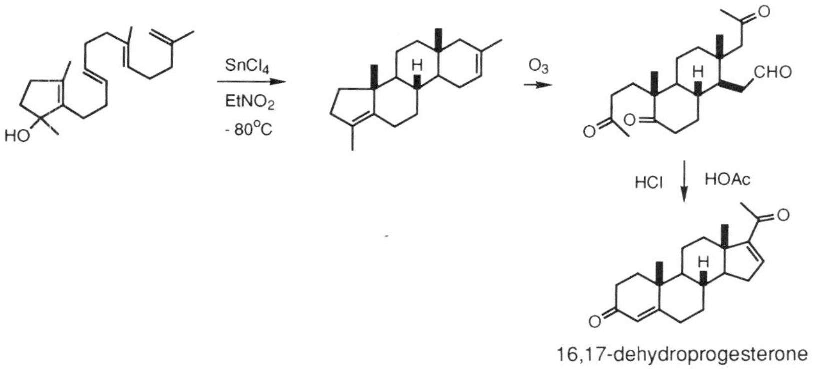

$SnCl_4$ / $EtNO_2$ / -80°C; O_3; HCl / HOAc

16,17-dehydroprogesterone

Remarkable advances in the understanding and hence the control of free radical reactions have been made in the recent past. Consequently, there has been burgeoning activity in the applications of such processes to organic synthesis. Like the cationic polyene cyclizations the reactive center can be relayed to a remote site attending each new bond formation; furthermore, free radical processes appear to be less demanding in terms of geometric alignment of the interacting groups (e.g. multiple bonds), therefore a greater variety of ring systems may be assembled by the method.

The statement above is adequately attested by synthetic routes to hirsutene [Curran, 1985] and silphiperfolene [Curran, 1987], two triquinane sesquiterpenes with a linear and an angular fusion, respectively.

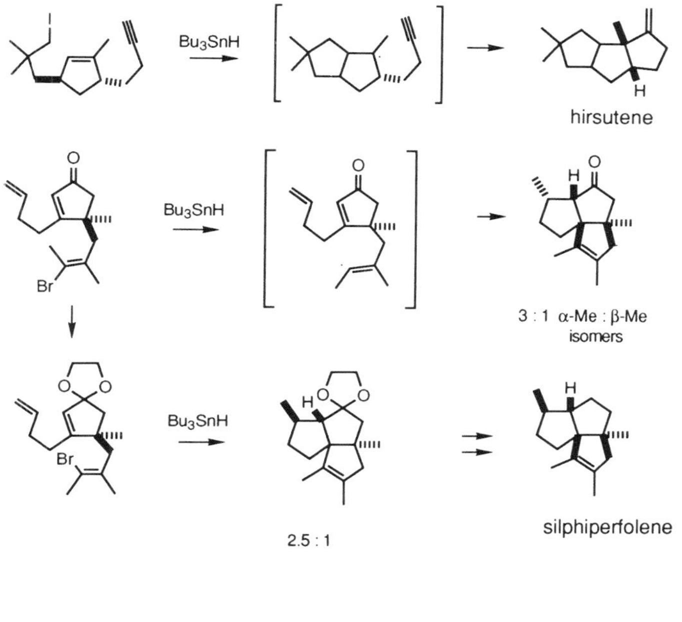

Bu_3SnH

hirsutene

Bu_3SnH

3 : 1 α-Me : β-Me isomers

Bu_3SnH

2.5 : 1

silphiperfolene

A method for the synthesis of [*b*]-fused quinoline system is initiated by addition of a free radical containing an ω-alkynyl group to an aryl isonitrile. This annulation can be considered as *vic*-dialkylation with respect to the alkyne linkage. As shown in an application to a formal synthesis of camptothecin [Curran, 1992a], this is a very efficient method.

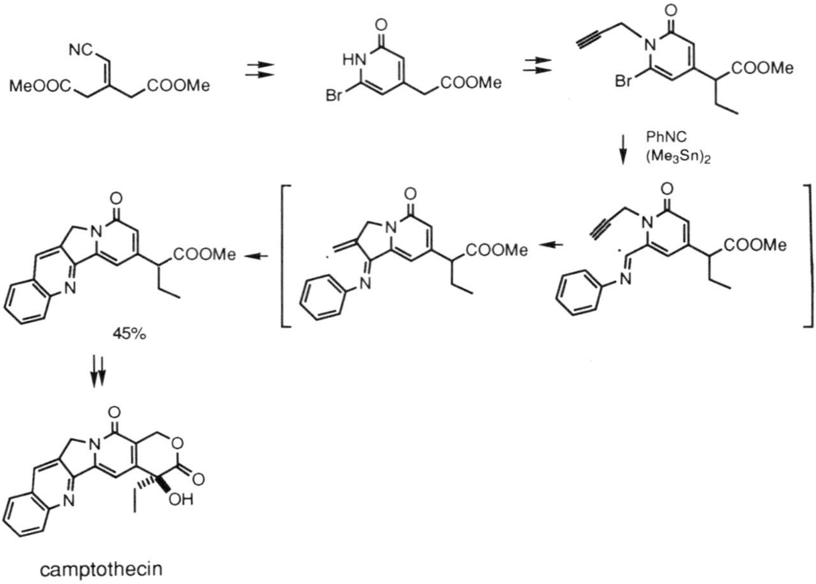

camptothecin

Bridged ring systems are also accessible by free radical processes. A synthesis of velloziolone [Snider, 1992a] attests to the value of such an approach.

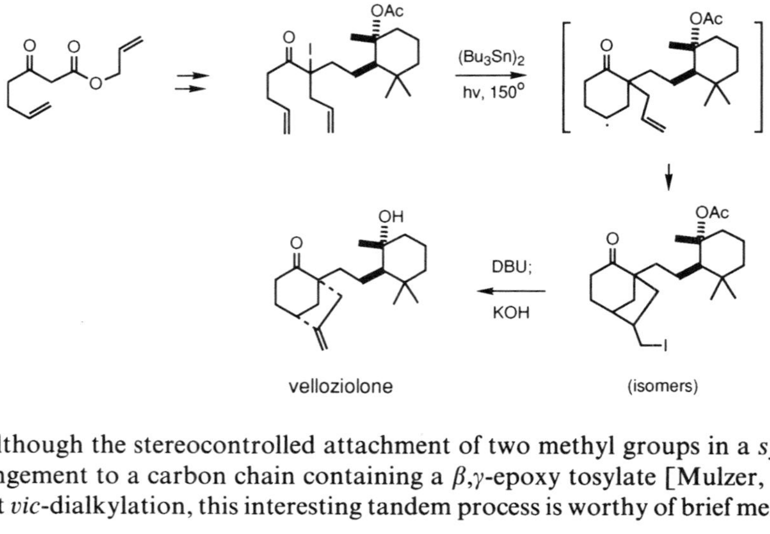

velloziolone

Although the stereocontrolled attachment of two methyl groups in a *syn*-1,3 arrangement to a carbon chain containing a β,γ-epoxy tosylate [Mulzer, 1990] is not *vic*-dialkylation, this interesting tandem process is worthy of brief mention.

The epoxide moves one carbon over on reaction with a methylcuprate reagent as a result of ring opening and internal displacement; the new epoxide undergoes electrophilic attack by the same reagent, the regiochemistry of the latter step presumably being determined by steric effect of the newly appended methyl group.

TsO O OH $Me_2Cu(CN)Li_2$ O OH OH OH

4.5. DIELS–ALDER REACTION AND RETRO–DIELS–ALDER REACTION

The Diels–Alder reaction [Carruthers, 1990] enjoys unmatched popularity in its applications to synthesis, owing to the formation of two σ-bonds in one step, besides inherent regio- and stereo-selectivity. The great increase in power of the Diels–Alder reaction in tandem with another reaction is to be expected, and the chemical literature contains ample prototypes of various possibilities.

Most obviously, a Diels–Alder reaction occurs when a diene or dienophile is generated in the presence of an existing partner under appropriate conditions. Intramolecular reactions are particularly favorable, and special molecular frameworks can often be erected by such a method.

Domino Diels–Alder reactions are featured in the elegant syntheses of dodecahedrane [Paquette, 1978] and pagodane [Fessner, 1983]. The dodecahedrane work involves the same dienophilic atoms for both steps, that is the diagonal carbon atoms of dimethyl acetylenedicarboxylate, which become trigonal in the intermediary stage. On the other hand, in the pagodane synthesis the dienophile for the second Diels–Alder reaction originates from the diene.

E E = COOMe -78° – 23° 3 h MeOOC COOMe dodecahedrane

MA O O O O O O

pagodane

While cyclobutadiene apparently undergoes dimerization only, dimethyl 1,2-cyclobutadienedicarboxylate liberated from its iron tricarbonyl complex gives the timer and tetramer in addition to the dimer [Mehta, 1992]. The oligomers are the products of consecutive Diels–Alder reactions with the monocyclic diester acting the diene.

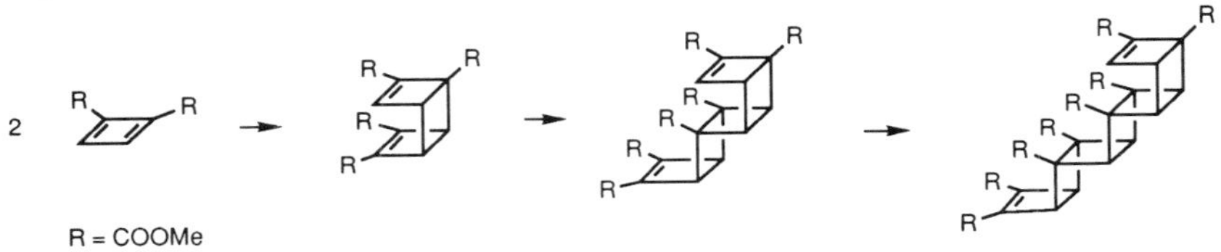

R = COOMe

Since the Diels–Alder reaction is thermally induced, other pericyclic reactions preceding or following it would form a tandem readily. In this regard the retro-Diels–Alder/Diels–Alder reaction sequence is quite well known. In terms of utility the following syntheses are representative: coronafacic acid [Ichihara, 1980], lysergic acid [Oppolzer, 1981a],barrelene [Zimmerman, 1969].

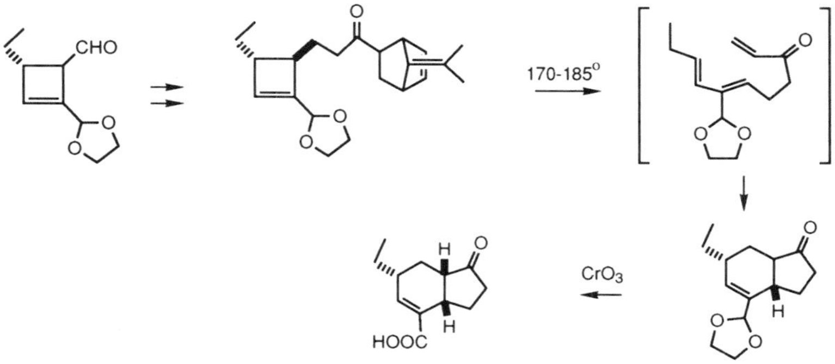

170-185°

CrO_3

coronafacic acid

200°

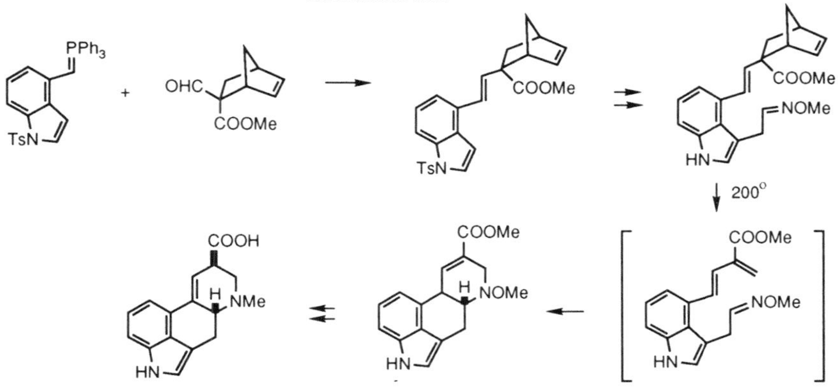

lysergic acid

Δ

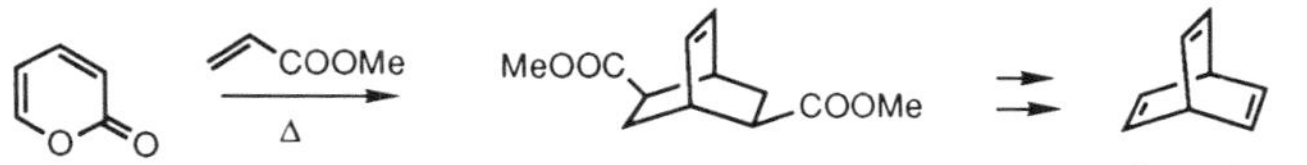

barrelene

A frequently exploited reaction sequence involves Diels–Alder reaction following a cheletropic elimination of sulfur dioxide from a 3-sulfolene. An interesting synthesis of a molecule containing the iceane skeleton [Hamon, 1982] involves a Diels–Alder reaction followed by sulfur dioxide extrusion and an intramolecular Diels–Alder reaction.

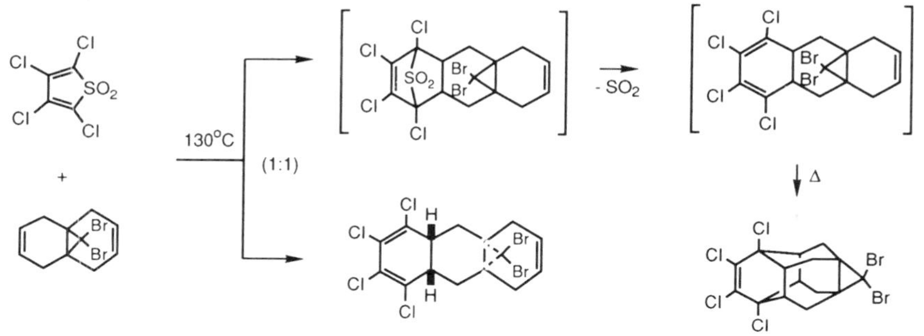

The cheletropy/Diels–Alder reaction tactic permits preservation of a sensitive conjugate diene during manipulation of other parts of the molecule. Routes to (+)-estradiol [Oppolzer, 1980] and lupinine [Nomoto, 1985] are exemplary. Note a great advantage, that the sulfolenes enable alkylation at C-2 and the sulfur dioxide extrusion method is suitable for the generation of the very reactive *o*-quinodimethanes. Another feature worth noting in the estradiol synthesis is the cyano substituent in the aromatic ring which controls the alkylation site; an oxygen function would reverse the regioselectivity.

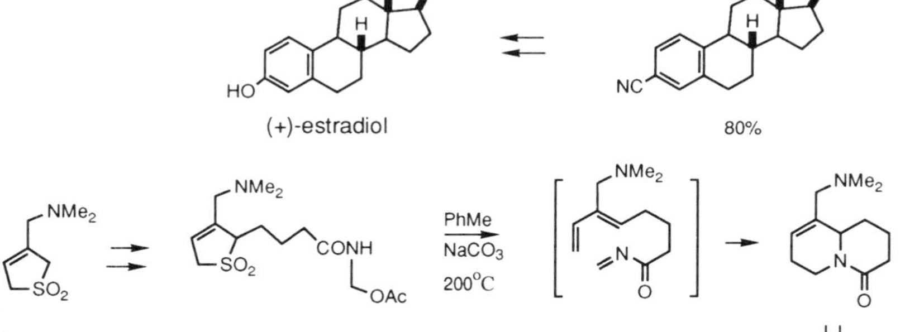

(+)-estradiol

80%

lupinine

Another effective tactic for the preparation of tetralin derivatives via interception of *o*-quinodimethanes is through generation of the latter species by pyrolytic cleavage of benzocyclobutenes. Many groups of researchers have employed this tactic in the elaboration of A-aromatic steroids. An interesting variant of this method is shown in a synthesis of (+)-chenodeoxycholic acid [Kametani, 1981] in which the benzocyclobutene unit forms part of the CD-ring component of the target molecule.

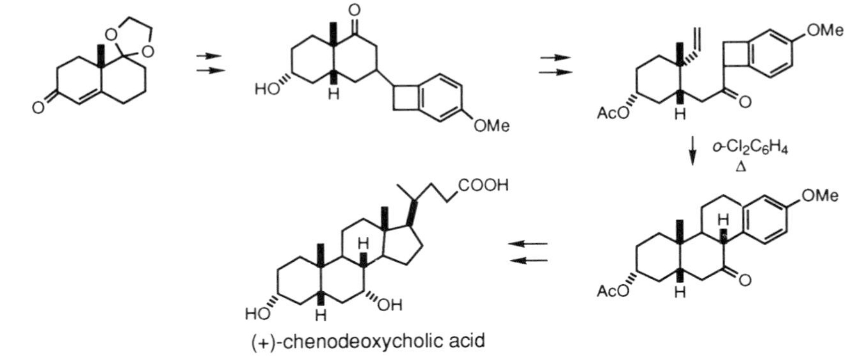

(+)-chenodeoxycholic acid

Electrocycloreversion of cyclobutenes prior to the Diels–Alder reaction actually requires lower temperatures, as shown in a synthetic approach to coronafacic acid [Jung, 1981]. The two-staged thermolysis is probably unnecessary. An intermolecular version of this method is involved in a synthesis of adriamycinone [Boeckman, 1983].

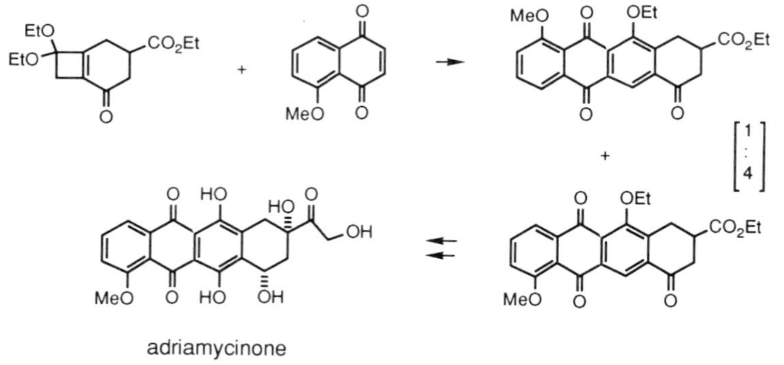

coronafacic acid

adriamycinone

A most spectacular pericyclic reaction sequence apparently mediates in the formation of the endiandric acids in nature. Indeed, various members of this family can be generated *in vitro* from the proper polyene carboxylic acid derivatives. To obtain endiandric acid-A methyl ester, heating briefly a hexaene precursor in toluene at 100°C induces two electrocyclizations to give, initially a cyclooctatriene, and then a bicyclo[4.2.0]octadiene (endiandric acid-E methyl ester), which is followed by an intramolecular Diels–Alder reaction. Endiandric acid-B and -C are similarly formed.

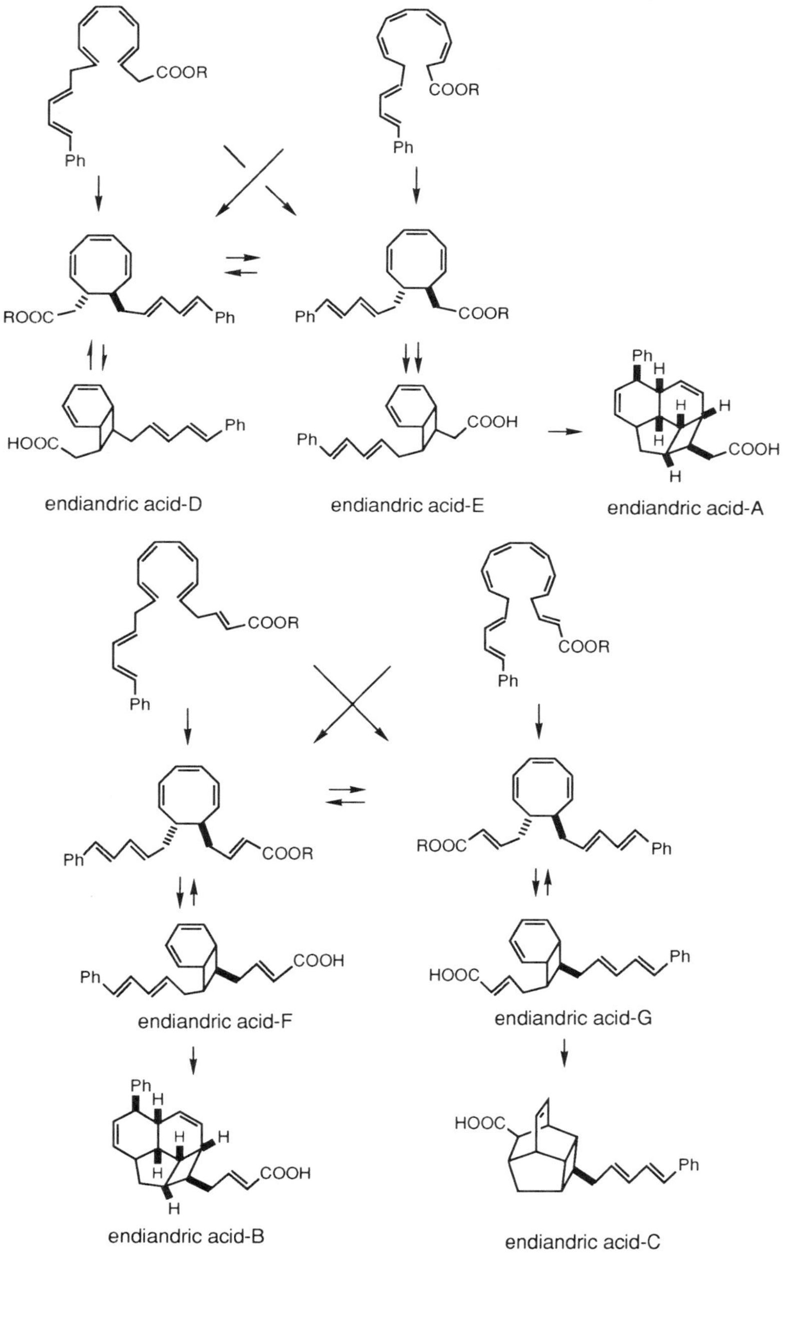

endiandric acid-D

endiandric acid-E

endiandric acid-A

endiandric acid-F

endiandric acid-G

endiandric acid-B

endiandric acid-C

2,3-Dimethylenoindoline shows comparable diene activity to *o*-quinodimethanes. It has been demonstrated that *in situ* trapping of such species is a valuable procedure for skeletal construction of many indole alkaloids. For example, a concise route to dehydroaspidospermidine [Gallagher, 1982] is predicated on *N*-acylation to generate the diene unit and an intramolecular Diels–Alder reaction to establish four rings of the pentacyclic alkaloid.

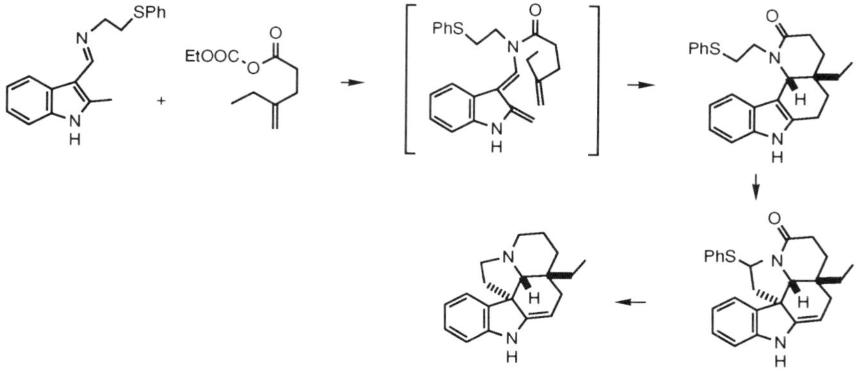

dehydroaspidospermidine

Secodine-type compounds generated *in vitro* undergo cycloaddition readily. Thus the unfoldment of the diene unit by an elimination process is a crux of a scheme (with many variations) for the synthesis of aspidosperma alkaloids, including vincadifformine [Kuehne, 1978, 1979a,b].

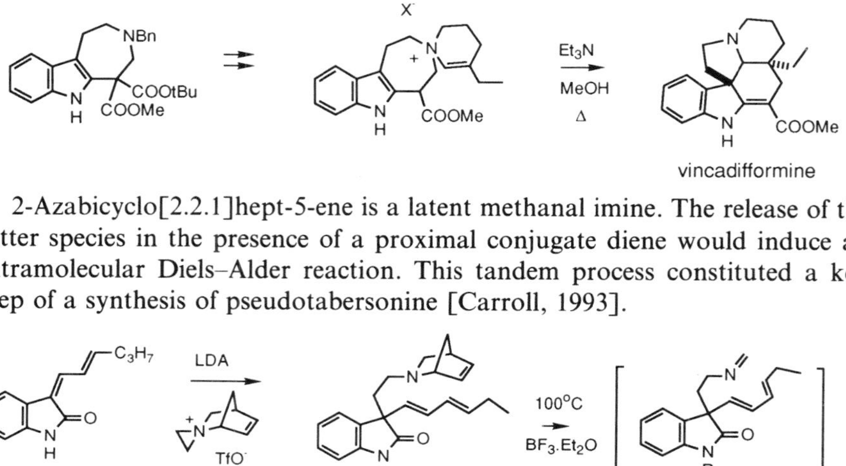

vincadifformine

2-Azabicyclo[2.2.1]hept-5-ene is a latent methanal imine. The release of the latter species in the presence of a proximal conjugate diene would induce an intramolecular Diels–Alder reaction. This tandem process constituted a key step of a synthesis of pseudotabersonine [Carroll, 1993].

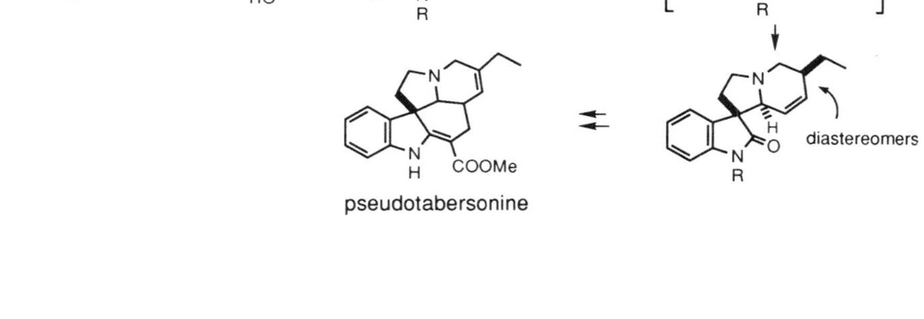

pseudotabersonine

The Diels–Alder/retro-Diels–Alder reaction sequence is also valuable in synthesis. The latter step is rendered facile when a stable small molecule (e.g. $CH_2{=}CH_2$, HCN, CO_2, N_2) is eliminated from the adduct. Synthetic routes leading to daunomycinone [Krohn, 1979], ligularone [Jacobi, 1984], reserpine [S.F. Martin, 1987b], *cis*- and *trans*-trikentrin-A [Boger, 1991] involve such reaction sequence.

- $CH_2{=}CH_2$

daunomycinone

Δ

- HCN

ligularone

0-25°
PhH

Ac_2O, NaOAc
160°

LiOH

cis-trikentrin A

A highly efficient synthesis of corannulene [L.T. Scott, 1991] depends on the formation of a fluoranthene intermediate which involves a tandem Diels–Alder reaction/cheletropic elimination/retro-Diels–Alder reaction sequence.

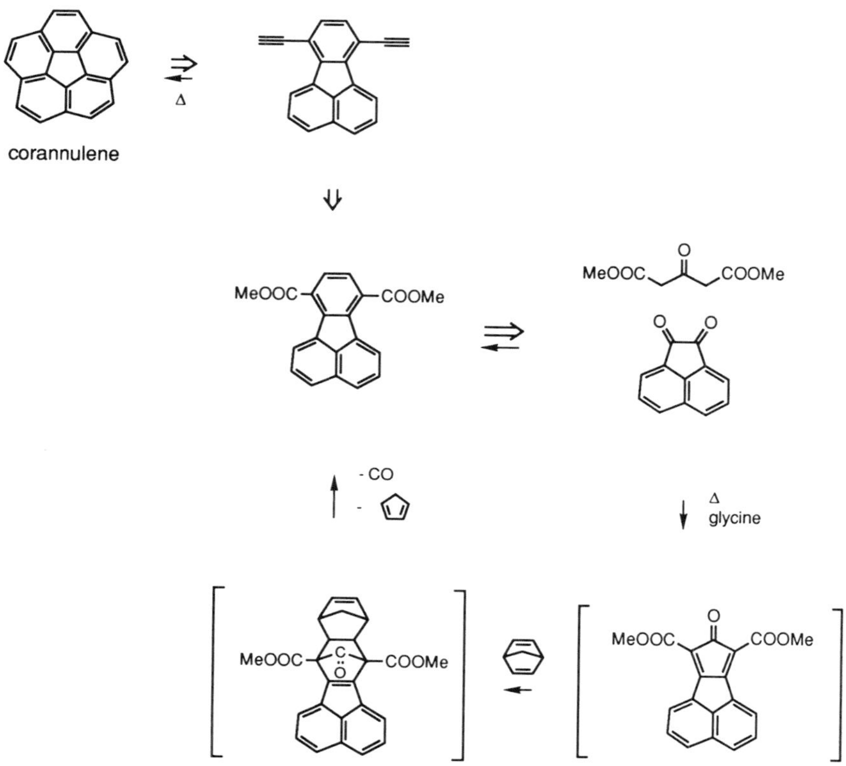

4.6. OTHER PERICYCLIC REACTIONS

As indicated above concerning the biosynthesis and synthesis of the endiandric acids, electrocyclic reactions are not only of mechanistic significance. Naturally tandem processes which are terminated by an electrocyclic reaction are known.

Consecutive electrocyclic reactions have played a critical role in synthesis of (+)-occidentalol [Hortmann, 1973]. In this route photoisomerization of a bicyclic ester precursor with a *trans*-ring juncture to two *cis*-fused diastereomers proceeds via conrotatory opening to a monocyclic triene, which undergoes a thermally induced disrotatory cyclization at −20°C. The two isomers are convertible to (+)-occidentalol and 7-epi-(−)-occidentalol, respectively.

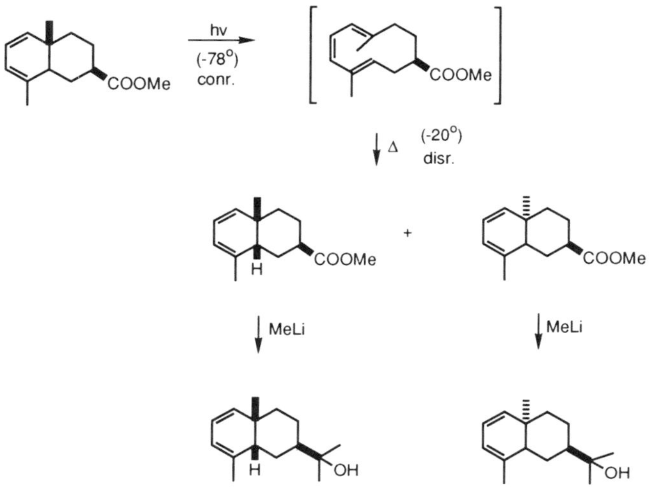

hv (-78°) conr.

Δ (-20°) disr.

MeLi

MeLi

7-epi-(-)-occidentalol

(+)-occidentalol

Resorcinol derivatives are formed at the end of a very interesting pericyclic reaction tandem. Some details are shown with an outline of an approach to maesanin [Danheiser, 1990]. Thus [2 + 2]-cycloaddition of a conjugate ketene with an alkoxyacetylene results in a vinylcyclobutenone which is subject to electrocyclic opening to afford a ketene with extended conjugation. Electrocyclization of the latter species gives rise to the aromatic compound.

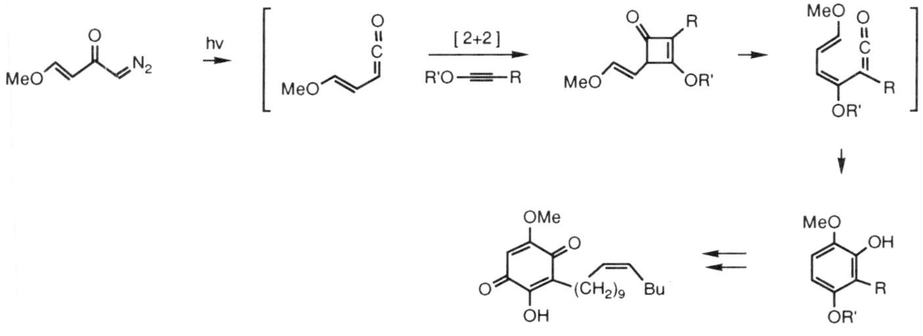

hv

[2+2]

R'O—≡—R

maesanin

The ene reaction and the Conia version have found many synthetic applications. The occurrence of an intramolecular ene reaction on pyrolysis of angularly methylated tricyclic compounds obtained from photocycloaddition of cyclobutene and cyclohexene derivatives was probably unanticipated. However, such products are desirable precursors for terpenes, such as calameon [Wender, 1980].

An ene reaction in tandem with Claisen rearrangement serves to create the five-membered ring of a dehydroestrone precursor [Mikami, 1990]. The preceding Claisen rearrangement joins the AB-element to an enyne chain and unveils the thermally reactive subunit.

An intramolecular ene reaction involving an acylnitroso component as enophile is the key step of a synthesis of mesembrine [Keck, 1982]. The acylnitroso compound is unstable and the conditions for its generation by a retro-Diels–Alder reaction are conducive to the ene reaction.

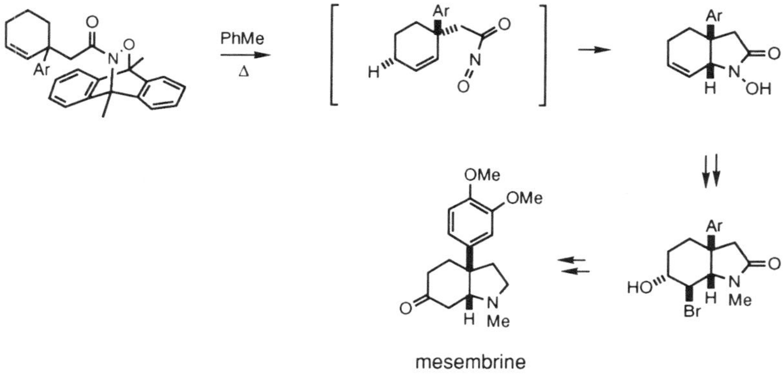

mesembrine

The metallo-ene reaction is one with an allylmetal as the ene component. However, it proceeds well only in the intramolecular version [Oppolzer, 1989]. Its versatility lies in the fact that ring formation is accompanied with transfer of the metal atom to another site of the molecule, and accordingly, further functionalization at this other site becomes possible. One type of the metallo-ene reactions is exemplified by a key step in a synthesis of (+)-khusimone [Oppolzer, 1982b]. This concise route employs a *vic*-dialkylation to erect a chiral 2,3-disubstituted cyclopentanone which can be converted into the magnesio-ene reaction substrate in a few steps. Carboxylation of the bicarbocyclic reaction product affords a properly functionalized molecule containing all the carbon atoms of the synthetic target.

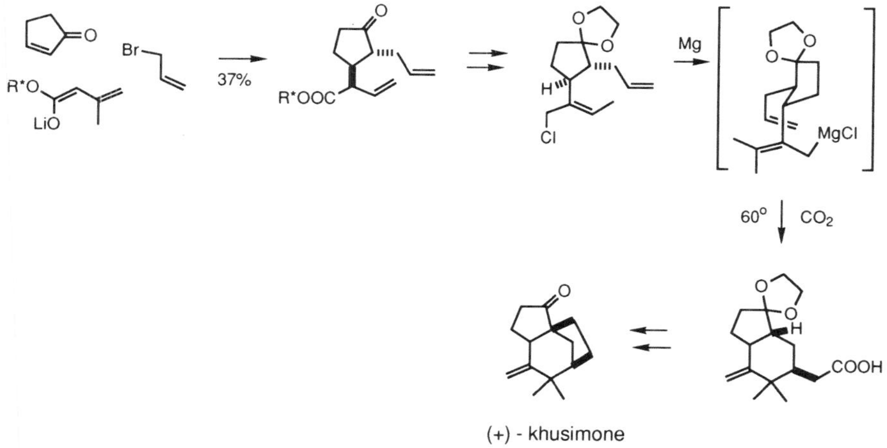

(+) - khusimone

The Conia reaction is an ene reaction in which the ene component is an enol. A double Conia reaction giving rise to a propellane system has been reported [Drouin, 1975].

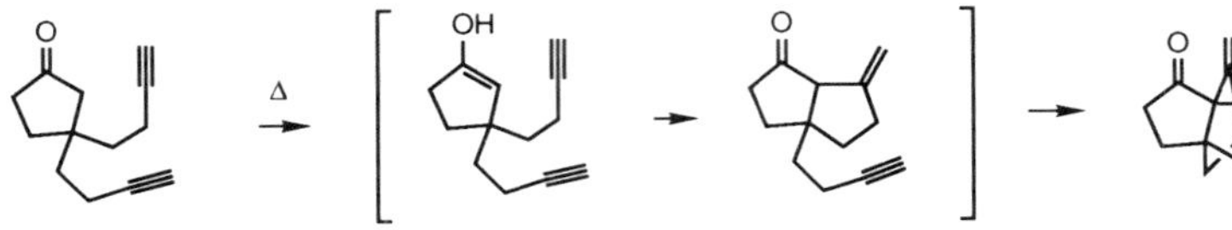

4.7. SIGMATROPIC REARRANGEMENTS

Rearrangements are powerful reactions for skeletal modification if such transformations are well defined. Sigmatropic rearrangements continue to be a reliable group of synthetic reactions because their pathways have been thoroughly clarified. Tandem reactions involving one or more sigmatropic rearrangements are even more valuable.

The combination of Claisen and Cope rearrangements, both belonging to the thermally allowed [3.3]-sigmatropy, is an excellent chain elongation method, as attested by the one-step elaboration of *β*-sinensal [Thomas, 1969] from myrcenol.

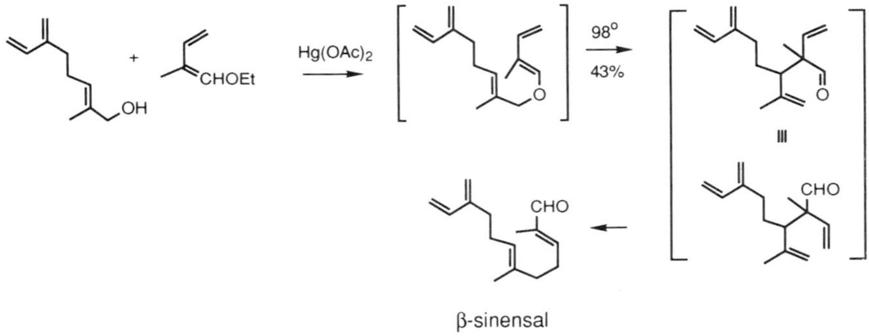

Hg(OAc)2 98° 43%

β-sinensal

With careful consideration substrates can be designed to accomplish the construction of molecules with three contiguous asymmetric centers in one step employing the tandem Cope–Claisen rearrangement. Such is the case of a methylenecyclopentane formation *en route* to pseudoguaianolide sesquiterpenes [F.E. Ziegler, 1982].

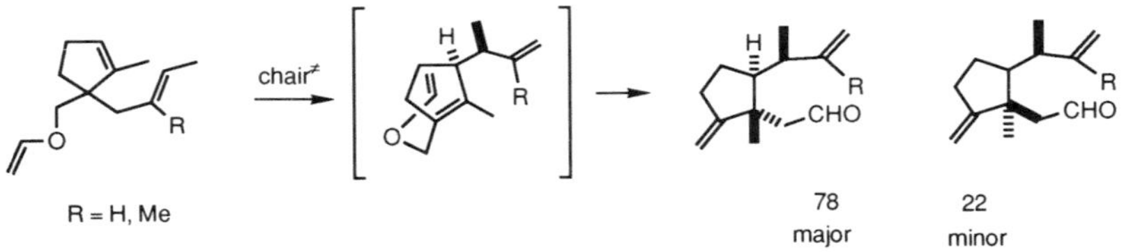

R = H, Me   chair≠   78 major   22 minor

An aromatic Claisen rearrangement of a 7-dimethylallyloxy coumarin, followed by two Cope rearrangements and prototropic shift, rapidly transforms it into gravelliferone [Cairns, 1987; Massanet, 1987]. Only because there are two available pathways for the Claisen rearrangement the yield of the target molecule is modest.

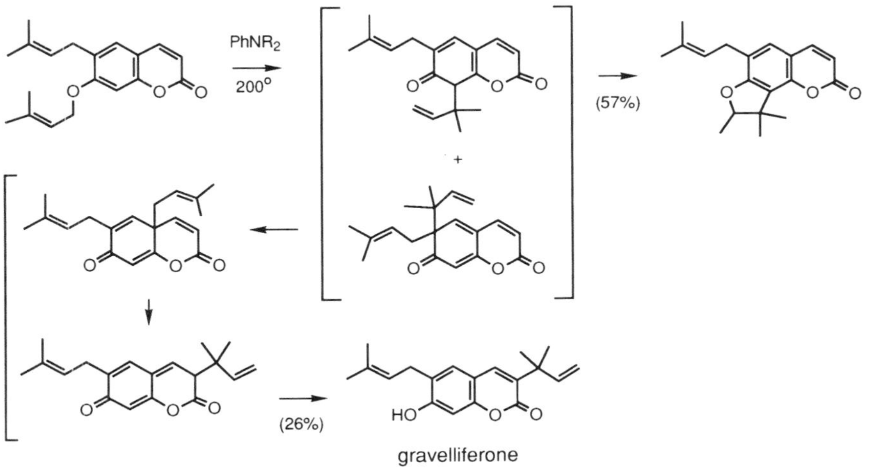

PhNR$_2$ 200° (57%) (26%)

gravelliferone

Allyl benzocyclobutene-1-carboxylates are thermally converted into 4-allyl-isochroman-3-ones. This result is the sum of an electrocyclic opening, electro-cyclization involving the ester carbonyl group, and a Claisen rearrangement. Such products have been identified as synthetic intermediates of several physiologically active substances, such as physostigmine [Shishido, 1986].

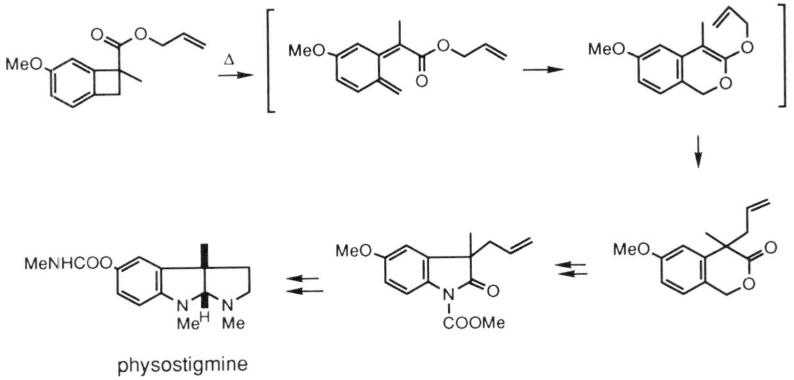

physostigmine

A Brook/Claisen rearrangement tandem constitutes an important process in the attachment of the eight-carbon sidechain of ophiobolin-C [Rowley, 1989].

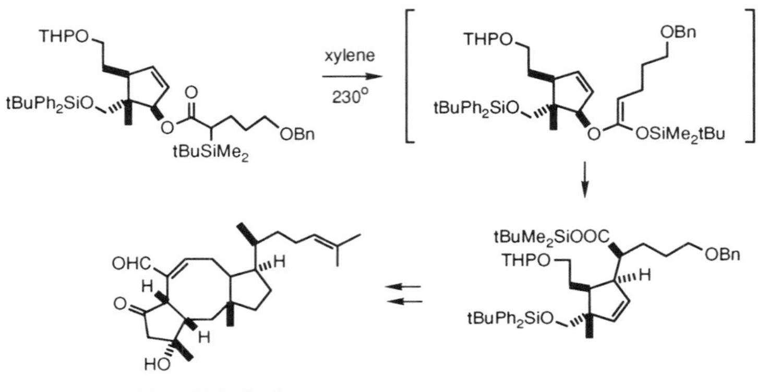

(+)-ophiobolin-C

It has not been possible to synthesize the germacrene framework by an intramolecular coupling of a bisallyl bromide with nickel carbonyl. The products contain an elemane-type skeleton due to a facile Cope rearrangement that ensues [Corey, 1969a).

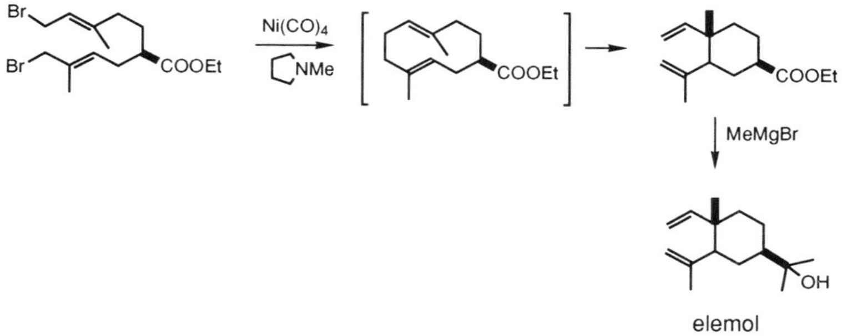

elemol

4.8. IONIC REARRANGEMENTS

The majority of skeletal rearrangements induced by electron-deficient centers are tandem reactions because multiple bond-forming, bond-breaking, and bond-migrating steps are involved. But since these rearrangements are difficult to control, synthetic steps based on rearrangements are usually incorporated with close analogy to precedents. However, when a model is established, an efficient approach to an intriguing molecule may present itself. A case in point is the formation of the tricyclic framework of longifolene [Volkmann, 1975].

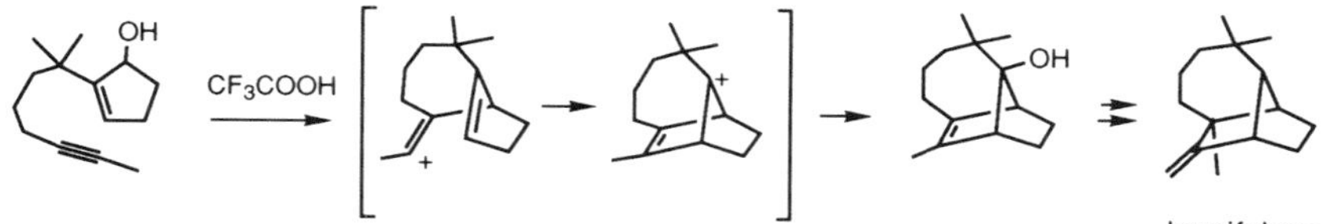

longifolene

The extensive 1,2-rearrangements of spirocyclic systems containing cyclobutane moieties have been exploited in a synthesis of isocomene and modhephene [Fitjer, 1988]. The cascade reaction is also the rationale for design of a [6.5]coronane synthesis [Wehle, 1987].

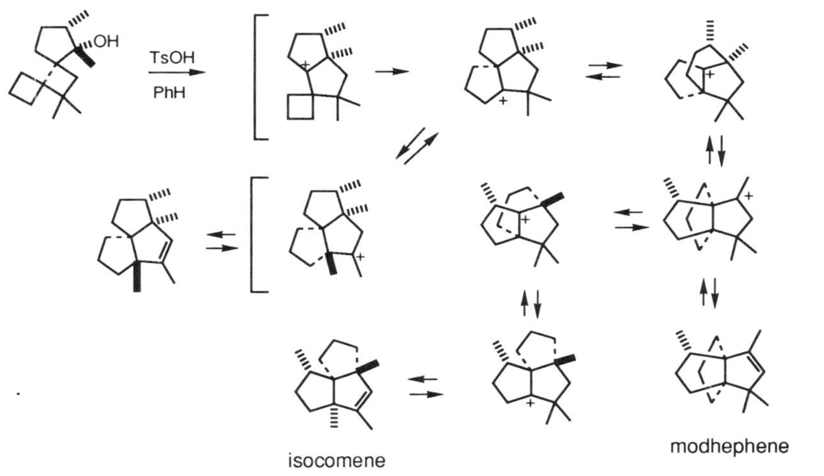

TsOH
PhH

isocomene

modhephene

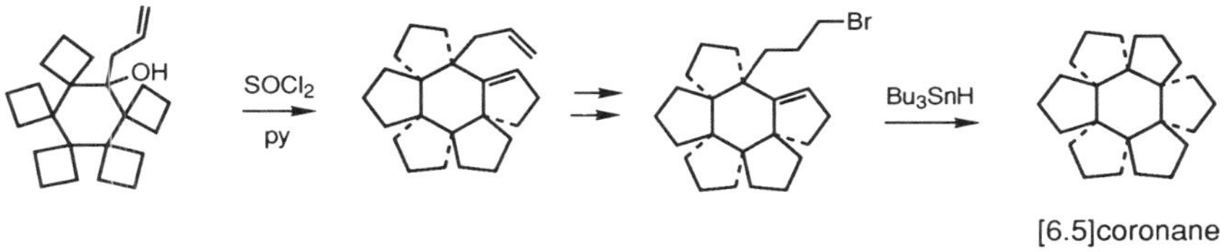

$SOCl_2$
py

Br

Bu_3SnH

[6.5]coronane

The skeletal rearrangement which transforms a 1-oxygenated 3-vinylbicyclo[2.2.2]oct-2-enes into octalones, as shown by the preparation of a synthetic intermediate of nootkatone [Dastur, 1974], consists of a fragmentation and recyclization.

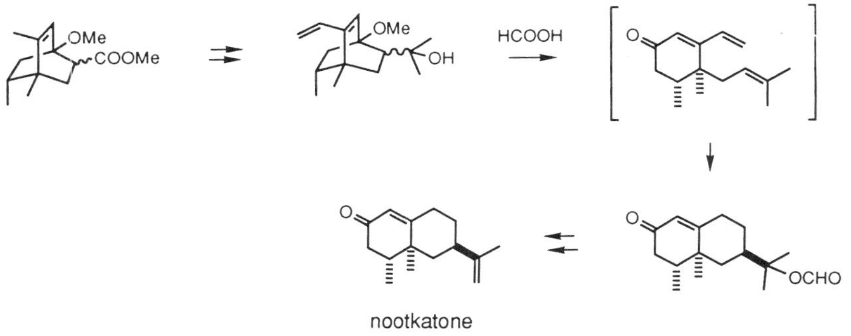

OMe
COOMe

OMe
OH

HCOOH

OCHO

nootkatone

Impressive tandem intramolecular transacylation of a lactam that contains a polyamine sidechain has been named "zip reaction" [Kramer, 1978]. In one operation a 53-membered macroheterocycle resulted.

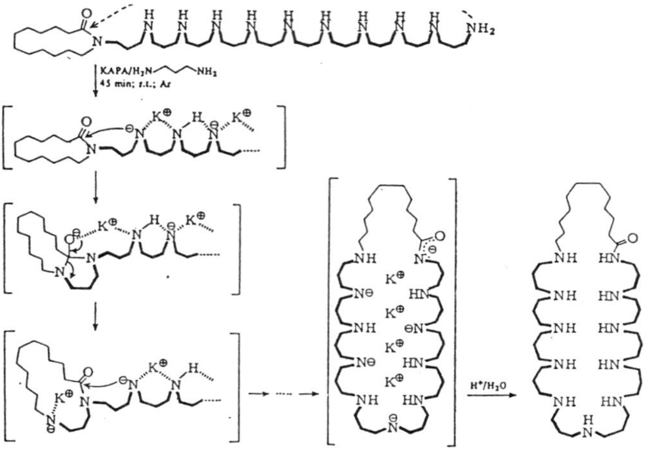

5

CYCLIC ARRAYS FOR STRUCTURAL AND STEREOCHEMICAL MANIPULATIONS

The progress of organic synthesis parallels the development of ring construction concepts and techniques. However, at certain stages of this development chemists began to consider the use of cyclic structures as scaffolds for erection of the more flexible chain segments which are characterized by the existence of variform conformations. Accordingly, the ease of five- and six-membered ring formation in combination with the well-established conformational effects of such systems has had enormous impacts on synthetic tactics.

The preparation of *o*-di-*t*-butylbenzene was not trivial. One approach [Barclay, 1962] in earlier days contemplated 1,1,4,4-tetramethyl-β-tetralone as a precursor, with relatively straightforward ring cleavage to a dicarboxylic acid and then reduction down to the hydrocarbon stage: the problem was solved.

OTs OTs COOH COOH O

Transitory annulation [T.-L. Ho, 1988] should be seriously entertained whenever the access of a medium or large ring system is involved. Furthermore, the most expedient way to assemble an acyclic array with multiple stereocenters is frequently via cyclic precursor. The tactic may include formation and selective cleavage of three- and four-membered rings.

The abundant 1,5-cyclooctadiene, 1,5,9-cyclododecatriene and analogs, which are cyclic oligomers of butadiene/isoprene, are unique precursors of certain natural products. For example, 1,5-cyclooctadiene has been the source of the sex pheromone of female pink bollworm moth [R.J. Anderson, 1975],

1,5-dimethyl-1,5-cyclooctadiene has been converted into iridomyrmecin [Mathews, 1975], and the all-(*E*) 1,5,9-trimethyl-1,5,9-cyclododecatriene into the cecropia juvenile hormone JH-III [Odinokov, 1985]. Of course cyclododecanone is a useful starting material for synthesis of macrocyclic ketones and lactones including muscone.

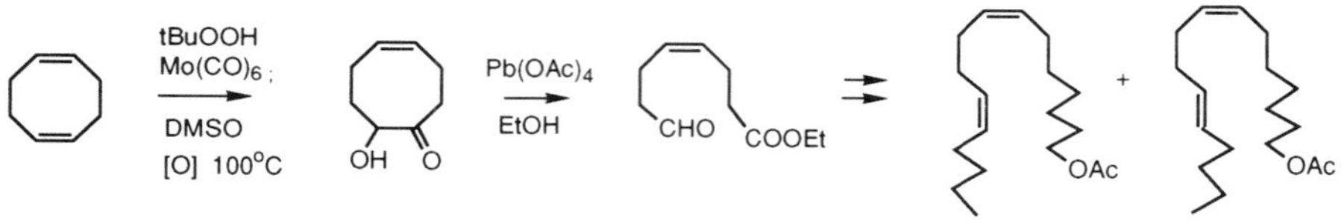

tBuOOH $Mo(CO)_6$; DMSO [O] 100°C; OH O; $Pb(OAc)_4$ EtOH; CHO COOEt; OAc; + OAc

gossyplure

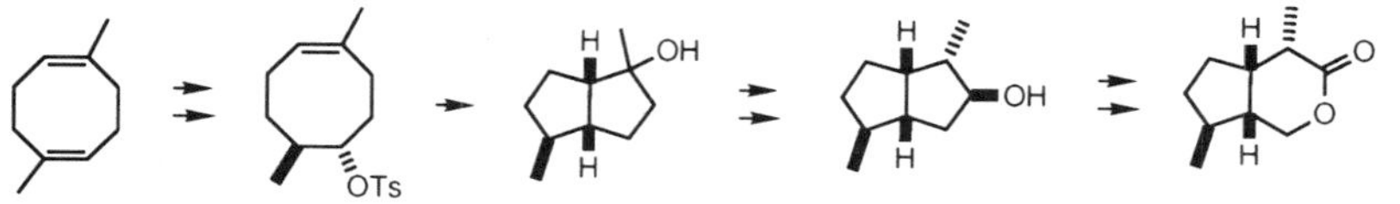

OTs; H OH H; H OH H; H O O H

iridomyrmecin

O_3; MeOH NH_4Cl; $(MeO)_2CH$ O; OHC COOMe; COOMe; O COOMe

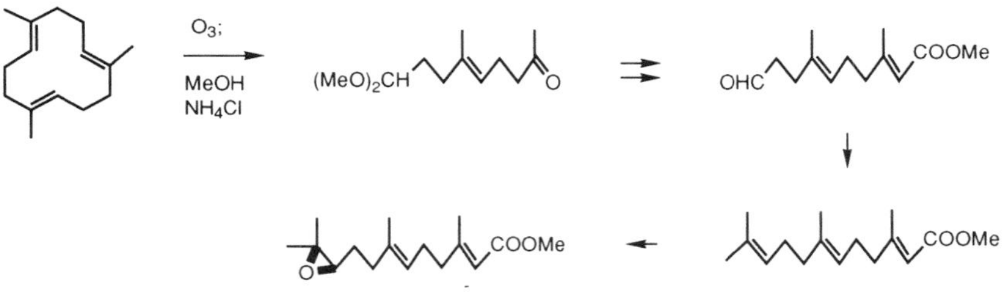

JH-III

The benzene ring is a source of a six carbon atom chain that may have special utility. For example the derived cyclohexadienones undergo photocleavage and such products have been employed in synthesis, e.g. chondrillin and plakorin [Snider, 1992b], and crocetin dimethyl ester [Quinkert, 1977].

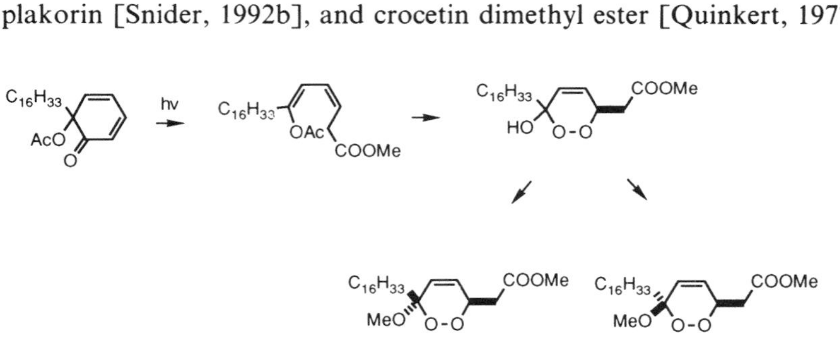

$C_{16}H_{33}$ AcO O; hv; $C_{16}H_{33}$ OAc COOMe; $C_{16}H_{33}$ HO O-O COOMe

$C_{16}H_{33}$ MeO O-O COOMe; $C_{16}H_{33}$ MeO O-O COOMe

chondrillin plakorin

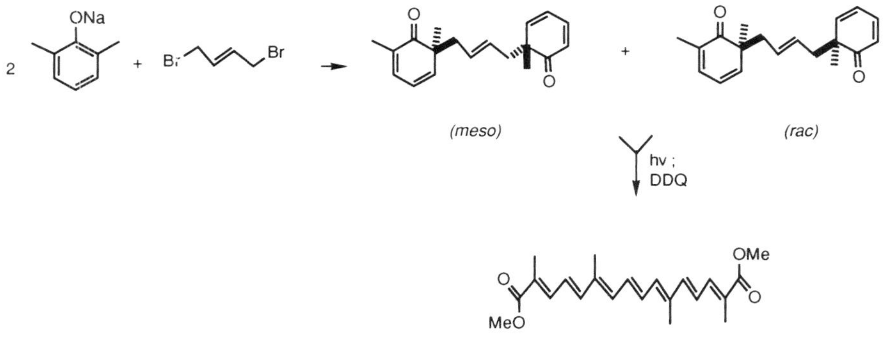

(meso)

(rac)

hv ; DDQ

crocetin dimethyl ester

Readily available polycycles are potential precursors of simpler cyclic systems. Naturally, the prerequisite for their use is that selective ring cleavage may be implemented. R.B. Woodward was perhaps the first chemist who systematically extolled the virtue of this tactic by adroit exemplification in many of his elegant syntheses.

A synthesis of quinine [Woodward, 1945] starting from 7-hydroxyisoquinoline is via *cis*-3,4-disubstituted piperidine derivatives. This approach solves the stereochemical problem by incorporating the two sidechains of the piperidine ring in a saturated six-membered ring which in turn is derivable from catalytic hydrogenation.

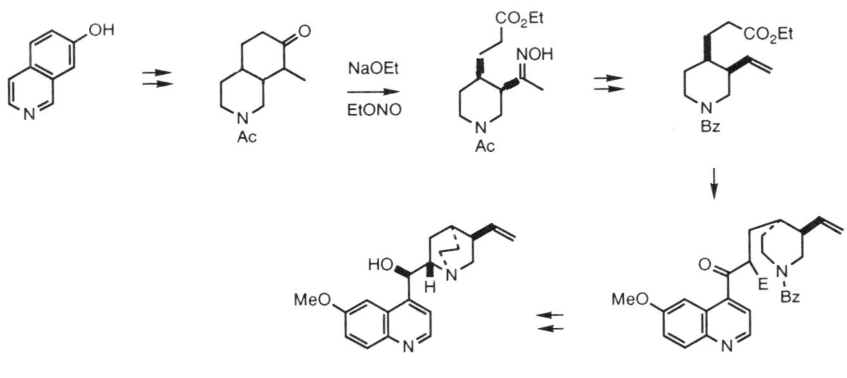

NaOEt / EtONO

quinine

E = COOEt

Ingenuity in the design for elaboration of strychnine [Woodward, 1963] is strikingly demonstrated by employing 2-veratrylindole to begin the synthesis. The veratryl group not only provides the necessary blockade of the Pictet–Spengler cyclization at C-2 of the indole nucleus, but also the latent elements for *cis*-muconic ester which would participate in the formation of ring-III and ring-IV.

OMe OMe NH2 NTs COOEt Ac H O3; HCl, MeOH COOMe

strychnine

A stereochemical problem pertains to the assemblage of chlorophyll-*a* [Woodward, 1960] in which the transformation of a porphyrin system into a chlorin or purpurin system is considered as a driving force for relieving unfavourable steric interactions among substituents in the vicinity of C-15 (γ-atom). Thus, the synthesis was designed on the basis of porphyrin construction and subsequent passage into a purpurin. In the latter process carbocyclization served to tetrahedralize the two adjacent stereogenic centers as required in the final product, but the additional ring was almost immediately cleaved.

NHAc COOMe MeOOC hv, air CHO Mg phytylOOC

chlorophyll-*a*

Derivation of three interconnected azacycles from elements partially provided by a cyclopentene via its cleavage product is the key feature of an ajmaline synthesis [Masamune, 1967]. Because of the local symmetry, stereochemical problems pertaining to generation of diastereomers were not encountered. Thus the hemiaminal formation and hence the subsequent Pictet–Spengler cyclization was predicted by *trans* disposition of the two vicinal sidechains of the ensuing heterocycle.

EtOOC O N Me BzO NHBz N Me BzO CHO CHO NHBz N Me OBz Bz N CHO N Me OH OBz Bz N CHO N Me HO N OH N H Me

ajmaline

The deceptively simple approach to cantharidin by a Diels–Alder route based on furan and dimethylmaleic anhydride was thwarted by the unreactivity of the addend combination. Using dimethyl acetylenedicarboxylate as dienophile many steps are required to reduce the ester groups (down to the methyl groups) and to add two vicinal carboxyl groups from the *exo* side [Stork, 1953]. With respect to accomplishing the latter requirement another Diels–Alder reaction was implemented, further ring cleavage and degradation were also involved.

O COOMe COOMe O COOMe COOMe O CHO CHO

$C_5H_{11}N$ HOAc

O O O O O CHO

cantharidin

A synthesis of dimethyl betamate [Büchi, 1978] was carried out from intermediates in which the two carboxylic groups were latent in a two-carbon bridge. The sensitive chromophore was delivered when the bridge was severed in the last step. This scheme also accommodated an elimination step (instead of oxidation) that could not occur before the severance (Bredt's rule).

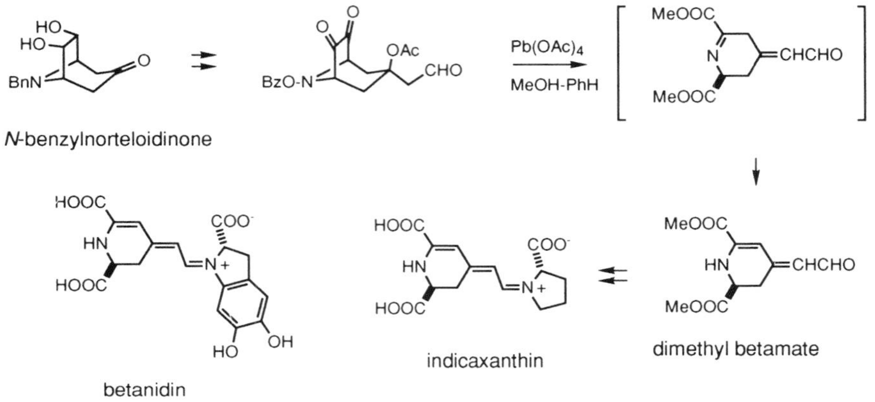

N-benzylnorteloidinone

$Pb(OAc)_4$ / MeOH-PhH

betanidin

indicaxanthin

dimethyl betamate

The following example indicates the utility of a bridged bicyclic triene in an elaboration of the Prelog–Djerassi lactone [Masamune, 1975] which is an intermediate of methynolide. Two *cis*-methyl groups were established from the two-carbon bridge via oxidative scission and subsequent manipulations.

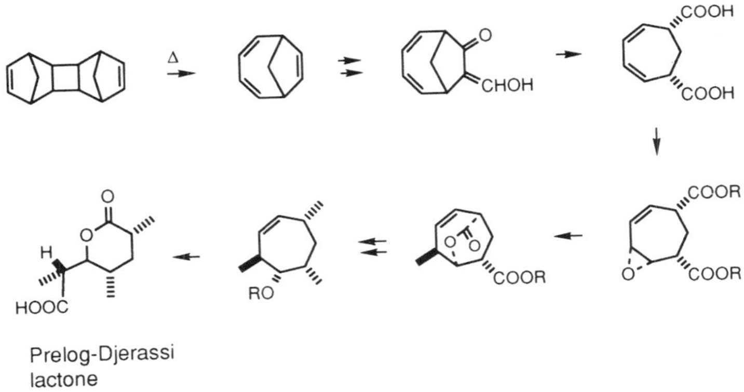

Prelog-Djerassi lactone

An expedient method for the elaboration of the *cis*-1,3-dimethylcyclopentane subunit of a molecule is via cleavage of a bicyclo[2.2.1]heptene precursor. Such a precursor is often available from a Diels–Alder approach, therefore an efficient reaction sequence may be developed. An example is shown in approaches to *cis*-trikentrin-B [Yasukouchi, 1989] and the herbindoles [Muratake, 1992].

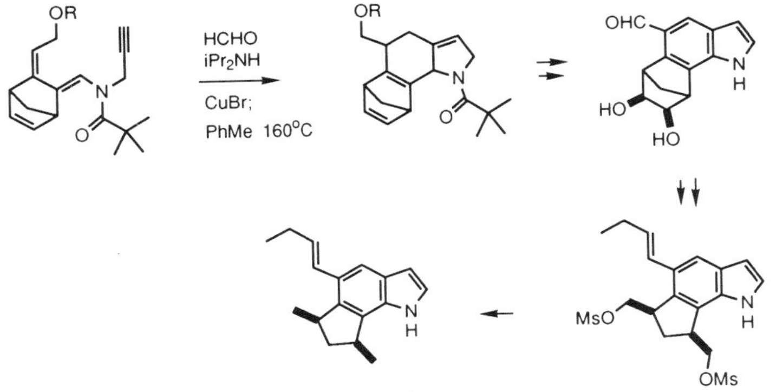

cis-trikentrin-B

Following the same scheme, *cis*-2,5-disubstituted tetrahydrofurans are readily prepared. Two varieties of dienophiles for furan are shown in synthetic routes of showdomycin [T. Sato, 1978; Just, 1980]. Tetrachlorocyclopropene can also be used as the C_3 component [Gensler, 1975].

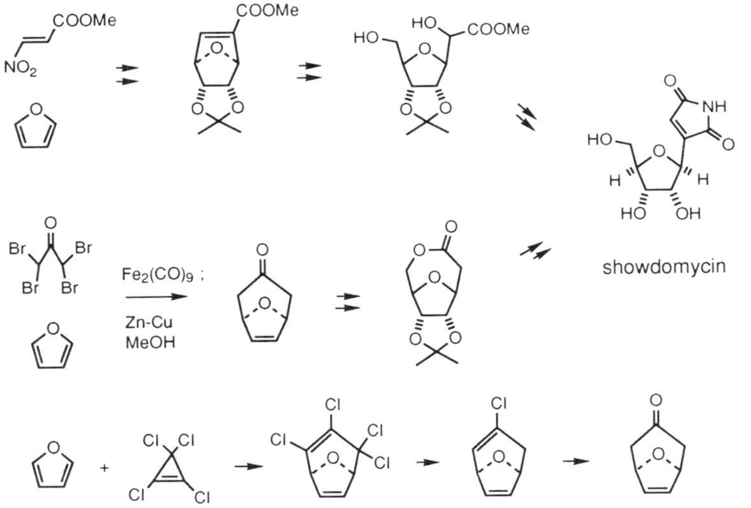

showdomycin

The key issue of substituent affixation during synthesis of (−)-bactobolin is the relative configuration of the contiguous asymmetric centers. Apparently the establishment of four of these centers was much more facile through manipulation of a bridged lactone system [Weinreb, 1989].

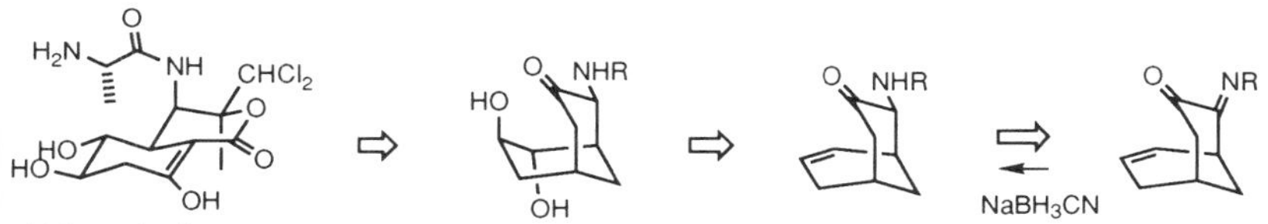

(-)-bactobolin

The solution of a stereochemical problem with the acid of ring formation is witnessed in a synthesis of geissoschizine [Benson, 1979]. Lactamization rendered the (*Z*)-isomer less stable, and that isomer underwent isomerization *in situ* so that a homogeneous product was obtained.

HOOC + HOOC

$(CF_3CO)_2O$

MeOOC OH

geissoschizine

While conformational change via lactonization was a key to isomerize a 3-epimeric precursor of reserpine (see Section 9.1.), the functionalization of the E-ring at the dicarbocyclic stage also benefited from the formation of a δ-lactone prior to acetolysis of an epoxide [Woodward, 1958]. Note that the direction of a *trans*-diaxial opening of the opoxide ring in the lactone is completely opposite to that of the hydroxy acid.

HOAc

DCC

HOAc

(corresponds to reserpine stereochemistry)

5.1. ALKYLATIONS IN CYCLIC SYSTEMS

Regio- and stereocontrolled alkylations are often required synthetic operations, in which different situations may demand special tactics. The simplest functional group interchange involving a *gem*-dimethyl group and a carbonyl is bridged by the spiroannulated cyclopropane, the C=O to CMe_2 conversion being achievable by the sequence of Wittig, Simmons–Smith reactions and hydrogenation (or more directly by treatment of the carbonyl compound with titanium reagents such as Me_2TiCl_2 [Reetz, 1985], $cp_2Ti{=}CH_2$ [Brown-Wensley, 1983]). Consequently it is possible to develop plans for the synthesis of longifolene [Oppolzer, 1978] and cedrene [Horton, 1984] based on an intramolecular deMayo reaction and intramolecular Michael reaction, respectively.

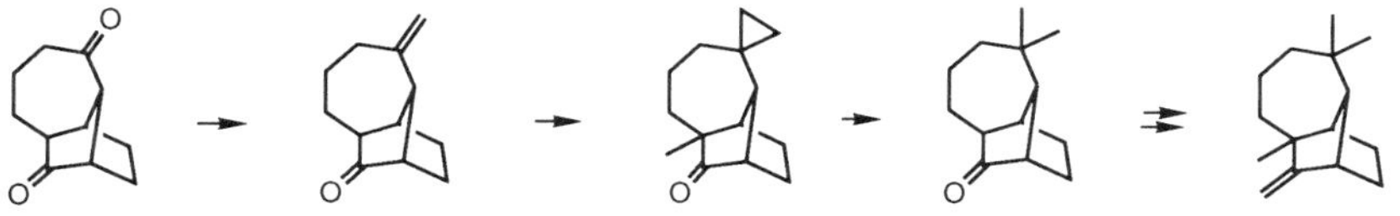

longifolene

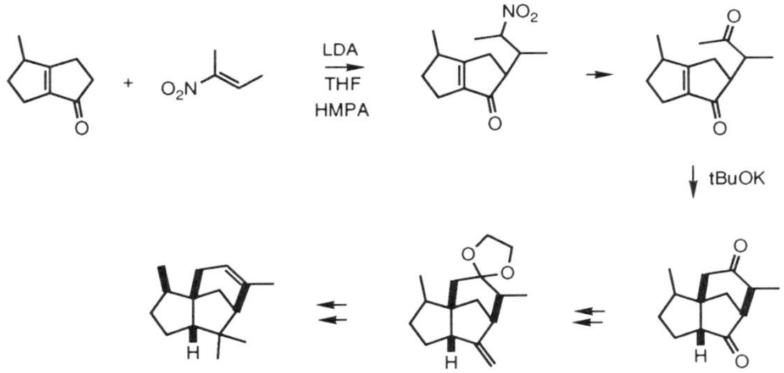

NO_2 LDA THF HMPA O_2N tBuOK

α-cedrene

Manipulation of fused cyclopropanes can result in either methylated ring systems or enlarged rings. The indirect methylation is found in the syntheses of (−)-valeranone [Wenkert, 1978a], sinularene [Antezak, 1985], isocaryophyllene [Bertrand, 1974], and 9-pupukeanone [Schiehser, 1980], each of them requiring introduction of a methyl group at a particular position.

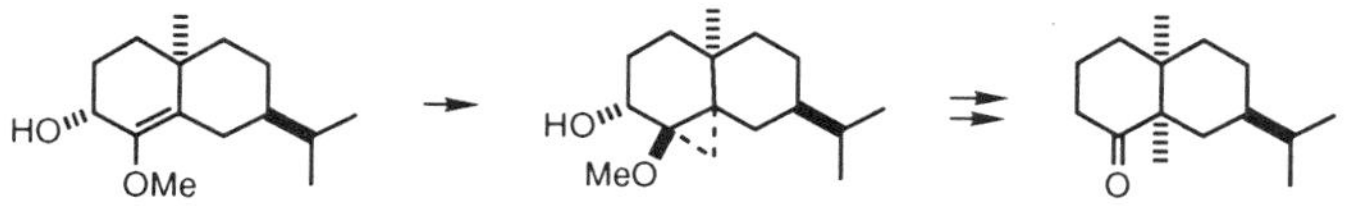

HO OMe MeO

(-)-valeranone

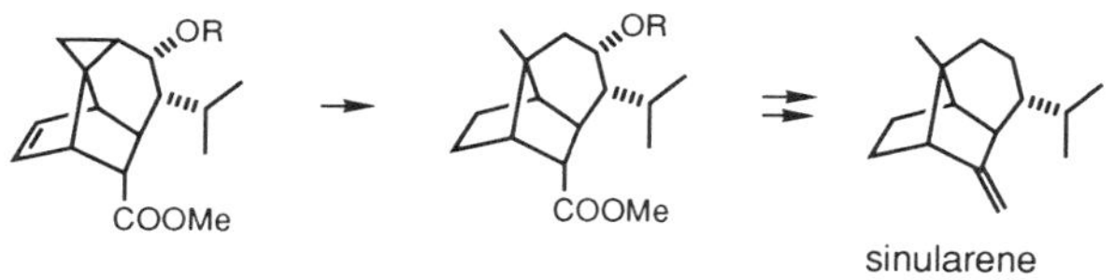

OR COOMe

sinularene

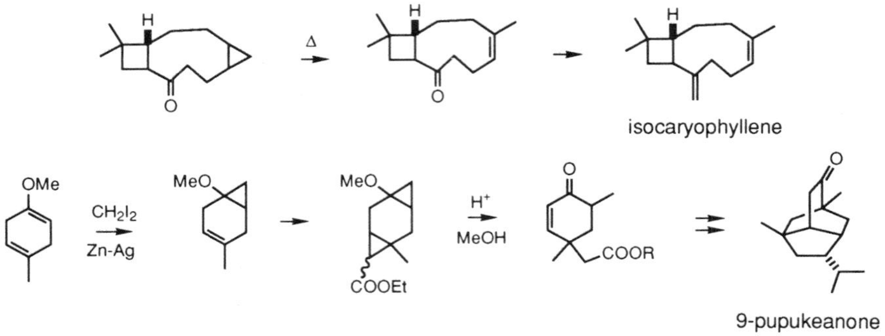

isocaryophyllene

9-pupukeanone

Cyclopropanation as an intervening step for alkylation has merits in the assembly of certain structural units. 1,6-Diketones such as that required in a synthesis of α-chamigrene [Iwata, 1979] are readily acquired by this method. In this instance the cyclopropanation is followed immediately by a fragmentation. On the other hand, a stepwise alkylation is exemplified in a route to α-cuparenone [Wenkert, 1978b]. The conversion of gibberellin-A_7 into antheridic acid [Furber, 1987] involving a 1,2-shift of a two-carbon bridge is also via stepwise process in which a cyclopropane intermediate is formed and cleaved.

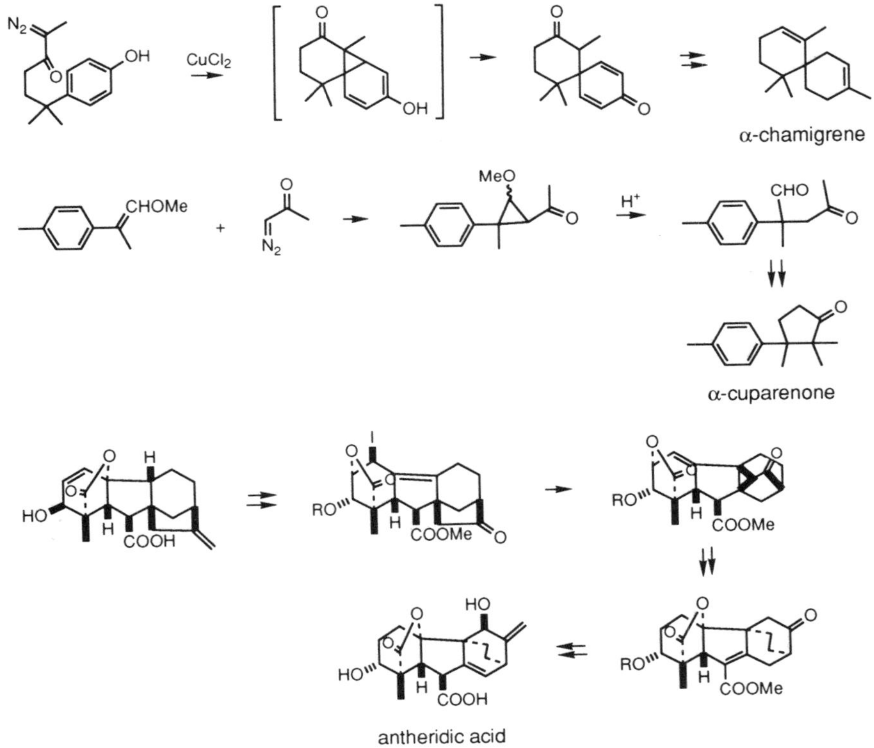

α-chamigrene

α-cuparenone

antheridic acid

Due to the great significance of functionality increase in a synthetic step the cleavage of cyclopropyl ketones effected by nucleophiles leading to γ-substituted ketones is well appreciated. The manipulation should be viewed in the perspective of a general method for cyclopropyl ketone formation consisting of cycloaddition of acylcarbenoid species to alkenes. Utility of the reaction sequence is witnessed in many syntheses, including chrysomelidial [Kon, 1980] and grosshemin [Rigby, 1987]. It is noteworthy that implementation of a cyclopropanation protocol is an important operation in the stereocontrolled introduction of the secondary methyl group in a synthesis of modhephene [Wrobel, 1983].

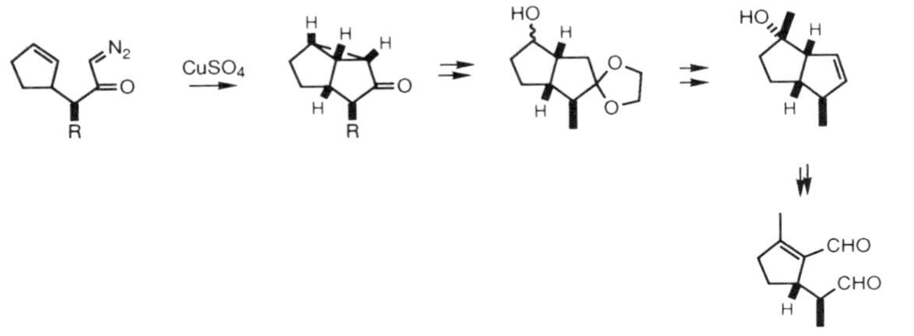

$CuSO_4$

chrysomelidial

$BF_3.OEt_2$ / Ac_2O

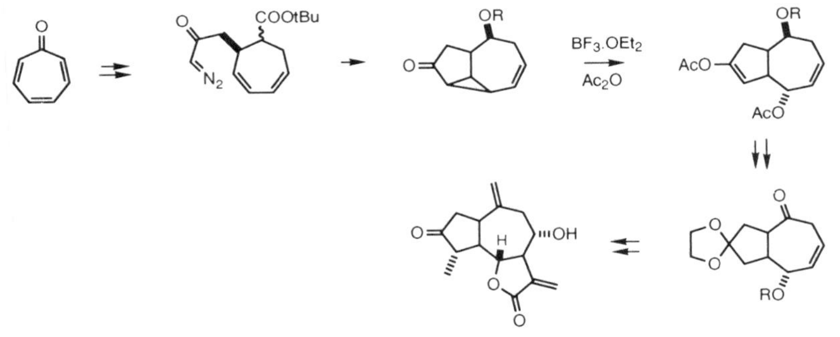

grosshemin

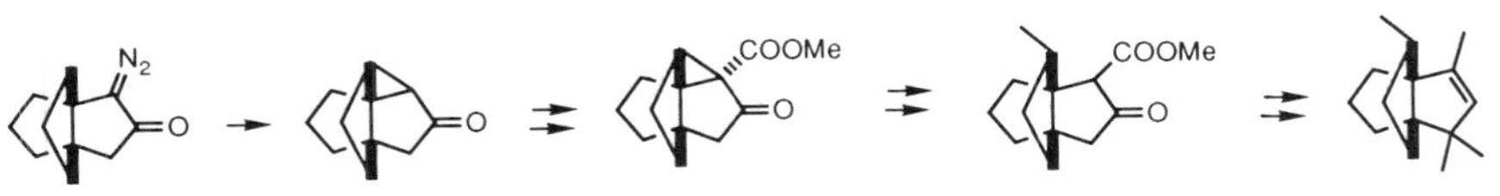

modhephene

The photochemical oxadi-π methane rearrangement is a very useful method for skeletal construction. A properly substituted bicylco[2.2.2]oct-7-ene-2,5-dione has been transformed into a tricyclic isomer containing a diquinane nucleus which is also framed by a cyclopropane. The latter unit determines the regio- and stereochemistry of alkylation of its adjacent ketone, thereby facilitating the synthesis of coriolin [Demuth, 1984].

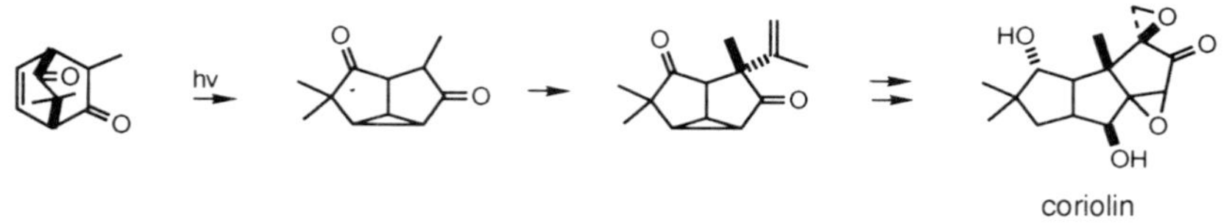

coriolin

Deliberate cyclopropanation to activate a special carbon atom and provide an entry to a different ring system is involved in a route to the prostaglandins, such as PGA_2 [Ali, 1980].

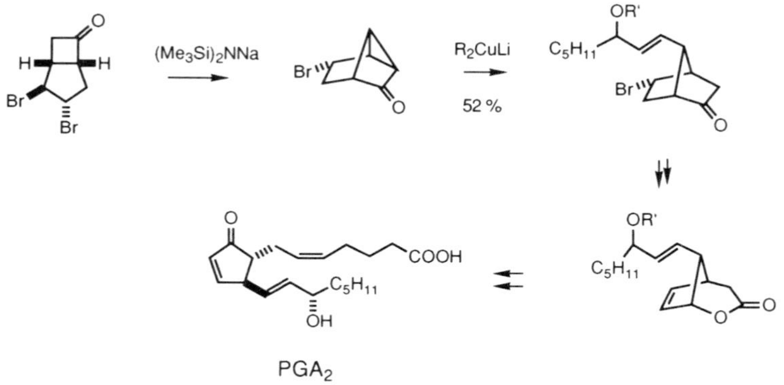

PGA_2

Stereocontrol afforded by a cyclopropane ring is seen in an elaboration of α-cyperone [Caine, 1974]. Stereoelectronic effects have been a sore point in the Robinson annulation approach from dihydrocarvone as epi-α-cyperone is the predominant product. When dihydrocarvone is converted into 2-caranone the electrophile can react from the face opposite to the fused dimethylcyclopropane ring resulting in a *cis* relationship between the angular methyl group and the isopropenyl sidechain which would be unraveled from the cyclopropane subunit.

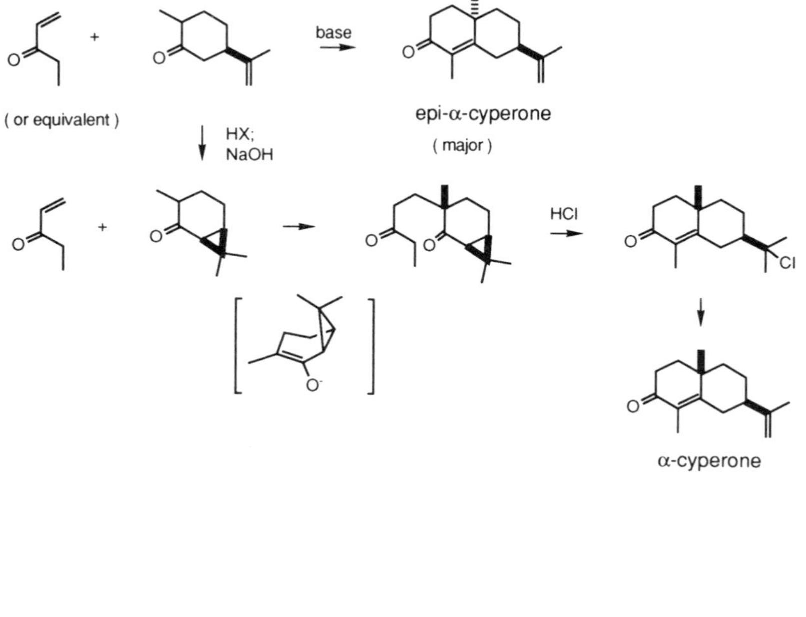

epi-α-cyperone (major)

α-cyperone

Stereocontrol by a cyclopropane is also an important aspect during the attachment of the isopropenyl group to a synthetic intermediate of eremophilone [F.E. Ziegler, 1977]. In this case the conformation of the Michael acceptor unit is fixed and the conjugated double bond is shielded on one side by a ketal.

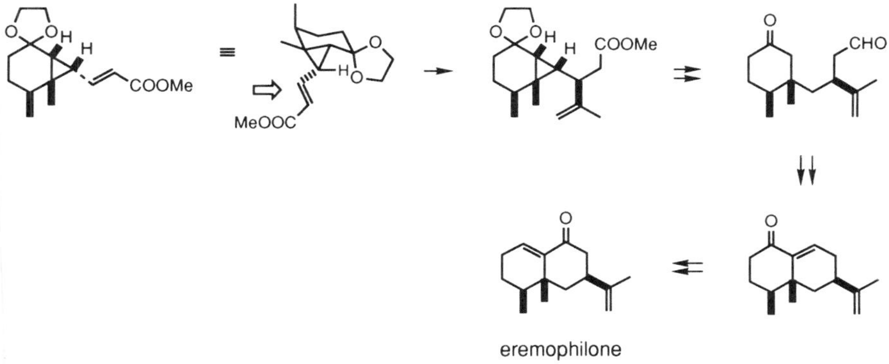

eremophilone

(The stereocontrolled introduction of oxygen functionalities at C-6 and C-8 during a synthesis of vernolepin/vernomenin [Isobe, 1978] is relevant. Cyclopropanation blocked formation of a double bond between C-6 and C-7 thus ensuring the generation of one isomer, and fixed the molecular conformation to direct the subsequent functionalization).

Reductive β-alkylation of conjugated cycloalkenones can be effected via [2 + 2]-cycloaddition and subsequent cleavage of the four-membered ring at the bond attaching to the α-carbon of the ketone group. The process involving thermal cycloaddition with an ynamine is illustrated in the assembly of an intermediate for *erythro*-juvabione [Ficini, 1974]. The ring cleavage is preceded by kinetic protonation of the aminocyclobutene portion.

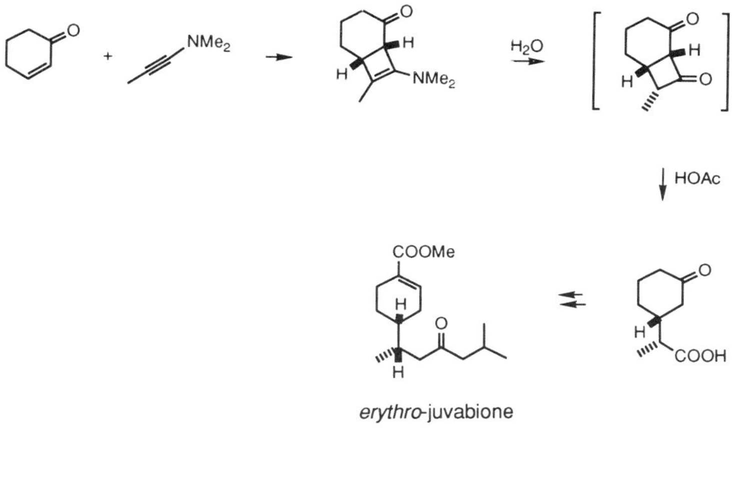

erythro-juvabione

A comparison with the preparation of an intermediate of *threo*-juvabione [Larsen, 1979, Kitagawa, 1983] is instructive. Kinetic protonation of enolates incorporated into certain bridged ring systems achieves the same results.

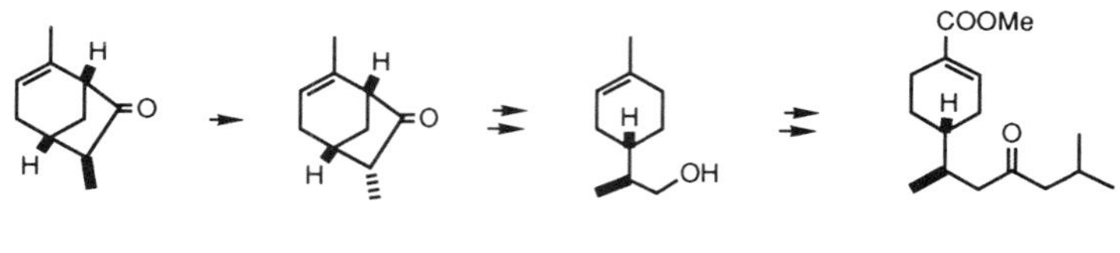

COOMe

H O

OH

threo-juvabione

The steroid D-ring and sidechain with correct configuration at C-17 and C-20 can be assembled via ynamine-enone cycloaddition, controlled hydrolysis and cleavage of the resulting bicyclic enamines [Desmaële, 1983]. Under thermodynamically controlled conditions the carboxyl unit thus generated corresponds to the C-20 methyl group.

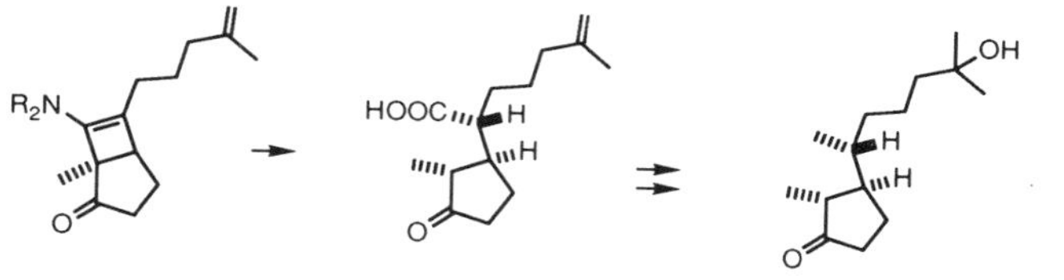

R_2N

HOOC H H

OH H H

By far the most popular method for the indirect alkylation consists of photo-cycloaddition with allene as the first step. It is very useful for the construction of bridged ring systems such as those existing in atisine [Guthrie, 1966] and 2-desoxystemodinone [Piers, 1985].

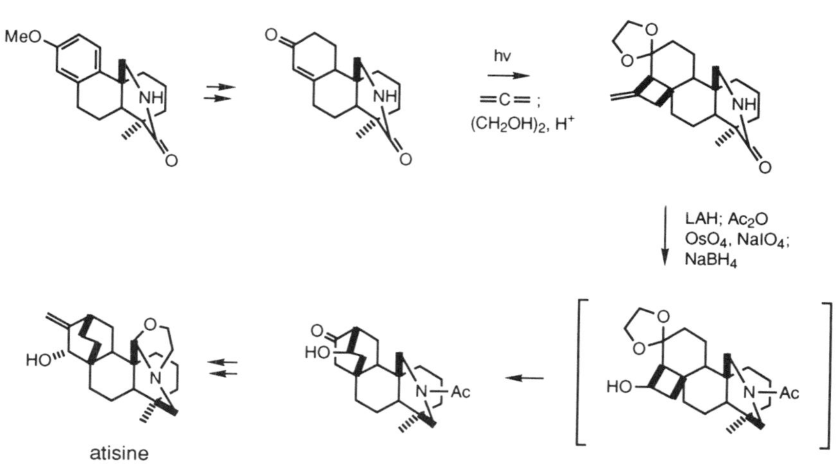

MeO NH O

O NH O

hv

=C= ;

$(CH_2OH)_2$, H^+

O O NH O

LAH; Ac_2O

OsO_4, $NaIO_4$;

$NaBH_4$

HO O O N-Ac

O HO N-Ac

HO O N

atisine

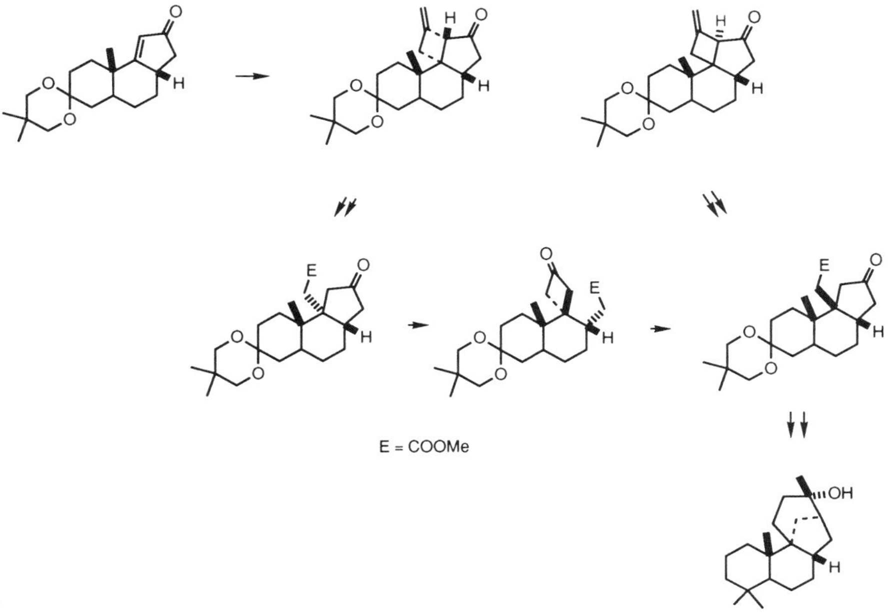

E = COOMe

2-desoxystemodinone

Two major products arising from the photocycloaddition of a tricyclic enone are convertible into the desired keto ester. It is noted that the *trans*-C/D isomer can be equilibrated via a spirocyclic cyclobutanone by a tandem twofold Claisen condensation and its retrograde reaction.

The combination of other photocycloaddition reactions with selective opening of the cyclobutane constitutes methods of great synthetic potential. It is interesting that an intramolecular photocycloadduct obtained at $-50°C$ fragments on photolysis at higher temperature. This observation together with the discovery of an $AgNO_3$-catalyzed epimerization of the isopropenyl chain has enabled a rapid synthesis of acorenone-B [Manh, 1981].

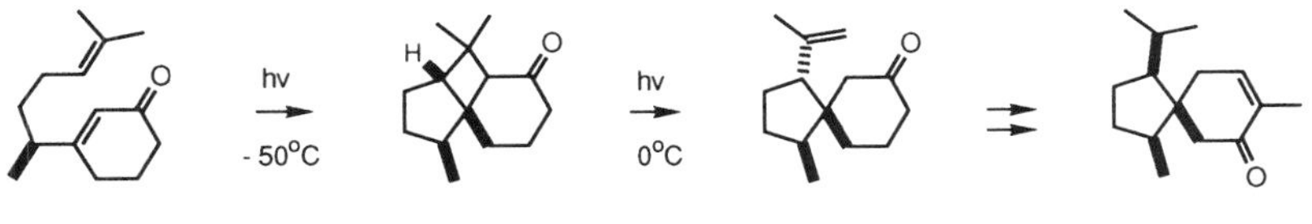

hv, - 50°C; hv, 0°C

acorenone-B

Cyclobutanols obtained from photocycloaddition of enols and derivatives to alkenes may be caused to fragment to provide substances difficult to access. Thus an approach to trichodiene [Yamakawa, 1976] exploits the presence of a fragmentable β-hydroxy ketone subunit in the head-to-head photocycloadduct.

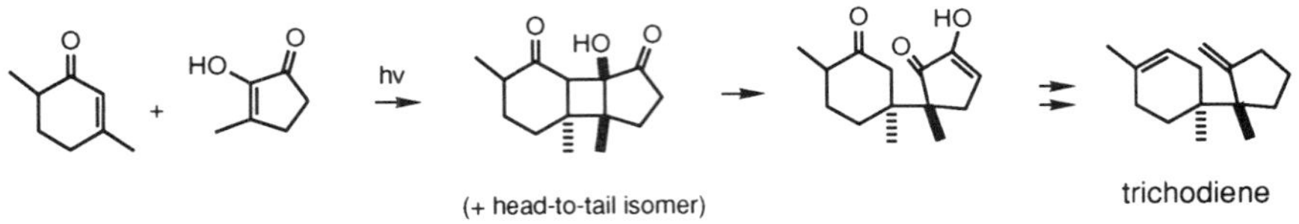

(+ head-to-tail isomer)

trichodiene

A combination of intramolecular ketene-alkene [2 + 2]-cycloaddition and selective cleavage of the cyclobutanone unit constituted the key features of a route to methyl dehydrojasmonate [S.Y. Lee, 1988]. Cyclization and functionalization at an exocyclic site were accomplished simultaneously.

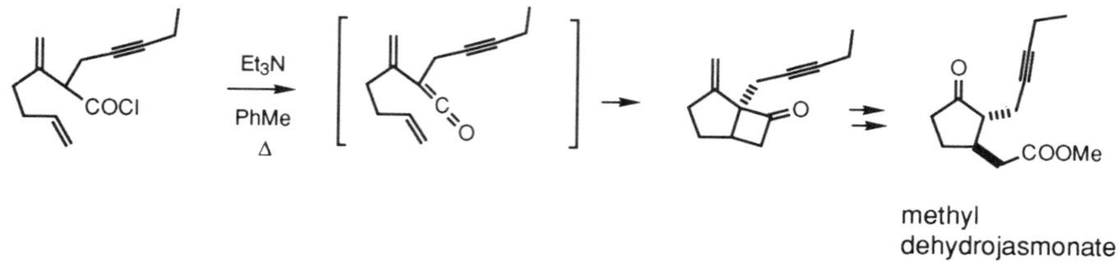

Et_3N, PhMe, Δ

methyl dehydrojasmonate

The stereochemical challenge of reserpine lies in the concentration of five asymmetric carbon atoms in a cyclohexane ring. A brilliant response [Woodward, 1958] to the synthetic challenge was initiated by a Diels–Alder reaction with which three contiguous asymmetric centers were erected. The enedione system derived from the *p*-benzoquinone dienophile was degraded in later steps to give an aldehydo-ester unit for heterocyclization (D-ring formation). It must be emphasized that the stereoselective introduction of the two other asymmetric centers in ring-E was under control with the help of conformational factors. Such control would be more difficult if the D-ring elements were not in a cyclic assembly.

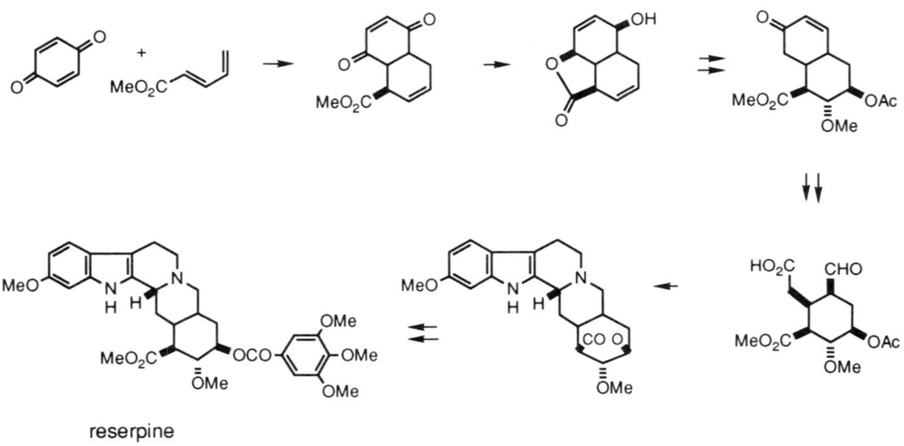

reserpine

In an early synthesis of yohimbine [van Tamelen, 1958] the nontryptamine portion was created from the Diels–Alder adduct of butadiene and *p*-benzoquinone. The quinone represents the E-ring moiety and the butadiene provided C-3, 14, and the carboxylate pendant atom. While a one-carbon fragment was extended from one of the original ketone groups, another such fragment was crafted out of an octalone intermediate. In essence the butadiene molecule furnished two chains to the E-ring which were necessary for the elaboration of the alkaloid.

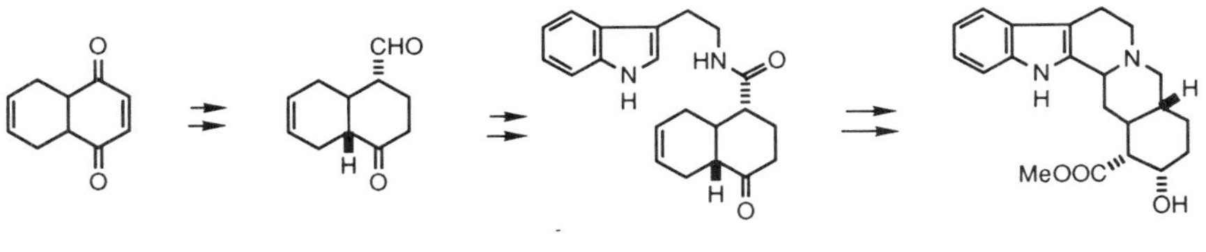

yohimbine

A neat approach to gephyrotoxin [Hart, 1981] involved the Diels–Alder adduct of butadiene and 2-cyclohexenone which, after epimerization of the ring juncture was derivatized into a succinimide at the ketone site. The cyclohexene moiety was cleaved and converted into two vinyl chains, and the one proximal to the succinimido group participated in heterocyclization. Noteworthy of this route is the exploitation of local symmetry, in the same spirit as that underlying the yohimbine synthesis.

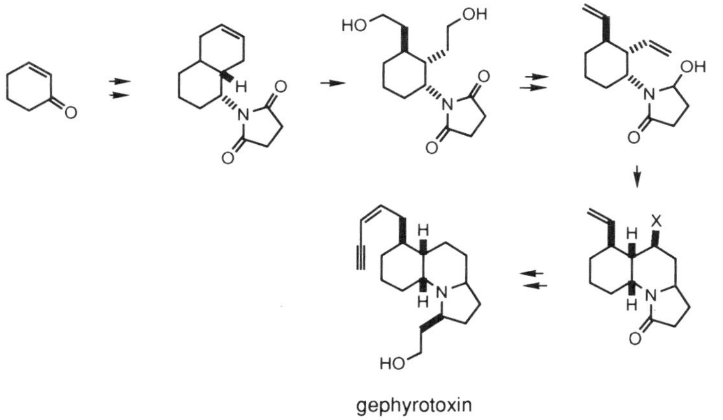

gephyrotoxin

Elaboration of terramycin [Muxfeldt, 1968] started from a Diels–Alder reaction of juglone acetate and 1-acetoxy-1,3-butadiene. However, only two skeletal carbon atoms originating from the diene were retained for the subsequent condensation reactions in the process of assembling the AB-ring moiety.

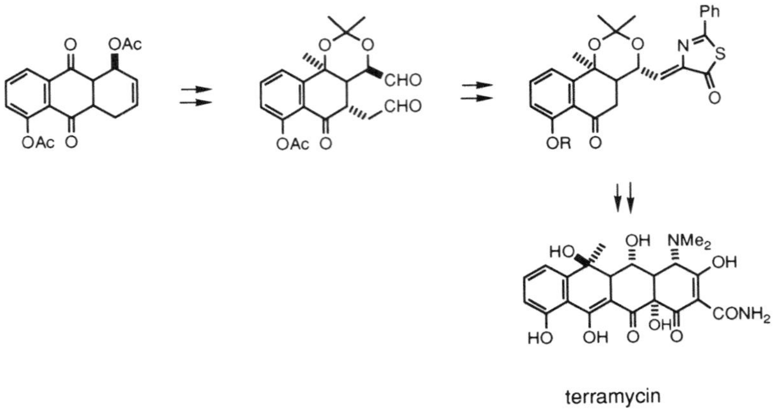

terramycin

The synthesis of more complex 4-substituted 2-cyclohexenones can be accomplished by Diels–Alder reaction of dihydroanisole derivatives (Birch reduction of the anisoles, and *in situ* isomerization) and fragmentation of the adducts or their modified forms. Suitable intermediates for the synthesis of *threo*-juvabione [Birch, 1970] nootkatone [Dastur, 1974], solavetivone [Murai, 1981], *inter alia*, have been obtained. These three examples illustrate the succinct assembly of monocyclic, condensed and spirocyclic ring systems with good stereocontrol.

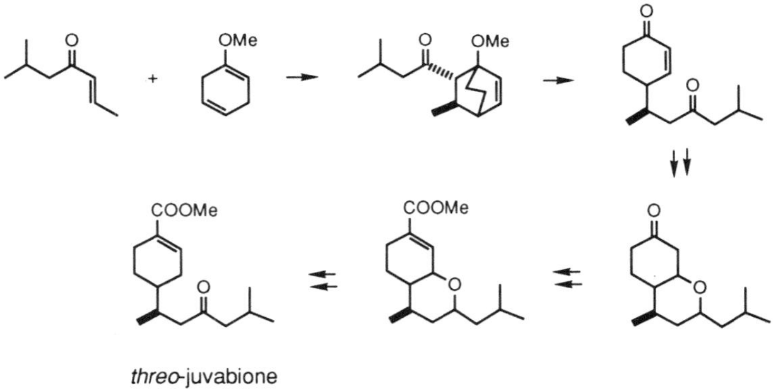

threo-juvabione

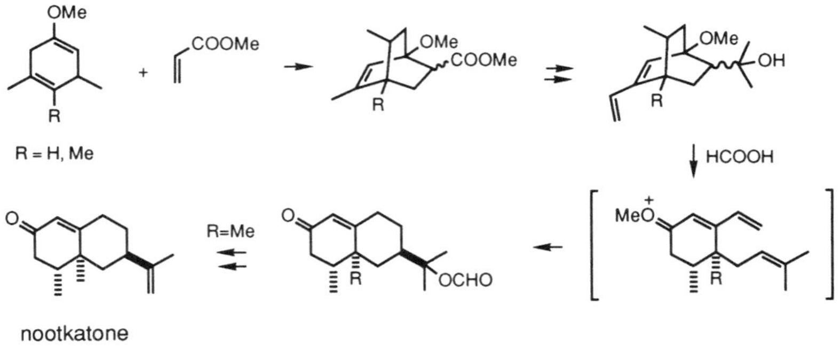

nootkatone

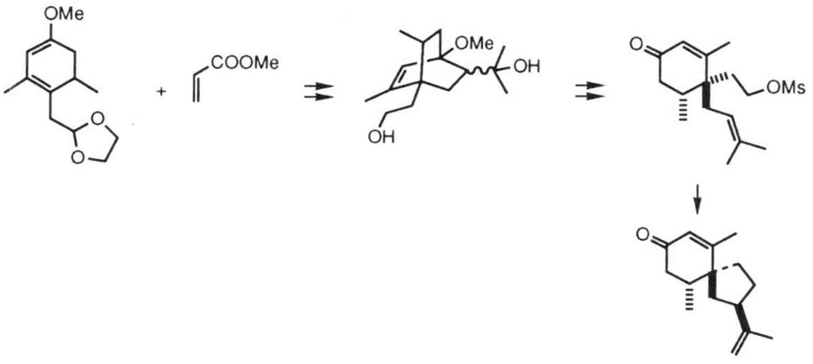

solavetivone

Pertaining to synthesis of trichodermol [Still, 1980f] and β-bazzanene [Kodama, 1980] is a process in which the fragmentative generation of a cyclohexenone occurs after much more complex transformations (e.g. ring contraction, etc.) of the part derived from the dienophile. However, the same advantage of stereocontrol in establishing two asymmetric centers on a bicyclic array is taken.

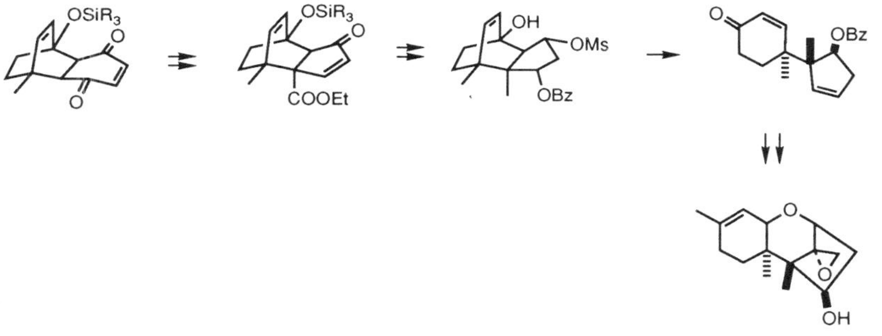

trichodermol

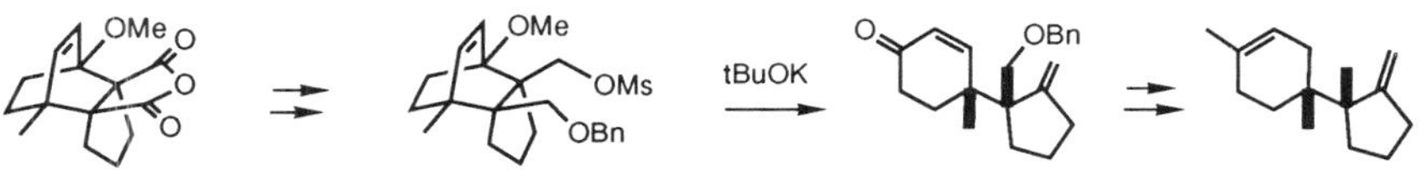

β-bazzanene

A circumstance analogous to the juvabione synthesis is that pertaining to an elaboration of chrysomelidial [Hewson, 1985] from a cyclopentanone derivative. Although the overall structural change consists of homologation of the sidechain ketone and the conversion of the β-keto ester function into a β-methyl enal, the advantage of conducting the tranformation via bicyclic intermediates is apparent. The stereochemical issue of establishing the two adjacent asymmetric centers is thereby laid to rest. In the bicyclo[3.3.0] octane system *exo*-oriented substituents (e.g., methyl) are more stable.

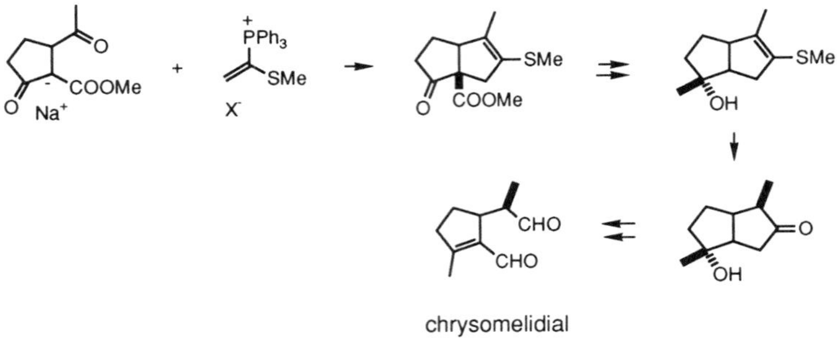

chrysomelidial

For the preparation of a 1-menthen-9-ol [Bartlett, 1981] which is convertible to *threo*-juvabione, stereocontrol over two adjacent centers may be extracted from a fused γ-lactone. *exo*-Methylation followed by reductive C-O bond cleavage led to the (*SR*, *RS*)-acid which was promptly reduced to the alcohol.

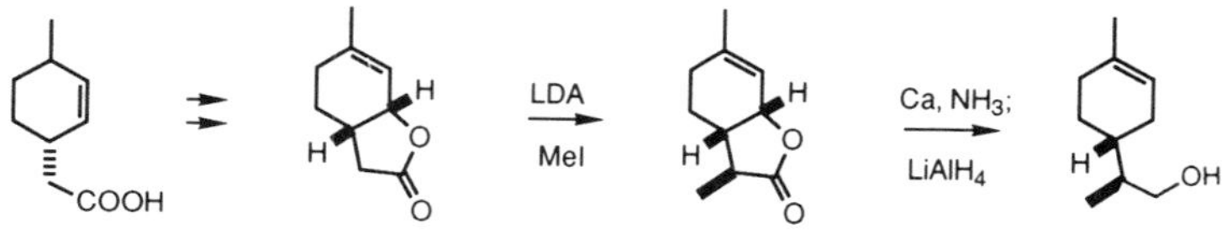

In cyclic skeletons 1,2-asymmetric inductions are more readily accomplished. Thus the stereocontrol of the two adjacent quaternary carbon centers during elaboration of trichodiene [Welch, 1980] is manageable when the formation of the five-membered ring is delegated to a Dieckmann condensation of a *cis*-fused lactone ester in which the stereochemical problem is already solved. The bicyclic lactone avails itself of stereoselective alkylation at the α-position, and the situation dictates an initial presence of a methyl group so that the carbon chain which would be modified to a four-carbon ester has a *trans* relationship with the angular substituents. The same principle is applied in the case of calonectrin [Kraus, 1982], although the alkylation is intramolecular and the donor is a ketone enolate.

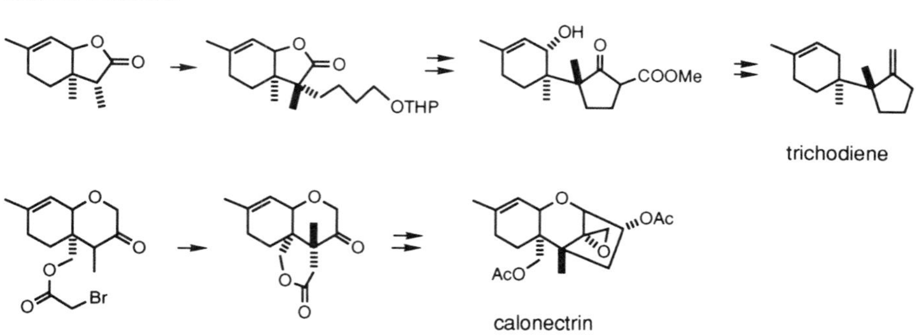

trichodiene

calonectrin

Norcamphor presents itself as an excellent building block for synthesis by virtue of its availability and steric manipulability in functionalization. The lactone derived from norcamphor also shares such characteristics, and it has been employed in the elaboration of many products. The kinetic *exo*-alkylation products of the lactone can be epimerized as desired, since the *endo*-substituent is equatorial. The following synthesis of emetine [Takano, 1982a] is representative of a body of work which exploits the structural and stereochemical attributes of a bridged ring system.

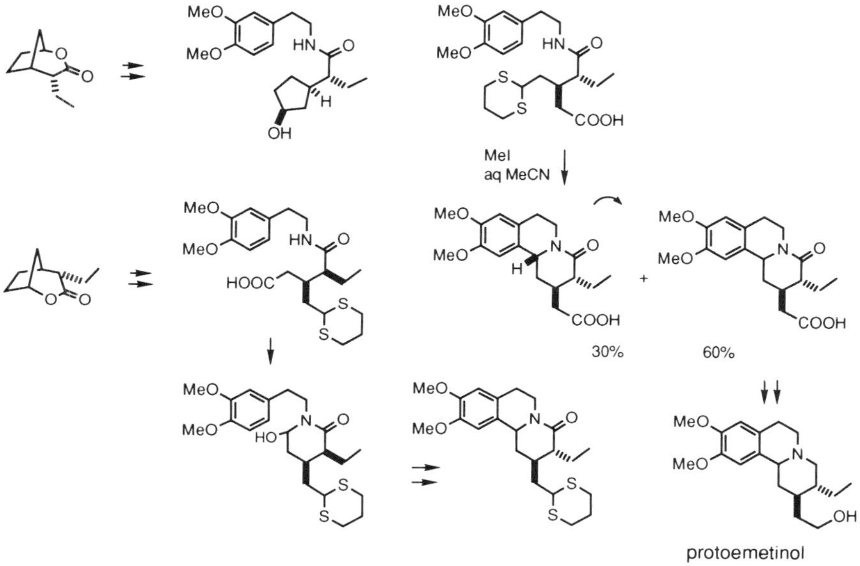

MeO MeO HN O OH H MeO MeO HN O S S COOH MeI aq MeCN MeO MeO HN O HOOC S S MeO MeO N O H COOH + MeO MeO N O COOH 30% 60% MeO MeO N O HO S S MeO MeO N O S S MeO MeO N OH

protoemetinol

(For a review of using camphor as building block and ring cleavage methodologies in natural product synthesis, see T.-L. Ho [1992a].).

Another important building block for synthesis of cyclopentanoids is norbornadiene which is readily converted into *exo*-5-halo-*anti*-7-carboxy-2-norbornanone via a Prins reaction, oxidation, and hydrohalogenation. Not only has this compound served in the synthesis of prostaglandins [Bindra, 1973; Peel, 1974], it has been elaborated into pseudoguaianolide sesquiterpenes [Grieco, 1977, 1980], steroids [Grieco, 1979b; Trost, 1979a], and macrolide antibiotics [Grieco, 1979a]. At least one of the five-membered rings is cleaved after playing its role of stereoregulation.

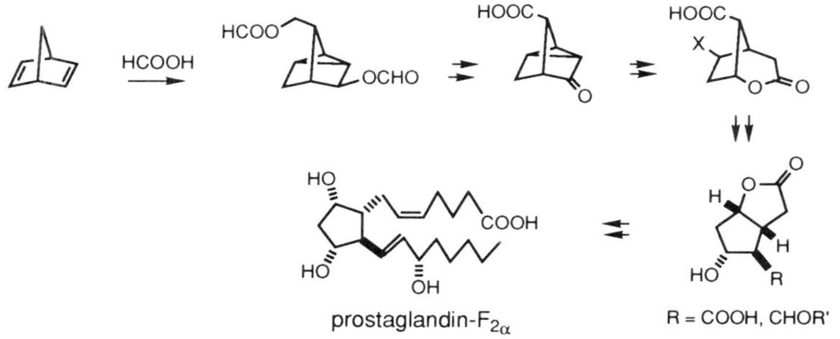

HCOOH HCOO OCHO HOOC O HOOC X O O HO COOH HO OH H O O H HO R

prostaglandin-$F_{2\alpha}$

R = COOH, CHOR'

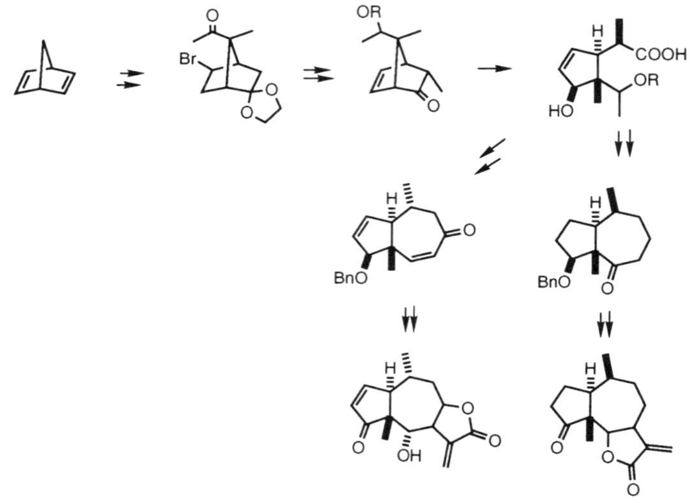

helenalin

damsin

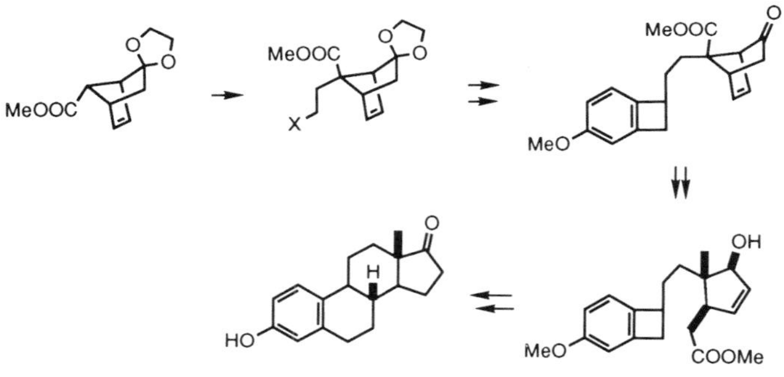

estrone

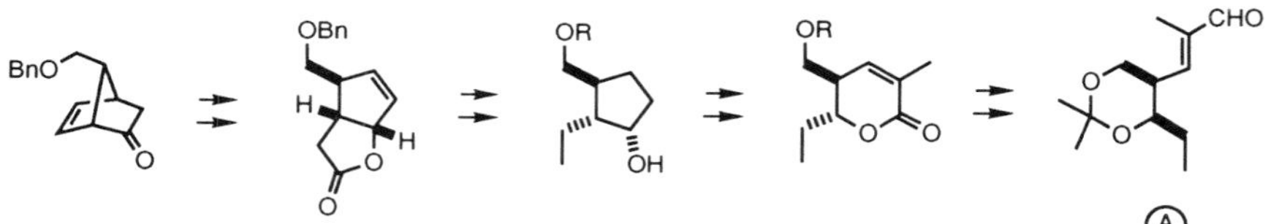

(A)

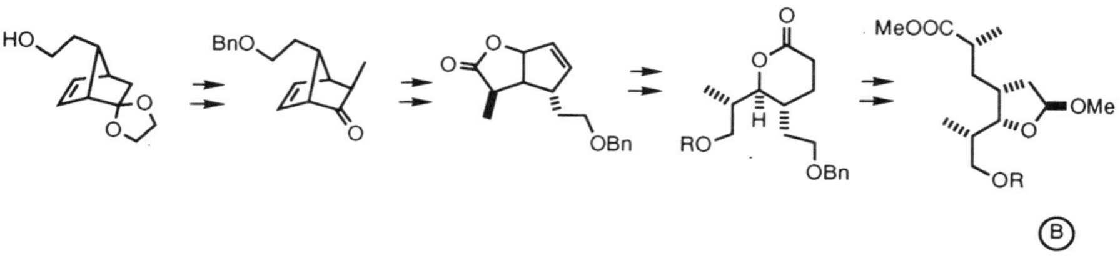

(B)

tylonolide hemiacetal

A silyl-directed aldol condensation (*anti*:*syn* >97:3) using a (*S*)-prolinol chiral auxiliary has been reported [Myers, 1990]. The geometry of the ketene *N*,*O*-acetal is fixed in a ring system. When the two alkyl groups on silicon are locked in a small ring, i.e. silacyclobutane, the reaction rate is enhanced, and *syn*-selectivity is observed [Myers, 1992a]. The mechanistic implication is that the initial complex of the two reactants undergoes pseudorotation to place the two alkyl substituents on silicon in the alternative apical and equatorial sites before C—C bond formation. This process is driven by angle strain of the silacyclobutane.

anti : *syn*
2.5 : 97.2

A high *anti*-selectivity is observed in the aldol condensation of 4-tetrahydrothiopyranone [K. Hayashi, 1991]. The ring effect is remarkable since reaction of the 3-pentanone (*E*)-enolate shows *anti*:*syn* aldol ratio of 54:46 [Dubois, 1975].

M = $SnMe_3$....

anti (96 : 4) *Syn*

A method for activation of cyclic amines toward reaction with Grignard reagents is by photooxidation [Pandey, 1991]. It is essential to provide stable intermediates and the trapping with a hydroxyl group to form oxazine derivatives serves the purpose for a selective synthesis of *cis*-2,5-disubstituted pyrrolidines and *cis*-2,6-disubstituted piperidines.

In an approach to coriolin [Danishefsky, 1981] there is a special design to effect regioselective reductive alkylation of an enedione system as a prelude to construct the third ring. This is a case in which synthetic strategy determines the tactics: the regiochemistry is of concern owing to the tetrasubstituted pattern of the double bond, such that alkylation at either the ring juncture or the desired position is possible. Even if the alkylation occurs at the desired position it would create a new stereochemical problem in reducing the double bond which must now be in a new position. The alkylation introduces a new asymmetric center, and *a priori*, it is difficult to conceive a simple method to achieve a stereoselective reduction. The solution involves a Diels–Alder reaction which is regioselective, being controlled by the methyl groups (*ortho*-adduct preferred) of the two addends. The new ring can be elaborated further and cleaved to generate the required sidechain. Note that the Diels–Alder reaction also provides a solution to the stereochemical problem. The stereoselectivity is conferred by the single asymmetric center of the diquinane; the molecule is predisposed to form a product with the more stable *cis*-diquinane framework.

OSiMe3 OH OH coriolin

(Note tactic of partial sacrifice of cyclic elements has been implemented in the syntheses of vernolepin/vernomenin [Danishefsky, 1977b] and pentalenolactone [Danishefsky, 1979].)

The very reactive siloxydienes readily undergo cyclocondensation with various aldehydes to provide 2,3-dihydro-4-pyrones. As a high degree of diastereoselectivity is often observed with aldehyde containing an α-asymmetric center, a chiral β-hydroxy acid is accessible via this condensation and oxidative cleavage of the resulting heterocycle, as shown in a synthesis of Boc-statine [Danishefsky, 1982].

BocNH + OMe OSiMe3 $ZnCl_2$ BocNH O_3; H_2O_2 BocNH OH COOH

Boc-Statine

It is appropriate to consider in this section the construction of a prostaglandin core from a substituted cyclopentaidene via Diels–Alder reaction [Corey, 1969d]. Essentially, the bicyclic array gives complete control over the formal alkylation and hydroxylation of the five-membered ring.

OR RO Cl CN RO O RO O O HOOC OR HO O O H H CHO AcO HO COOH HO HO

prostaglandin-$F_{2\alpha}$

Iridoid synthesis via formation of a bicyclo[3.3.0]octane allows execution of stereocontrol. For example, the *exo*-configuration of the methyl group on the lactone ring of isoiridomyrmecin can be readily achieved upon equilibration of the precursor [Sakan, 1960].

O O O OH OH O O

isoiridomyrmecin

Predicated by the methodology of asymmetric alkylation the precursor of the C-23 to C-32 subunit of ionomycin [Evans, 1990c] emerged as having an extra carboxamide pendant which must be degraded. Lactonization prepared the operation, when oxidative degradation of the sidechain by reaction of the acid chloride with peracid was found to be extremely inefficient.

BnO H O COXv OH PhMgBr H O O BnO O Tebbe; TsOH; O_3 OBn H O O OAc

ionomycin Ph$_3$P$^+$ I$^-$ H O RO OR

A synthesis of portulal [Tokoroyama, 1974; Kanazawa, 1975] addressed the problem of introducing the angular aldehyde group by considering its association with the vinylic methyl substituent. Although the annulation-cleavage tactic requires further degradation afterwards, its advantages outweigh the extra labor.

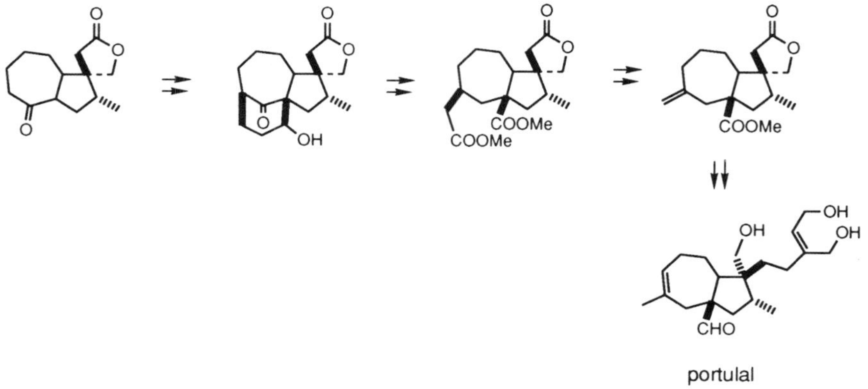

portulal

In connection with this tactic is how to approach hirsutic acid-C [Trost, 1979b]. In this case the rationale is to lock the two 1,4-related methyl groups in the form of a cyclohexene moiety so that they can be unraveled by ring cleavage and reduction. More importantly such an arrangement enables stereocontrol at five asymmetric centers, three of them being established during formation of the bicyclo[3.2.1]octane system, and the fourth one via an intramolecular Stetter reaction.

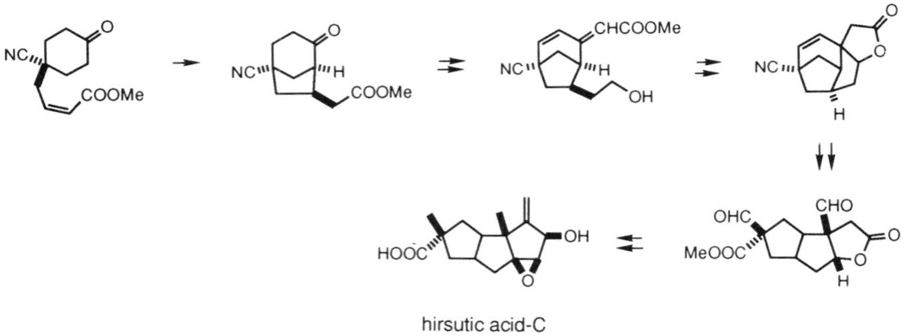

hirsutic acid-C

It should be noted that in the above synthesis the immolative ring, which acted as an original template, was kept for many steps. A simpler example of such utility is found in an elaboration of botryodiplodin [McCurry, 1973] from ethyl 2,6-dimethyl-4-oxo-2-cyclohexenecarboxylate.

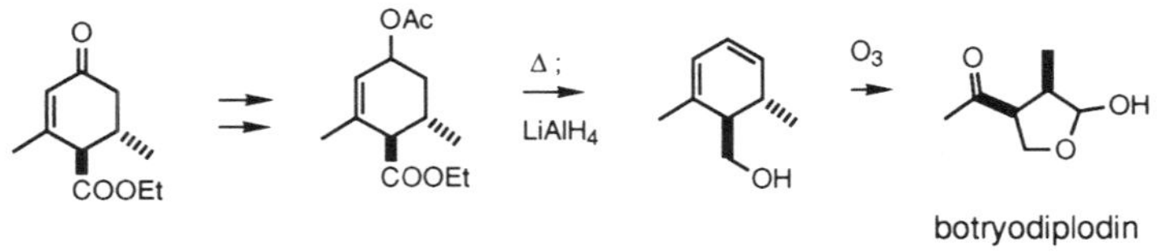

botryodiplodin

The juvenile hormone JH-I of the *cecropia* moth presents a stereochemical challenge in terms of the double bond geometry. One route based on the Grob fragmentation to stereospecifically generate the double bonds [Zurflüh, 1968] effectively translates the problem into that concerning the erection of asymmetric carbon atoms in a hydrindenone and a cyclopentanone. The substrate structure is predictable on the basis of the steric course of alkylation and the relationship of the angular ethyl group with the adjacent substituent which would be converted into a leaving group in the second-stage fragmentation.

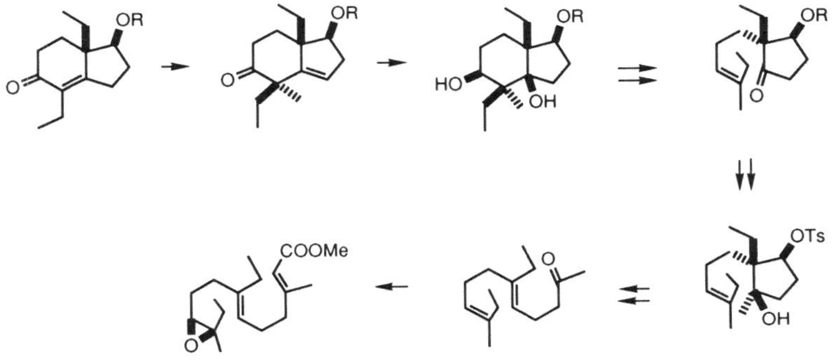

JH-I

p-Cresyl methyl ether provides the terminal 7-carbon unit of JH-I [Corey, 1968a] by a series of reactions including Birch reduction, selective ring cleavage, and deoxygenation of the primary alcohol. The (*Z*)-configuration of the double bond inherits from the cyclic precursor.

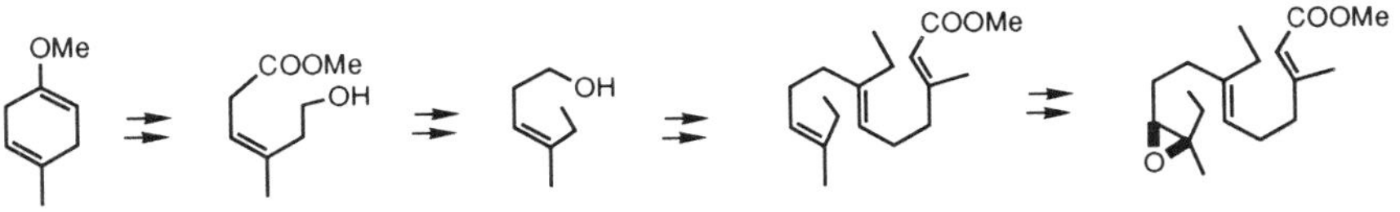

JH-I

In a synthesis of tetrahydrodicranenone-B [Moody, 1988] the dihydroaromatic compound was alkylated at an angular position, and one carbon atom from the six-membered ring was sacrificed immediately afterwards.

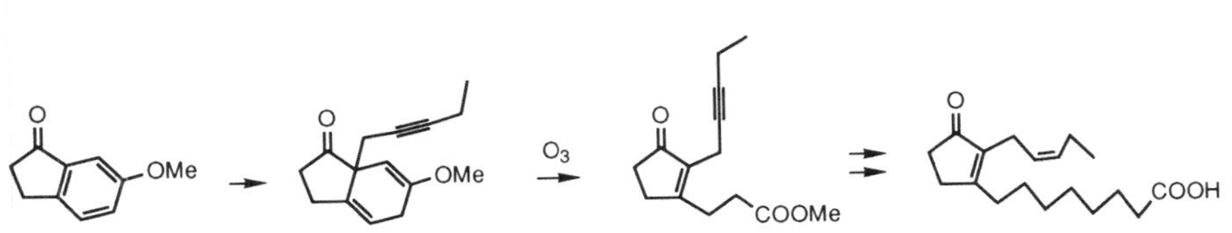

tetrahydro-
dicranenone-B

m-Disubstituted benzenes are latent 1,3-diketones as ozonolysis of the Birch reduction products would remove three of the ring carbon atoms. Synthetic operations based on this equivalency include an elaboration of the spiroketal subunit of milbemycin-β_3 [Holoboski, 1992]. A more elaborate structure for the synthesis of (+)-mycoticin [Poss, 1993] also was derived from 1,5-di(*m*-anisyl)-2,4-pentanedione.

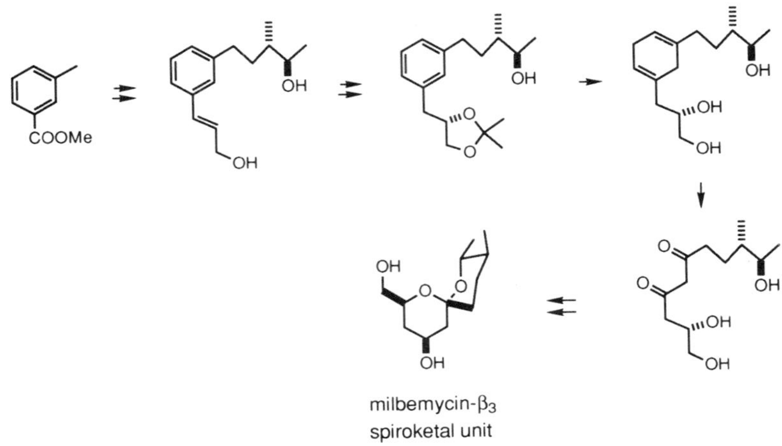

milbemycin-β_3
spiroketal unit

Analogously, statine derivatives of any desired configuration may be elaborated from an aryl ketone [Bringmann, 1990] via reductive amination with a chiral 1-phenylethylamine (available in both enantiomers), *N*-protection, Birch reduction, ring cleavage, and ketone reduction.

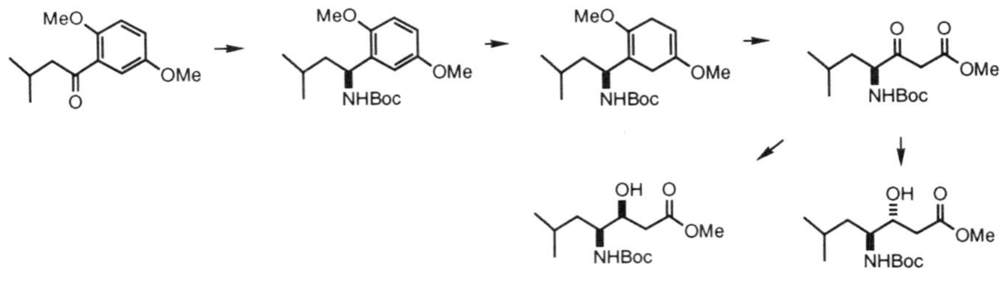

When submitted to Birch reduction and ozonolysis *m*-methoxystyrene oxides are converted into δ-hydroxy-β-keto esters from which 1,3-diols are obtainable by stereoselective reduction [Evans, 1991]. This reaction sequence should be valuable for the synthesis of polyketide natural products.

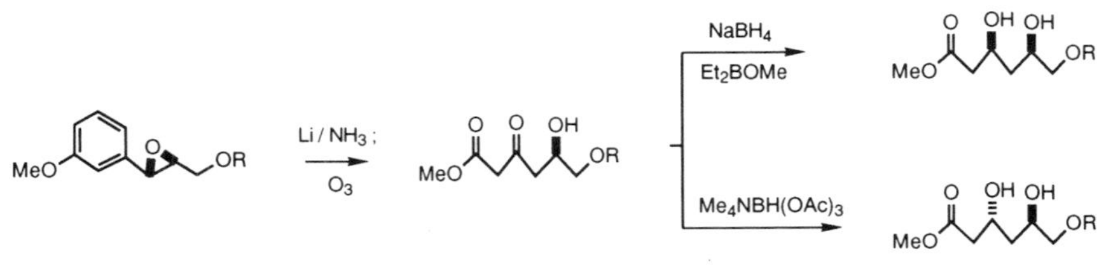

Stereocontrol of reactions can be exercised in acyclic carboxylic acids by temporarily incorporating them into macrocyclic templates [Still, 1984a,b]. For an approach to nonactic acid a kinetic methylation of a macrolactone derived from a 2,5-disubstituted furan which contains a carboxyl group and a hydroxyl function in the two sidechains, and a *p*-disubstituted benzene spacer was studied. It showed a very high diastereoselectivity, and it has been noted that when a larger ring, formed by increasing the spacer length, was enolized, the (*E*)-enolate is more stable than the (*Z*)-enolate, therefore kinetic methylation (of the *Z*-enolate) would lead to the *cis*-dimethyl isomer predominantly (ratio 25:1), whereas the thermodynamic controlled product (from the*E*-enolate) favors the *trans*-isomer (in a 4:1 ratio).

nonactic acid

iPr_2NLi >30 : 1
$(Me_3Si)_2NK$ 20 : 1

An assembly of the C-3 to C-9 segment of lysocellin demonstrated the same principle. Dimethylation of the dienolate of a chiral dilactone proceeded with >12:1 stereselection for the desired product. Peripheral hydroboration accomplished the goal. In the methylation step each entering methyl group is *cis* to the one associated with the carbinol stereocenter, probably due to local conformation of the (*Z*)-enolates.

lysocellin (C-3 to C-9)

LiTMP -78°C; MeI

BH_3; NaOOH

(>9 : 1)

(> 12: 1 in favor of indicated isomer)

α-Chlorohydrin derivatives can be alkylated, and with reductive decyanation an umpolung reaction of aldehydes is achieved. In the application of such a method to the synthesis of acyclic 1,3-polyol systems, stereocontrol is of paramount concern. A solution to the latter problem involves locking two oxygen functionalities in a 1,3-dioxane ring, so that alkylation leads to axial nitriles and the subsequent decyanation under thermodynamic control proceeds with retention of configuration. An assemblage of the C-12 to C-30 segment of roxaticin [Rychnovsky, 1992b] illustrates this tactic.

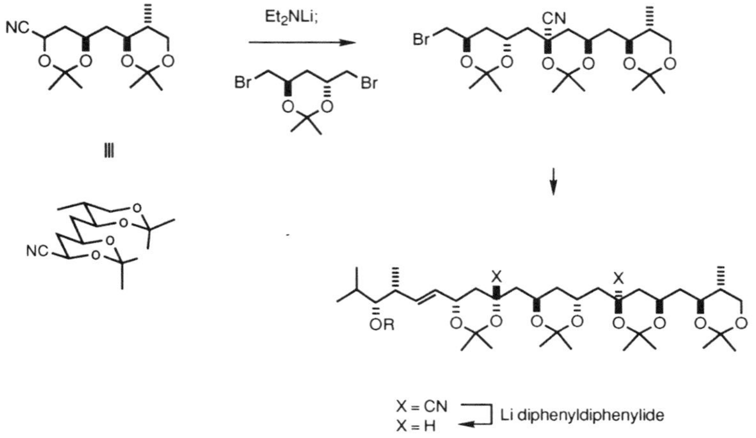

Et_2NLi;

X = CN
X = H
Li diphenyldiphenylide

5.2. DIALKYLATIONS VIA CYCLIZATIONS AND CYCLOADDITIONS

A frequently encountered synthetic step is to introduce stereoselectively two carbon chains in vicinal positions. There are many solutions in response to demand and existing structural constraints. From the following examples the reader should be able to gauge several situations.

A synthesis of podocarpic acid [Giarrusso, 1968] is based on the premise that the final stage involves cyclization of a B-seco precursor. In other words, this precursor is a cyclohexane derivative containing two carboxylic acid chains (one-carbon and two-carbon fragments) in a *cis-vic*-relationship, with an aryl group at the other adjacent ring carbon to the acetic acid pendant. A simple retrosynthetic analysis by association of the carboxylic chains indicates an ancestral role for a functionalized cyclopentane which permits regioselective ring cleavage. Stereochemically the five-membered ring must be *cis*-fused to the cyclohexane moiety, and such requirement is most accommodating (*cis*-hydrindane). Consequently, this route is easily reduced to practice. (Cf. a similar route to dehydroabietic acid [Ireland, 1966].)

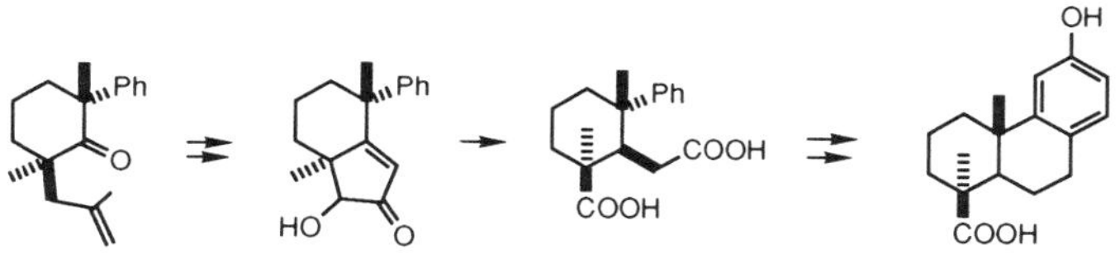

podocarpic acid

Bicyclo[3.3.0]octane-3,7-dione is very readily available via condensations of glyoxal with two equivalents of an acetonedicarboxylic ester, and decarboalkoxylation. Upon protection of one of the ketone groups the cyclopentanone moiety can be modified and cleaved to give, with proper manipulation at the other ring, loganin aglucone [Caille, 1984]. The important feature of this method is the establishment and maintenance of the ring juncture stereochemistry which also controls the steric course for the introduction of the secondary methyl group.

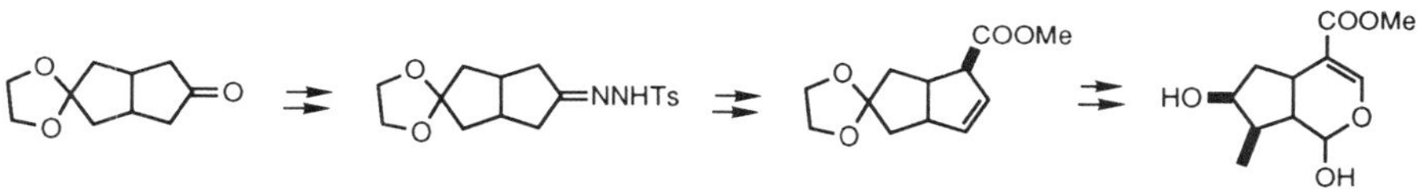

loganin aglucone

The [2 + 2]cycloadducts of cyclopentadiene and ketenes (esp. dichloroketene) are very useful synthetic intermediates. Numerous applications in the synthesis of cyclopentanoid substances which contain two or more sidechains have been disclosed. These substances include prostaglandin-$F_{2\alpha}$ [Newton, 1980], hybridalactone [Corey, 1984].

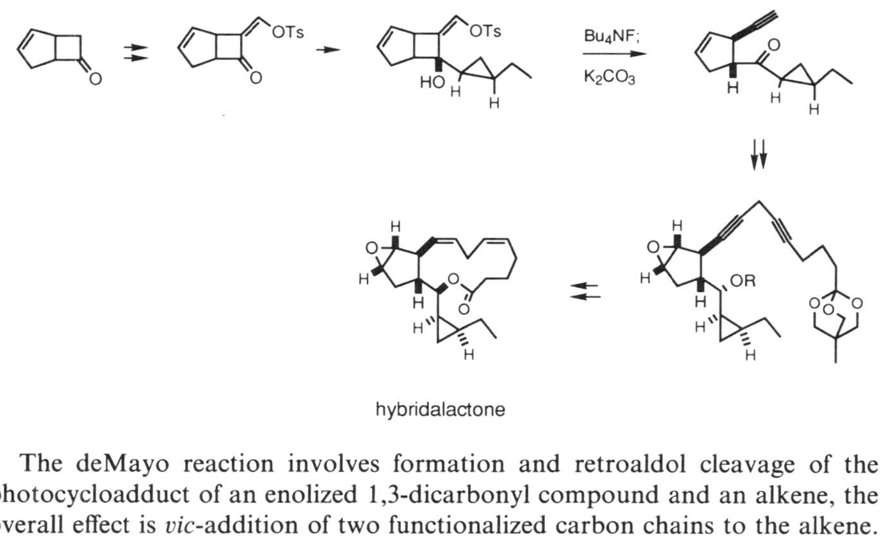

hybridalactone

The deMayo reaction involves formation and retroaldol cleavage of the photocycloadduct of an enolized 1,3-dicarbonyl compound and an alkene, the overall effect is *vic*-addition of two functionalized carbon chains to the alkene.

The cleavage step is thermodynamically driven, as the β-hydroxy carbonyl unit is incorporated in a highly strained cyclobutane ring. Numerous natural products, including loganin [Büchi, 1973; Partridge, 1973], hirsutene [Disanayaka, 1985], longifolene [Oppolzer, 1978], zizaene [Baker, 1981], daucene [Seto, 1985], and many others have succumbed to synthesis by using this method as the key step. These results, some of which are depicted in the following equations, demonstrate both inter- and intramolecular versions of the deMayo reaction, also it is possible to employ enolized dicarbonyl compounds and monoprotected forms.

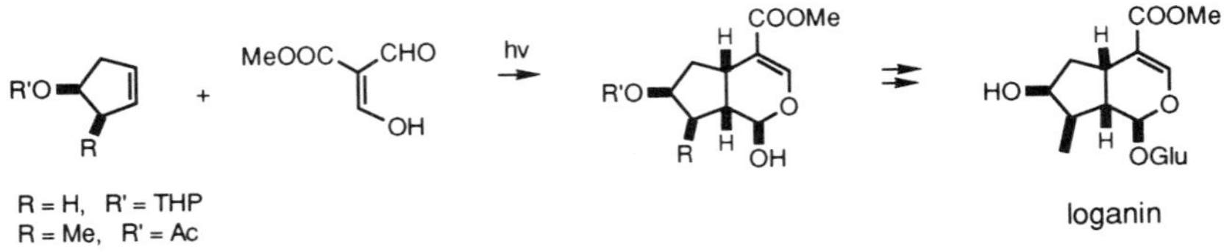

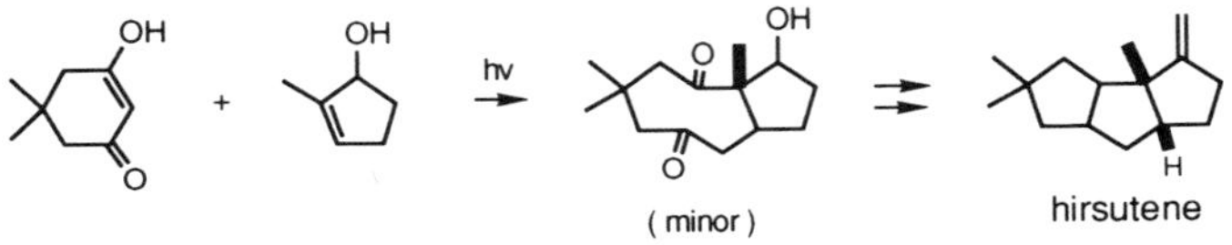

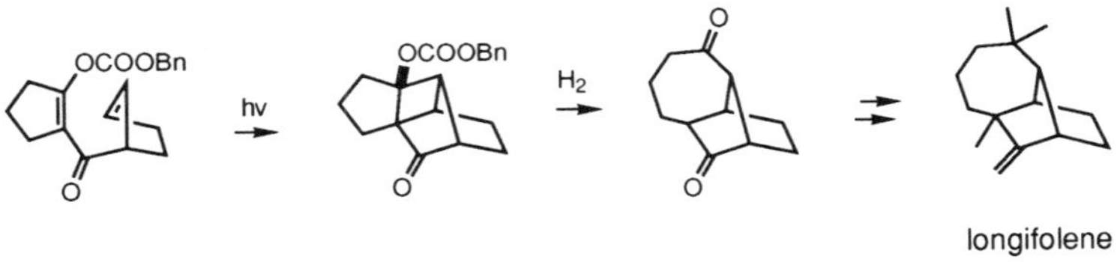

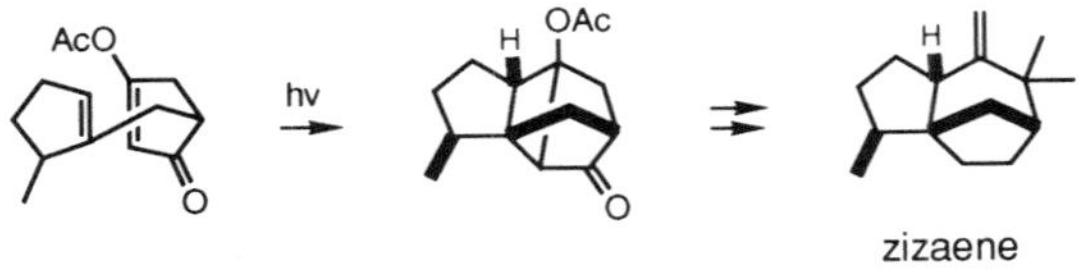

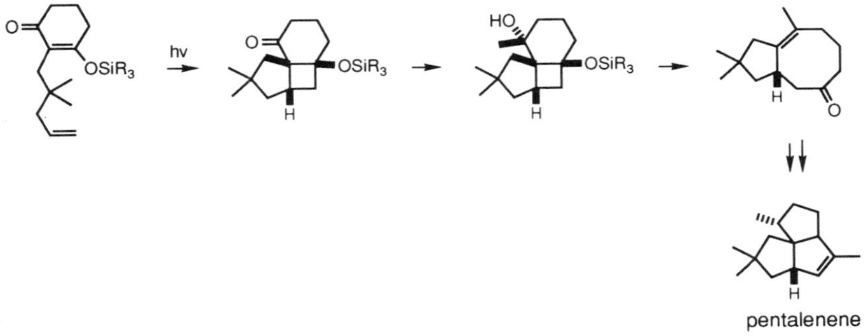

R = H, R' = THP
R = Me, R' = Ac

loganin

(minor)

hirsutene

longifolene

zizaene

pentalenene

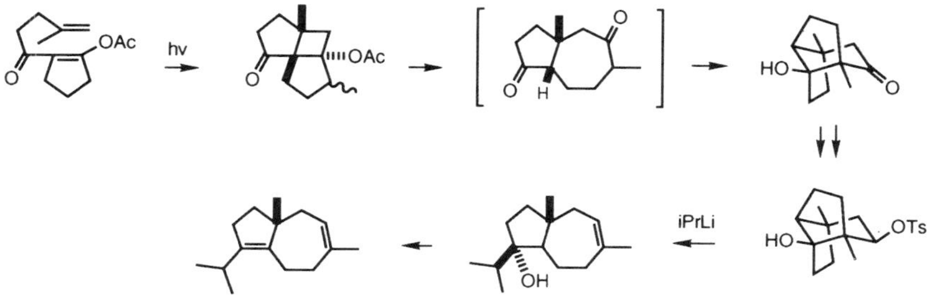

hv

iPrLi

daucene

When derivatized, an intrinsically unsymmetrical β-dicarbonyl compound would form one regioisomer from an intramolecular photocycloaddition (note situations in longifolene and daucene syntheses). A further advantage of using the protected addend is that it enables the modification of the free carbonyl group in the photoadduct (see syntheses of β-himachalene [Challand, 1967], pentalenene [Pattenden, 1984]). In the synthesis of daucene the retro-aldol cleavage was followed by another aldol condensation. Fortunately, the process served to differentiate the two ketone groups so that the new aldol was readily transformed into a 1,3-diol monotosylate, which on treatment with isopropyllithium accomplished a Grob fragmentation and the introduction of the missing sidechain. A similar, sterically enforced intramolecular aldol condensation also intervened, in tandem with a conjugate addition, during a synthesis of longifolene [McMurry, 1972].

(5*H*)-Furanones and (4*H*)-1,3-dioxin-4-ones are special photoaddends which are equivalents of β-oxo esters. They exhibit similar chemistry to those mentioned above, and their uses have elicited other methods for unraveling the hererocycle and the fused cyclobutane moiety of the adducts. Descriptions of usage of these compounds are found in syntheses of (−)-acorenone [S.W. Baldwin, 1982], occidentalol [S.W. Baldwin, 1982b], and *ent*-elemol [S.W. Baldwin, 1985].

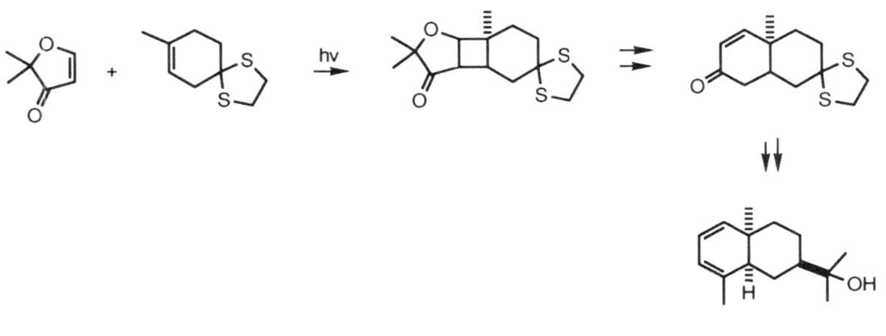

hv

occidentalol

A most fascinating exploitation of the intramolecular photocycloaddition of the dioxinones is the possibility of synthesizing *trans*-bicyclo[n.3.1]-alkanones

[Winkler, 1987, 1988]. A member of such ring systems is present in the diterpene ingenol.

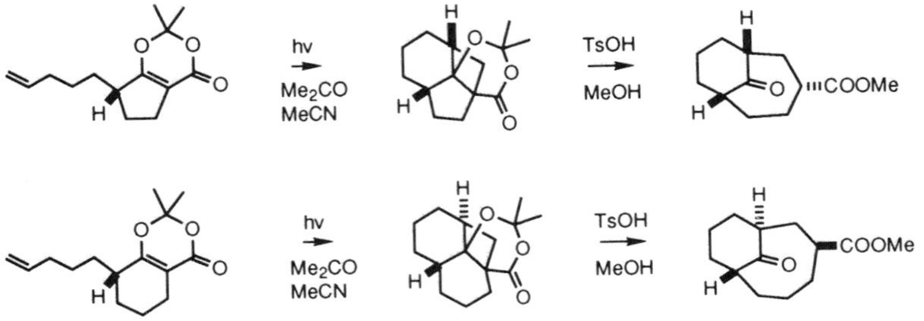

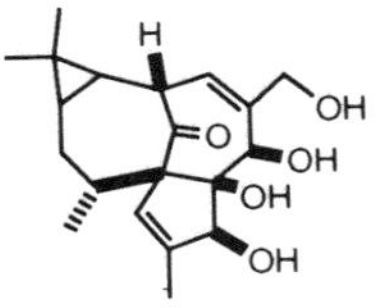

ingenol

vic-Diacylative annulation of an alkene may be achieved via [2 + 2]-photocycloaddition with a 1,2-disiloxyalkene and cleavage of the resulting di(siloxy) derivative. This method is best suited for the attachment of a five- or six-membered ring to an existing cycloalkenone as shown in its application to the syntheses of sesquiterpenes possessing hydrazulene and decalin skeletons. The following equations synopsize routes to damsin [DeClercq, 1977] and maritimin [van Hijfte, 1984].

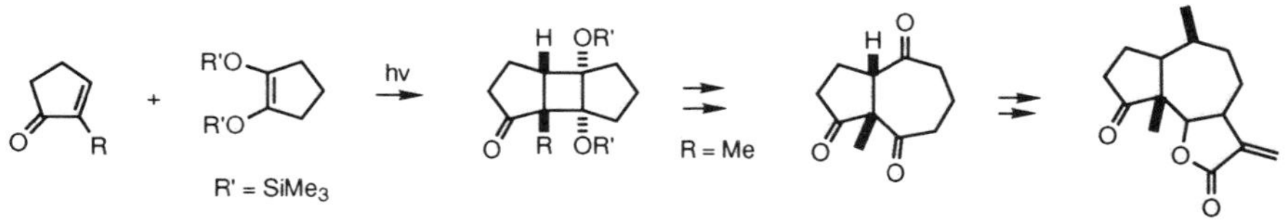

damsin

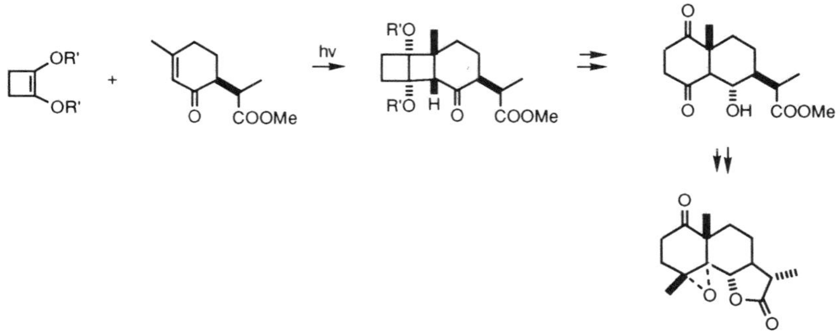

maritimin

An intramolecular Diels–Alder approach to the *trans*-octalin system is subject to stereocontrol by a γ-lactone lock [Davidson, 1985]. The conformational effect is that a competitive transition state leading to the *cis*-octalin is strongly disfavored.

COOMe H O O H xylene 140°C O O COOMe 75% O COOMe H O

A cyclohexene moiety created from a Diels–Alder reaction is easily partitioned into two functionalized carbon chains. Thus the combination of the cycloaddition and ring cleavage is a very effective protocol for *cis*-addition to an alkene of carbon segments which may be required for the elaboration of complex atomic arrays, as exemplified in a route to perhydrohistrionicotoxin [Ibuka, 1982].

COOEt OAc O O H O O O CHO O O COOMe HN HO HON HO

perhydrohistrionicotoxin

In a synthesis of pentalenolactone [Danishefsky, 1979] the δ-lactone moiety was created from a Diels–Alder adduct. The same tactic was successfully employed in the approach to vernolepin/vernomenin [Danishefsky, 1977b]. (Cf. coriolin synthesis in the previous section.)

O O O O O MeO Me_3SiO O O O O O O O O COOH O O COOH O O O

pentalenolactone

The stereochemical relationship of the three substituents on the pyrrolidine ring of domoic acid is agreeable with the elaboration from a Diels–Alder adduct of the hydroisoindole type. It is not surprising that a successful synthesis of domoic acid [Ohfune, 1982] has been accomplished in such a manner.

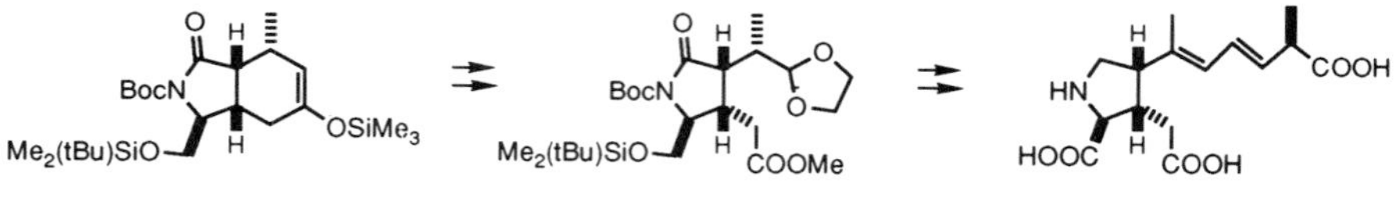

domoic acid

A method for reductive *gem*-dialkylation of cyclic ketones consists of spiroannulation and cleavage of the cyclobutanone unit. The released carbon chains are functionalized, and many synthetic operations are possible, such as elaboration of vetispiranes [Trost, 1975a].

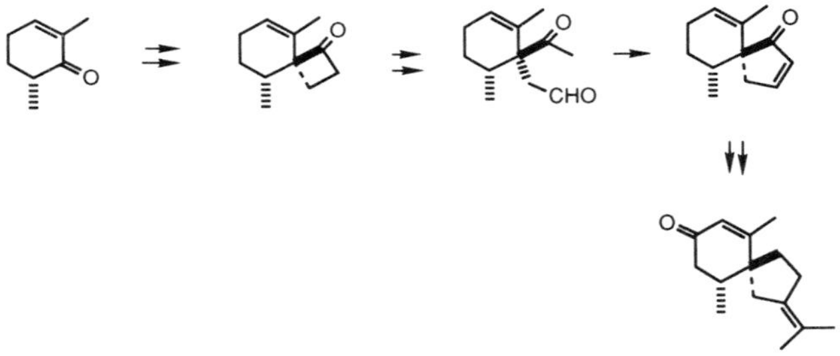

β-vetivone

5.3. COUPLING REACTIONS

Phenolic oxidative coupling is rendered more efficient and the relative stereochemistry of the new asymmetric center can be controlled by attaching the two aromatic rings in a cyclic skeleton [White, 1990].

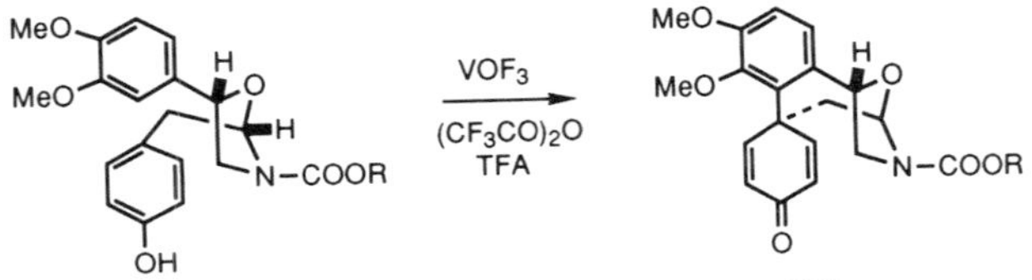

98%

5.4. RING EXPANSIONS AND CONTRACTIONS

5.4.1. Cleavage of Intercyclic Bonds

Cyclopropanated ring systems may be induced to undergo scission of the intercyclic C—C bond, resulting in ring expansion. In several natural product syntheses, for example cedrol [Corey, 1973], confertin [Marshall, 1976], longifolene [McMurry, 1972], and muscone [Stork, 1976], the triggering devices for the ring expansion are different, some of them leading to functionalized larger rings, others proceeding with concomitant formation of bridged or condensed carbocycles or heterocycles.

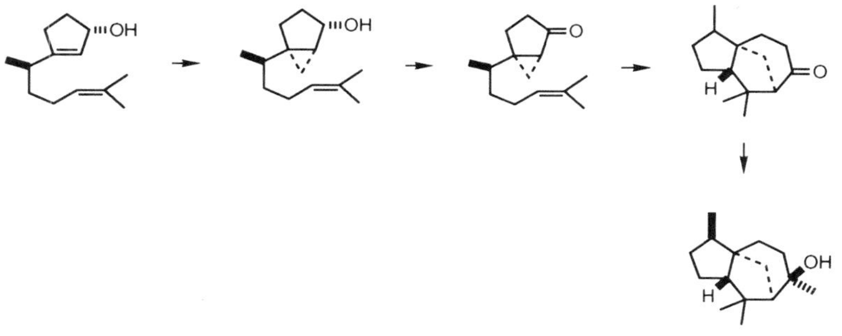

cedrol

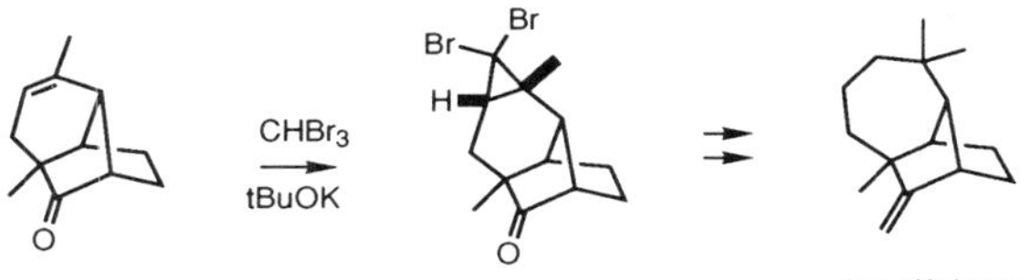

$CHBr_3$ / tBuOK

longifolene

$:CCl_2$

HCl, 66%

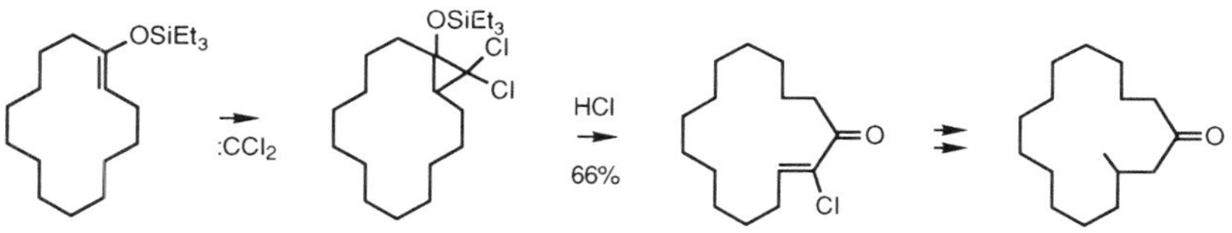

muscone

Intramolecular biaryl coupling to construct the tricarbocyclic skeleton of steganone failed [Magnus, 1985]. Fortunately, the formation of a seven-numbered ring proved feasible, the central ring was created indirectly from the accessible intermediate containing a fused cyclopropane.

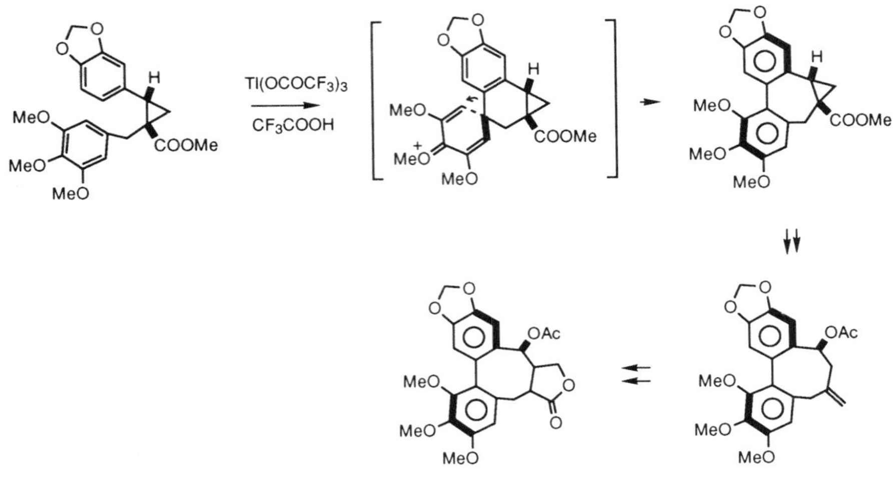

$Tl(OCOCF_3)_3$

CF_3COOH

steganone

Medium-sized rings are among the most difficult to construct; those containing defined functionalities are more so. Consequently the approach to caryophyllene and isocaryophyllene [Corey, 1964a] which is based on the assembly of an angular tricarbocycle and subsequent fragmentation of the hydrindane moiety is admirable. The fragmentation generated the γ,δ-cyclononenone whose (Z,E)-configuration depends on the relative stereochemistry of the angular methyl substituent and tosyloxy group at the adjacent carbon atom, that is *cis* → (E) and *trans* → (Z). It is worth mentioning the other novel feature of this synthesis: regioselective photocycloaddition.

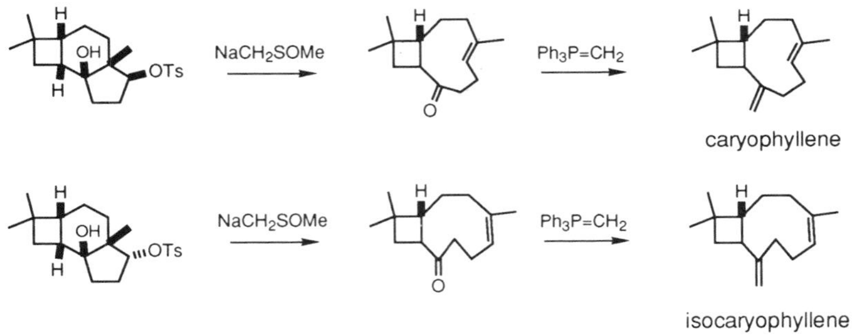

$NaCH_2SOMe$

$Ph_3P{=}CH_2$

caryophyllene

$NaCH_2SOMe$

$Ph_3P{=}CH_2$

isocaryophyllene

By means of a Grob fragmentation a monocyclic trienone precursor of periplanone-B has been obtained from a methyleneoctalindiol [Cauwberghs, 1988]. The diol was readily prepared from an intramolecular Diels–Alder reaction of furan and an allenyl ketone followed by reduction of the ketone group and cleavage of the bisallylic ether.

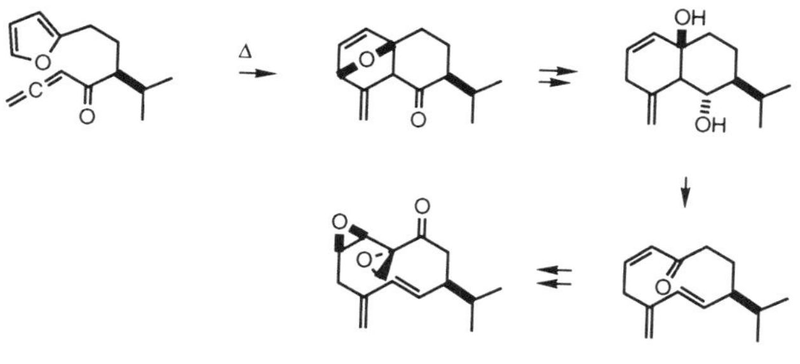

periplanone-B

Other syntheses employing the decompartmentation tactic include hedycaryol [Wharton, 1972], α- and β-longipinenes [Miyashita, 1974], and globulol [Marshall, 1974]. Even the latter two series of compounds are not monocyclic, and their access through the cyclodecadienes is definitely superior.

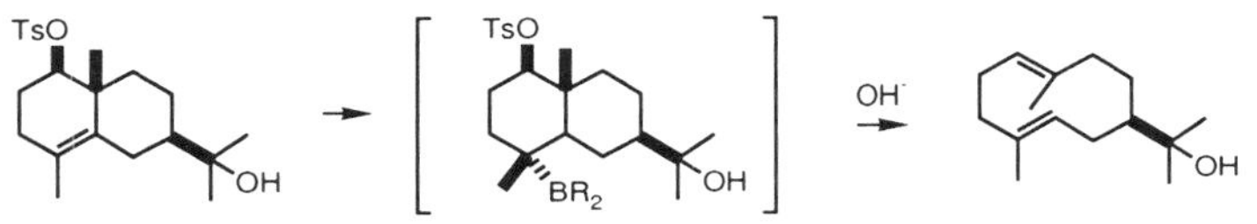

hedycaryol

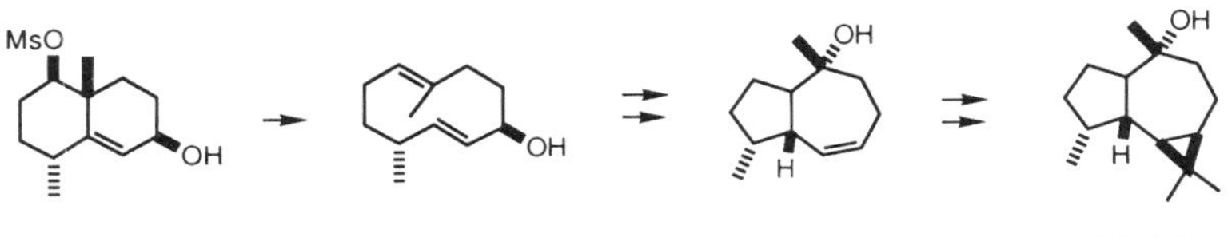

globulol

A conceptually concise route to guaiol [Buchanan, 1973] is based on annulation of a bicyclo[3.2.1]octanone template, and fragmentation to reach the hydrazulene skeleton. The bridged ring intermediate is inherent from an intramolecular aldol condensation, and it contains a one-carbon oxo bridge as latent ester (to be generated by retro-Claisen fission). Unfortunately yields of the various steps are low.

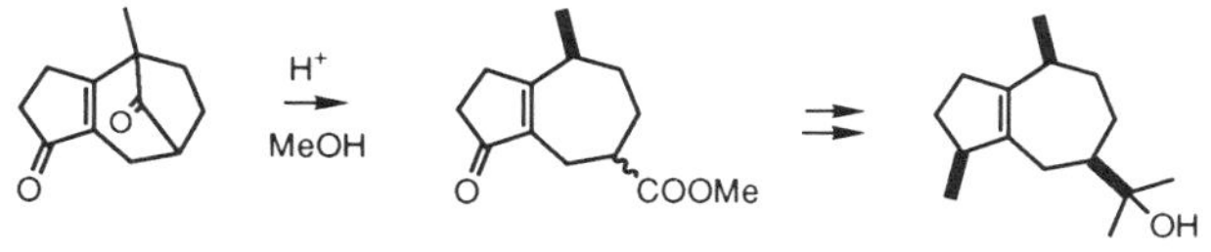

guaiol

Several di- and sesterterpenes possess a tricyclic 5:8:5-fused ring system which presents certain difficulties in synthesis, especially concerning the eight-membered ring. Attention must be paid to the four asymmetric centers at the ring junctures, for which the tactic involving fragmentation of a bridged ring system is most assuring. Again, manipulation of substituents in a precursor containing five- and six-membered rings is relatively simple. Thus, a synthesis of ceroplastol-I [Boeckman, 1989a] exploited such characteristics.

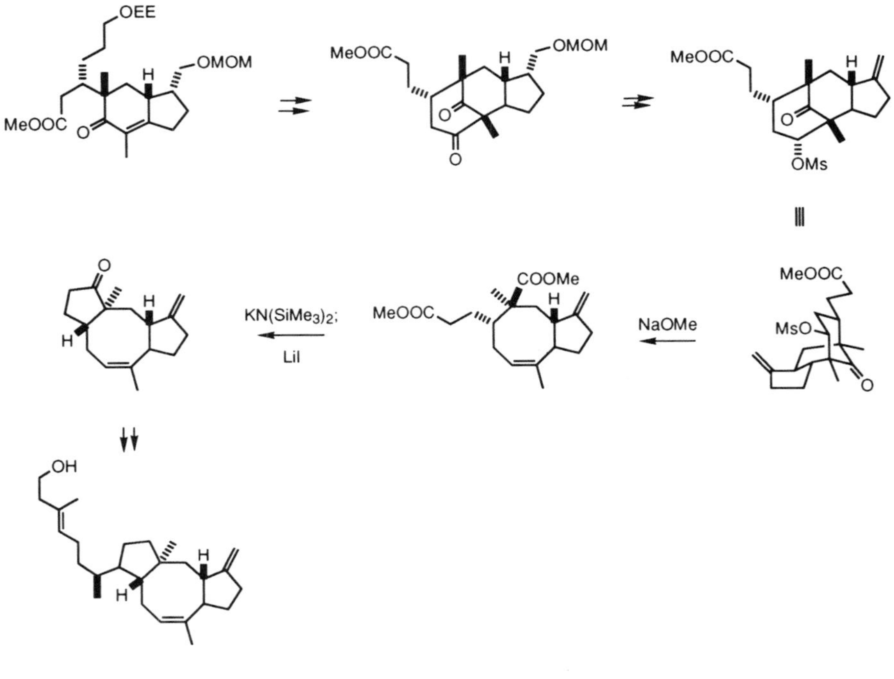

OEE OMOM MeOOC O H MeOOC OMOM H O O MeOOC H O OMs

O H H KN(SiMe3)2; LiI MeOOC COOMe H NaOMe MeOOC MsO O

OH H H

ceroplastol-I

Another example based on bridged ring template fragmentation is the access of (+)-sanadaol [Nagaoka, 1987, 1988].

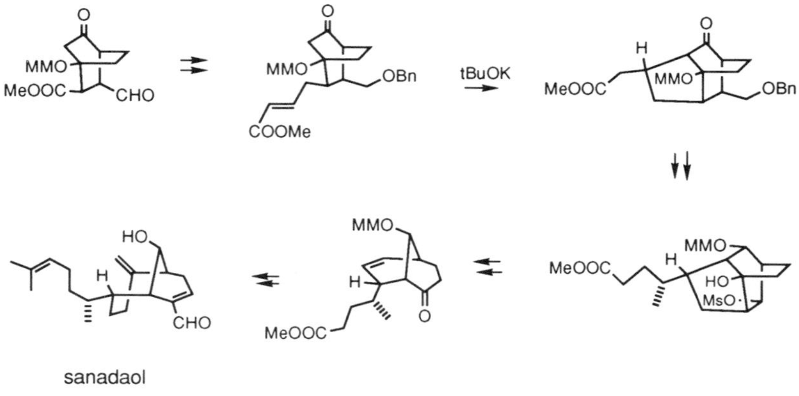

O MMO MeOOC CHO O MMO OBn COOMe tBuOK H O MeOOC MMO OBn

HO H CHO MMO H O MeOOC MeOOC H MMO HO MsO

sanadaol

It is even more challenging to introduce directly a bridge across the hydrindanone ring system in synthetic quest for pleuromutilin. The task was facilitated by the anchoring technique according to which a tricyclic system containing only six- and five-membered rings was formed as a template, and then a condensed ring was attached intramolecularly such that the product possessed all the framework carbon atoms and an extra C—C bond. The molecule was also provided with necessary functionalities for modification into a fragmentable species [Gibbons, 1982].

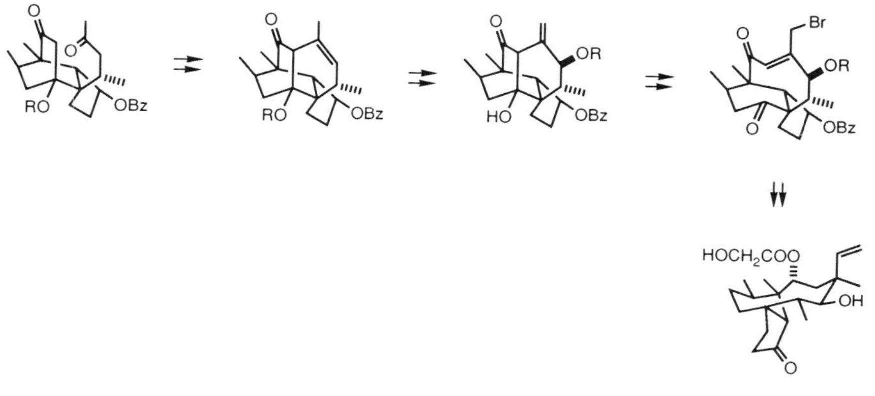

pleuromutilin

Another case of medium ring construction is that involved in the synthesis of byssochlamic acid [Stork, 1972b]. A dibenzocyclononatriene intermediate emerged from a Beckmann fragmentation of a bridged ring precursor.

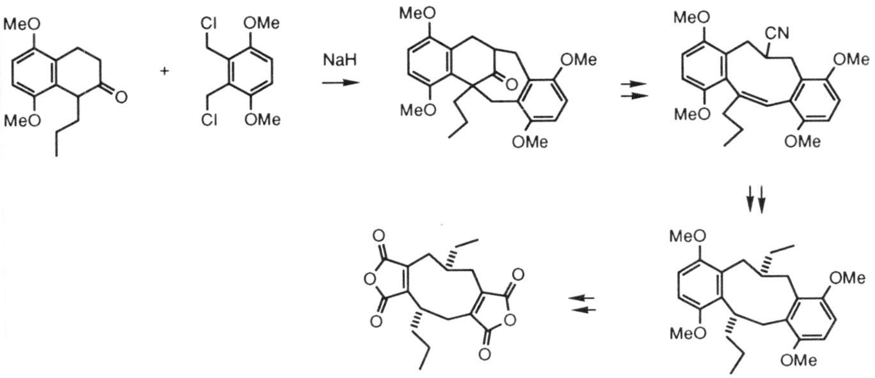

byssochlamic acid

As mentioned in a previous paragraph, an excellent method for the elaboration of *trans*-bicyclo[n.3.1]alkan-(n + 6)-ones (e.g. n = 4, 5) consists of an intramolecular [2 + 2]photocycloaddition and fragmentation. The bridgehead stereochemistry reflects the transition state strain of the cycloaddition step. Thus a *trans*-bicyclo[4.3.1]decan-10-one was obtained as a minor product [Winkler, 1988], in contrast to the exclusive formation of the homologous system [Winkler, 1986] to the exclusion of the *cis*-isomer.

In addition to the assortment of methods for ring expansion indicated above, this very important concept of ring construction has been developed into many more varieties.

Ynones in which the two functionalities are separated by two or more skeletal atoms are now accessible from conjugated cycloalkenones via epoxidation and fragmentation [J. Schreiber, 1967; Tanabe, 1967]. The effectiveness of the method makes it worthwhile to construct such ynones in a somewhat circuitous way. Sometimes the protocol is recommended for ring expansion, for example in a synthesis of muscone [Eschenmoser, 1967].

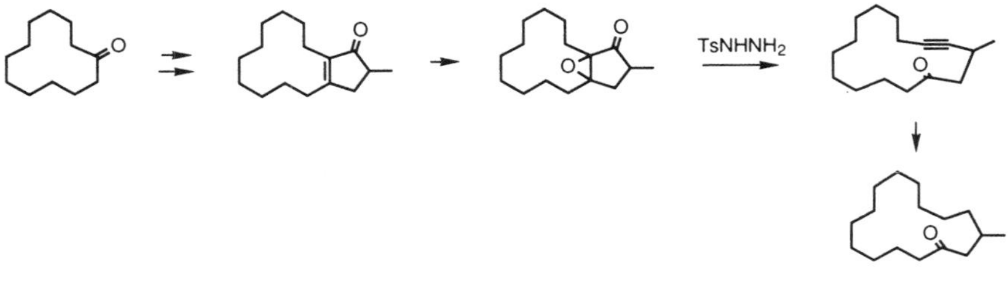

TsNHNH$_2$

muscone

Linear triquinanes are readily acquired by sequential photochemical and thermal transformations of the Diels–Alder adducts of cyclopentadienes and *p*-benzoquinones. Naturally, this protocol has been employed in natural product synthesis such as that of hirsutene and $\Delta^{9(12)}$-capnellene [Mehta, 1986].

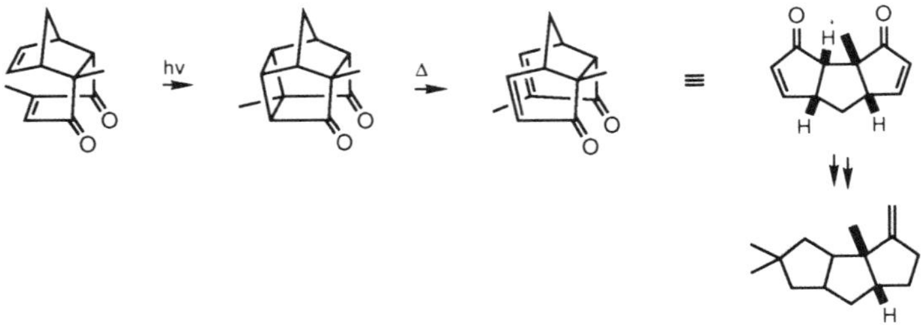

hν Δ

hirsutene

hν Δ

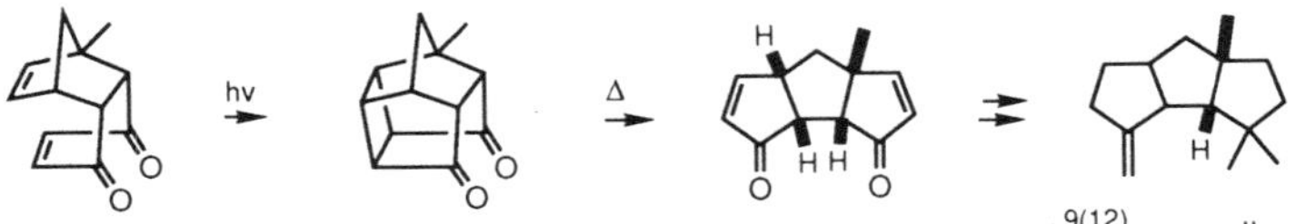

$\Delta^{9(12)}$-capnellene

An analogous reaction sequence has been applied to a synthesis of byssochlamic acid [White, 1992]. In this case the intramolecular [2 + 2]photocycloadduct is even more strained so that the thermal cleavage requires a much lower temperature.

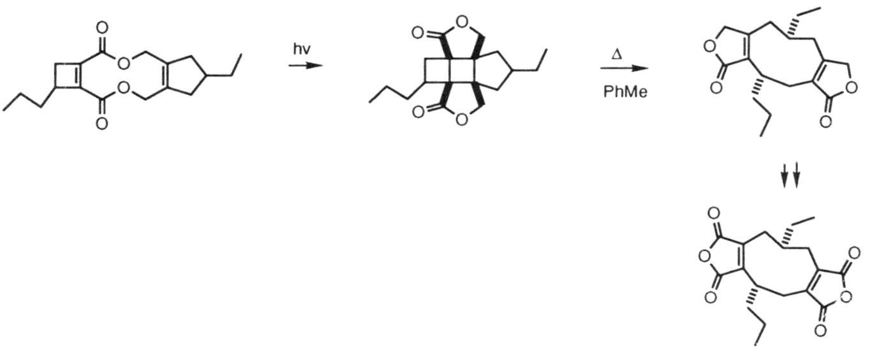

hv

Δ

PhMe

byssochlamic acid

A synthesis of dactylol [Feldman, 1990] demonstrates the transitory annulation tactic in which an extra ring is cleaved to generate two functionalities.

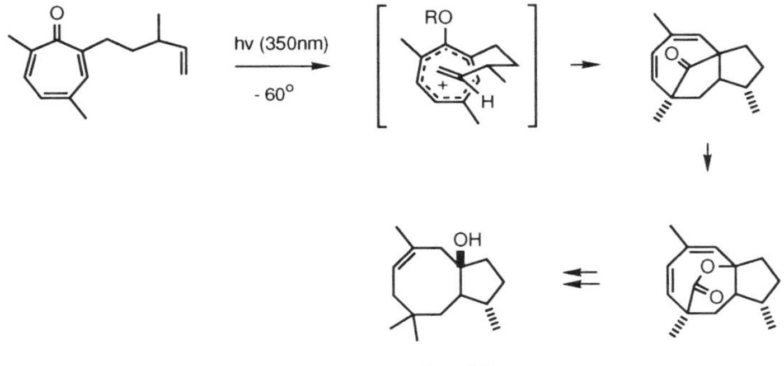

hv (350nm)

- 60°

dactylol

Tautomerization is the key to synthesis of 1,6-bridged cycloheptatrienes, including spiniferin-I [Marshall, 1983] and 1,6-methano[10]annulene [Vogel, 1964b]. An entry into the tropolone ring of colchicine [J. Schreiber, 1961] via intramolcular alkylation of a cyclohexadienedicarboxylic ester also featured tautomerization of the norcaradiene product.

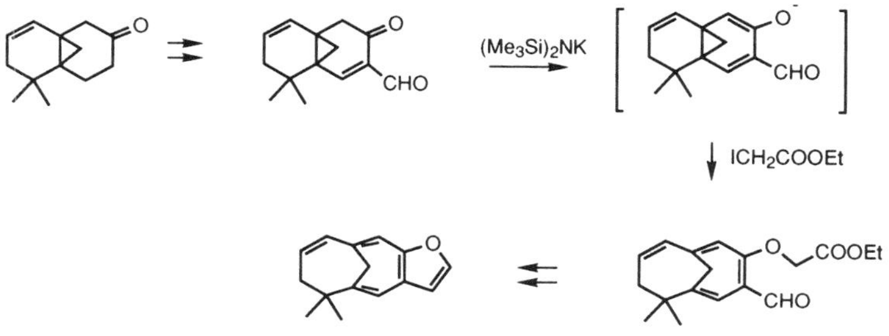

$(Me_3Si)_2NK$

ICH_2COOEt

spiniferin-I

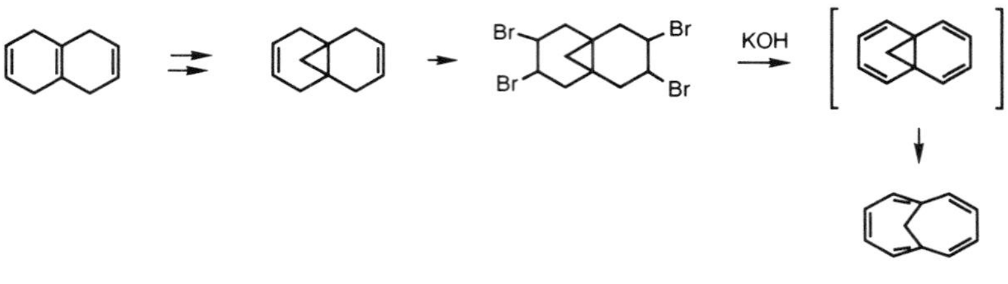

1,6-methano[10]annulene

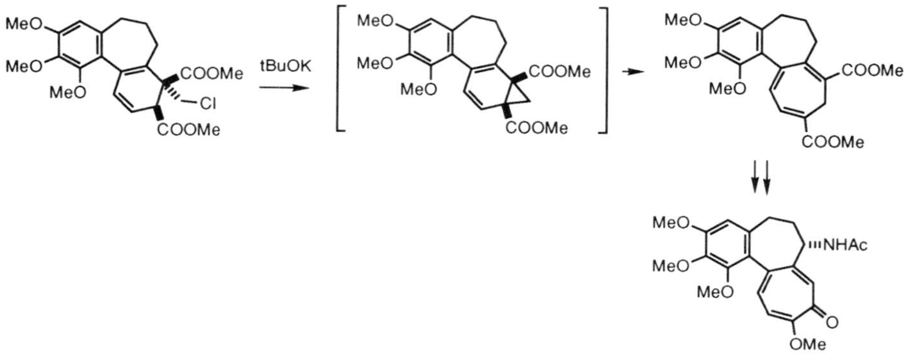

colchicine

An analogous process is that which followed the dehydrogenation of a bridged α-chloro sulfoxide, ultimately the product was used to complete a synthesis of the pyrene isomer, azuleno[2,1,8-*ija*]azulene, by a Ramberg–Bäcklund-type reaction [Vogel, 1984].

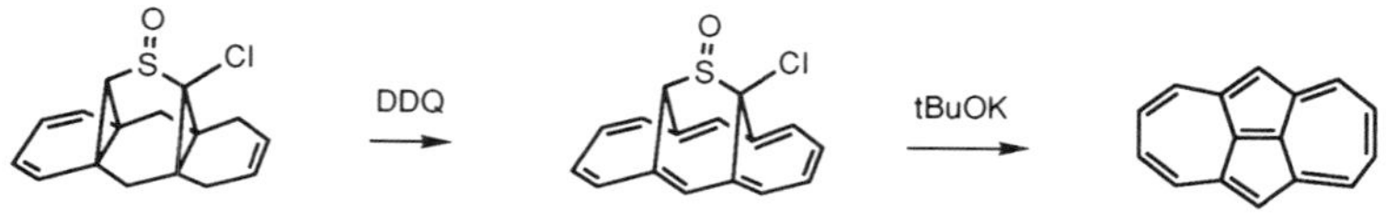

The central C—C bond of succinic acid can be cleaved via reaction of succinyl dichloride with phenylbis(trimethylsilyl)phosphine. This reaction is extendable to *trans*-1,2-cyclohexanedicarboxylic acid [Appel, 1983]. The ring opening was achieved by a 3,3-diphospha Cope rearrangement.

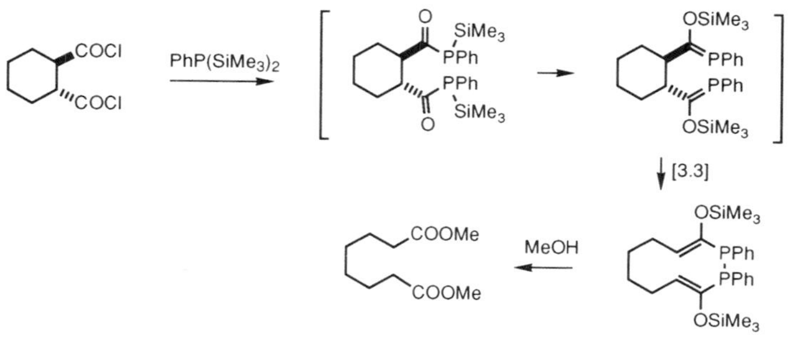

5.4.2. Oxy-Cope Rearrangement

The Cope rearrangement of 1,2-divinylcycloalkanes effects ring expansion by four carbon atoms. However, the equilibrium is not always in favor of the larger ring. In view of the tendency of many medium-sized 1,5-cyclodienes to undergo rearrangement in the opposite direction (cf. transformation of germacrenes to elemenes), incorporation of structural features to render the ring expansion irreversible is most desirable. Consequently the oxy and the anionic oxy version is the method of choice [Paquette, 1990a], and many intriguing variations on this theme have been developed. We need to examine a few syntheses to be able to appreciate the synthetic potential of this reaction: dactylol [Gadwood, 1986], periplanone-B [Still, 1979; S.L. Schreiber, 1984b], and pleuromutilin [Boeckman, 1989b].

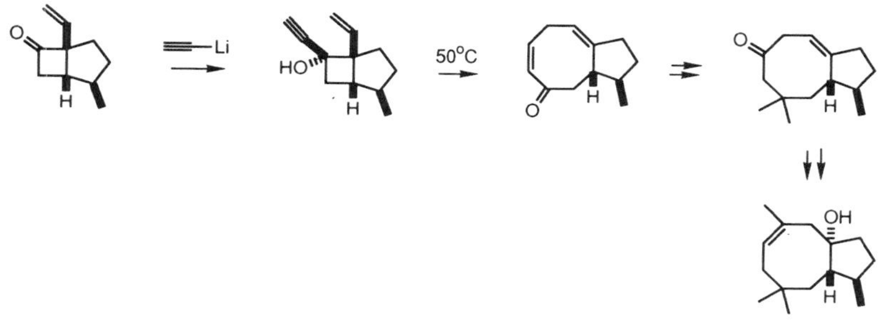

dactylol

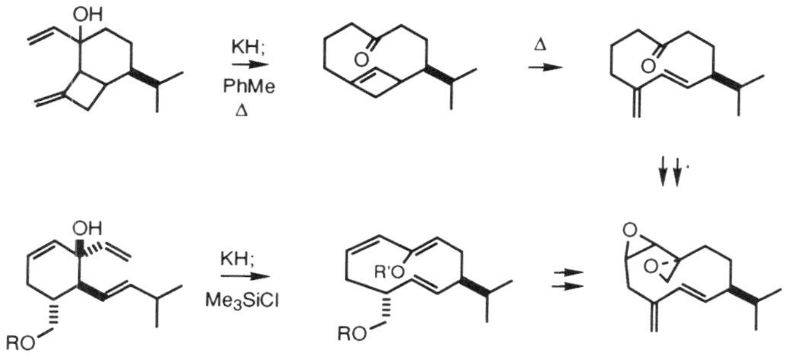

periplanone-B

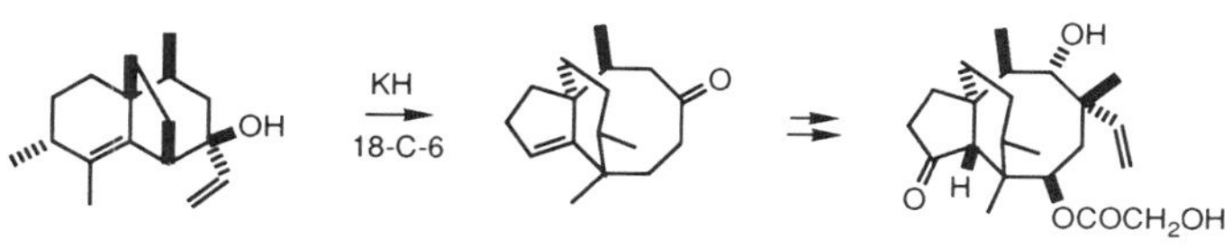

pleuromutilin

A double bond of a heterocycle such as furan is allowed as part of the 1,5-diene system that undergoes rearrangement. An example is seen in a synthesis of (+)-pallescensin-A [Paquette, 1992].

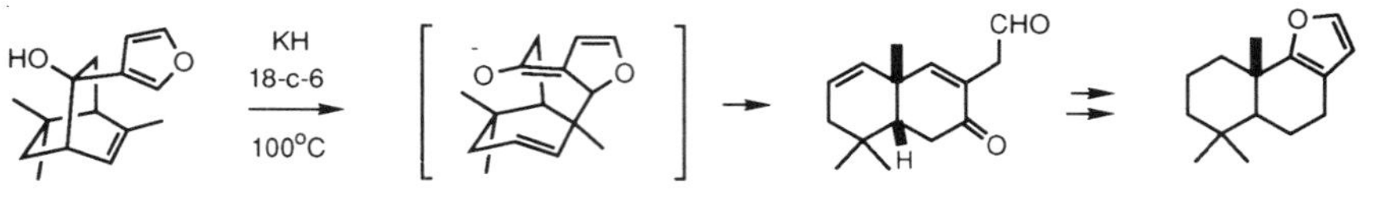

KH 18-c-6 100°C

(+)-pallescensin-A

Ring enlargement by eight carbon atoms in one step is possible. This method is illustrated in a synthesis of muscone [Wender, 1983]. It appears that two Cope rearrangements occurred in tandem [Wender, 1985].

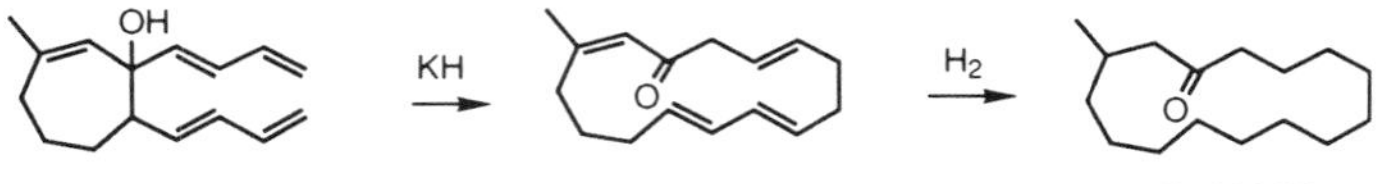

KH H_2

muscone

5.4.3. Ring Contraction

The use of a compound containing a larger ring to synthesize a cyclic molecule is dictated by the availability and structural attributes of the former compound, and convenient methodology. Perhaps the most well-known method is that which involves oxidative cleavage of a cyclohexene and intramolecular aldol reaction of the resulting dicarbonyl compound. This tactic was employed in the synthesis of cholesterol [Woodward, 1952], helminthosporal [Corey, 1965a], neosurugatoxin [Okada, 1992], and innumerable other substances.

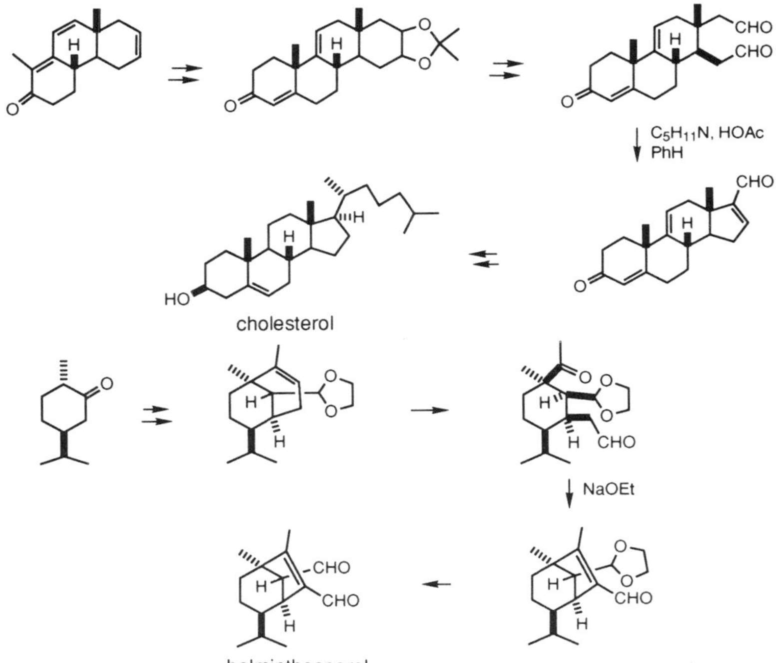

$C_5H_{11}N$, HOAc PhH

cholesterol

NaOEt

helminthosporal

As mentioned above, medium rings are inherently difficult to form by direct cylcization. Besides ring expansion methodologies one can consider an approach from the opposite direction. With a more easily available large ring compound the ring contraction process which has a much smaller entropy change is often practical. For example, a quantitative yield of the 9-membered carbocycle was obtained from a 13-membered lactam by an intramolecular transacylation during a synthesis of isocaryophyllene [Ohtsuka, 1984].

H H MeN O SO —LDA, -78°C→ H H O SO NHMe ⇒ H isocaryophyllene

Carbocycle construction via cyclic sulfones by pyrolytic extrusion of sulfur dioxide is particularly valuable in the synthesis of cyclophanes [Vögtle, 1979].

The use of [2.3]Wittig rearrangement in the synthesis of carbocycle is illustrated in a synthesis of haageanolide [T. Takahashi, 1987]. In the process a 13-membered oxacycle was contracted to provide an alcohol of desirable pattern, that is β-substituted with an isopropenyl group.

tBuCOO OH + OTHP Br → tBuCOO OTHP O ⇒ $OSiR_3$ O

↓ BuLi -78°C

haageanolide ⇐ $OSiR_3$ OH

It must be emphasized that not only the construction of medium rings may benefit from this tactic. As shown in the following, such ring segmentation is applicable to efficient elaboration of small and common rings. Thus, photochemical cyclization of eucarvone led to an intermediate of grandisol [Ayer, 1974], enolate Claisen rearrangement formed the basis of synthetic routes to *cis*-chrysanthemic acid [Funk, 1985], iridomyrmecin [Abelman, 1982], quadrone [Funk, 1986], and the bridged ring systems of taxane [Funk, 1988a] and ingenane [Funk, 1988b]. Heterocyclic compounds such as *N*-benzoylmeroquinene methyl ester [Funk, 1984] which is an intermediate of cinchona alkaloids, and (−)-kainic acid [Cooper, 1987] are also accessible by this method.

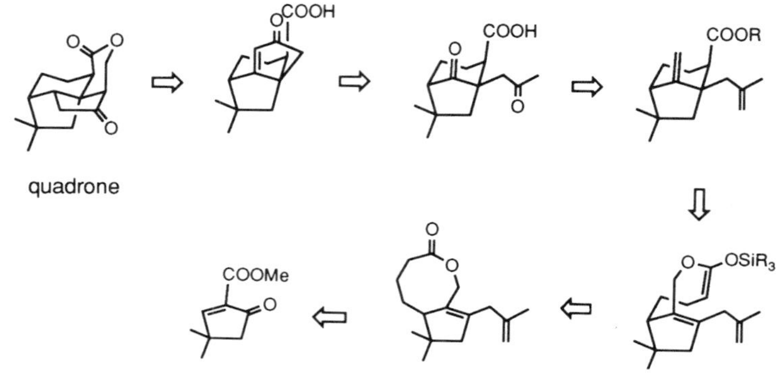

quadrone

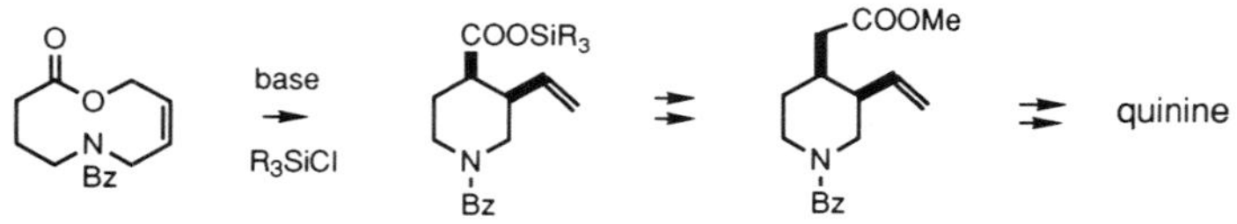

quinine

N- benzoylmeroquinene methyl ester

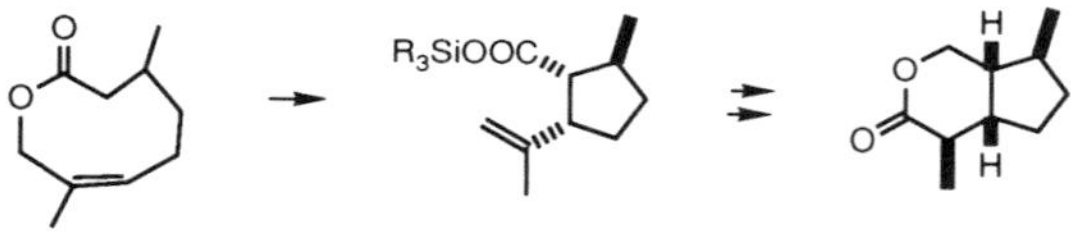

iridomyrmecin

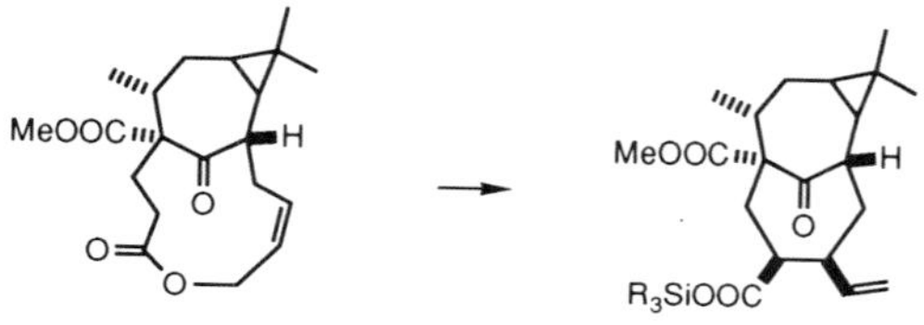

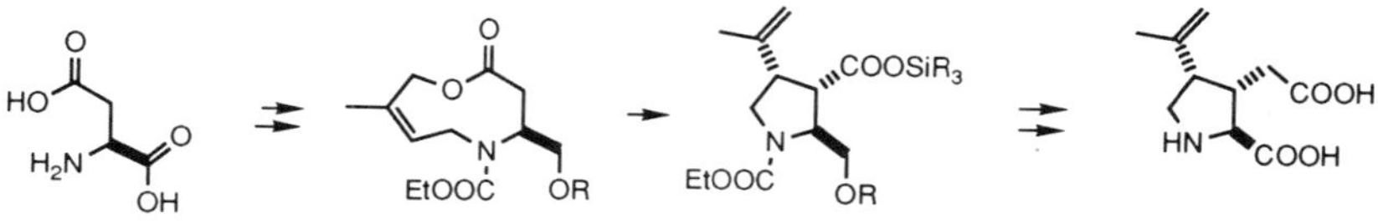

L-aspartic acid

α-kainic acid

Segmentation of an eight-membered ring is a central theme of synthetic approaches to iridomyrmecin [Mathews, 1975] and coriolin [Shibasaki, 1980].

Et$_2$NLi

OH

O

OSiR$_3$

RO

O

HO

O

O

O

HO

coriolin

RO

O

5.4.4. Cyclomutation for Skeletal Construction

We define the structural change as cyclomutation when a cyclic array serves as template for the construction of another ring and with the array itself dismantled afterwards. Usually the elements of the ring that undergoes destruction are converted into sidechains of the new ring but it would be interesting to include also certain transformations in which the liberated chains undergo ring formation anew. Both stereochemical and structural benefits accrue when this tactic is applied to proper systems. Some examples of this category given at the beginning of this chapter involve aboriginal mutating rings that are not formed during the synthetic operations.

In a synthesis of *erythro*-juvabione [Schultz, 1984] an intramolecular alkylation of 4-(3-iodopropyl)-3-methyl-2-cyclohexenone afforded a bicyclic compound. The relative stereochemistry of the two asymmetric centers of the synthetic target is easily established by hydrogenation of the product. The saturated ketone was then submitted to Baeyer–Villiger reaction so that the sidechain can be released.

I

O

O

O

OH

O

COOMe

O

erythro-juvabione

A bridged to spiro ring system exchange is featured in a synthesis of *ent*-hinesol [Chass, 1978]. Thus the base-promoted intramolecular condensation of a sulfone with a ketone triggers a fragmentation because a tosyloxy group is present in the γ-position of the alkoxide intermediate. The reaction results in the spirocyclic network of the sesquiterpene.

OTs O SO_2Ph NaH O SO_2Ph OH

ent-hinesol

2-(3-Oxoalkyl)cyclopentanones undergo rearrangement when subjected to acetalization conditions. This transformation, apparently proceeding through aldolization and fragmentation of the acetal, is of synthetic use. Synthetic approaches to bulnesol [M. Tanaka, 1988], acorenone-B [Nagumo, 1990], and trichodiene [M. Tanaka, 1991] are exemplary.

O O O O HO X O O OH OH X =

guaiol

O O OH OH $BF_3.Et_2O$ COO OH O

acorenone-B

The conversion of a bicyclo[3.2.1]octane system into the bicyclo[3.1.1]-heptane skeleton of α-*trans*-bergamotene [Larsen, 1977]is a sophisticated design. Concealing the tertiary methyl group as a carbonyl and anchoring it to the six-membered ring ensured reactivity and stereochemistry pertaining to the formation of the cyclobutane. The ketone group then served as the acceptor trigger for fragmentation to generate the cyclic double bond.

O MsO $OSiMe_3$ tAmOK O $OSiMe_3$ $CONH_2$

α-*trans*-bergamotene

A simpler operation is involved in the transformation of carvone into *trans*-chrysanthemic acid [T.-L. Ho, 1982a]. Hydrochlorination of the sidechain after removal of the conjugated double bond by a Kharasch reaction prepares the molecule to enter the bicyclic stage. Oxidative cleavage of the cyclohexanone moiety then begins.

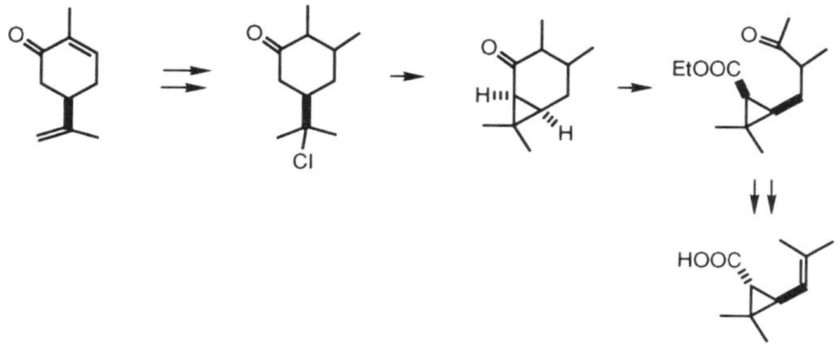

trans- chrysanthemic acid

A brief mention of a very powerful tactic for the elaboration of complex tetralin derivatives via intramolecular Diels–Alder reaction of *o*-quinodimethanes is in order. Such reactive intermediates can be generated from benzocyclobutenes on thermolysis. A concise and stereocontrolled route to estradiol [Kametani, 1978a,b] was developed.

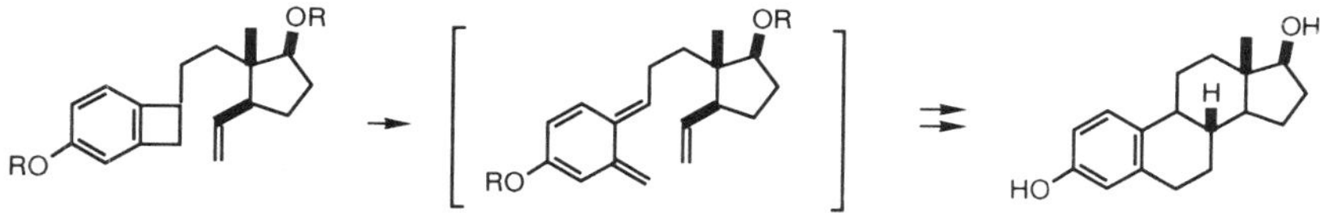

estradiol

Derivation of the angular methyl group and part of the methylenecyclopentane ring of $\Delta^{9,12}$-capnellene from the ethene bridge of a 1,6-annulated norbornene by reaction with Tebbe reagent is a novel transformation facilitating the synthetic effort [Stille, 1986]. Most gratifyingly the new alkenylidenetitanium species can react with an ester group at the adjacent carbon atom to form an alkoxycyclobutene. (Cf. hirsutic acid-C synthesis [Trost, 1979b] in which two aldehyde units generated by the cleavage of a cyclohexene were destined to become methyl groups, as referred to in the previous section).

3-Methyl-2-cyclopentenone and 3-methyl-2-cyclohexenone undergo photocycloaddition with ethene. The adducts are useful for elaboration of grandisol [Cargill, 1975; Zurflüh, 1970] because they contain two asymmetric centers corresponding to those of the insect pheromone molecule, and the two sidechains of grandisol are readily elaborated from the cycloalkanone moiety.

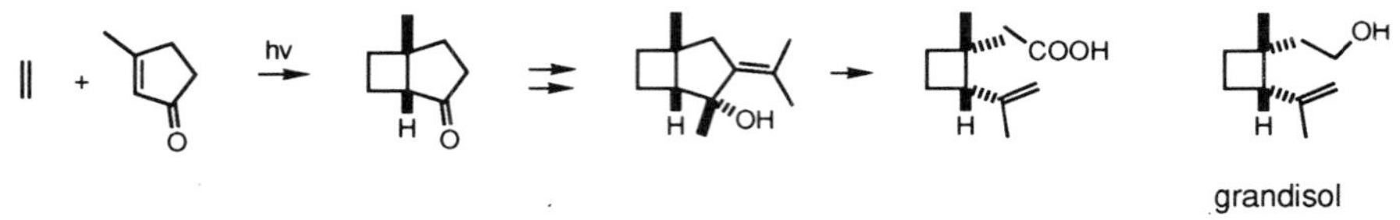
hv
COOH
OH
grandisol

An interesting approach to the 5:8:5-fused tricyclic system of several diterpenes and sesterterpenes is based on photocycloaddition and a Cope rearrangement [Jommi, 1991]. The latter step effects ring expansion from a four- to an eight-membered ring.

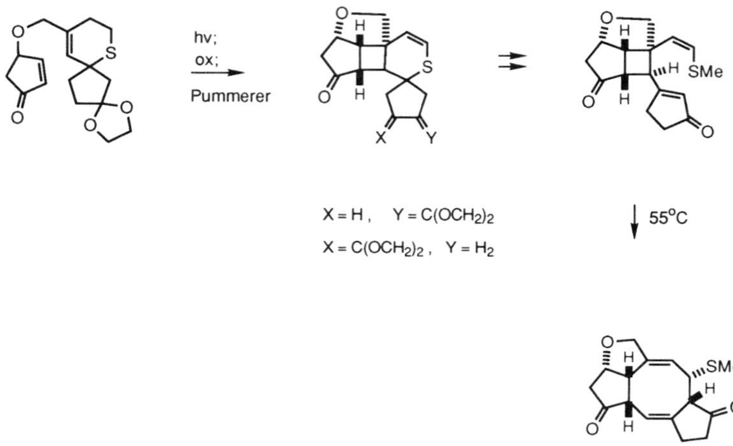
hv;
ox;
Pummerer
SMe
X = H, Y = C(OCH2)2
X = C(OCH2)2, Y = H2
55°C
SMe

Bicyclo[3.2.0]octan-6-ones are readily accessible from the cyclopentadiene-ketene adducts. While the use of such adducts in the elaboration of *vic*-disubstituted cyclopentanoids has been mentioned in a previous section, a more unusual application is the conversion into sativene [Sigrist, 1986].

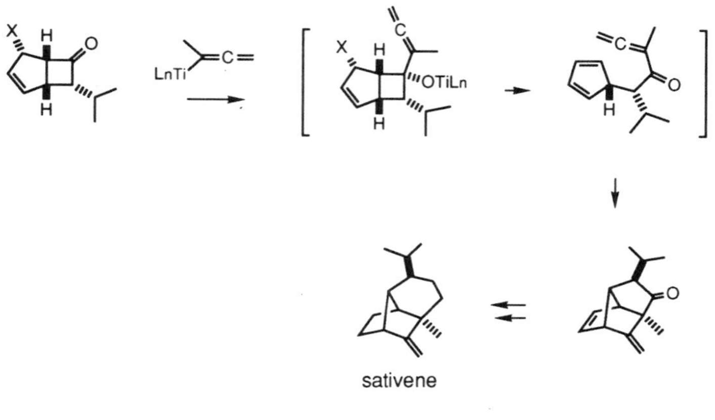
LnTi
OTiLn
sativene

While the most extensive use of benzocyclobutenes is in the generation of *o*-quinodimethanes (and *in situ* trapping, e.g. by dienophiles), a variant is shown in a synthesis of grandisol [Kametani, 1988] in which the aromatic ring was converted to an enone, permitting angular methylation. The cyclohexanone moiety was then elaborated into two carbon chains via ring cleavage.

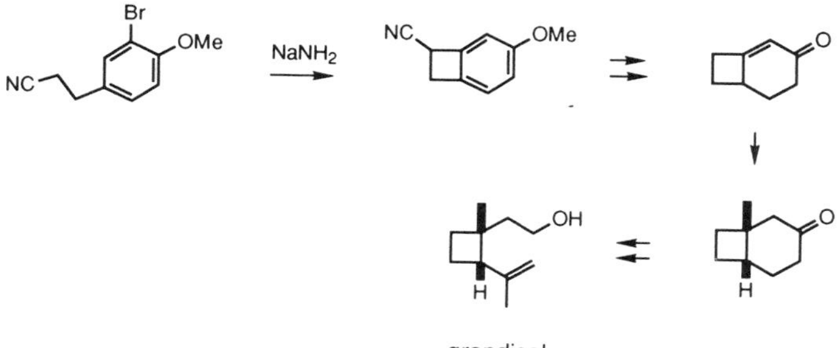

$NaNH_2$

grandisol

An unusually propitious entry into the pentalene system (although stabilized by donor and acceptor groups) from an azulene derivative is via two cycloaddition processes and a cycloreversion [Hafner, 1982].

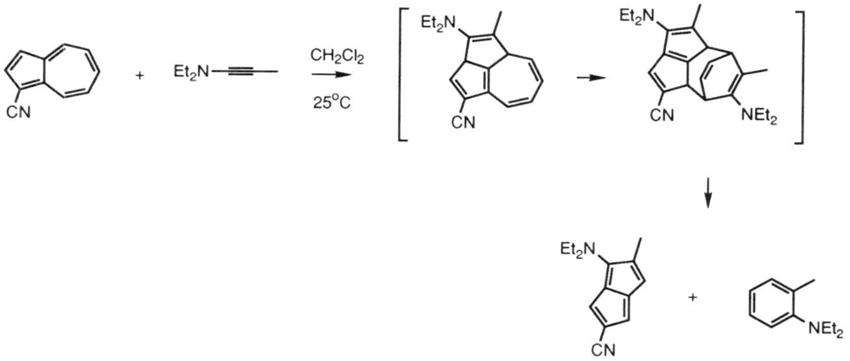

CH_2Cl_2, 25°C

A bridged ring system is more amenable to stereochemical manipulation than a condensed ring system, therefore the skeletal transformation of the former to the latter by a concerted process is a valuable methodology in synthesis. Based on the oxy-Cope rearrangement many 5:6- and 6:6- *cis*-fused carbocyclic and heterocyclic intermediates for synthesis (e.g. coronafacic acid [Jung, 1980], reserpiine [Wender, 1987]) have been acquired.

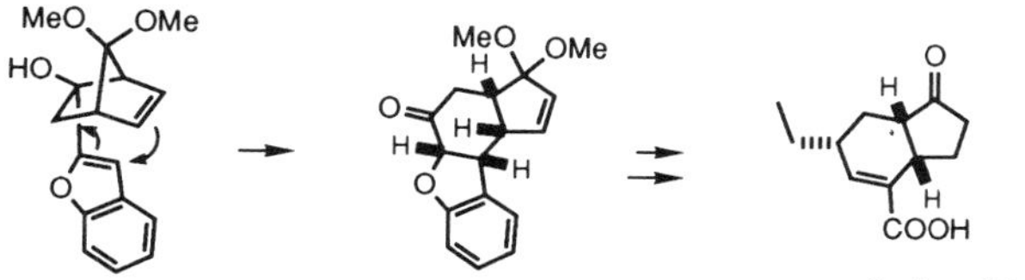

coronafacic acid

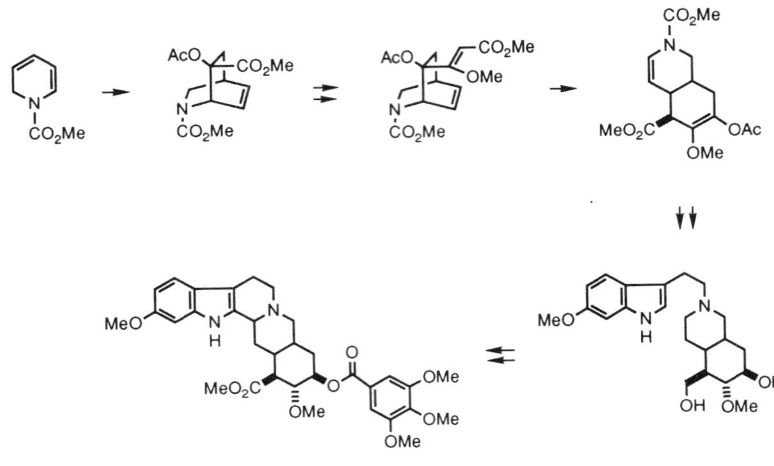

reserpine

The theoretically interesting triquinacene was first acquired from a tetracycle composed of two condensed bridged systems [Woodward, 1964]. By exploiting proximity effects a new cyclopentane unit was formed, but two existing five-membered rings were destroyed subsequently.

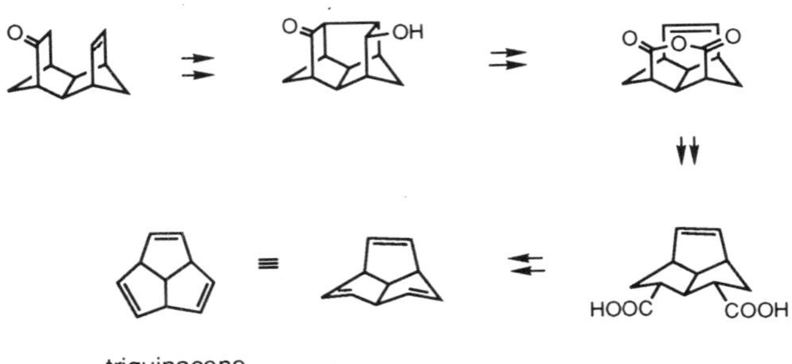

triquinacene

A retro-aldol fission served to unravel the ring system of velbanamine [Büchi, 1970a] from a more complex precursor which is more easily accessible. Note that cleavage of a bridging bond established a nine-membered ring.

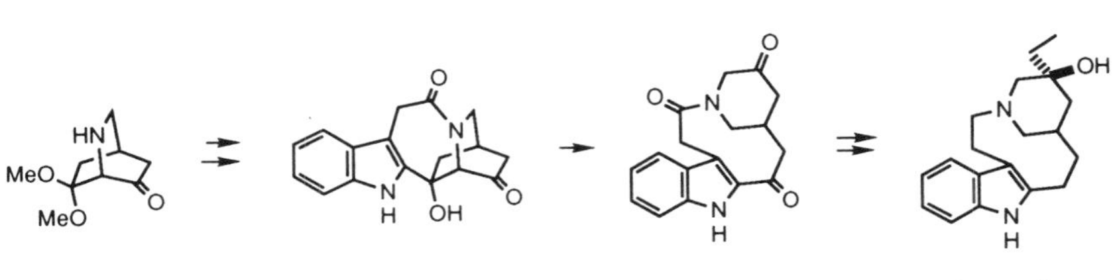

velbanamine

Quebrachamine possesses the same skeleton as velbanamine. Its synthesis via cyclomutation tactic has engaged two other different modes of reduction [Stork, 1963; Kutney, 1966].

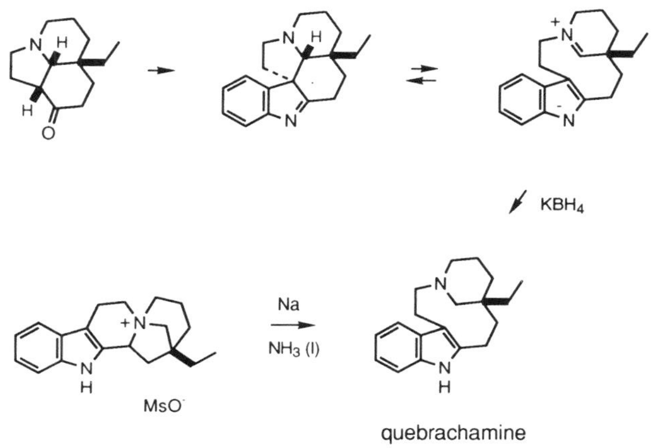

KBH_4

Na

NH_3 (l)

MsO⁻

quebrachamine

Formation of a medium ring from a bicyclic precursor may be followed by an alternative mode of cyclization. Thus the pyrrolizidine and indolizidine systems can be readily accessed from 1,3-cyclopentanedione derivatives [Ohnuma, 1983].

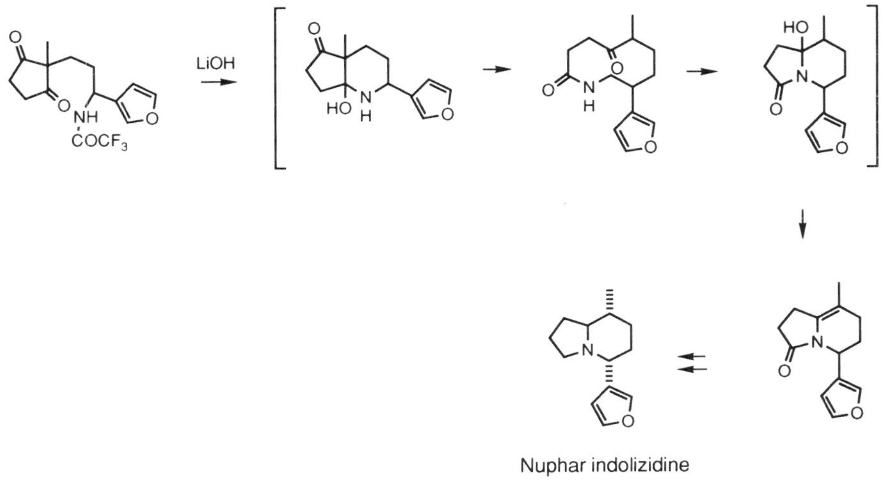

LiOH

$COCF_3$

Nuphar indolizidine

A significant transformation of one lactam into another via photo-rearrangement has been exploited in indole alkaloid synthesis [Ban, 1983]. Several structural types have become available from a common intermediate.

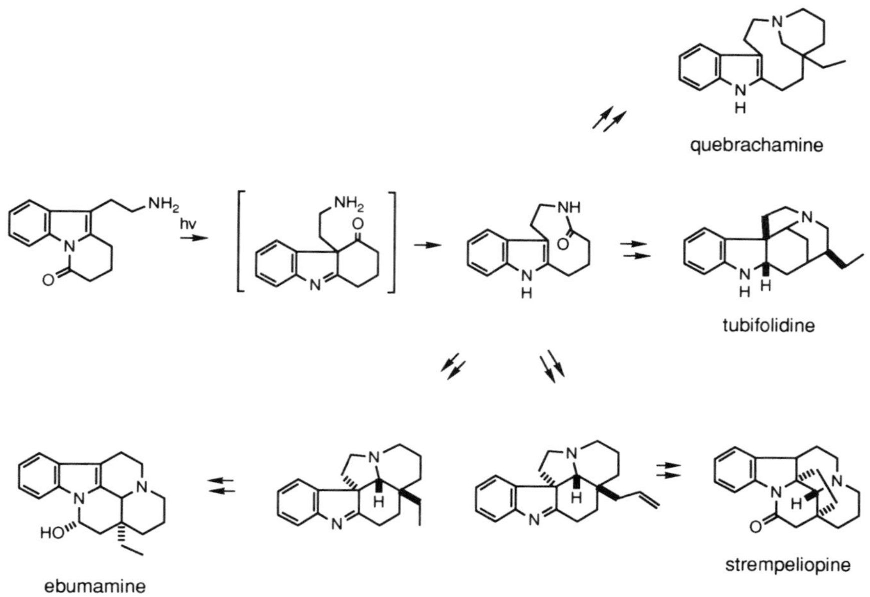

quebrachamine

tubifolidine

ebumamine

strempeliopine

The piperidine ring of *N*,4-diacetyl-2-(3-indolyl)piperidine served as a template in the formation of a carbazole *en route* to olivacine [Naito, 1986].

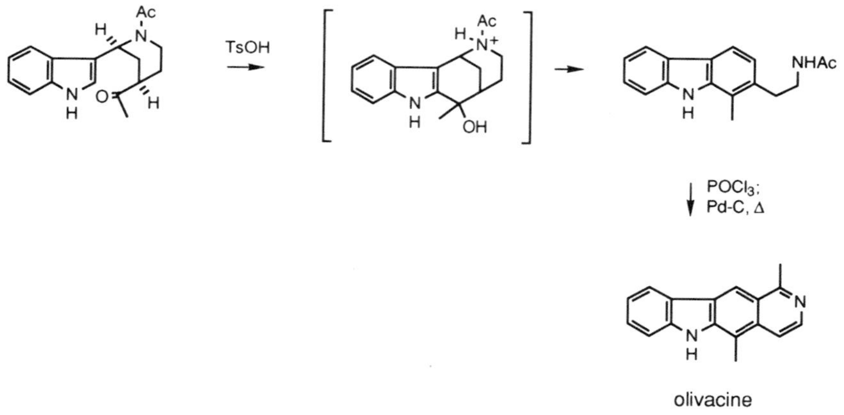

TsOH

$POCl_3$; Pd-C, Δ

olivacine

A series of fascinating transformations comprising ring formation and cleavage brought to bear on the creation of the AD-ring moiety of vitamin-B_{12} [Woodward, 1968]. The anisole ring of the 6-methoxyindole at the starting point supplied five carbon units containing a portion of ring-A and its three-carbon sidechain via Birch reduction and ozonolysis. The first carbocycle formed during the process underwent a Beckmann rearrangement at a later stage to furnish the other sidechain of ring-A and portion of ring-D. A few more ring formation/cleavage operations intervened before unraveling of the fully functionalized AD-ring moiety from a tetracyclic β-corrnorsterone which also involved fission of a carbocycle and a lactam. The tightly orchestrated appearance and disappearance of cyclic structures represents an unsurpassed act of molecular acrobatics.

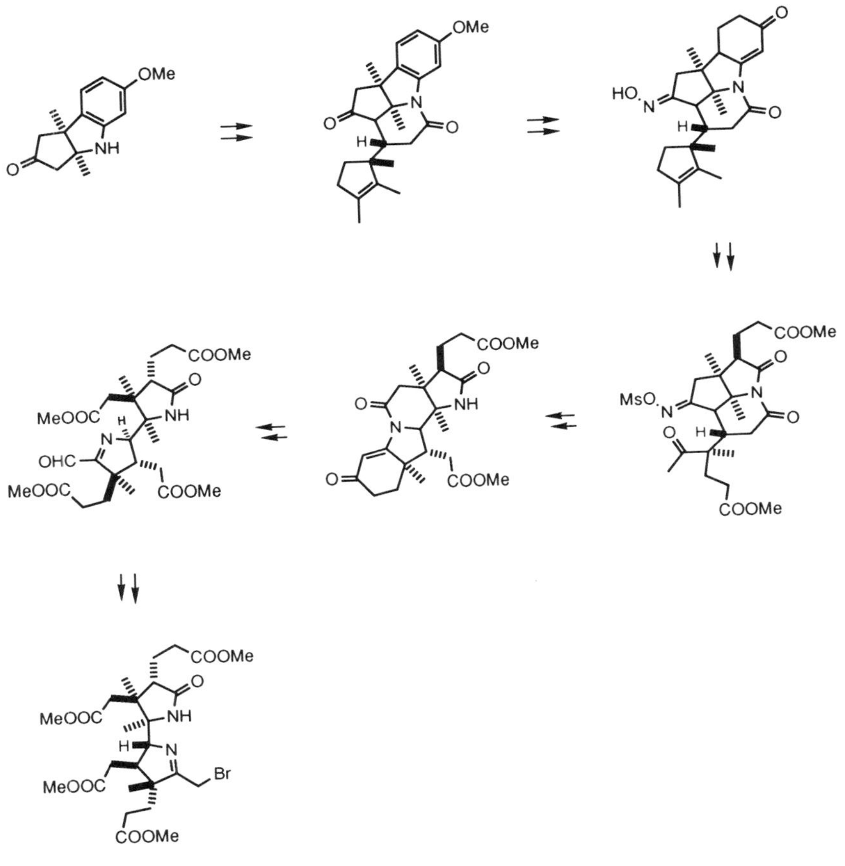

Two relevant changes are those involving formation of new ring types via Diels–Alder/retro-Diels–Alder reaction sequences starting with aromatic heterocycles (as the dienes) [T.-L. Ho, 1992b; Boger, 1987], and rearrangements as exemplified by conversion of a bicyclo[m.n]alkane to the [m-1,n + 1] isomer [T.-L. Ho, 1988]. Although these topics are not addressed here fully, the use of a symmetrical 1,2,4,5-tetrazine to provide two *para*-carbon atoms of the benzene ring of *cis*-trikentrin-A [Boger, 1991], and an approach to grandisol based on a rearrangement step [Wenkert, 1978a] are briefly delineated.

grandisol

5.5. MISCELLANEOUS

That stereoregulation of configurations is made possible in ring systems facilitates synthesis of the major component of olive fruit fly pheromone [Baker, 1982; Kocienski, 1983]. This compound is anhydro-1,3,9-trihydroxynonan-5-one in which the two terminal alcohols form a spiroketal with a ketone group. When the conformation specifies each of the two anomeric oxygen atoms to occupy an axial orientation the remaining hydroxyl group is equatorial. Accordingly, the structure represents the most stable isomer and its access is assured by hydration of an enone or equivalent under equilibration conditions.

HCl

THF, H_2O

20 : 1

olive fruit fly pheromone component

A similar tactic is involved in the assemblage of talaromycin-B [Kozikowski, 1984]. From a nitrile oxide cycloadduct a ketopentaol was created; having local symmetry in the two pairs of primary alcohols, the latter compound sought the most stable form during intramolecular dehydration (spiroketal formation). This reaction left a 1,3-diol unit and an isolated primary alcohol, the diol unit was promptly protected in the same operation, clearing the path toward talaromycin-B.

H^+

talaromycin-B

A tactic of good potential involves the formation of more than one ring in a process but retention of only one ring in the target molecule. An example is an intramolecular Diels–Alder reaction of a photoenol which served to establish the pyrrolidine ring of acromelic acid-E [Horikawa, 1993]. In this case a retro-Claisen fission accomplished the cleavage.

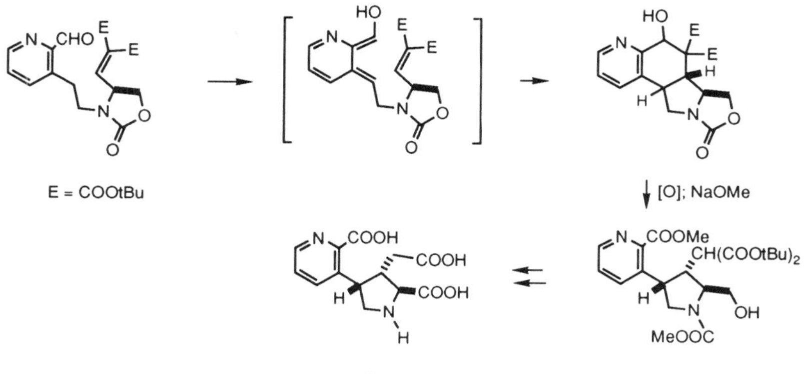

E = COOtBu

[O]; NaOMe

acromelic acid-E

The contrasting tactic to the ring expansion by cleavage of intercyclic bonds is compartmentization of a larger ring. In this regard is the creation of the pyrrolizidine skeleton from an azacyclooctan-5-one through reductive cyclization, and this technique permitted completion of a synthesis of isoretronecanol [Leonard, 1960]. A similar transannular cyclization is the photochemical conversion of α-allocryptopine to berberine [Dominguez, 1967].

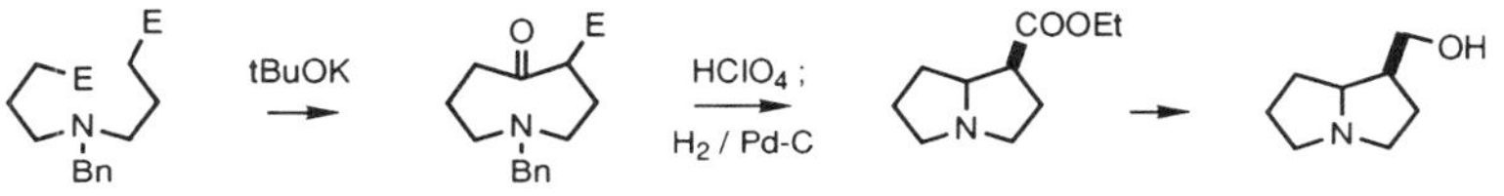

tBuOK

$HClO_4$; H_2 / Pd-C

isoretronecanol

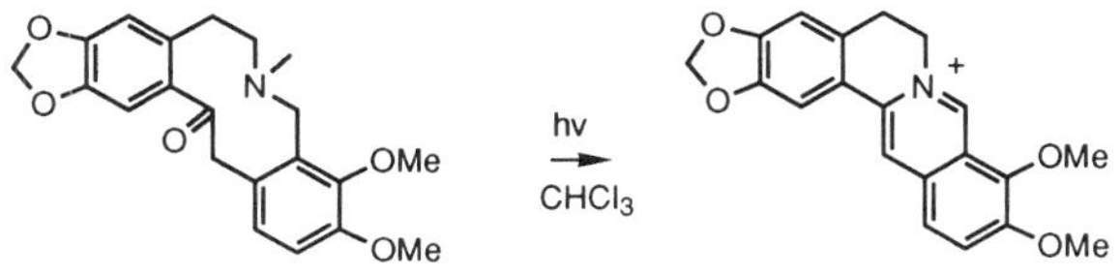

hν $CHCl_3$

α-allocryptopine

berberine

A synthesis of (+)-muscopyridine [Utimoto, 1982] demonstrated a mode of cyclic compartmentization with atom insertion. The cyclization step involved a Michael reaction and Schiff condensation.

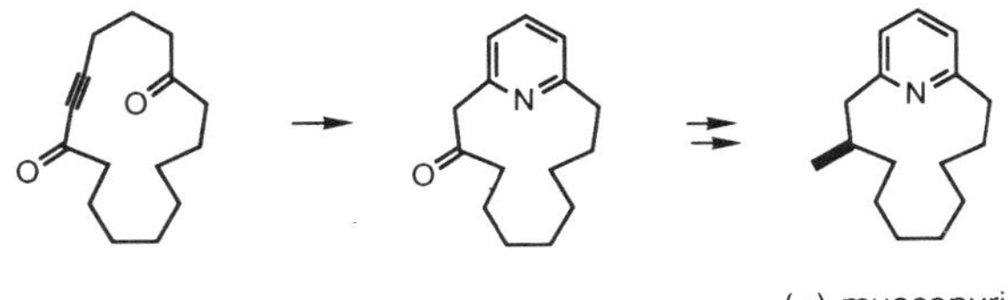

(+)-muscopyridine

In the carbocyclic area, intramolecular Diels–Alder reaction to form three rings from a macrocyclic triene [Marinier, 1988; T. Takahashi, 1988] has the greatest synthetic potential. Thus it is possible to construct the steroid skeleton with an A/B *cis*-ring juncture by thermal reaction of a cyclopentannulated 14-membered ring triene [T. Takahashi, 1988]. It should be noted that the transannular Diels–Alder reaction has been proposed as a key step in the biogenesis of several natural products, such as ikarugamycin [S. Ito, 1977].

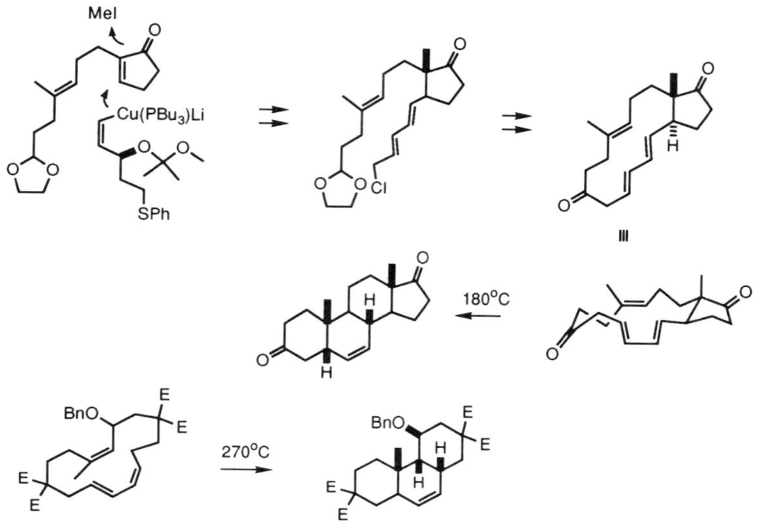

MeI

Cu(PBu$_3$)Li

SPh

Cl

180°C

270°C

5.6. CYCLIZATION WITH INSTALLATION OF A TRANSFERABLE CHAIN

This very powerful tactic is most readily appreciated by examination of the actual processes. Thus in a synthesis of 11-oxo steroids [Stork, 1981] the establishment of the B-ring was achieved by an intramolecular Diels–Alder reaction. Upon oxidative cleavage of the cyclohexene moiety of the adduct two critical carbonyl groups (at C-3 and C-11 of the steroid nucleus) were exposed. The Diels–Alder reaction provided a latent acetonyl unit (to be detached from C-11) to the pro-C-1 atom of the A-seco precursor.

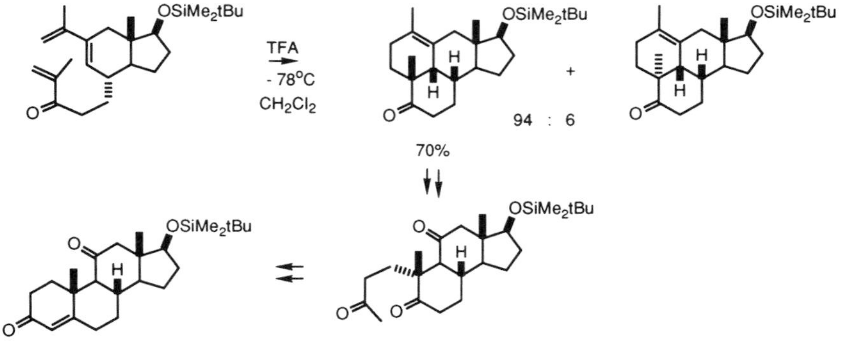

OSiMe$_2$tBu

TFA

- 78°C

CH_2Cl_2

94 : 6

70%

Similar tactics were implemented in a route to quadrone [Dewanckele, 1983; Schlessinger, 1983].

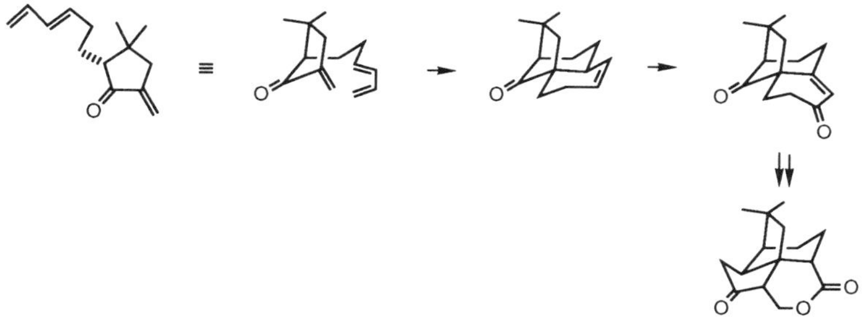

quadrone

In an elegant approach to (−)-picrotoxinin [Corey, 1979b] bridging of (−)-carvone set stage for the construction of the cyclopentane ring. The remaining two carbon unit of the bridge was subsequently redistributed to appear as the two lactonic carbonyl groups.

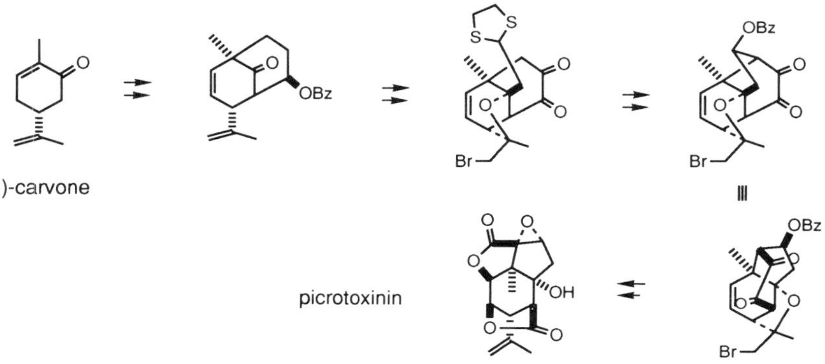

(-)-carvone

picrotoxinin

Simultaneous creation of two adjacent stereogenic centers of (−)-perhydrohistrionicotoxin including the spirocyclic carbon atom, while under control by the existing chirality at the α-carbon of the nitrogen atom [Winkler, 1989], greatly facilitated a synthesis of the bioactive substance. The reaction sequence employed here consists of intramolecular [2 + 2]-photocycloaddition and fragmentation.

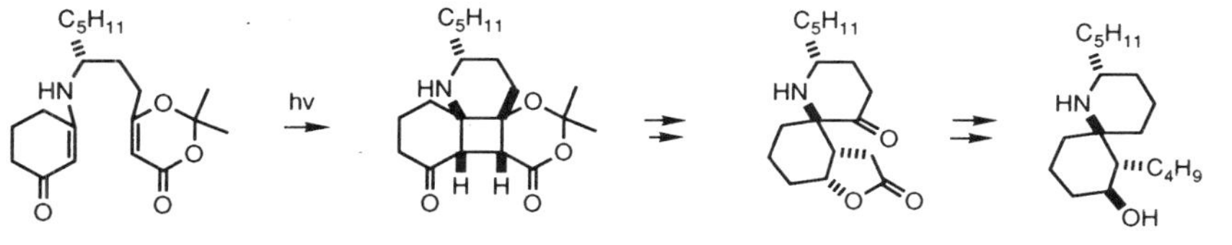

(-)-perhydro-
histrionicotoxin

A related tactic was employed to an approach to trichodiene [Harding, 1990]. A Nazarov cyclization served to establish the two adjacent quaternary carbon atoms, while subsequently the cyclic ketone group was removed.

trichodiene

5.7. HETEROCYCLES AS SYNTHETIC PRECURSORS

Heterocycles may be employed as templates for carbon framework construction because these substances are readily converted into aliphatic compounds with or without certain functionalities, and the timing of chain release can be regulated.We shall discuss some possibilities in the assembly of carbon chains based on manipulations of heterocycles.

A colchicine synthesis [Woodward, 1963–4] differs from all others in having the nitrogen atom present as a member atom of an isothiazole ring at the start. Substituents at C-3 and C-4 of this heterocycle were employed to extend carbon chains toward an aromatic ring and an alkadienoic acid chain, respectively. After formation of the benzosuberane system by an intramolecular reaction the unsubstituted carbon of the isothiazole ring was carboxylated, a process enabled by the intrinsic acidity of such carbon acid. The corresponding diester underwent Dieckmann cyclization to establish the other seven-membered ring. The heterocycle was maintained until this new ring was converted into a tropolone chromophore.

MeOOC N S ⇉ MeO MeO MeO E N S ⇉ MeO MeO MeO N S HOOC

MeO MeO MeO NHAc O OMe ⇇ MeO MeO MeO N S O ⇇ MeO MeO MeO N S COOMe COOMe

colchicine

N-Carbethoxymethyl-5-hydroxyethyl-4-methylthiazolium bromide is a useful 1,3-dipolar precursor for generating proline derivatives, amply demonstrated in its application to a synthesis of α-allokainic acid [Kraus, 1985].

OBn O (Et3N) O OBn COOH COOEt COOEt COOH HO H α-allokainic acid

3,5-Dimethylisoxazole is a latent acetonyl group and also an equivalent of 4-chloro-2-butanone by virtue of its ready chloromethylation. After reaction of the chloromethylated isoxazole with an enolate the heterocycle can be cleaved to reveal a methyl ketone which can then be subjected to intramolecular aldolization [Stork, 1967]. A route to ferruginol [Ohashi, 1968] exemplifies some advantages over conventional alkylating agents.

COOEt N-O COOEt OH OR ferruginol

2,3,5-Trimethylisoxazolium salts condense with aromatic aldehydes under mild conditions [Kashima, 1977]. The 1,3-diketones are liberated by treatment with sodium methoxide and hydrochloric acid, hence these heterocyclic compounds are a synthetic equivalent of 2,4-pentanedione 1,5-dianion.

In previous times furan derivatives were under-utilized due to the susceptibility of these compounds to polymerization. However, the difficulties have been largely overcome by careful control of ring cleavage reaction conditions, and the furan ring is now considered as a viable latent 1,4-dicarbonyl unit. For example, from 2-methylfuran, convenient syntheses of *cis*-jasmone [Büchi, 1966b],and of a precursor for 16,17-dehydroprogesterone [W.S. Johnson, 1970a, 1971] via cationic polyene cyclization have been described.

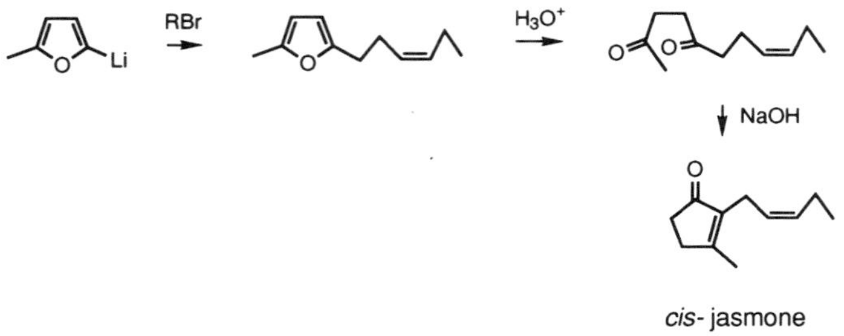

RBr → H$_3$O$^+$ → NaOH

cis-jasmone

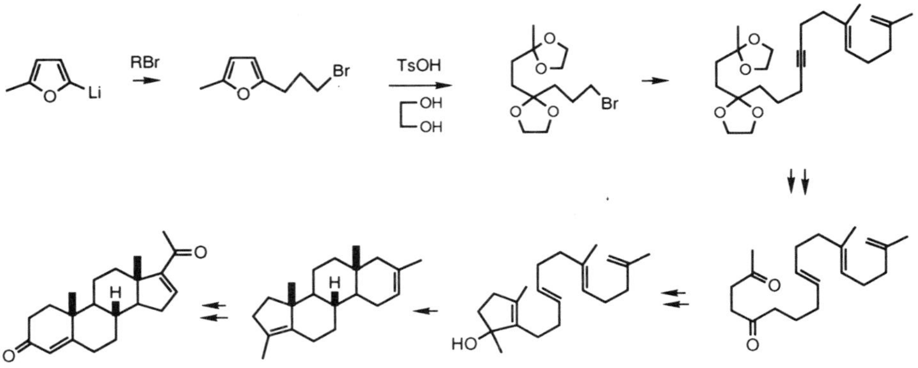

RBr → TsOH, HOCH$_2$CH$_2$OH → Br

Various γ-keto esters are available from furfural in a three-step method [Lewis, 1961; Lukes, 1961], i.e. Grignard reaction, ethanolysis, and hydrolysis. Acylation of a furan (with a free α-position) followed by formation of the tosylhydrazone and thermal decomposition of its sodium salt leads to conjugate enynone [Hoffman, 1971].

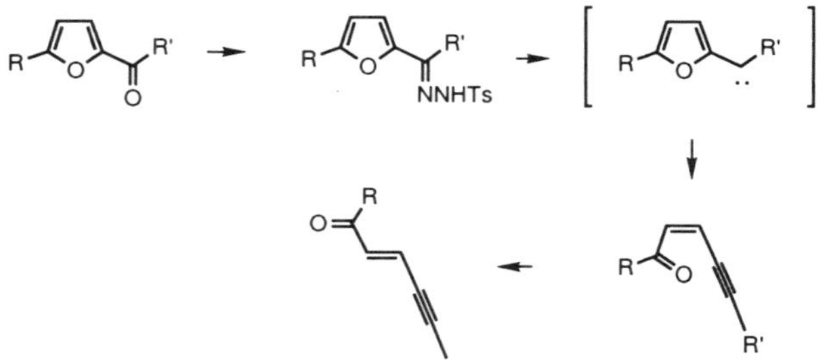

A double bond of the furan ring may participate in photoinduced electrocyclization [Ninomiya, 1983] and this behavior has been exploited in numerous alkaloid syntheses. A recent example concerns the access of eburnamonine [Naito, 1992] in which the dihydrofuran moiety of the photochemical product plays the role of a latent acetic acid sidechain. The same intermediate is apparently suitable for synthesis of cuanzine-type alkaloids.

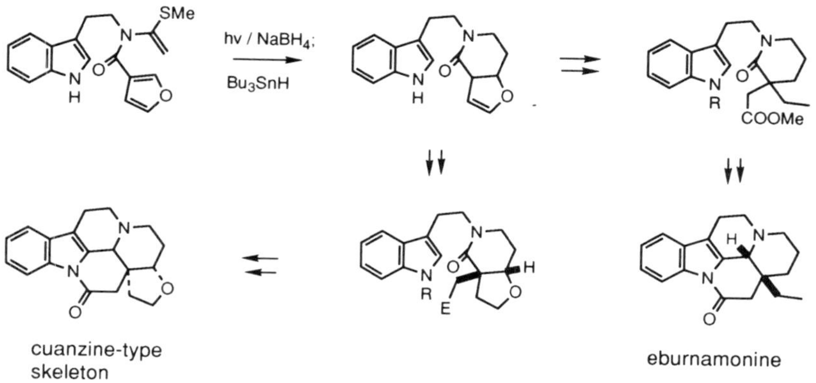

cuanzine-type skeleton

eburnamonine

Unusual synthetic opportunities are available from the Paterno–Büchi reaction adducts of furans and aldehydes. For example, the syntheses of avenaciolide [S.L. Schreiber, 1984a] and asteltoxin [S.L. Schreiber, 1984c] have been accomplished with the aid of this reaction.

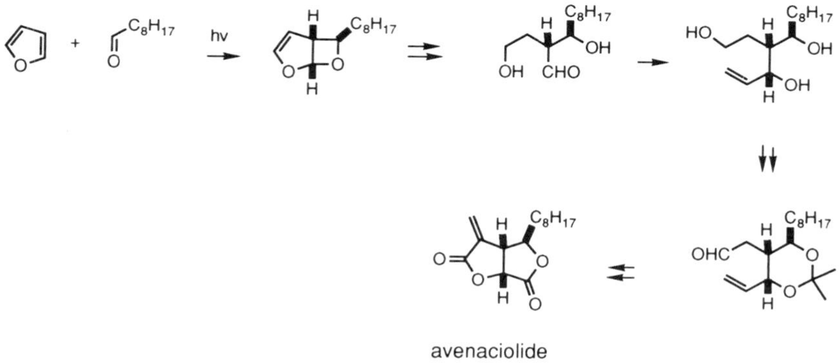

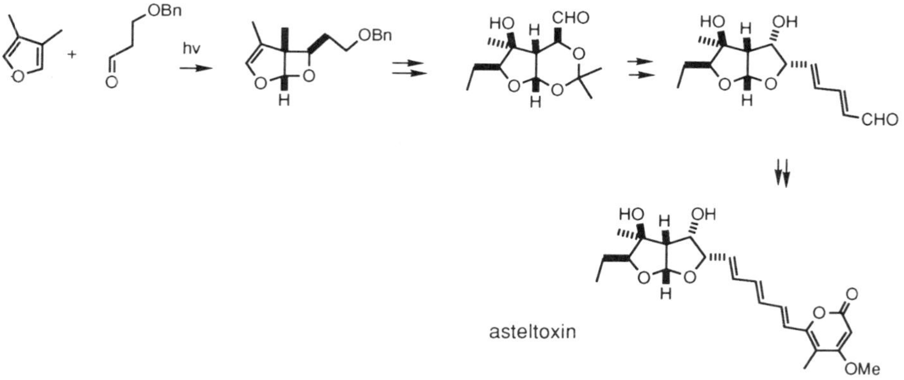

avenaciolide

asteltoxin

With a double bond serving as dipolarophile for nitrile oxides, a furan ring provides a template for the elaboration of aminopolyols [Müller, 1982].

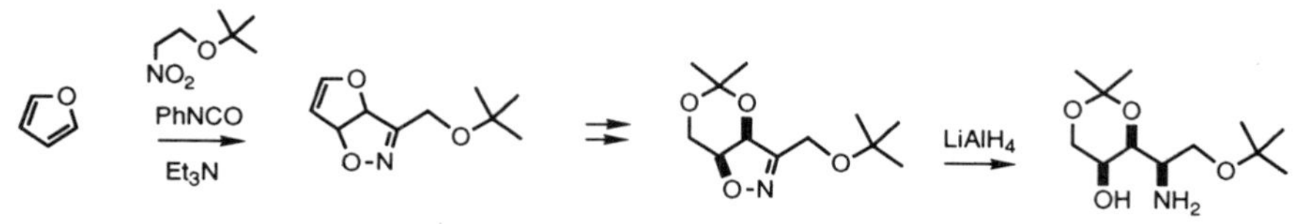

A route to the olivin-like tricyclic compound which was based on Diels–Alder reaction of a furan has been devised [Kraus, 1983].

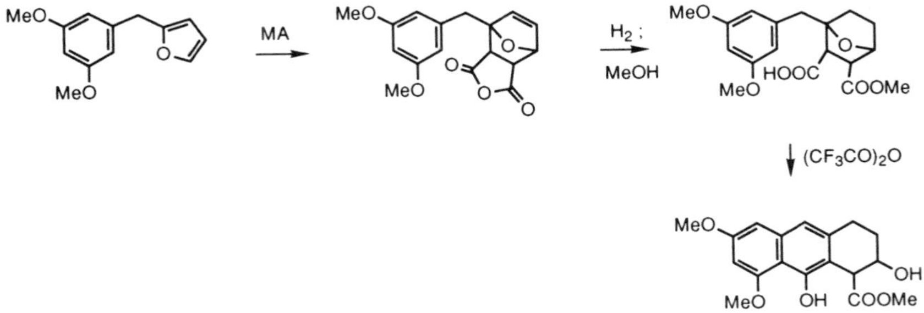

It is easily conceived that the dehydration of a 7-oxabicyclo[2.2.1]hept-2-ene leads to an aromatic ring. Occasionally a reaction sequence involving the preparation of such a compound and its aromatization has merit in synthesis, such as a route to the jatropholones [A.B. Smith, 1985].

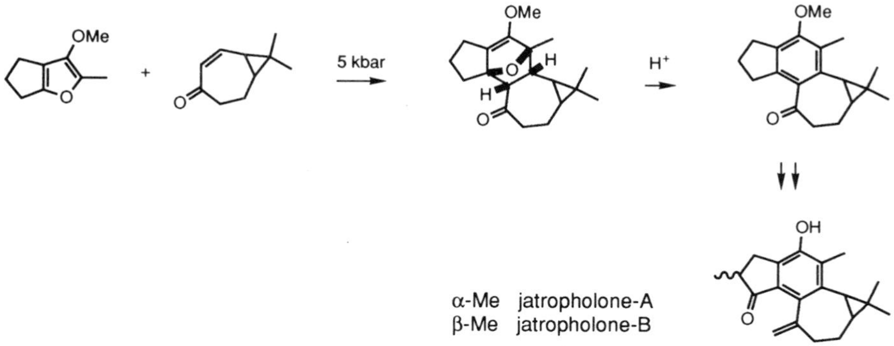

α-Me jatropholone-A
β-Me jatropholone-B

Methyl 3,6-oxa-1-cyclohexenecarboxylate, obtainable from the Diels–Alder adduct of furan and methyl 3-nitroacrylate via diimide hydrogenation and elimination, served as a dienophile in a reaction that yielded a precursor of (+)-compactin [Grieco, 1986]. The heterocycle was cleaved in a fragmentation step to generate a double bond and the oxygen substituent.

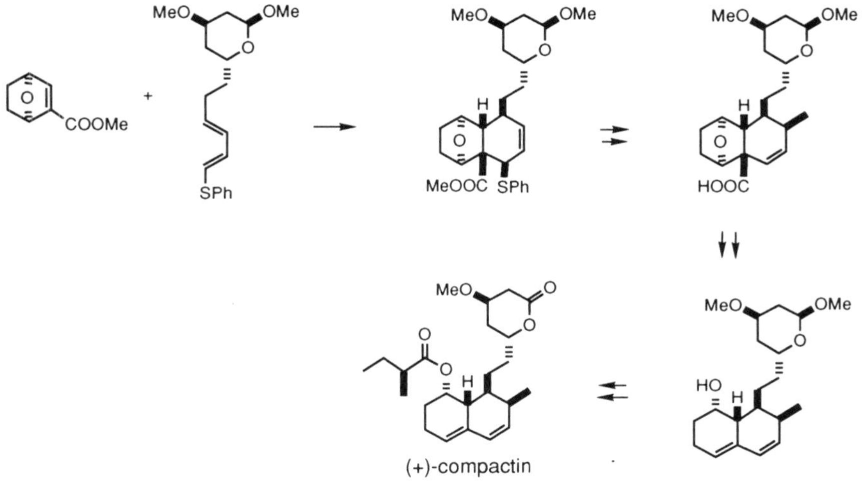

The facile formation of an 8-oxabicyclo[3.2.1]oct-6-en-3-ones by interception of 2-oxyallyl cations with furans has been exploited in the elaboration of the Prelog–Djerassi lactone [White, 1979], and nezukone [Takaya, 1978].

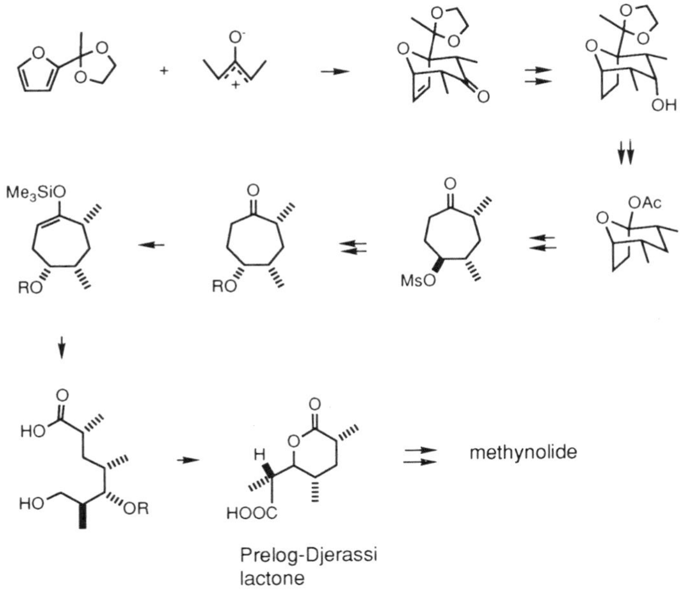

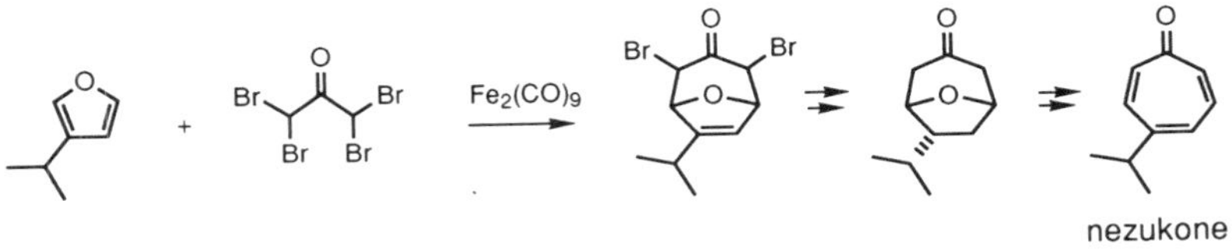

Note that an S_N2' methylation has been applied to a bicyclic alcohol to generate a cycloheptene which is a stereopentad synthon for rifamycin-S [Lautens, 1992].

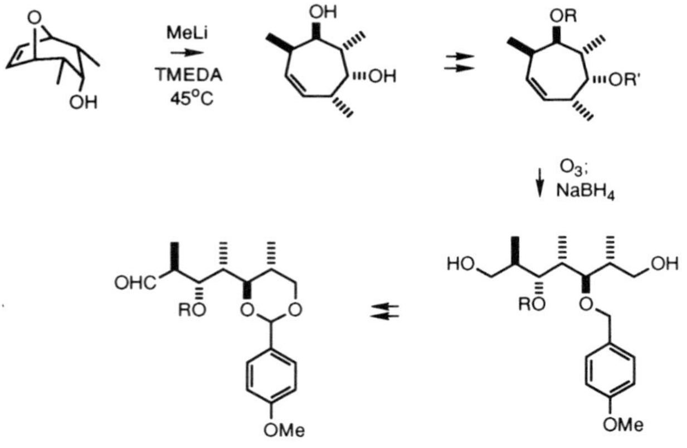

rifamycin-S segment

Oxidation of (2-furyl)carbinols gives rise to 2,3-dihydro-6-hydroxypyran-3-ones. This ring transformation is not only useful for carbohydrate synthesis, its extension is eminently suitable for the elaboration of various natural products ranging from disparlure [Kametani, 1990] to erythronolide-B [S.F. Martin, 1989].

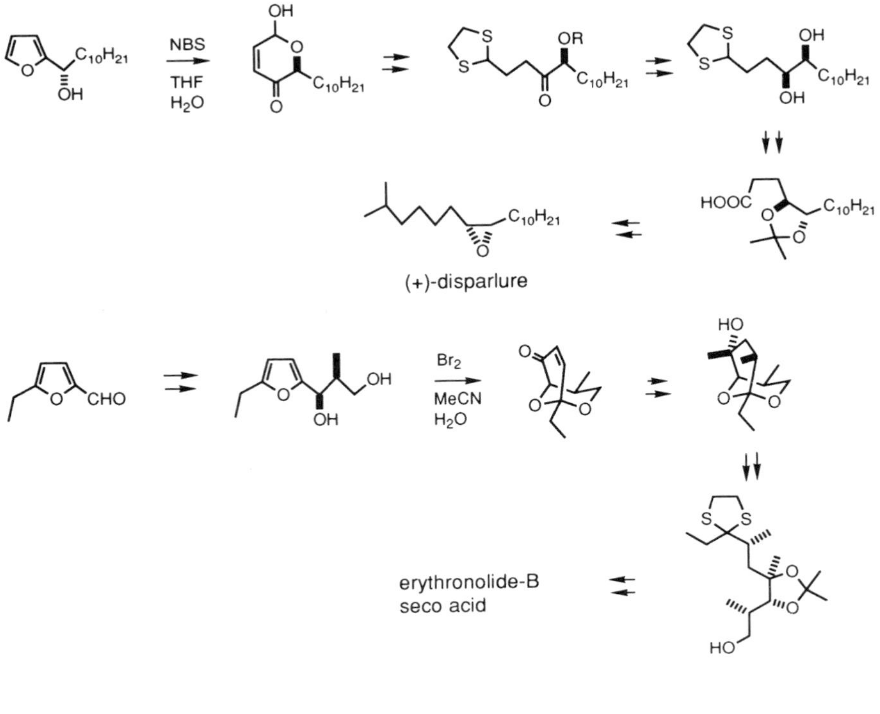

(+)-disparlure

erythronolide-B seco acid

2-Alkenyl-5-alkylidenyltetrahydrofurans are subjected to Claisen rearrangement to afford 3-cycloheptenones. This method allowed the preparation of karahanaenone [Demole, 1969].

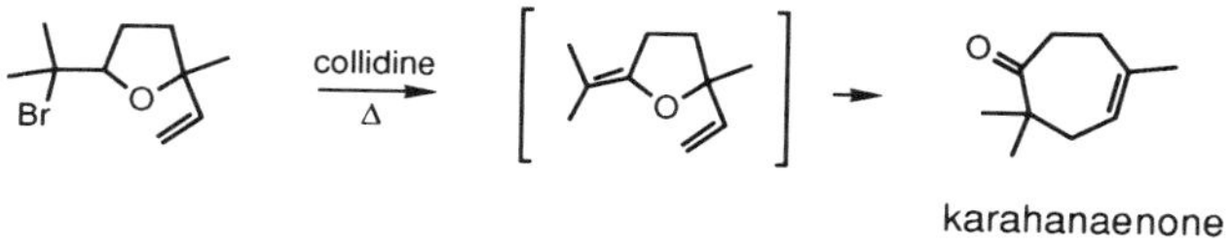

collidine
Δ

karahanaenone

The analogous rearrangement of 3,4-dihydro-2*H*-pyranylalkenes leads to substituted cyclohexene derivatives [Büchi, 1970b], thus representing a complementary method to the Diels–Alder reaction.

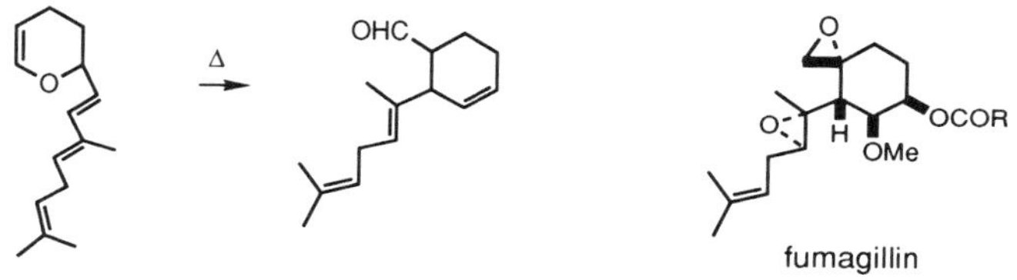

Δ

fumagillin

Intermediates for synthesis of erythronolide-A and -B have been acquired via the dioxanone-to-dihydropyran conversion approach [Burke, 1986b,1987a,b]. The dioxanones were prepared from chelation-controlled Grignard reactions, and the Ireland–Claisen rearrangement converted the dioxanones stereospecifically into the dihydropyran products useful for further elaboration of the functionalized segments of the antibiotic aglycones.

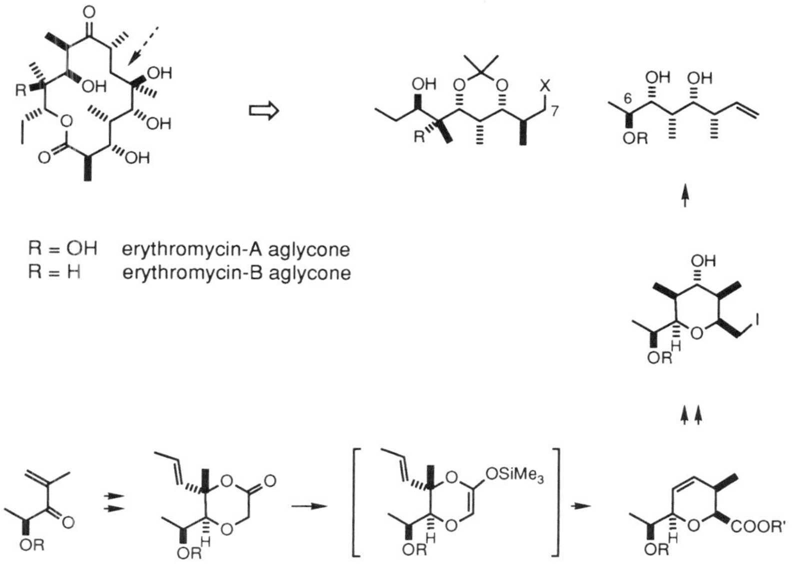

R = OH erythromycin-A aglycone
R = H erythromycin-B aglycone

R = OH , X = OBn
R = H , X = SPh

The same tactic proved successful in the construction of the left-half of the ionophore antibiotic indanomycin which is a trisubstituted tetrahydropyran [Burke, 1985].

indanomycin

The applicability of monosaccharides to the synthesis of chiral natural products [Hanessian, 1983] can hardly be overemphasized. Because of the enormity of this subject and consequently the difficulty in singling out any one or a few contributions, it is perhaps better to point out the potential of a less commonly utilized derivative, levoglucosenone, which is available from the pyrolysate of cellulose. With its rigid bicyclic framework,levoglucosenone offers excellent stereocontrol in its reactions. A synthesis of serricornin [M. Mori, 1982] was based on this property.

levoglucosenone

serricornin

Reduction of the pyridine ring to the tetrahydro stage and further cleavage of the C—N bonds would provide a carbon chain. If arrangement of substituents on the heterocycle is readily achievable the construction of a carbon skeleton may take this advantage. A simple illustration of this protocol is a synthesis of 1,4-diene precursors for insect pheromones from 2-alkylpyridines [Bac, 1982, 1986].

LDA; Me_3SiCl; MeI ; $NaBH_4$ → $SiMe_3$ → MeI ; CsF

The pyridine ring is also a latent 1,5-dicarbonyl unit. The conversion is easily achieved by reduction with sodium in liquid ammonia followed by mild hydrolysis. Since picoline undergoes alkylation readily, and 2-vinylpyridine behaves as a Michael acceptor, incorporation of the 1,5-dicarbonyl subunit into a complex molecule is tactically feasible by using this chemistry. A synthesis of D-homoestrone [Danishefsky, 1972] serves to demonstrate the concept.

tAmOK

EtONa

D-homoestrone

Porphobilinogen (PBG) which occurs in urine of patients suffering acute porphyria, is a direct precursor of blood and plant pigments, biliprotenoids, and vitamin B_{12}. A very efficient synthesis of PBG [Frydman, 1969] is based on the formation a tetrahydro-6-azaindol-5-one intermediate from a pyridine derivative, and subsequent cleavage of the lactam ring. Thus the two sidechains at C-2- and C-3 of the pyrrole nucleus were derived from the heterocyclic template.

porphobilinogen

Because β-lactams are now readily prepared, they have been utilized quite routinely as building blocks for synthesis. Thus in an approach to (−)-lankacidin-C [Kende, 1993] the pyruvamido substituent and the lactone carbonyl of the metabolite evolved from a β-lactam.

TBDPSO OH OTBS PMBO O N TBS I Bu$_4$NF ; MsOH HO NH$_3$+ H O O PMBO I HO O HN O O H O O HO OH

(-)-lankacidin-C

An expedient way for the preparation of 3,4-disubstituted β-lactams is through photoisomerization of 2-pyridones followed by oxidative cleavage of the resulting 2-azabicyclo[2.2.0]hex-5-en-3-ones [Brennan, 1981]. A more elaborate heterocycle dismutation involving opening of a Dewar pyridinium ion and formation of a hydropicolinic acid which occurred when 3-(4-pyridyl)-alanine was subjected to photolytic conditions has been exploited in a synthesis of decarboxybetanidin [Hilpert, 1985], the blue-violet pigment of a flower.

NH O hv NH O O$_3$; NaBH$_4$ OH OH NH O

N HOOC NH$_2$ hv NH$_3$, MeOH [+NH $^-$OOC NH$_2$ NH$_2$ CHO $^-$OOC NH$_2$ NH$_2$ $^-$OOC N+ H] (S)-cyclodopa

HO HO N+ COOH $^-$OOC N H

decarboxybetanidin

The discovery of the "zip reaction" makes the synthesis of macrocyclic polyamine lactam alkaloids much easier. For example, in an approach to chaenorhine [Wasserman, 1983b] a 13-membered triazacycle was constructed from a nine-membered aminolactam precursor, itself deriving from hexahydropyrazine and ethyl acrylate. The N—N bond of the first bicycle intermediate was cleaved reductively.

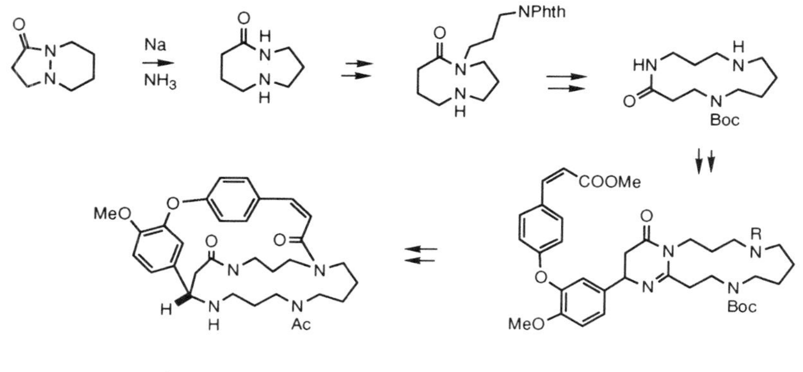

chaenorhine

The two above-mentioned transformations exemplify a tactic comprising manipulation of one heterocycle into another. Further representations are the synthesis of piperideines from oxazinones [Angle, 1989; Wang, 1991].

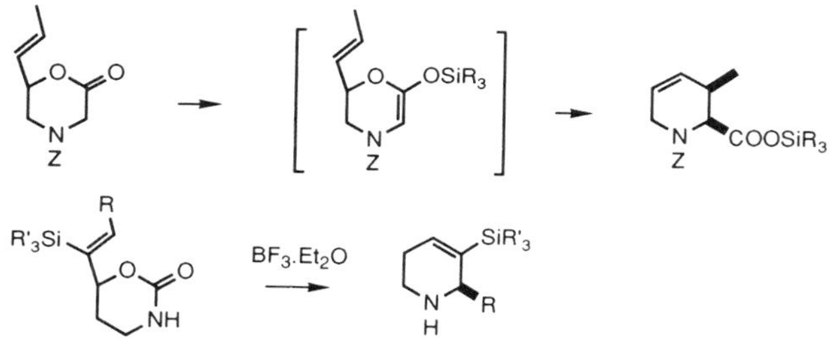

Most applications of sulfur-containing heterocycles as templates to synthesis have been targeted at sulfur-free products. Thiophene provides a four-carbon chain which can be extended in both directions, for example via acylation. Reduction in the presence of Raney nickel is the most expedient method for removal of the heterocycle [Wynberg, 1956]. 1,10-Diketones are obtained from bithienyl as shown below. Furthermore, cyclization of ω-thienylcarboxylic acids gives rise to condensed ring systems, and similar reduction leads to substituted cycloalkanones [Cagniant, 1953, 1955; Murad, 1973], including muscone [Catoni, 1980].

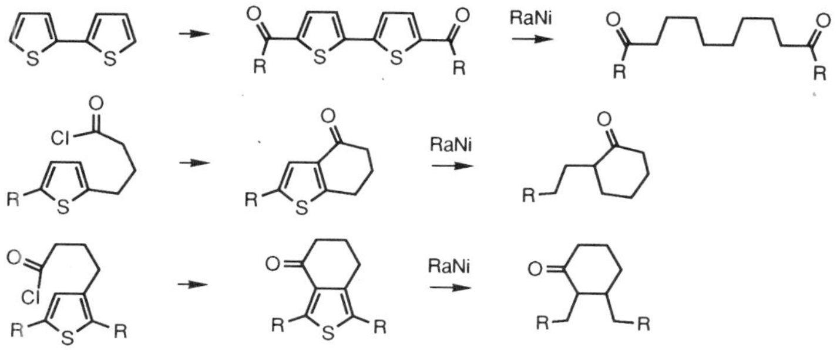

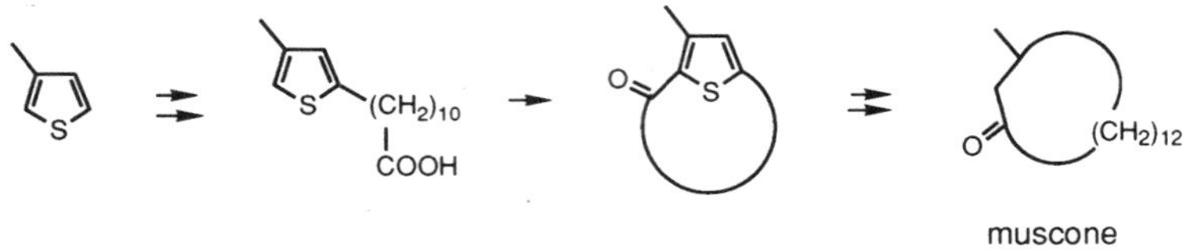

muscone

For synthesis of a long-chain carboxylic acid, thiophene is acylated, deoxygenated (by Clemmensen or Wolff-Kishner reduction), acetylated or succinylated to furnish a methyl ketone or a keto acid which is processed accordingly (haloform reaction or subjected to ketone reduction), before the thiophene ring is destroyed. Many other variations are possible.

By applying the Strecker synthesis to the proper formylthiophene a straight- or branched-chain amino acid is accessible [Goldfarb, 1962].

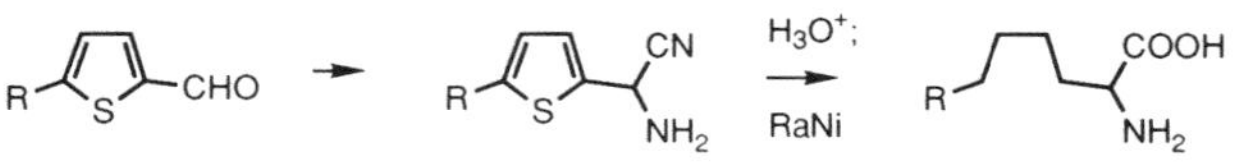

In a synthesis of xestospongin-A [Hoye, 1993] the difunctional key intermediate was assembled from a thiophene ring. At the final stage of the synthesis hydrodesulfurization released a tetramethylene unit in each of the chain segments separating the oxaquinolizidines. The choice of the thiophene template was perhaps because it facilitated the assembling process possibly including the macrocyclization, and it did not interfere with subsequently employed chemical reactions at other sites.

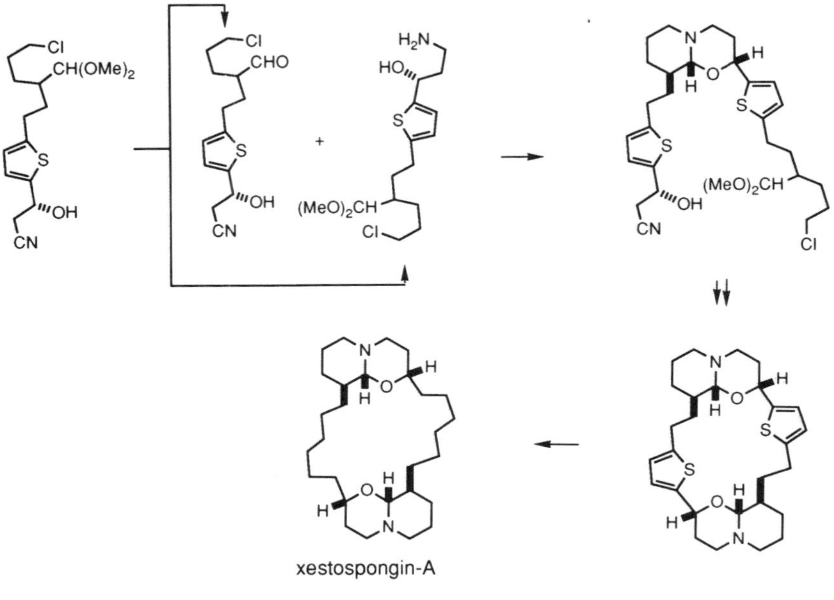

xestospongin-A

A recent development in the synthesis of (*Z*,*Z*)-1,3-dienes consists of nickel(O)-catalyzed Grignard reactions with thiophene (also with furan, selenophene, and tellurophene) [Wenkert, 1984a].

Sulfur-containing heterocycles may also provide better stereocontrol in the modification of carbon skeletons. Indeed, several 2,6-dideoxy L-sugars including mycarose, oleandrose, olivose, olivomycose, cymarose, digitoxose, are readily obtainable from methyl α-L-glucoside via 2-oxa-5-thiabicyclo[2.2.2]octane derivatives [Toshima, 1991a,b].

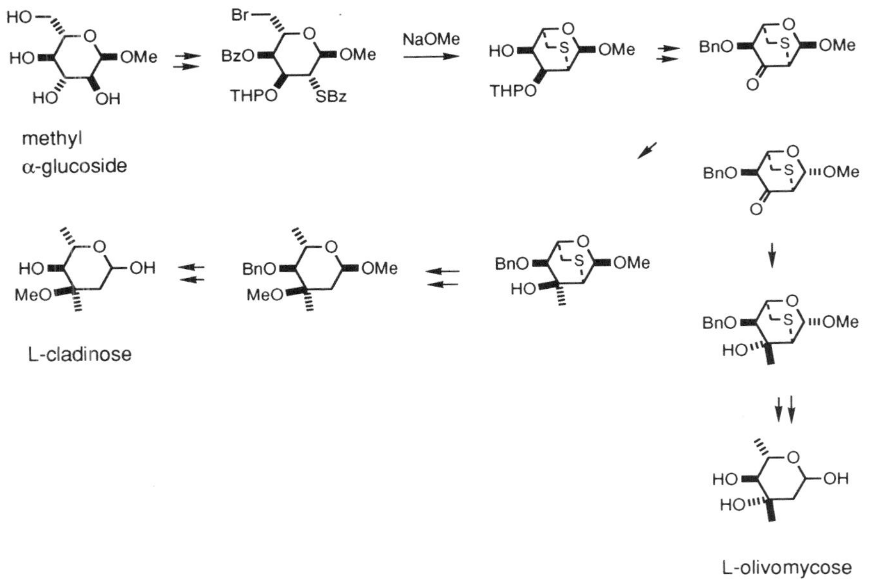

methyl α-glucoside

NaOMe

L-cladinose

L-olivomycose

An excellent synthesis design of the *Cecropia* juvenile hormone JH-I [Kondo, 1972; Stotter, 1973] consists of assembly of two thiopyran units, each of them serving as masked alkene segments of desired configuration. Cleavage of C—S bonds was accomplished by successive treatment with $Li/EtNH_2$ and Raney nickel.

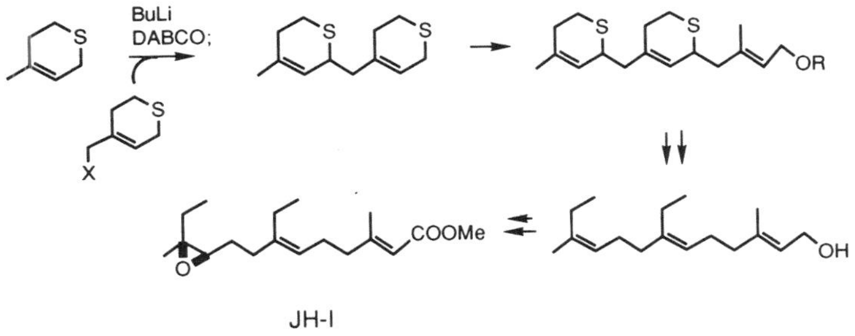

BuLi DABCO;

JH-I

The alkylation product of 5,6-dihydro-2*H*-thiopyran with epichlorohydrin provided a precursor of (*Z*)-5-octen-2-one [Torii, 1977], therefore *cis*-jasmone is readily acquired.

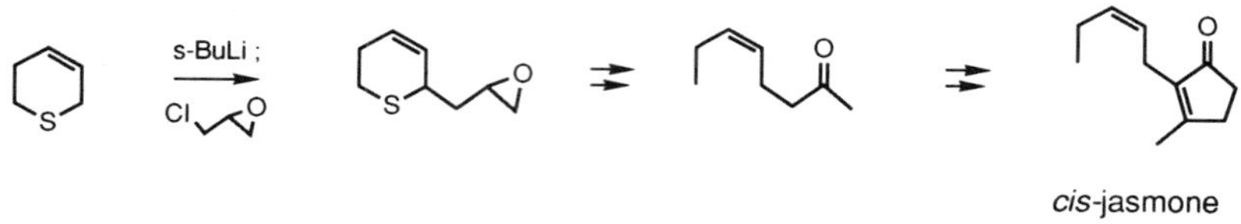

s-BuLi ;

cis-jasmone

In an approach to erythromycin [Woodward, 1981] which started from the preparation of erythronolide-A seco acid the design for chain assembly involved elaboration of a fused bisthiopyranone by intramolecular aldol reaction. When the carbon chain segments were assembled and properly adorned with functionalities desulfurization was carried out to break the heterocycles and reveal the methyl groups.

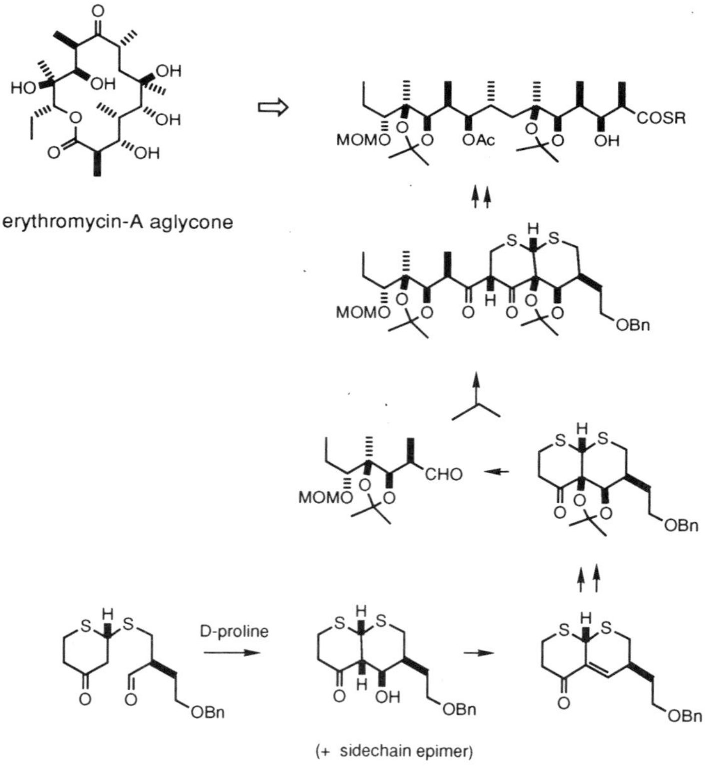

erythromycin-A aglycone

D-proline

(+ sidechain epimer)

Ring expansion via [2.3]-sigmatropic rearrangement of a sulfonium ylide provides a solution to many difficult synthetic problems. One would appreciate its value on examination of its application to synthesis of zygosporin-E [Vedejs, 1984b, 1988]. Reiterative use of this reaction further enlarges its role.

Ph N Bz O ⇉ SiMe3 Cl OAc H S N Bz O Ph O NaI K_2CO_3 MeCN Δ [SiMe3 OAc S + - H Ph N Bz O O]

zygosporin-E

A long-sought "direct" Diels–Alder route to cantharidin was realized [Dauben, 1980] by employing high pressure and cyclofixation tactic which decreases the rotational freedom of the two methyl groups in the dienophile by bridging them into a 3,4-dihydrothiophene ring.

pressure (major) RaNi cantharidin

A different application of a 2,5-dihydrothiophene derivative as dienophile to facilitate the Diels–Alder reaction and to ensure the stereochemical congruence to the target compound which is devoid of sulfur element is shown in an approach to a diester precursor of *trans*-perhydroindanone containing an angular methyl group [Stork, 1969]. The compound is a CD-ring synthon of steroids.

MeOOC S COOMe → EtO COOMe S COOMe; OH OH, RaNi → COOMe COOMe → COOMe

By means of an intramolecular Diels–Alder approach involving a furan ring and an allyl sulfide double bond it is possible to gain complete stereocontrol of three asymmetric centers during assemblage of nemorensic acid [Klein, 1985].

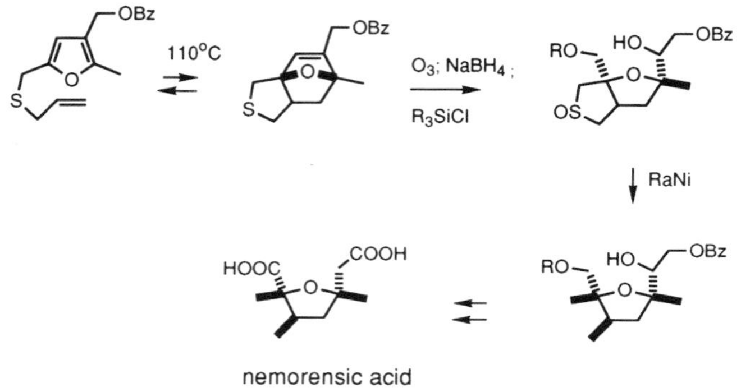

nemorensic acid

Thiacycloalkanonecarboxylic esters obtained from Dieckmann condensation of the aliphatic diesters are useful precursors of certain chiral β-hydroxy esters [Hoffmann, 1982c] in view of the enantioselective reduction of cyclic β-keto esters by baker's yeast.

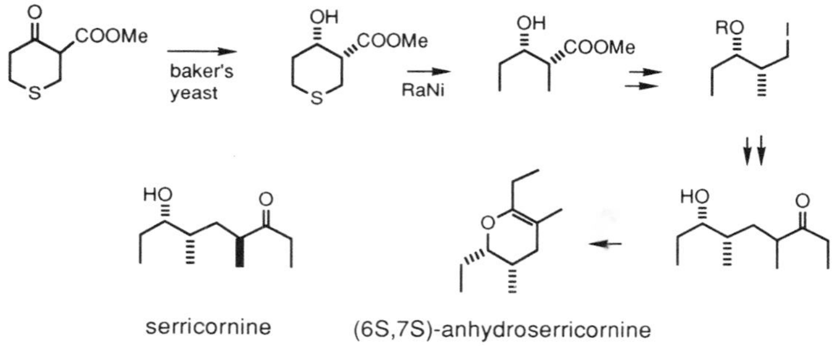

serricornine (6S,7S)-anhydroserricornine

Sulfur-containing heterocycles may be contracted by a Ramberg–Bäcklund reaction, and application of such a tactic in a synthetic route to a prostaglandin intermediate has been reported [Fujisawa, 1991].

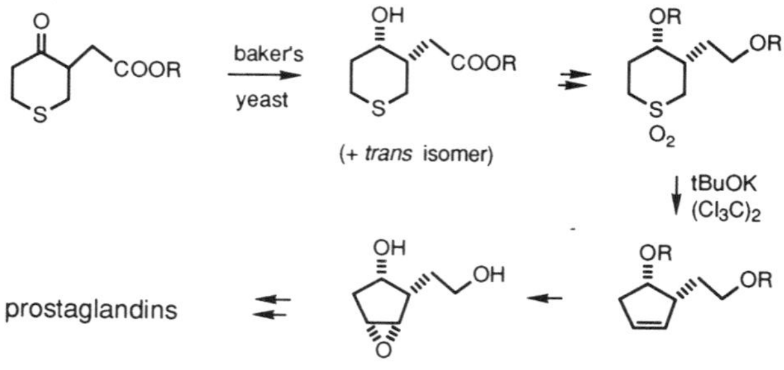

The dithioacetal function is a masked carbonyl. Cyclic dithioacetals which are readily introduced to an α-position of a ketone have an additional application in directing base-promoted C—C bond cleavage. Thus a synthesis of antirhine [T. Suzuki, 1986] could use a well-known 3-substituted cyclopentanone as a building block. It should be noted that the dithiolation occurred at C-5, and the product of ring cleavage has differentiable functionalities.

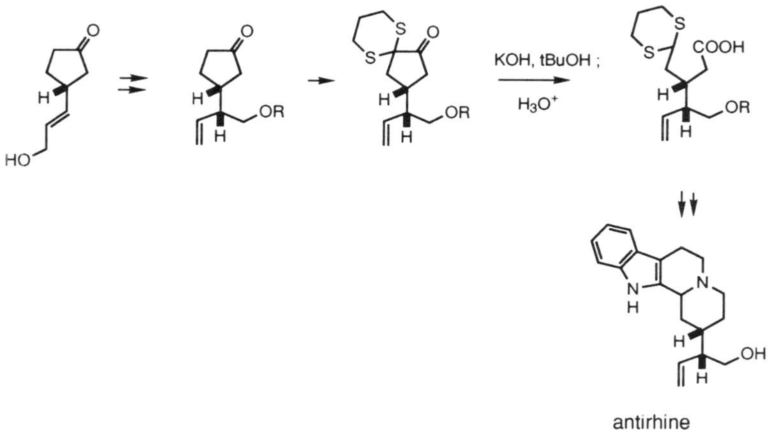

KOH, tBuOH ; H_3O^+

antirhine

One of the most intensively studied sulfur-containing heterocycles for synthetic manipulations is 3-sulfolene, the prototype of which is the cycloadduct of 1,3-butadiene and sulfur dioxide. However, the sulfolene moiety may be constructed *de novo* at an intermediary stage, as illustrated in a synthesis of elaeokanine-A [Schmittenhenner, 1980] in which a 3-acyl-2,5-dihydrothiophene was constructed from an (α-acyl)vinylphosphonate by reaction with α-thioacetaldehyde in a Michael/Horner–Emmons reaction tandem.

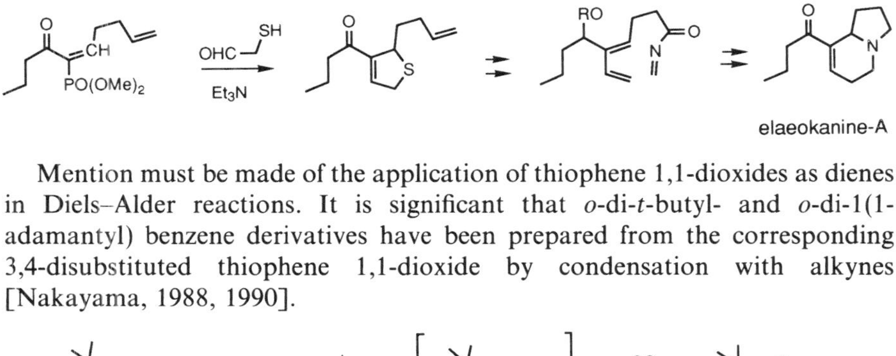

Et_3N

elaeokanine-A

Mention must be made of the application of thiophene 1,1-dioxides as dienes in Diels–Alder reactions. It is significant that *o*-di-*t*-butyl- and *o*-di-1(1-adamantyl) benzene derivatives have been prepared from the corresponding 3,4-disubstituted thiophene 1,1-dioxide by condensation with alkynes [Nakayama, 1988, 1990].

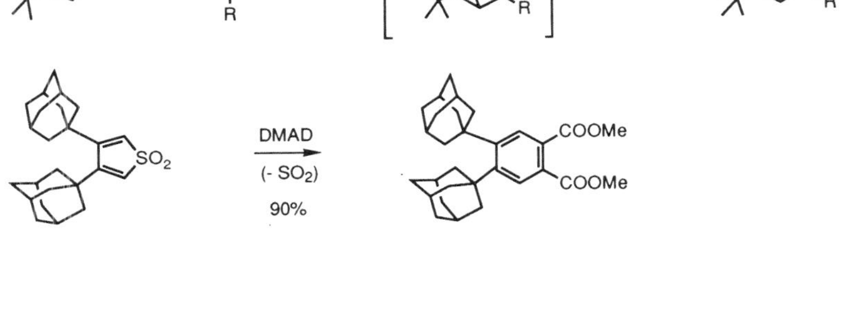

DMAD (- SO_2) 90%

The goal-directed transformation of a thiacycle to another one of different size is illustrated in a route to (+)-biotin [Baggiolini, 1982]. Thus a transannular 1,3-dipolar cycloaddition involving a macrocyclic vinyl sulfide was the key step that established the tetrahydrothiophene nucleus and the missing nitrogen atom.

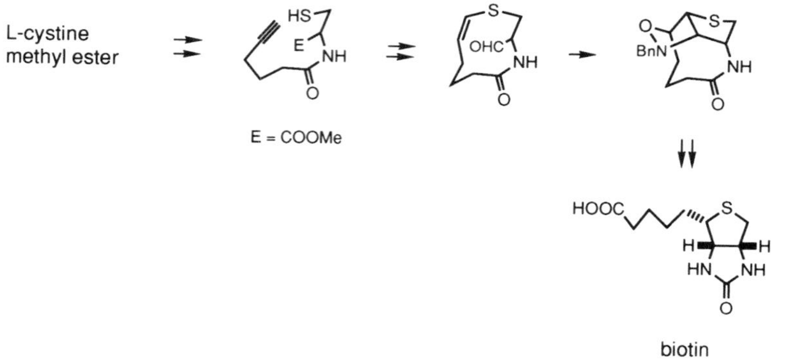

L-cystine methyl ester

E = COOMe

biotin

Condensation of 2,4-pentanedione with urea gives 4,6-dimethyl-2-hydroxypyrimidine which can be hydrogenated to the saturated cyclic urea and thence the *meso*-1,3-diamine [Hutchins, 1972]. Since hydroxypyrimidines prepared from various 1,3-diketones can be alkylated at the sidechain [Wolf, 1970], the preparation of *syn*-1,3-diamines is thereby greatly facilitated.

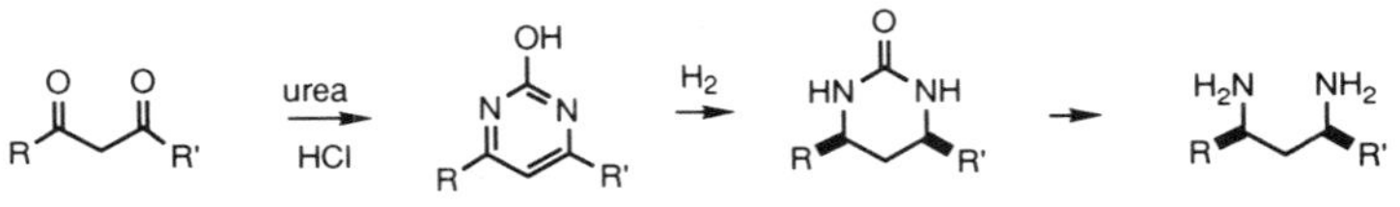

urea HCl H_2

Insertion of a methylene group between two of the double bonds of a conjugate diene may employ the tactic involving hetero-Diels–Alder reaction with an azodicarboxylic ester, followed by cyclopropanation, conversation of the hydrazine moiety into an azo linkage, and pyrolytic extrusion of dinitrogen [Berson, 1969].

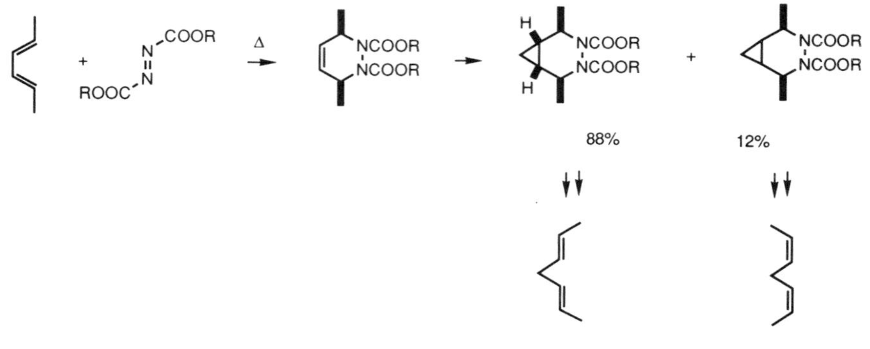

88% 12%

In tabtoxin there is a 1,6-relationship between the two amide carbonyl groups, and an intracircuit *cis*-1,4-amino (tertiary) alcohol unit. The structural features have been considered in a synthesis [J.E. Baldwin, 1983] which consists of a Diels–Alder reaction between ethyl 1,3-cyclohexadienecarboxylate and a niroso compound, conversation of the ester of the adduct into an aminomethyl group, oxidative cleavage of the double bond, and reductive scission of the N-O linkage. The latter two processes rendered a bicyclic intermediate into a chain with sterically obligatory functionalities.

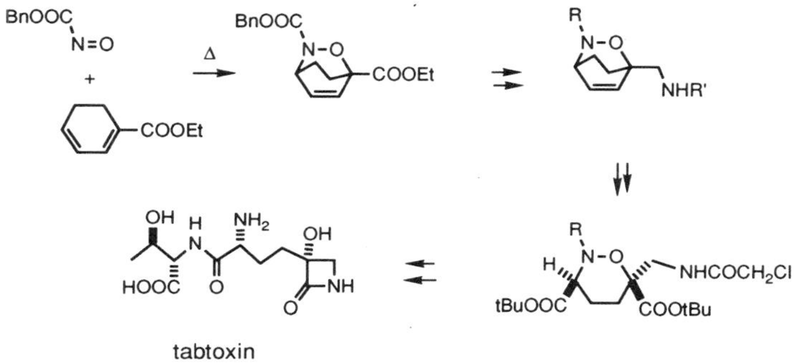

BnOOC N=O + COOEt → Δ BnOOC N-O COOEt ⇉ R N-O NHR'

⇊

tabtoxin ⇇ R N-O H NHCOCH₂Cl tBuOOC COOtBu

tabtoxin

The temporary formation of a 2,2-dimethyl-1,3-dioxane to provide stereocontrolled access to the *syn*-1,3-diol unit is a valuable synthetic tactic. The carbon chain constitutes substituents at C-4 and C-6 of the heterocycle unit and the three consecutive ring carbon atoms. As the *cis*-diequatorial arrangement of these substituents in a chair conformation is more stable, it can be reached through equilibration if such a process is allowed. Deketalization unveils the *syn*-diol system [S.L. Schreiber, 1987b; Nakata, 1989].

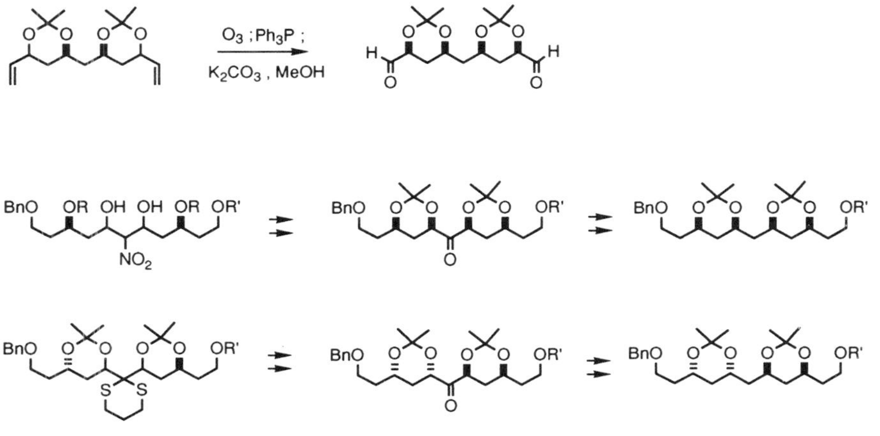

O_3 ; Ph_3P ; K_2CO_3, MeOH

BnO OR OH OH OR OR' NO_2 ⇉ BnO OR' ⇉ BnO OR'

BnO S S OR' ⇉ BnO OR' ⇉ BnO OR'

6

INTRAMOLECULARIZATION AND NEIGHBORING GROUP PARTICIPATIONS

Generally an intramolecular reaction is more favorable than the intermolecular counterpart owing to lesser demand in further entropy decrease. Such favorableness can be exploited in synthesis with respect to exemption from activation of reactants and to the defiance of normal regio- and stereoselectivity. Another common consequence is increase in yield of an intramolecular reaction as compared with the intermolecular version because side reactions are often minimized. Perhaps the most eloquent and convincing statement is that which was voiced by Eschenmoser [1970] in the description of the difficulty of condensing the B-ring and C-ring components during a synthesis of vitamin-B_{12}, and the circumvention by the formation of a divinyl sulfide followed by sulfide contraction. Indeed this necessity of intramolecularization tactic inspired the formulation and development of the powerful methodology. Another example is the general observation that the Pauson–Khand reaction, which typically gives cyclopentenones in yields ranging between 30–50% can be greatly improved when the alkene and alkyne components are connected by a tether of proper length.

6.1. INTRAMOLECULARIZATION

6.1.1. Facilitation of Reactions

Intramolecular catalysis facilitates many organic reactions. For example, the hydrolysis of methyl acrylate is accelerated by 16,000-fold in the presence of Cu(II) ion and *N*,*N*,*N′*-trialkylethylenediamine [Duerr, 1989]. The acceleration has been attributed to complex formation of the Michael adduct of the amine and the acrylic ester with the metal ion, and the metal-bond hydroxide attacks the nearby ester group. After hydrolysis the carboxylate ion does not participate strongly in chelation, allowing the free amino acid to undergo retro-Michael decomposition.

Simple alkyl esters higher than methyl esters are not cleaved by thiolate anions. However, 2-haloethyl esters undergo *O*-dealkylation on exposure to trithiocarbonate [T.L. Ho, 1974b] and ω-chloroalkyl esters are cleaved on treatment with sodium sulfide [T.L. Ho, 1974c]. The cleavage is the result of an intramolecular displacement after an initial reaction of the diphilic reagent at the haloalkyl carbon.

The difficulty for converting a hindered carboxylic acid into the phenylselenyl ester was overcome by intramolecular acylation [Ireland, 1985]. Thus a mixed anhydride was formed by reaction with phenyl dichlorophosphate and then reacted with phenylselenol in the presence of triethylamine. It is thought that the second reaction proceeded by a selenyl group hopping from phosphorus to the acyl carbon atom.

The remarkable dienophilic reactivity of monosubstituted alkenes in intramolecular Diels–Alder reactions has been repeatedly reported. There are successful applications of such reactions to synthesis of steroids and other complex natural products including gephyrotoxin [Y. Ito, 1983]. In the generation of the analogous 2,3-dimethyleneindole species by N_b-acylation of a 2-methyl-3-indolecarbaldehyde imine with an acylating agent containing a double bond, intramolecular Diels–Alder reaction occurs. This process constitutes the key step in a concise approach to indole alkaloids such as aspidospermidine [Gallagher, 1982].

SPh
EtOOC
PhS
aspidospermidine

The facile formation of carpanone [Chapman,1971] by Pd(II) oxidation of an *o*-propenylphenol is thought to proceed via an intramolecular hetero-Diels–Alder reaction.

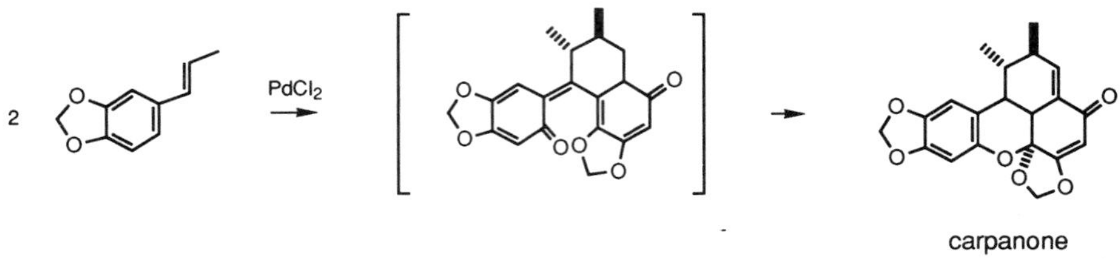

carpanone

The naphthalide lignan skeleton can be constructed by dehydrative dimerization of 3-arylpropynoic acid at below room temperatures, as exemplified in a preparation of helioxanthin [Holmes, 1970].

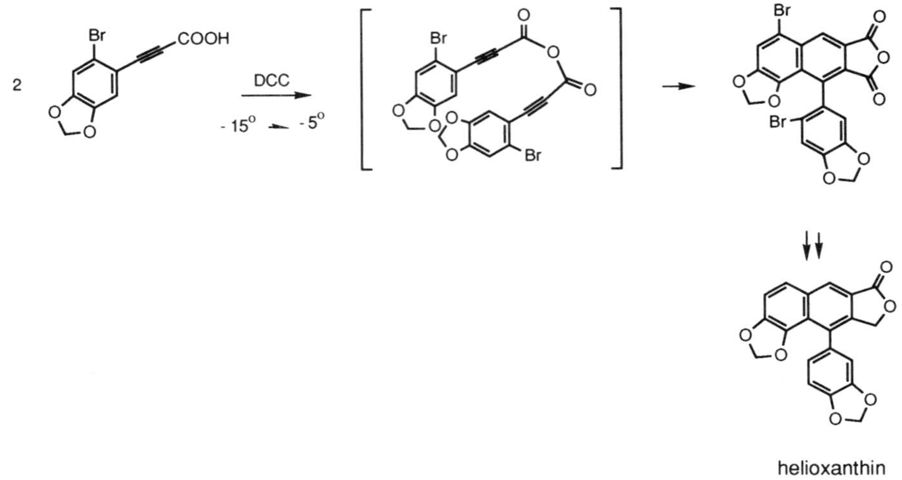

helioxanthin

Jasmine ketolactone contains a 10-membered lactone fused to a cyclopentanone unit. While the cyclization of the medium ring is traditionally a difficult reaction, a better prospect for the synthesis of jasmine ketolactone is to conduct an intramolecular cyclization by a transannular Michael reaction of a 13-membered ring precursor [Shimizu, 1992]. With a 1,5-dicarbonyl disposition and a *trans*-ring juncture the target molecule is eminently suitable for elaboration by the Michael reaction.

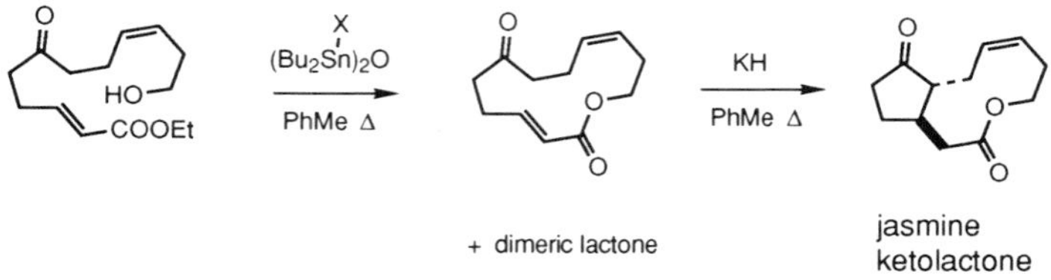

+ dimeric lactone

jasmine ketolactone

An amino group on a benzene ring when enveloped by the long bridge of an *m*-cyclophane is too hindered to undergo intermolecular alkylation. However, triansa compounds can be made [Schill, 1967] by an intramolecular reaction in which the two alkylating agents are joined to common carbon atoms in the form of a ketal. The ingenious exploitation of this chain anchoring tactic developed into a successful approach to catenanes [Schill, 1972].

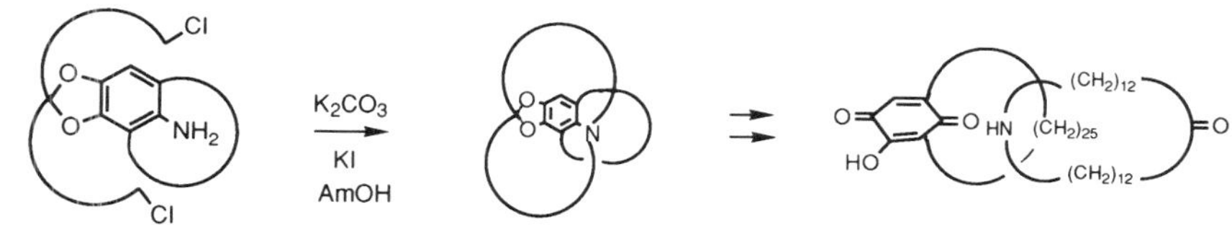

K_2CO_3 / KI / AmOH

The acid-sensitivity of a precursor of (−)-bertyadionol gave much trouble in the recovery of a ketone from the dithiane [A.B. Smith, 1986]. However, by oxidation of the latter moiety to a monoxide and treatment with acetic anhydride and triethylamine in aqueous tetrahydrofuran, a carbocationic intermediate could be trapped intramolecularly in a rate faster than the cuprit reaction of cyclopropane opening.

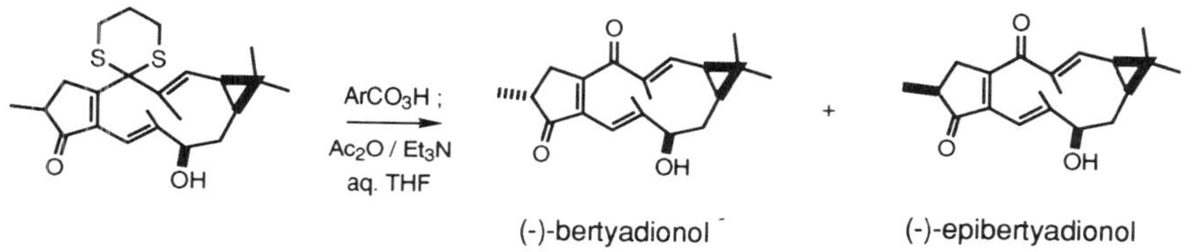

$ArCO_3H$; Ac_2O / Et_3N aq. THF

(-)-bertyadionol (-)-epibertyadionol

6.1.2. Stereocontrol

Control of double bond geometry in olefination reactions is not always easy, particularly when the immediate steric environments on both sides of the carbonyl group are the same. Pertaining to bryostatin synthesis it is obviously difficult to introduce stereoselectively the unsaturated ester pendants by a Horner–Emmons condensation with the 4-tetrahydropyranone units, but a tactic based on neighboring group participation provides a neat solution to the problem [Evans, 1990a]. Accordingly, an intramolecular condensation using a tether to control the stereochemistry of the product fulfills the need, the tether length being determined by consideration of its overall thermodynamic stability, minimal transannular interactions in the transition state, and stereoelectronic effects.

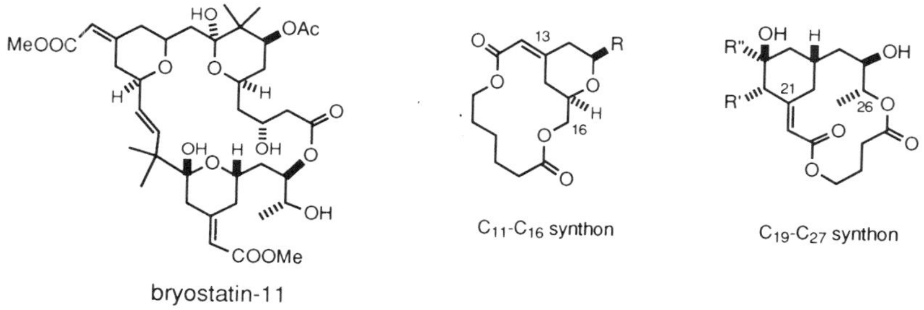

bryostatin-11 C_{11}-C_{16} synthon C_{19}-C_{27} synthon

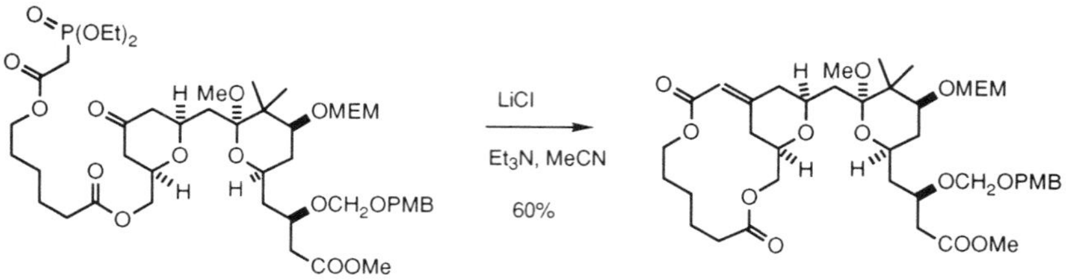

LiCl

Et$_3$N, MeCN

60%

The same is true in the case of calicheamicinone synthesis which must consider the stereochemistry of the trithioethylidene group in its introduction. A fine solution to this problem [Cabal, 1990] made use of an intramolecular Horner–Emmons condensation by anchoring the phosphonate chain to the secondary hydroxyl group.

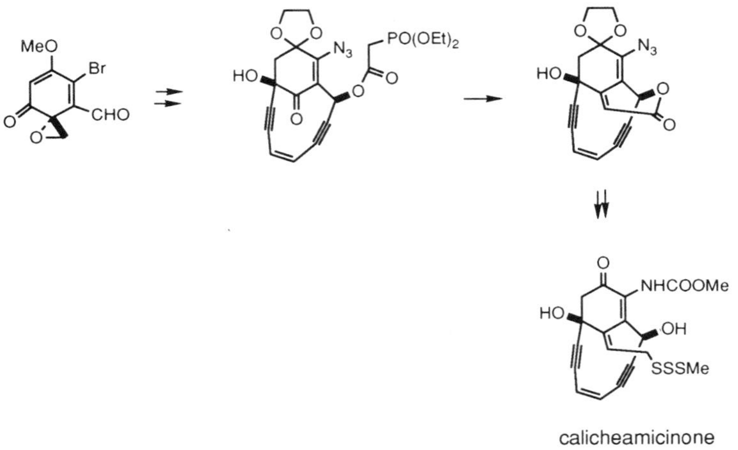

calicheamicinone

Formal hydration of a double bond to introduce an asymmetric center to the AB-ring subunit of daunomycinone has been accomplished via bromolactonization of a derivative containing a proline auxiliary [Terashima, 1978]. The intramolecularity ensured induction of absolute stereochemistry at the prochiral site.

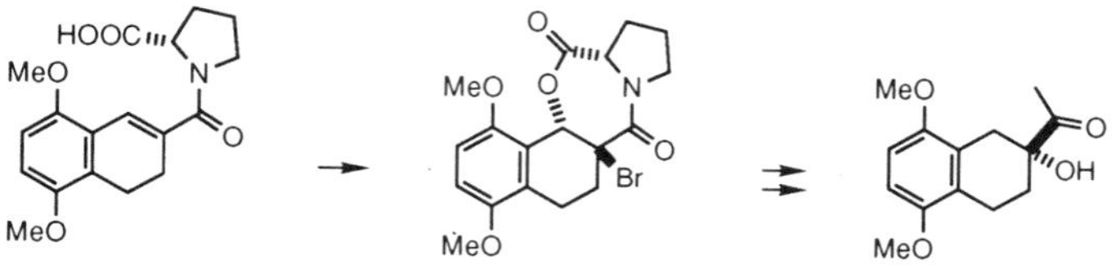

Stereoselective reduction or functionalization of the double bond of a homoallylic alcohol is feasible by hyrosilylation [Tamao, 1986; Hale, 1992]. Specifically, a silane is first linked to the oxygen atom to render the subsequent reaction intramolecular. The different selectivities achieved on a hydroxy diene (derivatives) is notable.

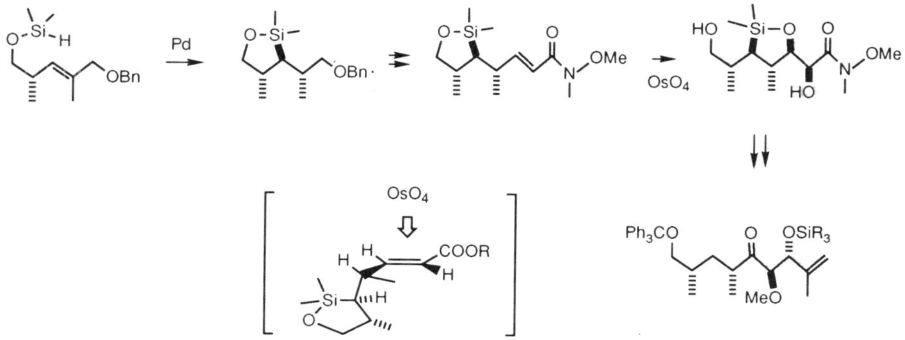

C-27 to C-33 segment
rapamycin

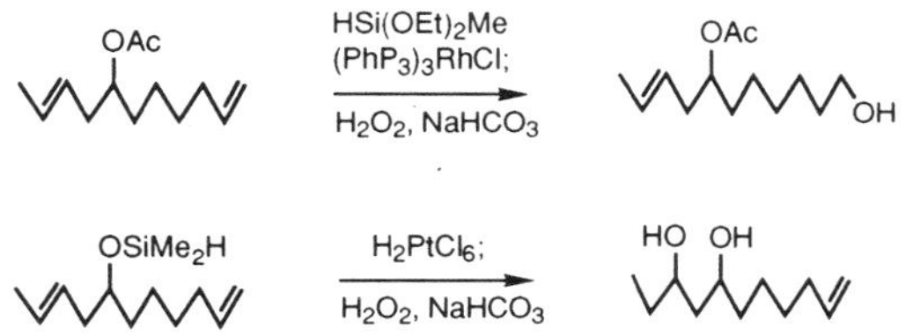

The similar hydrosilylation of allylamines which forms 1-aza-2-silacyclobutanes [Tamao, 1990] provides a method for stereoselective synthesis of 2-amino alcohols.

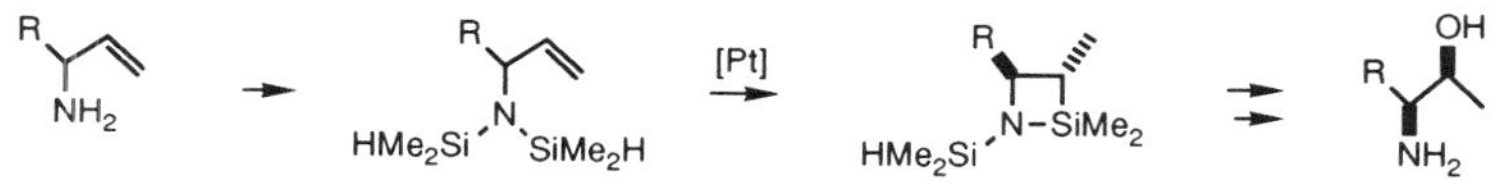

When the homoallylic alcohol is converted into a disilanyl derivative, 1,2-disilylation of the double bond occurs in the presence of a proper catalyst [Murakami, 1993]. Note that intermolecular disilylation of alkenes has not been successful.

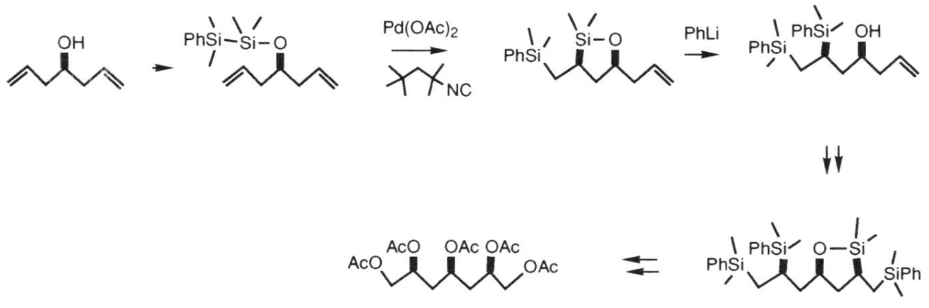

The hydroboration of double bonds separated by one or two skeletal atoms often shows stereoselectivity due to the intramolecular nature of the second stage hydroboration. Three reports on the synthesis of Prelog–Djerassi lactone [Morgans, 1981; Still, 1981; Yokoyama, 1991] described such a phenomenon. In the hydroboration of 1,4-diene, three contiguous asymmetric centers were elaborated. The stereoselectivity arose from avoidance of a transition state possessing severe $A^{1,3}$-strain.

$ThexBH_2$

NaCN
$(CF_3CO)_2O$;
NaOOH

Prelog-Djerassi
lactone

Hydroboration of dialkenyl carbinol derivatives using thexylborane proceeds stereoselectively to give mainly the *syn-anti* products [Harada, 1990]. These results are opposite to those of simple symmetrical dienes in which stereocontrol is exerted by equatorial orientation of substituents in the transition states.

A related problem concerns the elaboration of the vitamin-E sidechain in which the two chirality centers are separated by a propylene group, and it would be most expeditious to create these stereocenters by 1,5-asymmetric induction. The twofold hydroboration with a bulky monoalkylborane meets this challenge in that the transition state for the second, intramolecular reaction should bear semblance to the low-energy conformation of the mesocycle. A preference for the *meso*-diol formation from 2,6-dimethyl-1,6-heptadiene would be predicted. A *meso*:*dl* ratio of 15:1 was realized [Still, 1980].

An intramolecular Michael reaction constituted the critical step in a route to the AB-ring moiety of the taxane diterpenes [Nagaoka, 1984]. In this manner the stereochemistry of the quaternary centers between the B/C-ring juncture was established. This interesting approach also involved unraveling the bicyclo[5.3.1]undecane system from a tricyclic precursor.

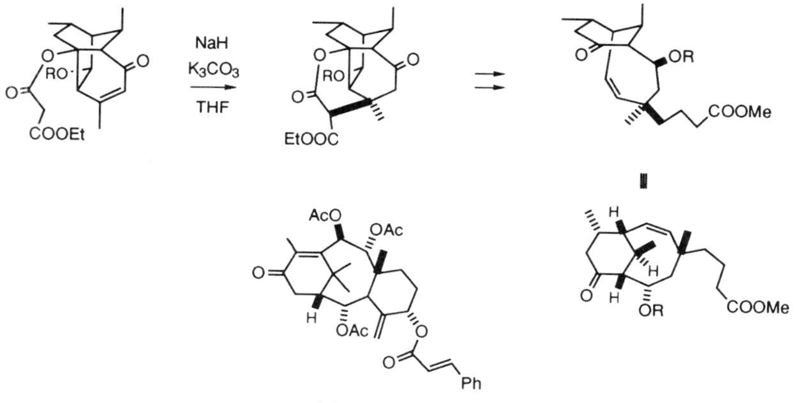

taxinine

Hydroformylation of alkenes with a dirhodium complex [Broussard, 1993] has been found to proceed much faster and give higher linear-to-branched aldehyde ratios than with the $Rh(acac)(CO)_2/0.82\ Ph_3P$ system. The last step being hydride transfer is more favorable in the case of intramolecular reaction. Excess alkene inhibits the reaction, presumably due to acylation of both metal centers.

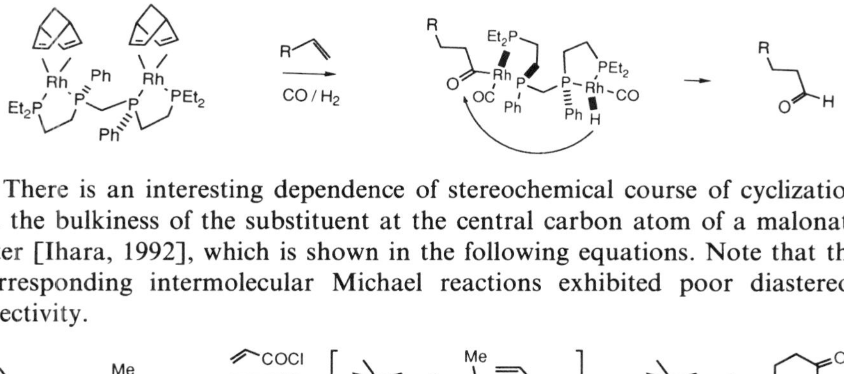

There is an interesting dependence of stereochemical course of cyclization on the bulkiness of the substituent at the central carbon atom of a malonate ester [Ihara, 1992], which is shown in the following equations. Note that the corresponding intermolecular Michael reactions exhibited poor diastereoselectivity.

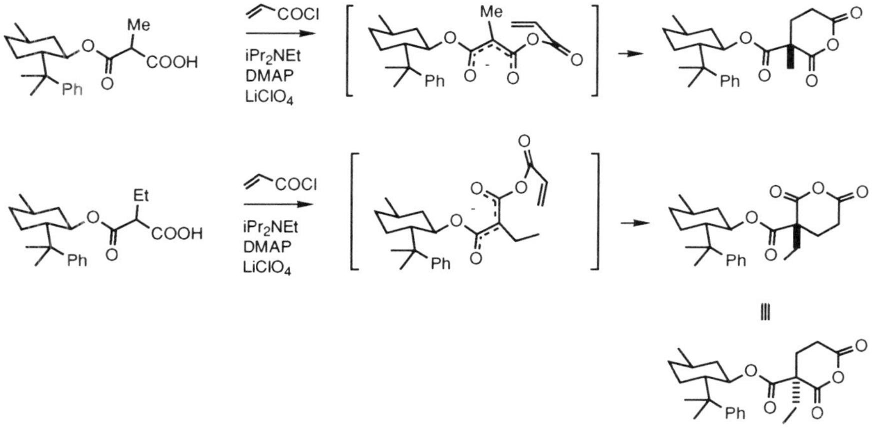

By means of an intramolecular addition by an incipient imide ion of an imino ester to a conjugated nitrile, the bridgehead nitrogen atom of a bicyclic intermediate for a potential synthesis of tetrodotoxin was installed [Alonso, 1993]. Similar stereocontrolled N-C bond formation can also be induced in unactivated double bonds, such as that involved in a synthesis of *N*-acetylristosamine [Pauls, 1983].

tetrodotoxin

O^1-methyl-
N-acetylresistosamine

A similar tactic proved to be critical for controlling the functionalization of the off-pyranose stereocenters during a synthesis of lincomycin [Knapp, 1990]. In fact, the epoxide precursor was prepared with the aid of neighboring group participation.

lincosamine
hexaacetate

A double Michael reaction between 3,5-dimethyl-4-methylene-2-cyclohexenone and dimethyl acetonedicarboxylate was the crucial step in a synthesis of occidentalol [Irie, 1978; Mizuno, 1980] because it led to a *cis*-decalindione in one step. The *cis* ring fusion resulted from steric constraint experienced in the cyclization step (intramolecular reaction). Although this ring formation process consists of two steps, the first one being intermolecular, its inclusion in the present discussion serves the purpose of emphasizing stereocontrol by intramolecularization.

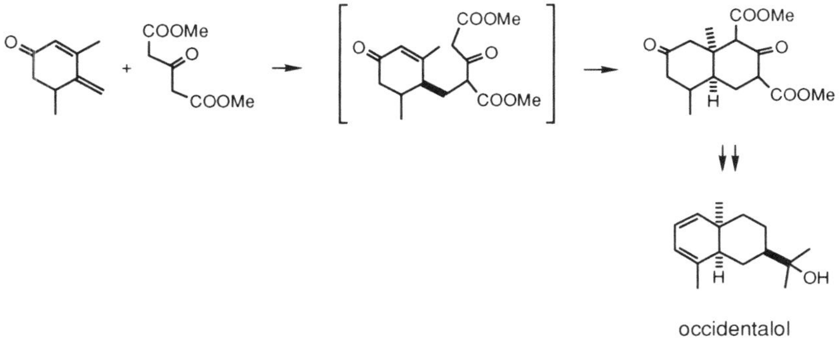

occidentalol

Annexation of both the lactone and A-ring to a tricarbocyclic intermediate to complete a synthesis of gibberellin-A_1 [Lombardo, 1980] called upon intramolecular Michael and aldol reactions. The significant feature of this approach is that the tricarbyclic intermediate has a strong steric bias so that attack on the ketone group by external reagents from the same side as the angular hydrogen is preferred. As a result the oxygen atom is forced to the α-face, and when it is esterified with a propanoyl group a subsequent intramolecular Michael reaction is sterically defined. Moreover, an aldehyde fashioned from the carbon chain previously introduced in the nucleophilic attack can now participate in an aldol reaction. While simultaneously creating two asymmetric centers in this latter condensation the products can have only one possible configuration at C-4.

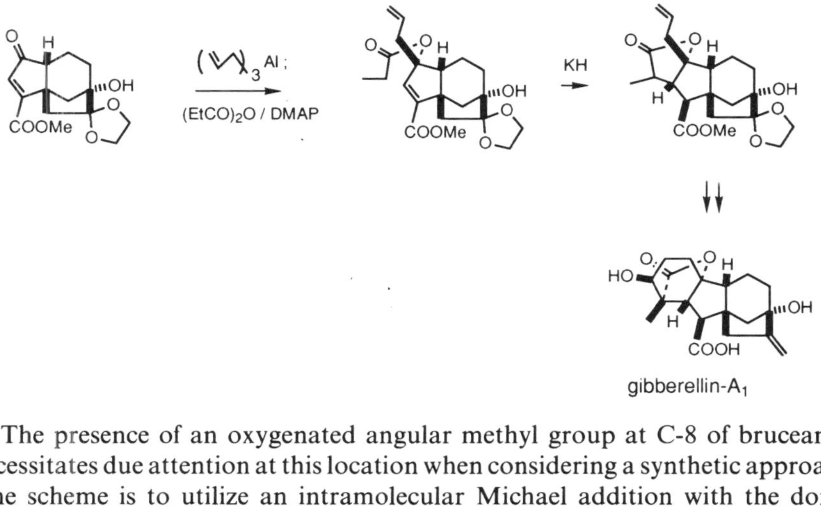

gibberellin-A_1

The presence of an oxygenated angular methyl group at C-8 of bruceantin necessitates due attention at this location when considering a synthetic approach. One scheme is to utilize an intramolecular Michael addition with the donor group attaching to the other angular substituent (between A/B rings). A model study has been carried out [Hedstrand, 1987].

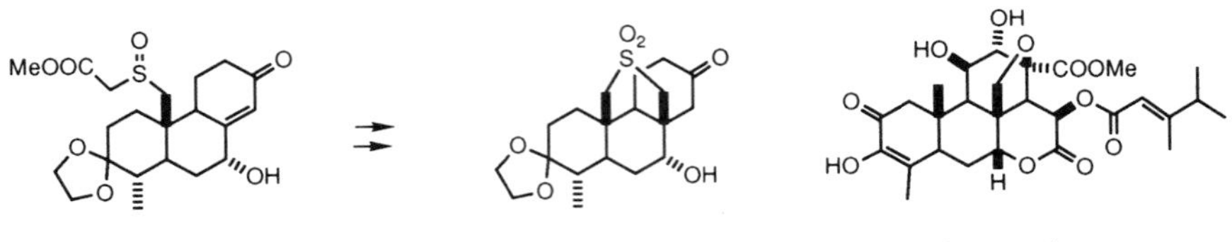

bruceantin

The generation of *cis*-fused ring systems from intramolecular alkylation is a well-known event. A recent example is related to a synthetic study on *cis*-clerodanes [Piers, 1991].

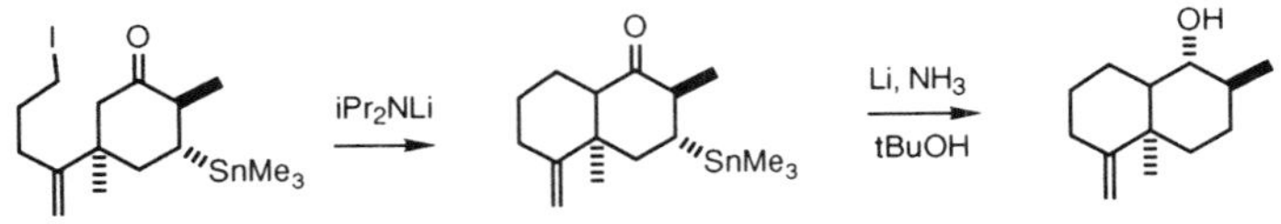

The central carbocycle of strychnine is the most highly concentrated in the content of asymmetric centers. In a synthetic perspective its bridging to the piperidine moiety is a fortunate feature because these two directly related centers necessarily arise in an unambiguous manner. Taking advantage of this arrangement a route passing through a seco compound has been devised [Magnus, 1992a], and the synthetic step involved correct establishment of two new asymmetric centers via a transannular Mannich reaction. Actually the cyclization was so facile that the seco compound, on purification by chromatography, was partially converted into the product, presumably via air oxidation.

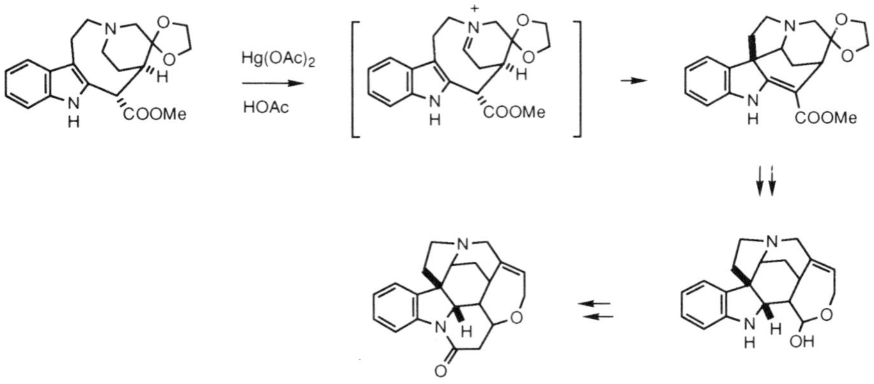

strychnine

Intramolecular Diels–Alder reactions often violate the Alder *endo* rule, consequently they can be exploited accordingly. For example, the following observation [Roush, 1982] formed the basis for a synthesis of the ionophore antibiotic X-14547A. Such stereochemical course is probably the result of an unsymmetrical transition state, in which a more advanced bond formation condition between the β-carbon of the dienophile and the internal terminus of the diene is established, and anticipates the thermodynamic stability of the developing five-membered ring. Furthermore, the alternative *syn* transition state suffers from steric interaction between the allylic hydrogen of the dienophile and a vinylic hydrogen of the diene. With a chiral auxiliary attaching to the dienophilic portion such a reaction generates a product of excellent enantioselectivity [Evans, 1984b; Oppolzer, 1985].

anti T.S. *syn* T.S.

MeOOC MenOAlCl$_2$ 23°C MeOOC H 72% MeOOC H H 0%

Both stereochemical and structural features demanded by selina-3,7(11)-diene are fulfilled by the intramolecular Diels–Alder approach [S.R. Wilson, 1978a]. Similarly, a synthesis of patchouli alcohol [Näf, 1981] reaped full benefit from this mode of elaboration.

OSiR$_3$ 200°C OSiR$_3$

selina-3,7(11)-diene

OH 280°C HO H$_2$ HO

patchouli alcohol

The steric course of the Diels–Alder reaction for the construction of the A-ring (C-2/C-3 and C-4/C-5 bond formation) of gibberellic acid [Corey, 1978a,b] is not readily discernible by casual inspection of molecular models. Consequently, the establishment of the C-5 configuration, which requires an acrylic ester to approach the diene from the α side, can be guaranteed only when an intramolecular reaction is employed. Of course the pendant destined to become the carboxyl group at C-6 must be epimeric (but equilibratable at a later stage), and when this pendant is in the form of a carbinol, its derivatization into an acrylic ester fulfills the need. Furthermore, the lactone product would be forced to undergo methylation stereoselectively in the desired sense.

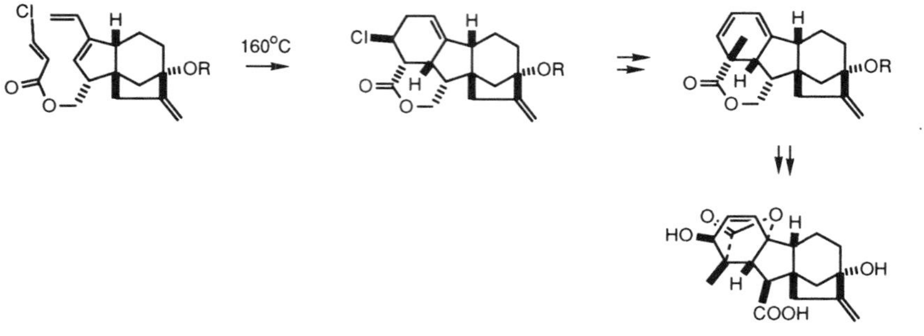

gibberellic acid

An intramolecular Diels–Alder reaction through linkage to the C-1 hydroxyl group was an excellent choice for establishing the stereochemistry of the quaternary angular carbon atoms of forskolin [Ziegler, 1985, 1987]. The same tactic was also employed in a synthesis of myrocin-C [Chu-Moyer, 1992] in which the quaternary carbon atom bearing the methyl and vinyl groups was created. The stereochemical consequence was in accordance with expectation as the dienophile attaching to C-7 of the decalin system was delivered from the α-side.

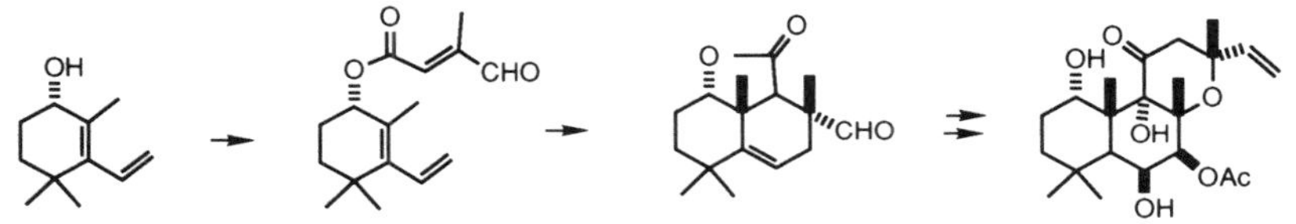

forskolin

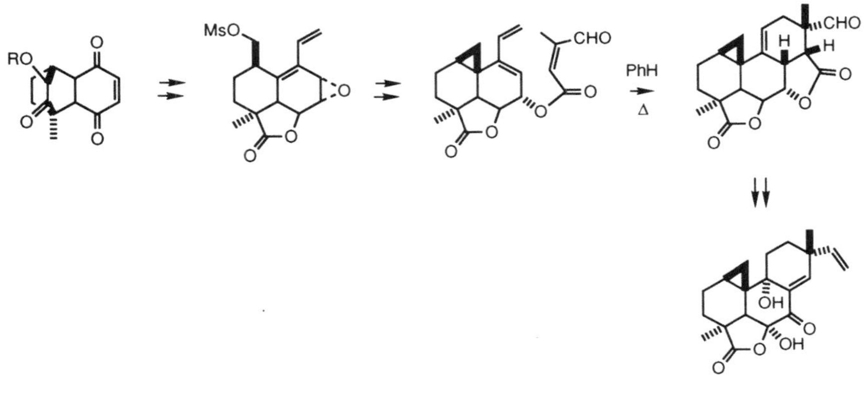

myrocin-C

A facile synthesis of sedridine [Uyehara, 1990] involved desilylative alkoxycarbonylation of *N*-trimethylsilyl-1-aza-1,3-butadiene and intramolecular

Diels–Alder reaction to provide a key intermediate. An exo transition state was apparently adopted.

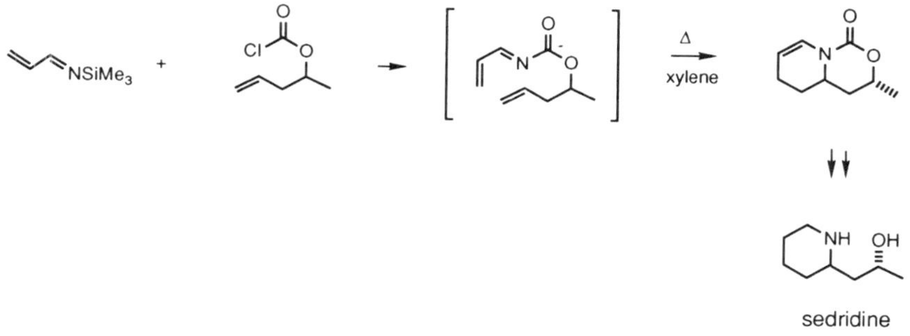

sedridine

For the synthesis of naphthyridinomycin by cycloddition of acrolein to a dipolar pyrazine-2.6-dione, a model study [Garner, 1989] showed that the problem of diastereoselection could be solved by temporary anchoring of the dipolarophile to the phenethyl hydroxyl group.

A synthesis of reserpine [Pearlman, 1979] enlisted an intramolecular [2 + 2]-photocycloaddition, thereby fixing configurations of the two asymmetric centers destined to become the C/D ring juncture atoms.

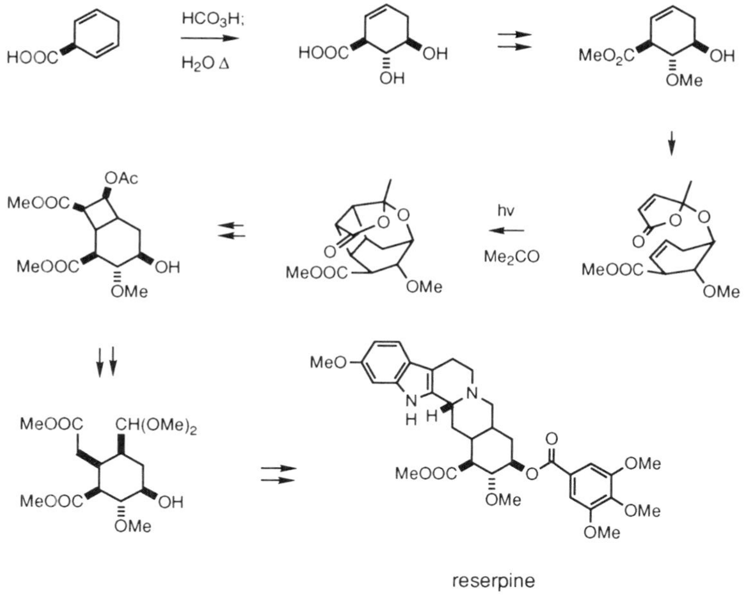

reserpine

An analogous tactic was employed to create three rings of the ginkgolides [Crimmins, 1989b] in a model study. These rings constitute the core of the complex terpenoids, and since they are spiro-related, stereocontrol during their establishment is of paramount importance. The effectiveness of this approach is evident. A total synthesis of the less complex congener, bilobalide, has been accomplished [Crimmins, 1992].

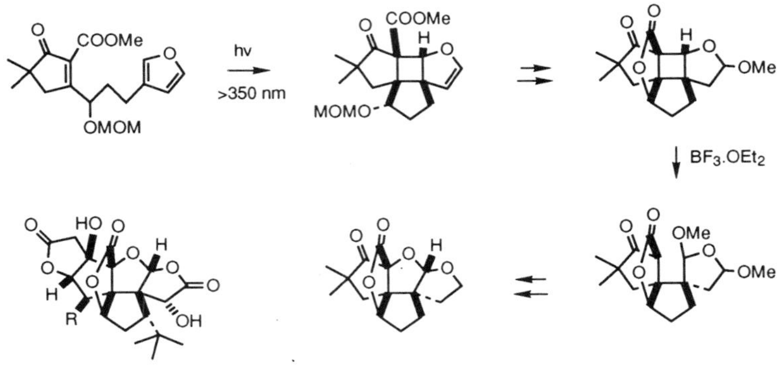

R = H ginkgolide-A
R = OH ginkgolide-B

Kainic acid is a proline derivative in which C-3 and C-4 are substituted and both substituents are *trans* to the carboxyl group at C-2 of the pyrrolidine ring. By means of an intramolecular [1,3]-dipolar cycloaddition involving an azomethine ylide to assemble the skeleton, the two nonepimerizable substituents can be installed in a (*Z*)-alkene precursor [Takano, 1988a]. The configuration of the secondary carboxyl group, being *trans* to the other sidechains, is more stable and equilibratable. The intramolecular cycloaddition ensures regiochemistry of the adduct.

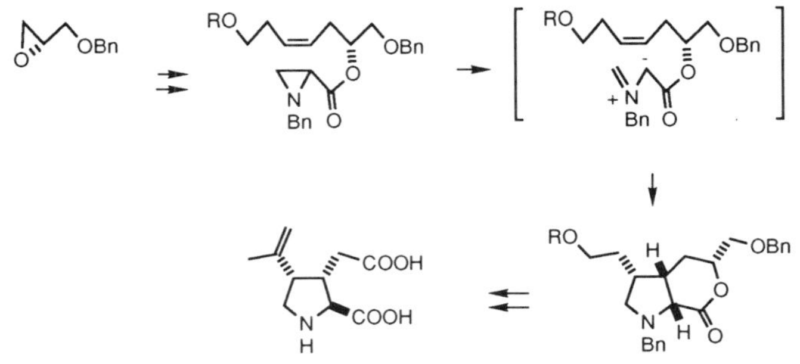

kainic acid

Reaction of allylic alcohols with α-methoxycarbonylnitrones in the presence of titanium isopropoxide proceeds via transesterification, (*E*,*Z*)-isomerization, and cycloaddition [O. Tamura, 1993].

Oxidative addition of an allylic epoxide to carbonyliron(O) species leads to the internal π-complex of allyloxycarbonyliron, the aminolysis of which occurs with transposition. An α-allyl-β-lactam is formed on Ce(IV) demetallation of the latter complex. This reaction sequence is amenable to application to a synthesis of thienamycin [Hodgson, 1985]. The required cis-2,3-disubstituted β-lactam readily arises from attack of the Lewis acid-activated complex by the amine to disengage the ester moiety from the opposite face which is followed by *N*-acylation to form a new complex, and subsequent oxidative demetallation. The intramolecular *C*-acylation in the last step apparently proceeds rapidly so that the stereochemistry of the product is predetermined upon initial complex formation and the introduction of the amino group.

OMe OMe Fe(CO)5 hv Fe(CO)3 CH(OMe)2 ZnCl2 BnNH2 (OC)3Fe NBn CH(OMe)2 CAN

HO H NH3 S COO⁻ CH(OMe)2 NBn

thienamycin

A problem existing in the synthesis of kaurene [R.A. Bell, 1966] from a hydrophenanthrene intermediate was the stereoselective introduction of a proper sidechain at C-8. A nice solution is by Claisen rearrangement, the steric course of which determined by the relative configuration of the allylic hydroxyl. Thus application of an intramolecular tactic removed any stereochemical uncertainty concerning reaction at a hindered position and the potential trouble of obtaining a mixture of products.

OH OMe OH OEt Hg(OAc)2 Δ CHO OH O

kaurene

The spirocyclic center of bakkenolide-A is not likely to be established directly by alkylation of a precursor containing a carboxylic acid/ester sidechain because the steric course would lead to an epimer. Accordingly, the problem was resolved

[Evans, 1973] by reversing the order of events in the introduction of the two chains. By employing an intramolecular [2.3]-sigmatropic rearrangement of a sulfur-stabilized carbene a functionalized one-carbon unit was placed from the less hindered *exo* face of the hydrindane skeleton, fulfilling the need for successful elaboration of the sesquiterpene lactone.

Br H S SMe H MeS S H O O H

bakkenolide-A

The β-anomer of a 2′-deoxyuridine derivative can be acquired via an intramolecular Vorbrügen coupling [Jung, 1993], using the 5-hydroxyl group to direct the stereochemical course.

The spirocyclic (+)-hydantocidin was also synthesized by such a tactic [Chemla, 1993].

OH O OMe OSiMe3 N N O O OAc OMe TfOSiMe3 O N N O O OMe NaOH HO O HN N O O OMe

An interesting chirality transfer from center to axis is involved in a synthesis of a naphthylisoquinoline alkaloid (−)-ancistrocladine [Bringmann, 1986]. The transfer occurred in the Heck-type coupling to give a lactone. The product was then converted into the target molecule by reduction, deoxygenation, and *N*-deprotection.

MeO OMe O O Br NBn MeO (Ph3P)2PdCl2 100°C MeO OMe O O NBn MeO MeO OMe HO NBH MeO

(-)-ancistrocladine

6.1.3. Regiocontrol

In 8,14-cedranediol the *endo*-orientation of the hydroxymethyl chain is an exploitable feature for fixing an oxygen function at C-8. In the form of a carboxyl group in a precursor containing a bicyclo[2.2.1]heptene core, it would participate in the ring expansion following the cycloaddition [Landry, 1983].

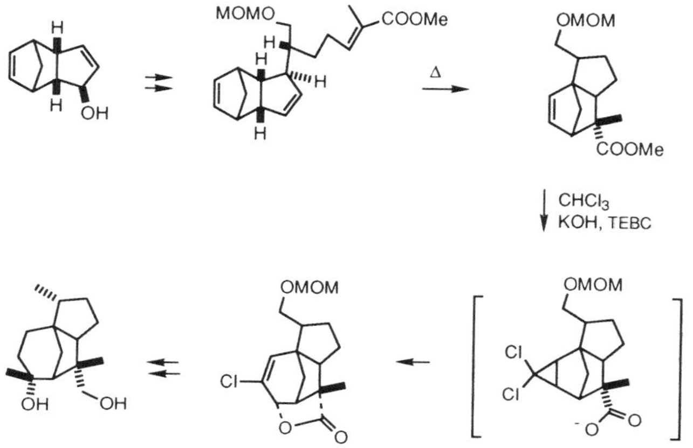

8,14-cedranediol

Chelidonine has been synthesized via intramolecular trapping of an *o*-quinodimethane derivative [Oppolzer, 1983c]. The presence of a secondary hydroxyl group in the hydroaromatic ring required a hereosubstituent at the corresponding position of the dienophilic moiety of the precursor, and a *β*-nitrostyrene served adequately in that capacity. It must be emphasized that the major intermolecular Diels–Alder reaction product is a regioisomer.

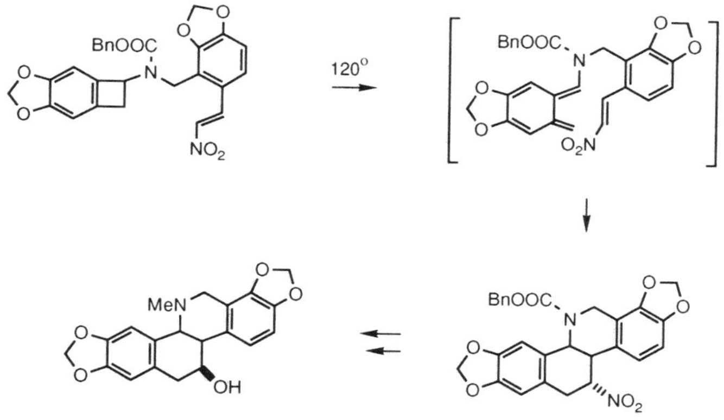

chelidonine

The strategy of synthesizing staurosporine aglycone through intramolecular nitrene insertion at C-1 of an existing carbazole ring system necessitates the acquisition of a 2-(*o*-nitrophenyl)carbazole. An excellent method for the preparation of such a carbazole is by exploiting the intramolecular Diels–Alder reaction [Moody, 1992].

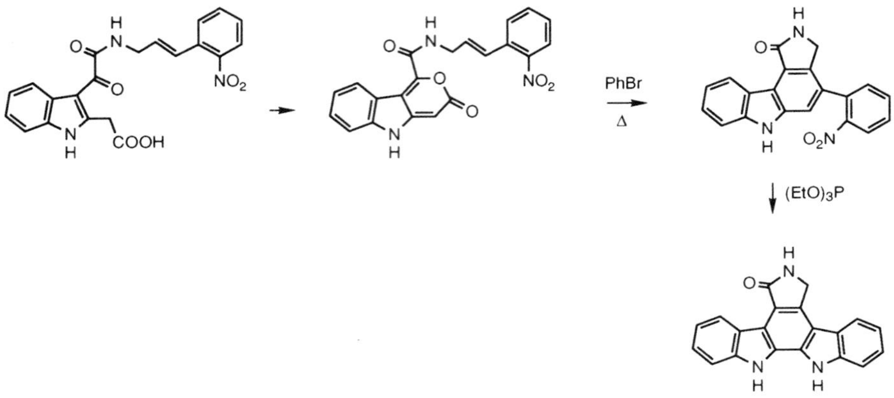

NO2 NO2 PhBr Δ O2N (EtO)3P

staurosporine aglycon

Intramolecular Diels–Alder reactions of substrates in which the diene and dienophile components are tethered by a silacarbon chain have become popular [Gillard, 1991; Shea, 1991; Stork, 1992a] due to the fact that the silicon atom can be converted into other functionalities at a later stage. Furthermore, the length of the tether determines with the substituent orientation of the adduct. Thus, it is possible to reverse the normal regiochemistry using a designed addend [Shea, 1991].

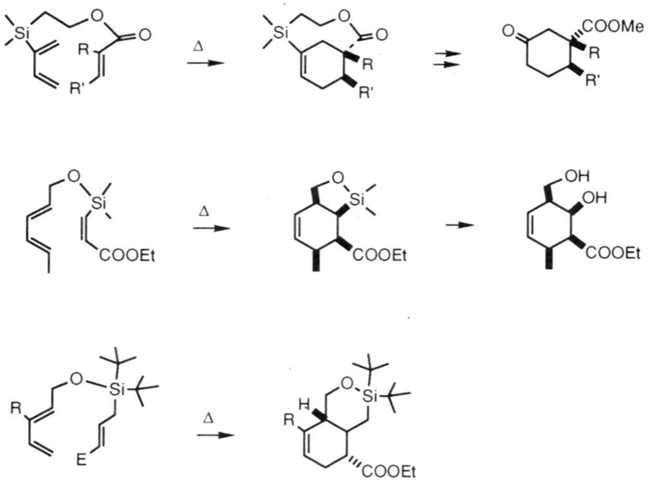

E = COOEt

A more labile but still useful device for joining a diene and a dienophile which bear properly situated hydroxyl groups is a boron atom. In forming a boronate to render the subsequent Diels–Alder reaction intramolecular, orientation of the adduct is defined [Narasaka, 1991].

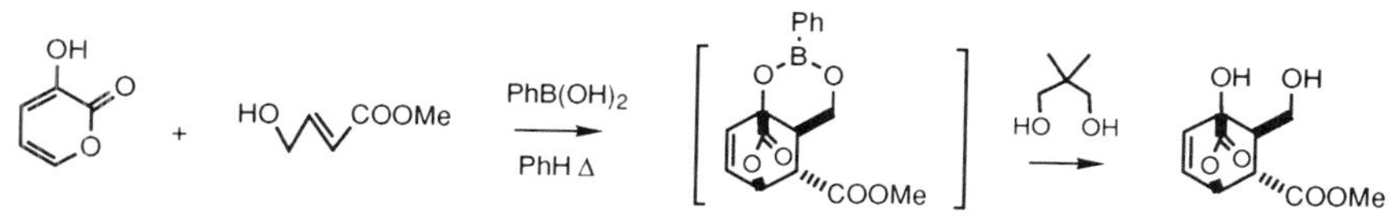

Pyrrolidine formation by [2 + 3]-cycloaddition of alkenes to iminium ylides as applied to the synthesis of eserethole [R. Smith, 1985] in an intramolecular version seems to be against natural arrangement of polar groups. It is a case of intramolecular enforcement of regiochemistry.

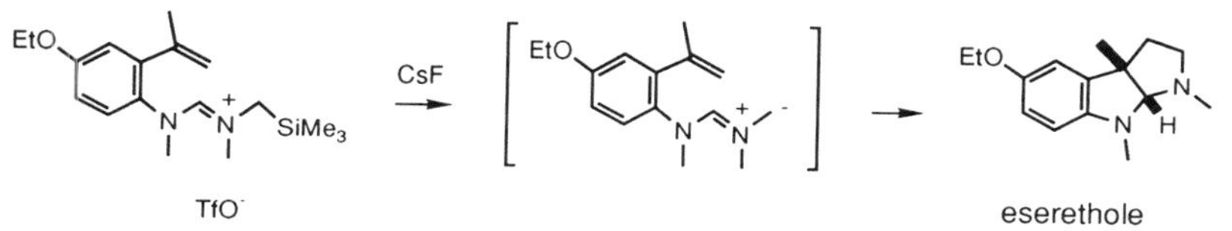

eserethole

Although it is obvious that an expedient route to stoechospermol would involve [2 + 2]-photocycloaddition, regiochemistry would be an issue. Such a control as well as that of absolute stereochemistry can be exercised by using a chiral butenolide derived from (*S*)-glutamic acid to form a photoreactive ester. Clearly, only the useful cycloadduct(s) were obtained [M. Tanaka, 1985].

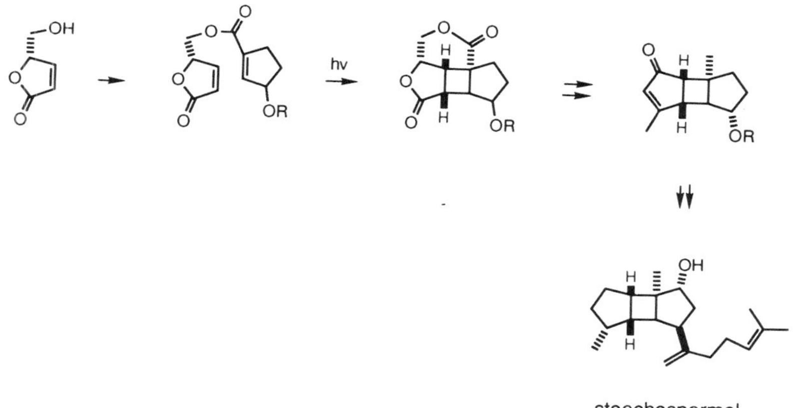

stoechospermol

The success of a synthesis of precapnelladiene [Inouye, 1993] via intramolecular photocycloaddition, ether cleavage, and retro-aldol fission depended on the first step which fixed the substitution pattern of the adduct. Equally important is that its anticipated stereochemistry of the relevant centers corresponds to those of the target molecule.

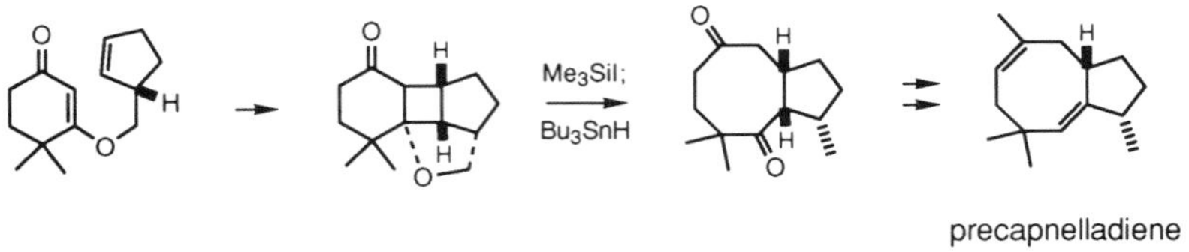

Me$_3$SiI; Bu$_3$SnH

precapnelladiene

Under normal conditions the preparation of the *syn-cis* photodimer is not easy. However, the linkage at the nitrogen atom of two molecules by a removable short tether such as (2*Z*)-buten-1,4-diyl restricts the intramolecular photo-cycloaddition to proceed in the desired manner [Burdi, 1992].

Cinnamoyl groups fixed in a threitol derivative undergo intramolecular photo-cycloaddition to furnish two truxinates in a 2:1 ratio [Green, 1976/1977]. The chiral induction was excellent.

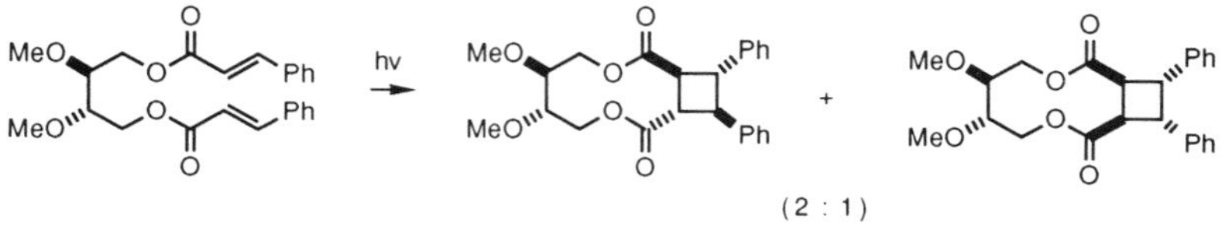

(2 : 1)

As a method for the synthesis of calomelanolactone [Neeson, 1988], the transition metal catalyzed cyclotrimerization of alkynes does not guarantee correct organization of the various substituents in the indane nucleus. However, when two of the *ortho* substituents are linked in the precursor, the desired effect can be achieved.

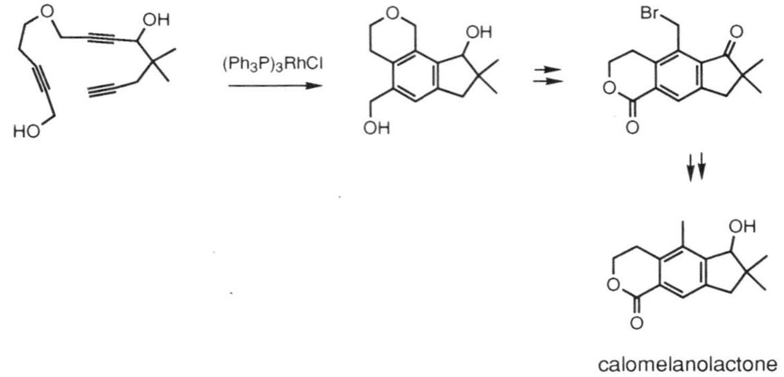

(Ph$_3$P)$_3$RhCl

calomelanolactone

The Dötz annulation is used to form a 1,4-naphthoquinone monoether by formally assembling the alkyne, an aromatic moiety, and carbon monoxide, with an aryl Fischer carbene complex to provide the two latter components. Regioselectivity regarding the incorporation of the alkyne can be achieved in such a reaction only if the substituents of the alkyne are quite different, and that is not expected from one bearing two alkyl groups. Consequently, the application of this reaction to a synthesis of deoxyfrenolicin [Semmelhack, 1985a] must rely on an intramolecular version.

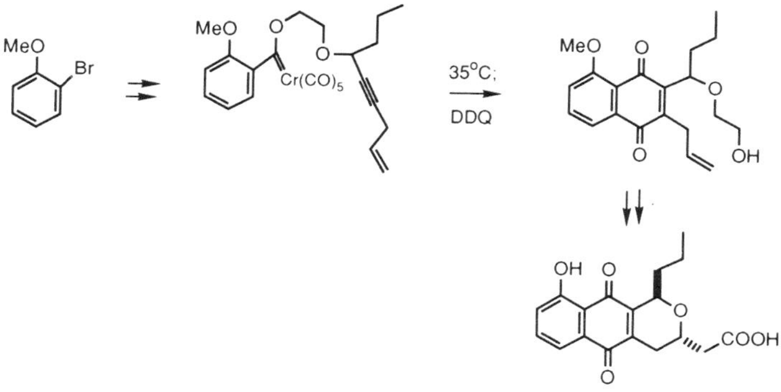

MeO Br → MeO O O Cr(CO)5 → 35°C; DDQ → MeO O O O OH → OH O O COOH

deoxyfrenolicin

Another access route to 1,4-naphthoquinones is via alkyne insertion to phthaloylcobalt species. For an efficient synthesis of nanaomycin-A [South, 1984] utilizing this method regiocontrol must be exercised, and the same intramolecularization tactic served well.

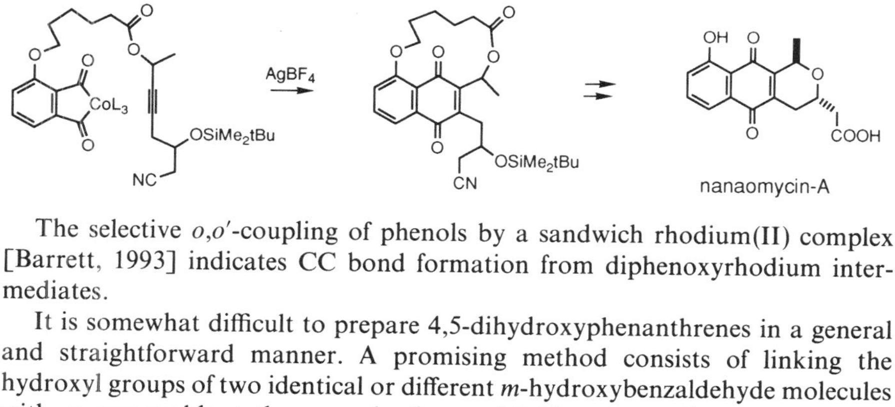

O O O O CoL_3 O $OSiMe_2tBu$ NC → $AgBF_4$ → O O O O O $OSiMe_2tBu$ CN → OH O O O COOH

nanaomycin-A

The selective *o,o'*-coupling of phenols by a sandwich rhodium(II) complex [Barrett, 1993] indicates CC bond formation from diphenoxyrhodium intermediates.

It is somewhat difficult to prepare 4,5-dihydroxyphenanthrenes in a general and straightforward manner. A promising method consists of linking the hydroxyl groups of two identical or different *m*-hydroxybenzaldehyde molecules with a removable tether, conducting a McMurry coupling, and then a photochemical electrocyclization under oxidative conditions [Ben, 1985].

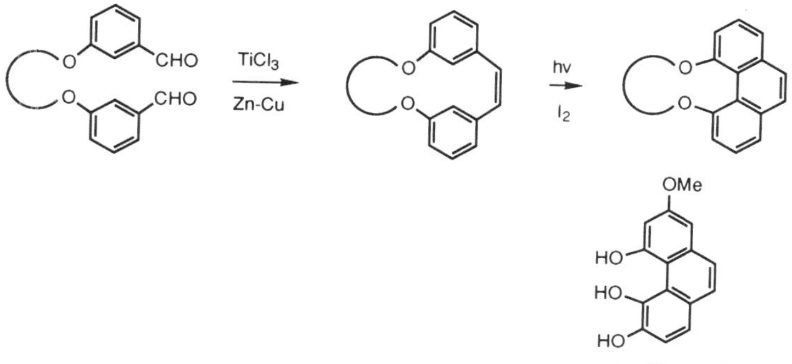

O CHO O CHO → $TiCl_3$ Zn-Cu → O O → hν I_2 → O O

OMe HO HO HO

cannithrene-II

Biaryl synthesis also benefits from intramolecularization. For example, by anchoring two molecules of an *o*-haloaroic acid to the C-2 and C-2′ of chiral 1,1′-binaphthyl and completing an Ullmann coupling, one diasteromer is produced [Miyano, 1986, 1988]. Another method is by using *o*-hydroxybenzyl alcohol to esterify the aroic acid so that the coupling product can be cleaved selectively by transacylating techniques [M. Takahashi, 1992]. Consequently, the two carboxylic acid groups can be differentiated.

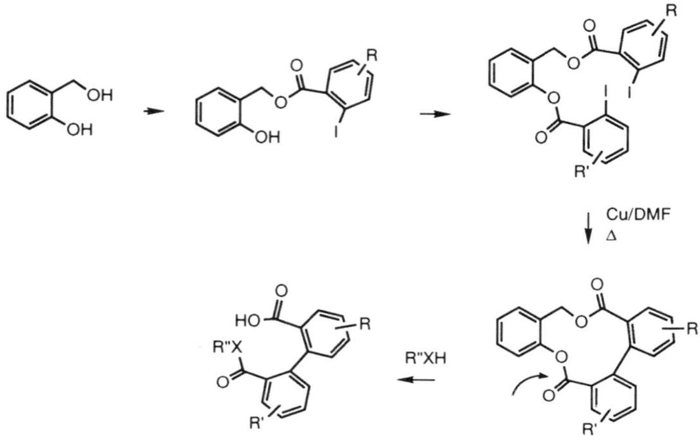
OH
OH
O
O
R
OH
I
Cu/DMF
Δ
R"XH
HO
R"X
R'

The elaboration of more complex biphenyls such as that constituting a portion of vancomycin by coupling methods must resort to intramolecularization [Rama Rao, 1992]. In this case an ester was submitted to a palladium-catalyzed C—C bond formation.

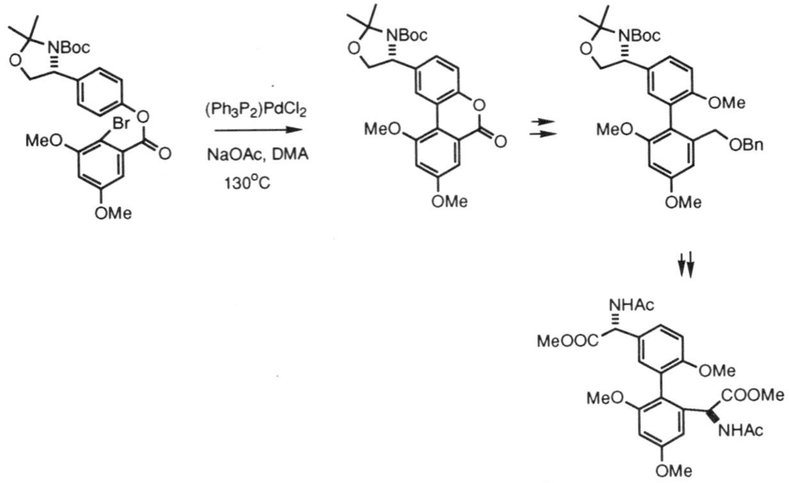
NBoc
Br
MeO
OMe
(Ph3P2)PdCl2
NaOAc, DMA
130°C
OBn
NHAc
MeOOC
COOMe

Using a heteroatom to link up two structural units and facilitate a crucial C—C bond formation constitutes an excellent synthetic tactic. An examination of the synthesis of dihydrojasmone [Hendrickson, 1985] and cuparene [Ishibashi, 1988] readily demonstrates this point.

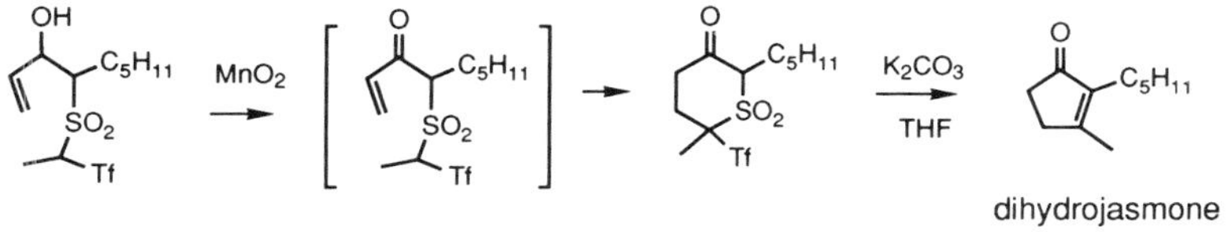

OH C5H11 SO2 Tf MnO2 O C5H11 SO2 Tf O C5H11 SO2 Tf K2CO3 THF O C5H11

dihydrojasmone

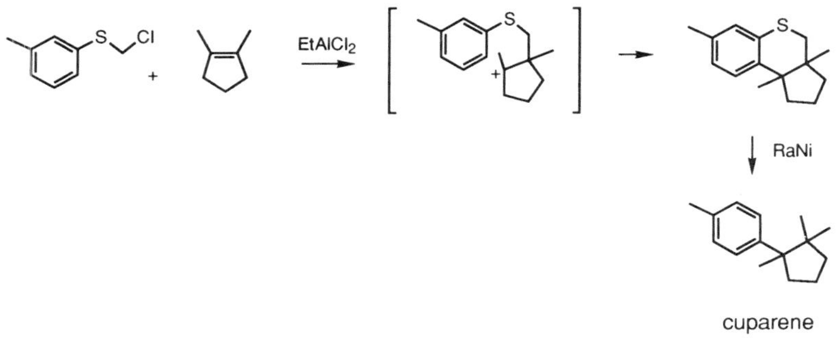

S Cl + EtAlCl2 S + S RaNi

cuparene

6.2. NEIGHBORING GROUP PARTICIPATIONS

Neighboring group participation [Capon, 1976] accounts for regioselectivity and stereospecificity of many reactions. It encompasses all intramolecular reactions and all reactions involving nonelectrostatic through-space interactions between groups in the same molecule. Most of the participating groups are nucleophiles, including n- and π-donors. If the participation leads to an enhanced reaction rate, the group is said to provide anchimeric assistance.

Sometimes the reactivity of a substrate is modified by a neighboring functionality. A case in point is the α-hydroxyl group of a tricyclic aldehyde, whose essential role is to enable the occurrence of an intramolecular ene reaction for the construction of a bridged ring system *en route* to 2-deoxystemodinone [White, 1987]. The corresponding deoxy analog fails to afford any ene product either thermally or in the presence of a Lewis acid. The exact role of the hydroxyl group is not clear.

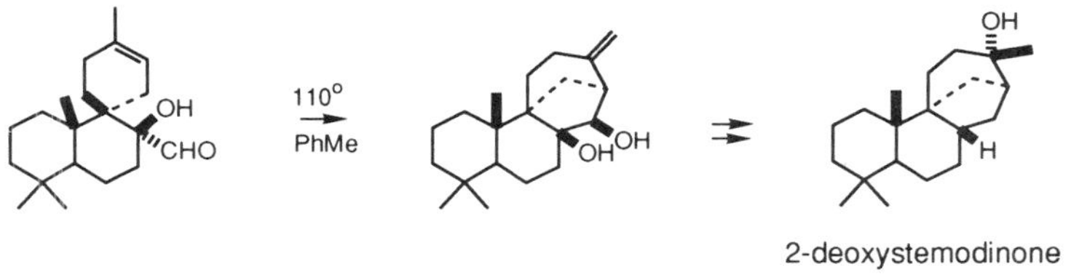

OH CHO 110° PhMe OH OH OH H

2-deoxystemodinone

An umpoled activation of a secondary amine at is α-position by neighboring group participation involves *N*-*o*-iodobenzylation and treatment with samarium(II) iodide [Murakami, 1992]. A free radical generated at the α-carbon of the amine via hydrogen abstraction by the initially formed phenyl radical, reacts with the samarium reagent to form a carbon nucleophile.

Sml2
THF
HMPA
RR'C=O

Medium-sized rings are difficult to create by cyclization reactions due to entropy factors and nonbonding interactions. Frequently, the undesirable intermolecular reactions predominate. In light of this general trend it is of interest to note that the skeleton of quebrachamine has been formed in 87% yield by treatment of a C-seco carboxylic with polyphosphoric acid [Ziegler, 1969]. A reasonable mechanism indicates *N*-acylation prior to ring closure. The process describes a relay which has a favorable course because the transacylation involves two atoms separated by five bonds.

PPA
87%

quebrachamine

6.2.1. Chemoselectivity

Allylic and homoallylic alcohols undergo epoxidation selectively by *t*-butyl hydroperoxide in the presence of a metal catalyst such as $VO(acac)_2$ [Sharpless, 1979]. For example, geraniol reacts with this reagent to furnish the 2,3-epoxide, whereas it is preferentially epoxidized by a peracid at the other double bond. Furthermore, high diasteroselectivity is often observed in reactions with the metal system, because they apparently proceed via chelate transition states.

The greatly enhanced chemo- and regio-selectivity of this epoxidation has enabled implementation of a scheme for converting farnesol into dendrolasin [E. Lee, 1982].

farnesol

tBuOOH

VO(acac)2

dendrolasin

An outstanding development is the employment of tetraisopropyl titanate as catalyst in conjunction with a chiral tartaric ester [Finn, 1985, Rossiter, 1985] to replace vanadyl acetylacetonate. Asymmetric inductions in the epoxidation of a wide range of allylic alcohols are extremely high. Furthermore, the ready availability of both (+)- and (−)-tartaric acid esters enables obtention of either enantiomeric epoxide.

Existing chirality at the carbinolic center of an allylic alcohol has a profound influence on the reaction course. As only one of the enantiomers undergoes rapid epoxidation, leaving the other enantiomer practically untouched, an effective procedure for kinetic resolution for such substances becomes available [V.S. Martin, 1981]. Modified catalyst systems have also been found for the epoxidation of allylic alcohols with an opposite enantiofacial selectivity [Lu, 1984].

(S) + tBuOOH → L-(+)-DIPT, $(iPrO)_4Ti$, fast → *erythro* + *threo* (98 : 2)

(R) + tBuOOH → L-(+)-DIPT, $(iPrO)_4Ti$, slow → (38 : 62)

In a synthesis of the *cecropia* juvenile hormone JH-I [Yasuda, 1979] from farnesol the directed epoxidation of the two allylic alcohol systems was crucial because deoxygenation of the bishomo tetraol followed a defined steric course. (*E*,*Z*)-isomers would result even if stereocontrol is not maintained in both reactions.

OAc ArCO$_3$H ; Et$_2$NLi ; Ph$_3$CCl , TsOH OH OH OTr O OH OH OTr O OAc COOMe

JH-I

Methylenation of a carbonyl group with a 1,3-dimolybdacyclobutane species is facilitated by a neighboring hydroxyl group [Kauffmann, 1989]. A similar phenomenon is the acceleration of replacing a vinylic bromine atom with the methyl group by reaction with transition metal reagents [Kauffmann, 1991].

O OH O Cl O Mo Mo O Cl O OH

HO R Br Me$_4$MLi$_2$ Me O-M R Br [M] O R Br HO R

6.2.2. Stereocontrol

The recognition of stereocontrol by a proximal hydrogen donor (e.g. OH) in an alkene during epoxidation with a peroxycarboxylic [Albrecht, 1957; Henbest, 1957b] has had impact on the development of other synthetic methodologies. The *syn*-epoxidation of allylic alcohols can be explained in terms of a hydrogen-bonded transition state which is consistent with a highly negative entropy of activation. As expected from the intramolecular hydrogen bonding, there is very little dependence on solvent and peracid.

R O H O O H O → O H O + RCOOH

Generally, a hydroxyl group in the allylic position is more proficient than the homoallylic analog in directing *syn*-epoxidation. (However, this trend is not universal. For example, *cis*-6-methyl-2,4-cycloheptadienol undergoes epoxidation with peracid at the distal double bond [Pearson, 1992], apparently via a hydrogen-bonded complex in which the ring assumes a crown-like conformation).There are numerous uses of this directed epoxidation in synthesis. An example is the formation of a β-epoxide from (+)-α-cyperone in its transformation into cyperolone [Kikino, 1966]. The ring contraction step via rearrangement of the epoxide which is not directly accessible from the enone.

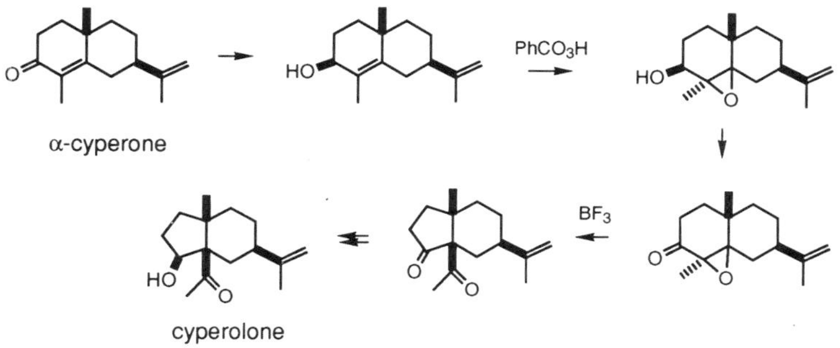

$PhCO_3H$

BF_3

α-cyperone

cyperolone

A molybdenum-catalyzed epoxidation with *t*-butyl hydroperoxide [Chamberlain, 1991] may have involved interaction of both a homoallylic alcohol and an allyic ester with the metal.

Aranorosin has been synthesized [McKillop, 1993] using the 4-hydroxyl group (unfolded in situ from a hemiacetal) of a 2,5-cyclohexadienone to direct the base-catalyzed twofold epoxidation.

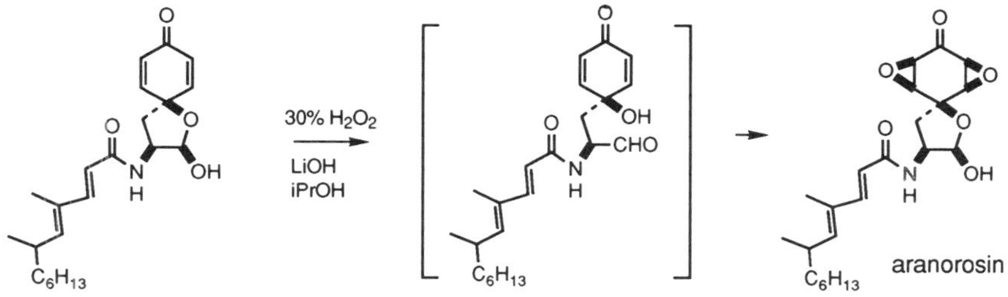

30% H_2O_2
LiOH
iPrOH

aranorosin

Of relevance is the existence of a similar directing influence by an allylic alcohol [Winstein, 1961] in the Simmons–Smith cyclopropanation reaction [Simmons, 1958, 1973], although it must be emphasized that the corresponding ether is also effective (however, there are differences with respect to asymmetric induction by a C_2-symmetric disulfonamide [H. Takahashi, 1992], and cinnamyl methyl ether undergoes cyclopropanation to afford a racemate). Complexation in the Simmons–Smith reaction apparently employs the allylic oxygen atom as a ligand for the zinc species. For 2-cycloalkenol cyclopropanation a pseudoequatorial hydroxyl group has a better orienting effect than a pseudoaxial one [Chan, 1968]. The striking change of stereoselectivity between 2-cycloheptenol and 2-cyclooctenol has been ascribed to the relative accessibility of the π-faces [Poulter, 1969].

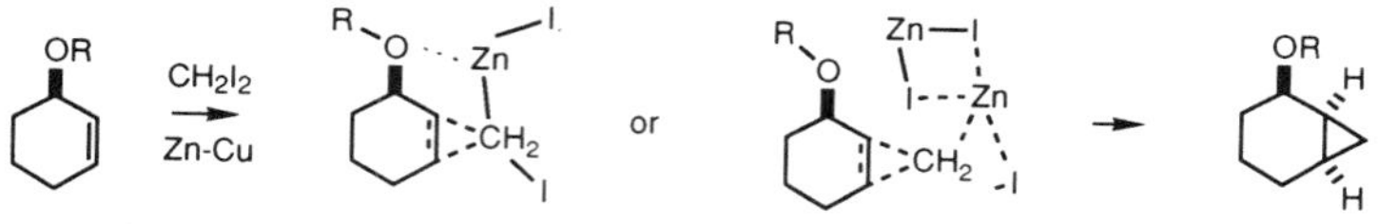

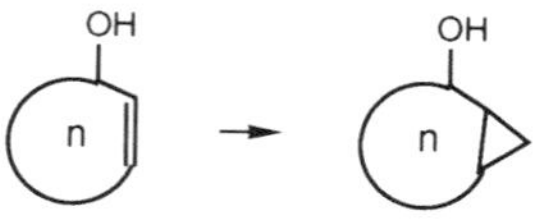

n	epimer distribution syn	anti
5	99	-
6	99	-
7	90	10
8	0.5	99.5
9	0.05	99.95

In the synthesis of thujopsene it is imperative to place the cyclopropane ring *cis* to the angular methyl group. A reaction sequence involving a hydroxyl-directed Simmons–Smith reaction was devised for this purpose [Dauben, 1963], and although the reaction gave the product in a 23% yield, the other diastereomer has not been observed.

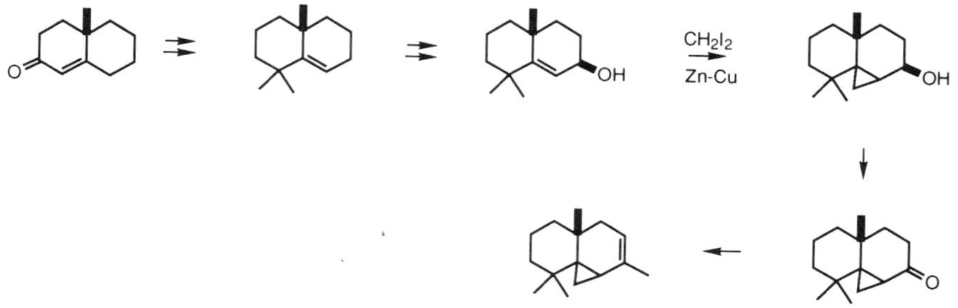

thujopsene

The combination of lactonization and ring expansion constitutes a key operation in a synthesis of confertin [Marshall, 1976]. The reaction involves ionization of a cyclopropylcarbinol with carboxyl group assistance. The formation of a *trans*-fused lactone indicates a concerted process.

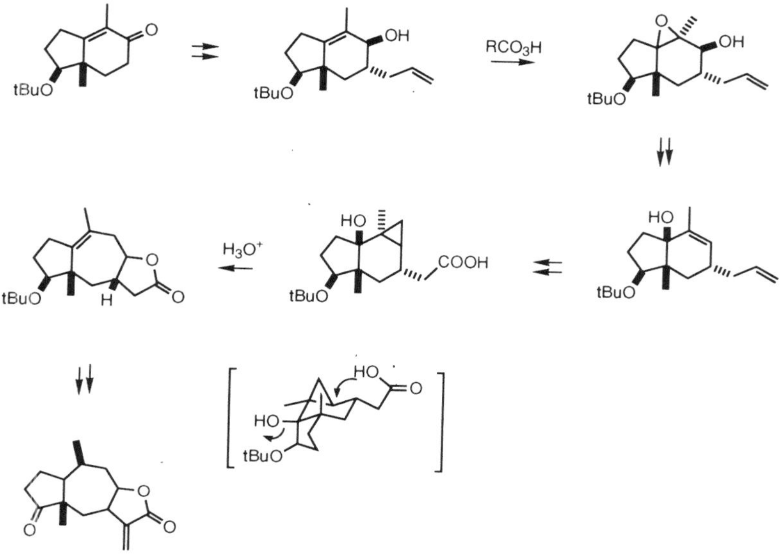

tBuO tBuO OH RCO$_3$H tBuO OH

tBuO H O O H$_3$O$^+$ HO tBuO COOH HO tBuO

HO HO O tBuO

confertin

An allylic alcohol can be cyclopropanated in an enantioselective manner when it is attached to the anomeric centre (as β-anomer) of a sugar derivative containing a free 3α-hydroxyl group which acts as stereoconductor [Charette, 1991]. Both antipodes are available from D-glucose and L-rhamnose, respectively.

Chiral 2-substituted cyclopropanols are available by diastereoselective Simmons–Smith reaction of certain 1-alkenylboronic esters with subsequent oxidation [Imai, 1990]. Chelation of the reagent by the substrates must be responsible for the stereochemical consequences.

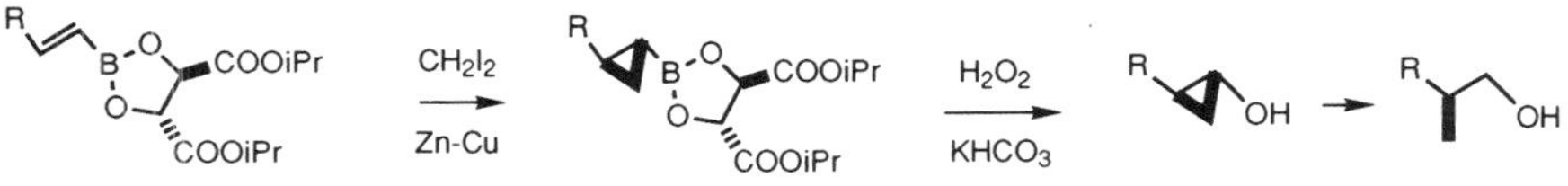

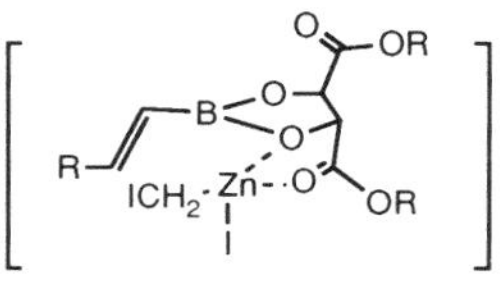

R COOiPr COOiPr CH$_2$I$_2$ Zn-Cu R COOiPr COOiPr H$_2$O$_2$ KHCO$_3$ R OH R OH

O OR B O O R ICH$_2$ Zn O OR I

The generation of a carbon chain with alternating hydroxyl and methyl substitution patterns [Collum 1986] via oxymercuration of cyclopropylcarbinols provides a complementary method to fulfill such need.

HO R $\xrightarrow[LiAlH_4]{(CF_3COO)_2Hg;}$ HO OH R 45 : 1 (inv. : ret.)

HO R $\xrightarrow[LiAlH_4]{(CF_3COO)_2Hg;}$ HO OH R 57 : 1

An interesting template-directed epoxidation is the formation of a 17α,20α-epoxide of a 3α-arylacetoxy steroid on treatment with *t*-butyl hydroperoxide and molybdenum hexacarbonyl [Breslow, 1977b]. Presence of a *t*-carbinol substituent in a particular position of the aromatic ring is crucial. (For more remote functionalization processes, see Section 6.2.3.)

$\xrightarrow[Mo(CO)_6]{tBuOOH}$

The high stereoselectivity (ca. 10:1) for the introduction of the 7α-hydroxyl group in galisubinone-D by solvolysis of the derived bromide [Kende, 1981; Confalone, 1981] is probably due to hydrogen bonding of the attacking water molecule by the existing tertiary hydroxyl group. This behavior is quite different from that of the closely related daunomycinone series from which the *trans*-diols are usually formed in preponderance.

galirubinone-D $\xrightarrow{NBS}$ (bromide) $\xrightarrow{aq. THF}$ [intermediate] → aklavinone

A synthesis of β-acorenol [Iwata, 1985a] is distinguished by stereocontrol of the spirocyclic center using a nascent tertiary hydroxyl group. The cyclic ether was then cleaved in the same step of diene reduction.

β-acorenol

One of the earliest recognitions of neighboring group direction of C—C bond formation is the addition of organometallic species to a proximal multiple bond in certain cyclic systems [Wittig, 1963; Klumpp, 1967; Eisch, 1979].

A method for stereocontrolled functionalization of the triple bond of propargylic alcohols involves formation of oxaaluminacyclopentene intermediates. An ideal case is that which enabled synthesis of a diene intermediate for the *cecropia* juvenile hormone JH-I [Corey, 1971b].

The high stereoselectivity in the transmetallation of a bicyclic tin compound [Newman-Evans, 1985], in contrast to the usual trend in loss of configuration for other cases, is attributable to the effect of the *syn*-hydroxyl group. Under the same conditions the epimer does not under exchange of lithium for tin.

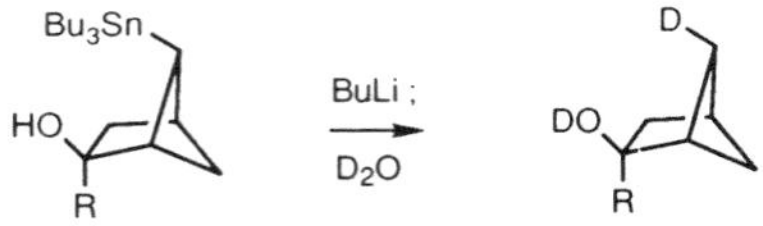
Bu3Sn
HO
R
BuLi ;
D2O
D
DO
R

A hydroxyl group in an alkyl sidechain of an aromatic ring has strong directing effect on the π-complexation of a tricarbonylchromium group [Uemura, 1986, 1987]. The hydroxyl group may situate at thebenzylic position or beyond (at least to three carbon atoms).

Intramolecular carbametallation of enynes mediated by palladium species is subject to control by an allylic substituent [Trost, 1989a]. An allylic alcohol favors the generation of 1,3-dienes.

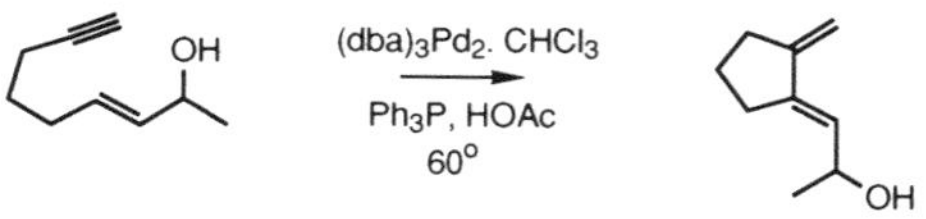
OH
(dba)3Pd2. CHCl3
Ph3P, HOAc
60°
OH

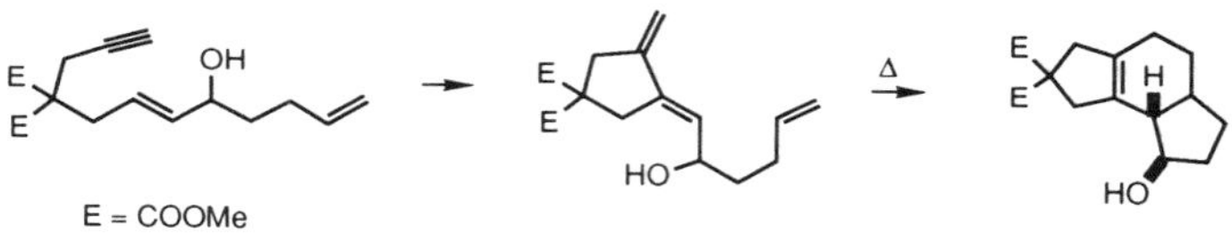
E
E
OH
HO
Δ
H
HO

E = COOMe

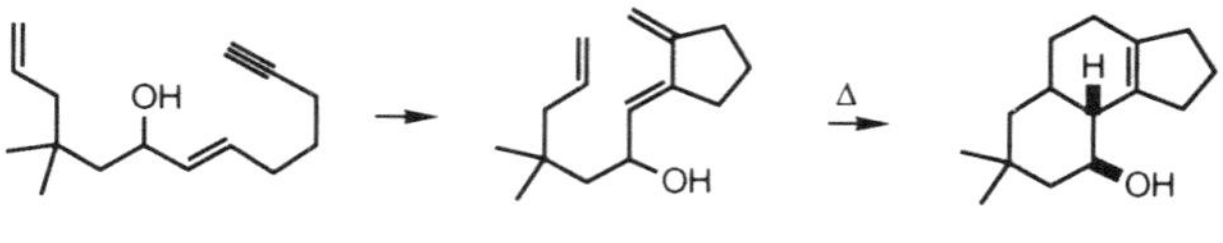
OH
OH
Δ
H
OH

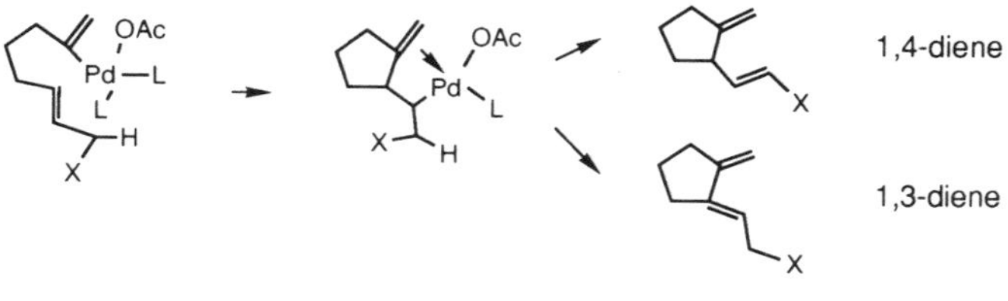
OAc
Pd—L
L
H
X
OAc
Pd
L
X
H
X
X
1,4-diene
1,3-diene

It has been observed that a pendant alkyne acts as a ligand for the Pd(O) catalyst to direct the intramolecular coupling of 1,1-diiodo-1-alken-5-ynes

[Nuss, 1993], permitting a rapid construction of the neocarzinostatin chromophore. The coupling of 1,1-dihaloalkenes with alkynes was previously shown to be (*E*)-selective, i.e. the (*E*)-halogen is removed.

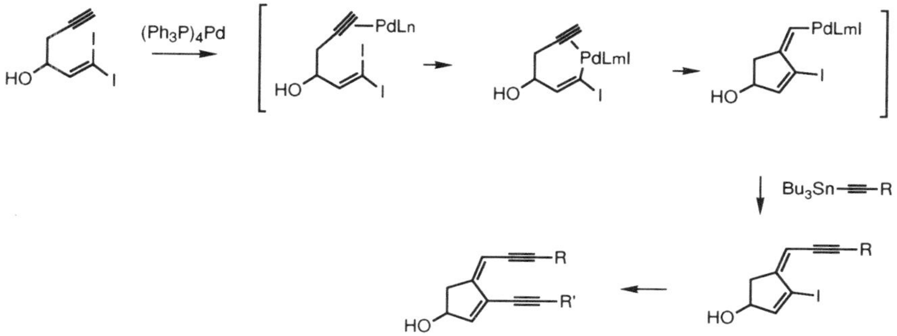

Transpositional reduction of allyl formates occurs in the presence of Pd(O). A mechanism involving antarafacial displacement of the ester group by the metal species, capture of the formate ion by palladium, and return of a hydride ion from the formyloxy group to the organic ligand manifests an S_N2' process with configuration inversion. This method can be used to establish stereochemistry of ring juncture [Mandai, 1992].

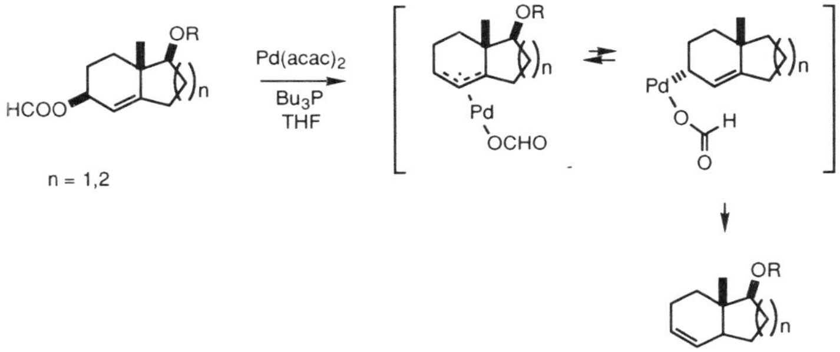

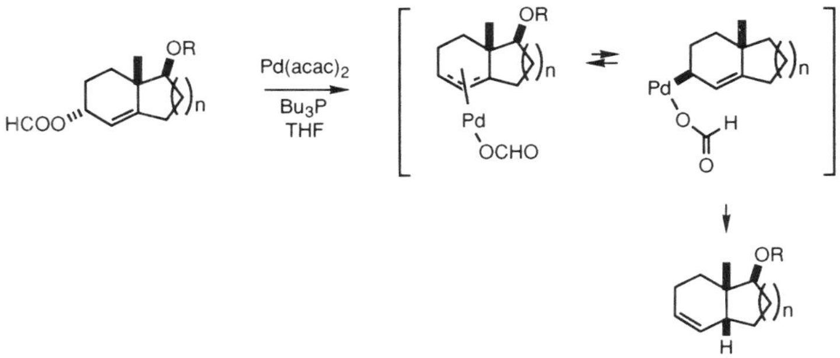

The stereochemical course of rhodium-catalyzed hydroformylation can be subjected to ligand mediation. Thus (2-cyclohexenyl)methyl diethyl phosphite gives the *cis* 2-substituted cyclohexanecarbaldehyde [Jackson, 1992], but interestingly the major product (ratio 9:1) from the same reaction of the (2-cyclohexenyl)ethyl homolog is the *trans*-isomer. (For regiocontrol of hydroformylation, see Section 6.2.3.) The hydroboration of (2-cyclohexenyl-oxy)diphenylphosphine with catechol-borane in the presence of tris(triphenyl-phosphine)rhodium(I) chloride is analogous, it proceeds regio- and stereoselectively [Evans, 1992c] in terms of neighboring group participation.

PPh_2 BH $(Ph_3P)_3RhCl$ PPh_2 [Rh]−$B(OR)_2$ H PPh_2 [Rh] H $B(OR)_2$ [Rh] PPh_2 $B(OR)_2$ OAc OAc

Synthetically the establishment of the angular hydroxyl group between A/B rings of phorbol requires careful planning and experimentation. A solution as demonstrated in a model study is that involving intramolecular delivery of a 1-propynyl group to a ketone precursor via temporary anchoring on an adjacent hydroxymethyl sidechain [Shigeno, 1992].

HO H OR O $2CeCl$ H O O Cl Ce RO HO H OR HO $C_{13}H_{27}COO$ OAc H OH O OH OH H OR O OH

phorbol

Catalytic hydrogenation of indenes and dihydronaphthalenes is subject to stereocontrol by existing substituents on the saturated ring carbon atom(s). This behavior causes problems in a synthesis of secopseudopterosin aglycone [McCrombie, 1991] based on such a reduction. However, the desired *trans*-1,4-disubstituted tetralin has been acquired by employing an intra-molecular hydrosilylation technique.

Crucial to both syntheses of modhephene [Kraus, 1991] and cuanzine [Palmisano, 1991] is the stereoselective reduction of a trisubstituted double bond. While heterogeneous catalytic hydrogenation led to compounds having the incorrect configuration, homogeneous hydrogenation with an iridium catalyst afforded the desired products, apparently due to precoordination of the metal with the oxygen functionality of the substrates, enforcing entry of hydrogen from the *exo* side. This directed hydrogenation was previously demonstrated in the stereoselective synthesis of *cis*- and *trans*-hydrindanone derivatives [Stork, 1983].

Diimide reduction of a double bond in a hydrophenanthrene system has been shown to be directed (stereoselection >99:1) by a proximal hydroxyl group or the lithium alkoxide [Thompson, 1977]. The corresponding sodium and potassium alkoxides are incapable of exerting the same effect.

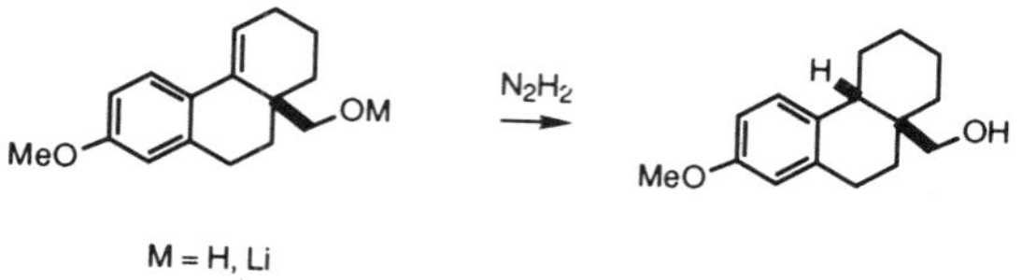

M = H, Li

Contrasteric delivery of reagent to a molecule has been observed when another functionality intervenes in the reaction course. In reduction with a complex metal hydride often a hydroxyl group in the proximity of the reaction site forms an alkoxymetal hydride reagent to serve as reducing agent. Thus, the α-pyridone ring of a hexacyclic intermediate for strychnine synthesis [Woodward, 1963] was converted into a dihydro derivative by such a mechanism, the hydride ion attacked the iminium species from the more hindered *endo*-face. (Note the α-pyridone moiety apparently undergoes reduction in the pyridinium 2-oxide form, thus the carbonyl group is protected as the enolate ion against reduction. Compare the removal of the other lactam carbonyl in the same molecule).

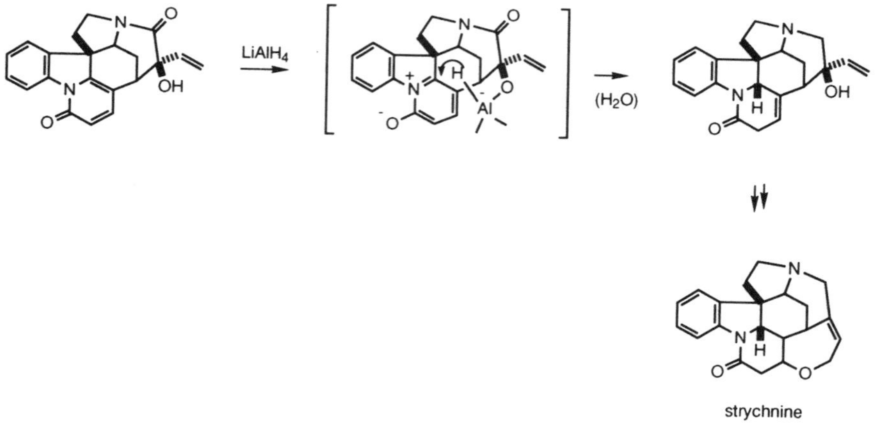

strychnine

(Note the stereoselective hydration at C-17 of the aspidosperma skeleton [Danieli, 1984] apparently occurs with the participation of the basic nitrogen atom acting as hydrogen bond acceptor of a water molecule.)

In a synthesis of (+)-brefeldin-A [Solladie, 1993] a *trans*-3,4-disubstituted cyclopentanone was submitted to triacetoxyborohydride reduction which led to a secondary alcohol having the desired absolute configuration, as a result of neighboring group participation.

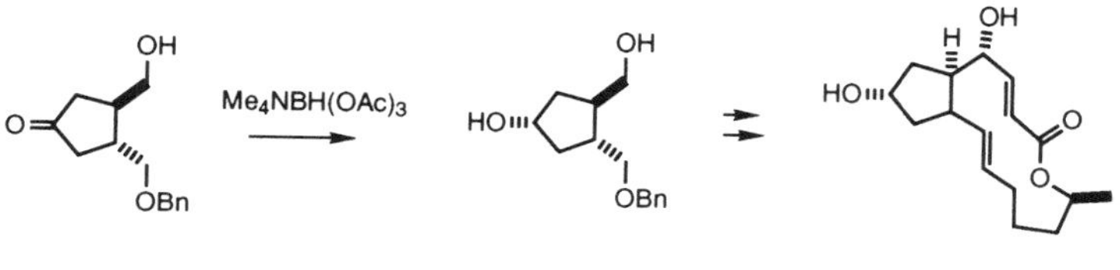

brefeldin-A

In a model study for synthesis of phorbol, three new stereocenters (in the C-ring) were created in one step of $LiAlH_4$ reduction [Magar, 1992]. As expected the first reagent attacked from the less hindered *exo*-face, however, the subsequent cleavage of the epoxide was accomplished regioselectively and stereoselectively by intramolecular delivery of a hydride ion from the oxyaluminate species.

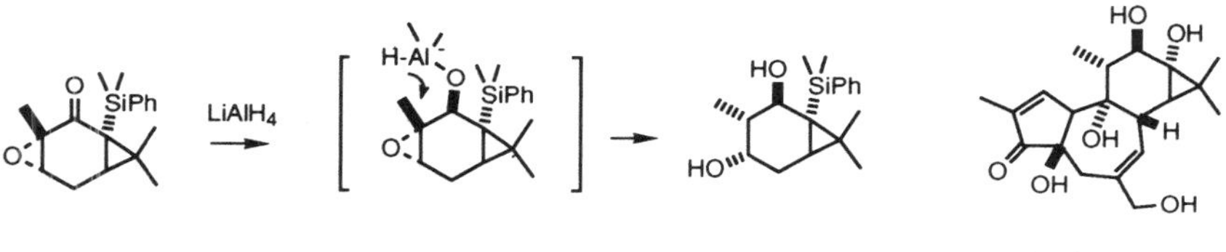

phorbol

A synthetic intermediate with the correct sidechain of stoechospermol was acquired by a reduction process [Salomon, 1984]. Participation of hydroxyl group(s) is evident in the Michael-type reaction of the cyclopentylidenemalonic ester. This tactic was used previously to establish the C-7 configuration of vernolepin/vernomenin [Isobe, 1978] (with $NaBH_3CN$ *suspension*, then BH_3), and the stereoselective reduction of certain γ-acetoxy α,β-unsaturated nitriles on the periphery of a hydrazulene skeleton for the synthesis pseudoguaianolides [Lansbury, 1982].

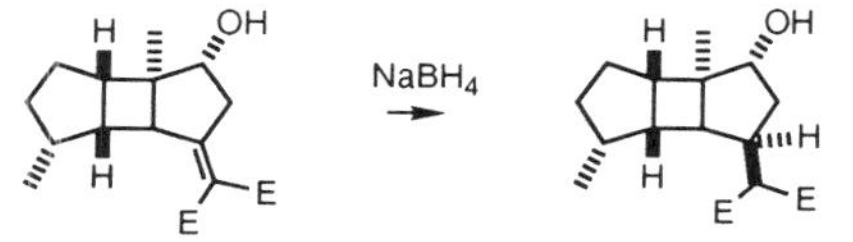

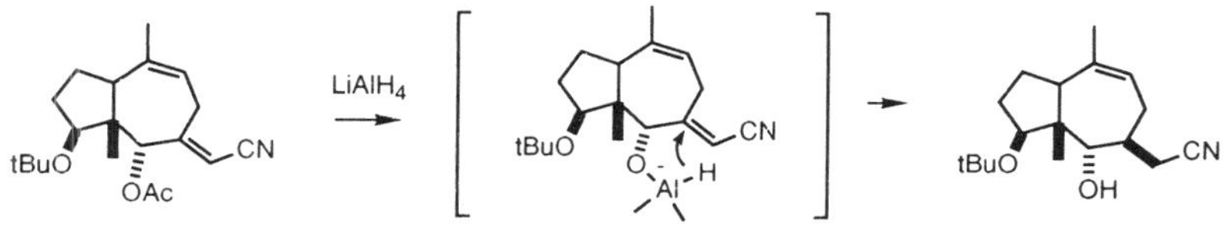

The hydroxyl-directed reduction of β-hydroxy oximino ethers [D.R. Williams, 1992] is interesting because of its dependence on the geometry of the oxime, e.g. (Z)-oxime to 1,3-*anti* product, (E)-oxime to 1,3-*syn* isomer.

OH N OBn — $Me_4NBH(OAc)_3$ / MeCN / HOAc → OH NH OBn

anti- (95:5)

OH N·OBn — $Me_4NBH(OAc)_3$ / MeCN / HOAc → OH NH OBn

syn- (85:15)

Remindful of these reductions is the semihydrogenation of propargylic alcohols by reaction with lithium aluminum hydride which has been utilized in a synthesis of vitamin-A [Attenburrow, 1952].

Lithium-liquid ammonia reduction of homonuclear octalones in which the conjugated double bond is exocyclic to the other cyclohexane ring is subject to thermodynamic control, resulting in predominantly the *trans*-decalones. This general trend finds exception when a hydrogen donor is present on the same side as the angular substituent, because intramolecular protonation of the enolate β-carbanions by the hydrogen donor is favorable. For example, 19-hydroxy-Δ^4-3-keto steroids give mainly AB-*cis* ketones [Knox, 1965], the proton source being the angular hydroxymethyl group. A similar protonation by a water molecule held by lithium carboxylate over the β-face of an octalone accounts for the appearance of the *cis*-decalonecarboxylic acid as major reduction product in "wet" ammonia [McMurry, 1978].

HOOC — Li, NH_3 (H_2O) → HOOC ... H major

Li, NH_3 → HOOC ... H

[Li O O O-H H O H]

Internal protonation at the β-carbon atom during Li/NH_3 reduction by a hydroxyl group accomplished simultaneous asymmetric induction at two remote sites, and thereby simplifies an access to solavetivone [Iwata, 1981a].

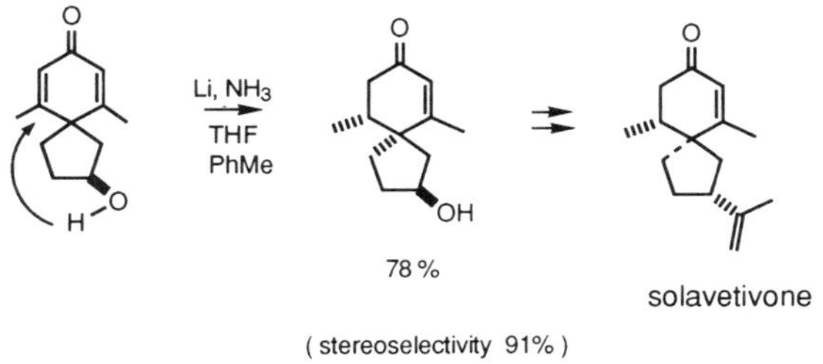

78 %

solavetivone

(stereoselectivity 91%)

The same technique was applied to the construction of a tricyclic intermediate for the stemodane diterpenes [Iwata, 1984]. Note that the stereoisomeric precursor for the aphidicolane diterpenes could be prepared from the same hydroxydienone by forming a cyclic ether before the reduction.

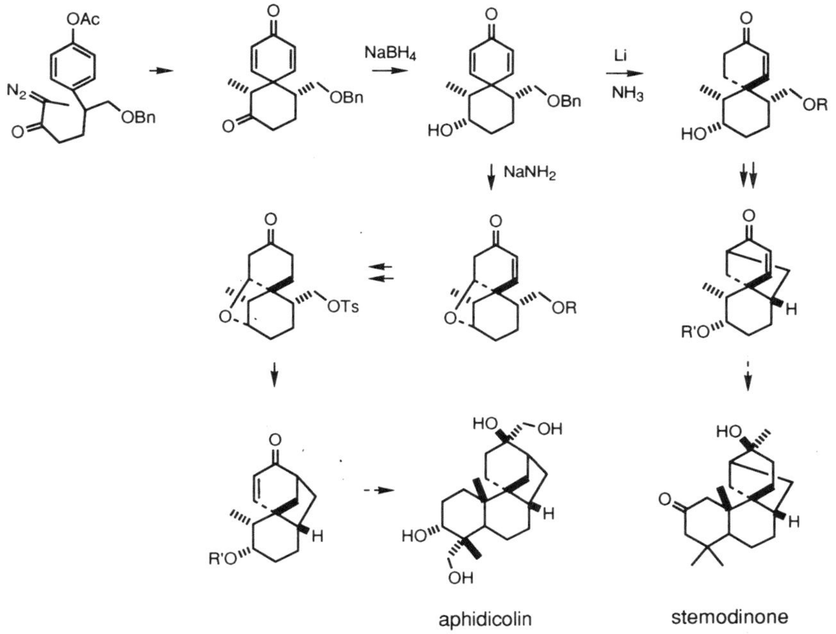

aphidicolin

stemodinone

Neighboring group participation by way of bromolactonization to oxygenenate pro-C-5 with the desired stereochemistry (relative to the adjacent quaternary carbon atom) was crucial to the elaboration of the C-1/C-9 fragment of erythronolide-B [Corey, 1978c,d]. The subtarget carboxylic acid is an asymmetrical molecule, but this particular structural limitation was dispelled.

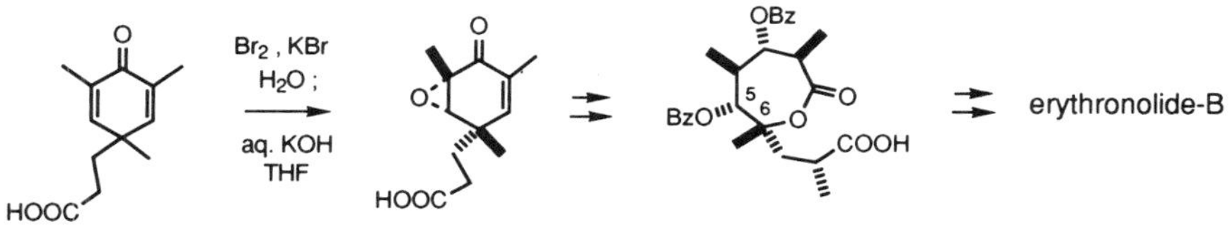

erythronolide-B

Iodolactonization provided a solution to the stereochemical problem concerning the hydration of a cyclopentene intermediate in a synthesis of sesbanine [Bottaro, 1980].

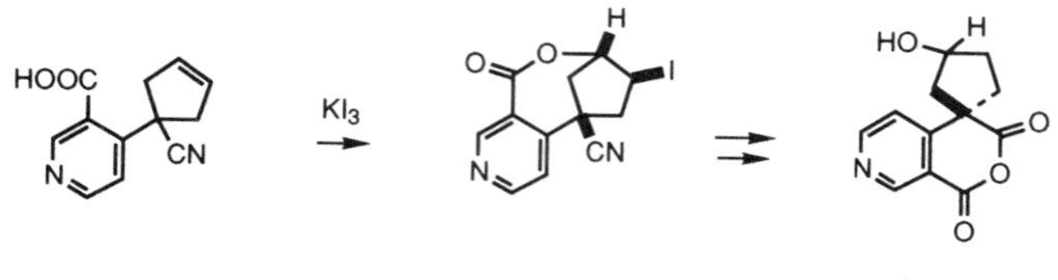

sesbanine

Addition of organometallic reagents to (Z)-unsaturated sulfones containing an α-trimethylsilyl substituent is subject to stereocontrol by a hydroxyl group at the γ- or δ-position [Isobe, 1980, 1981, 1986b]. Exactly because of the neighboring group participation the configuration of the double bond is most important, the (E)-isomers afford C $\rightarrow$ O silyl transfer without achieving the addition. This process played an important role in the total synthesis of okadaic acid [Isobe, 1986a].

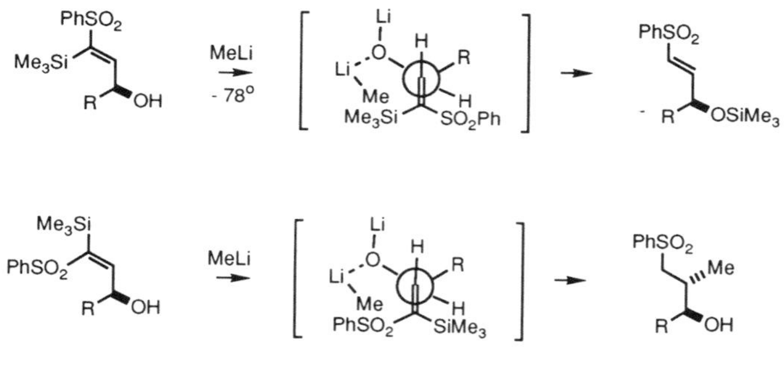

vic-Dihydroxylation of olefins via osmylation and permanganate oxidation is *syn*-selective, whereas *anti*-dihydroxylation results when the alkenes are treated with hydrogen peroxide and formic acid. Overall *anti*-addition can also be achieved by the method of Prevost, in which he reagent is a combination of iodine and silver benzoate (1:2 molar ratio). The initial products are *trans*-1,2-iodobenzoates which undergo nucleophilic substitution with participation of the neighboring ester group. Since the second reaction consists of a double inversion, the diester products are still *trans*. Saponification of the diesters yields *vic*-diols without further change of stereochemistry. This process has a complement in the Woodward method using equimolar iodine and silver acetate in aqueous acetic acid. The presence of water prevents or greatly reduces the neighboring group participation in the displacement step, resulting in *cis*-diol monoacetates. Hydrolysis affords the *cis*-diols.

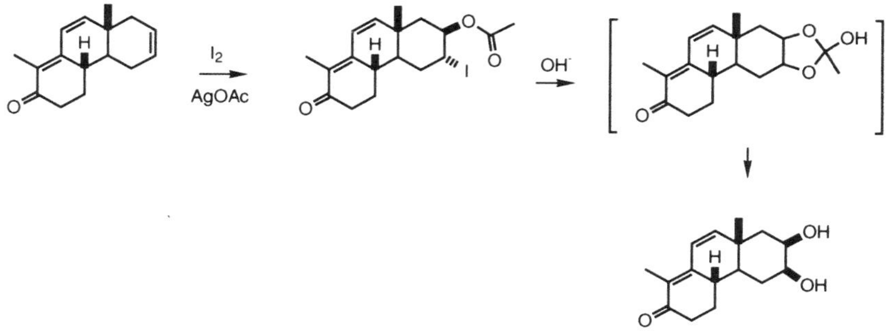

The normal reaction of dichlorocarbene with 7-*t*-butoxybicyclo[2.2.1]hepta-2,5-diene occurs almost exclusively at the *syn* double bond [Kwantes, 1976]. This contrasteric effect manifests the intervention of the oxygen atom in the cycloaddition.

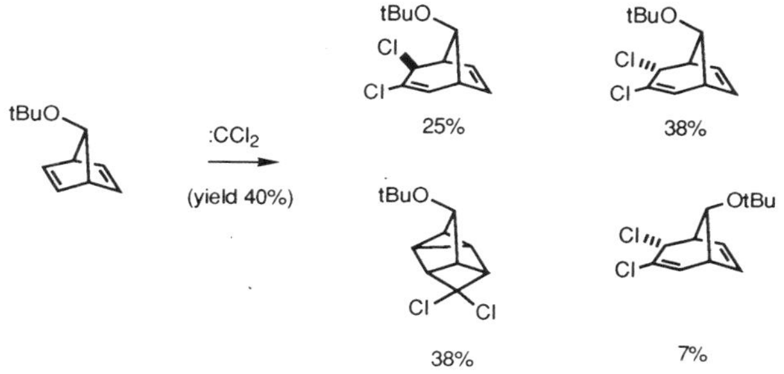

A route to methyl *O*-methylpodocarpate [Node, 1989] via B-ring formation is under stereocontrol by the ester group. The tertiary carbenium ion intermediate is stabilized by interaction with the ester C=O and such an interaction requires an axial orientation of the latter group and the steric shielding forces an approach of the aromatic ring from the opposite side.

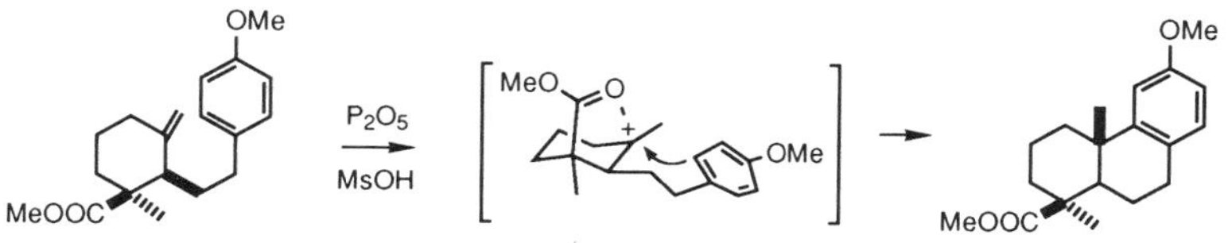

The synthesis of the aromatic 15,16-dioxo-*syn*-1,6:8,13-bismethano[14]-annulene is a challenging task. The stereocontrol problem was solved by intramolecular carbene addition to provide intermediates containing extra but cleavable links [Balci, 1981].

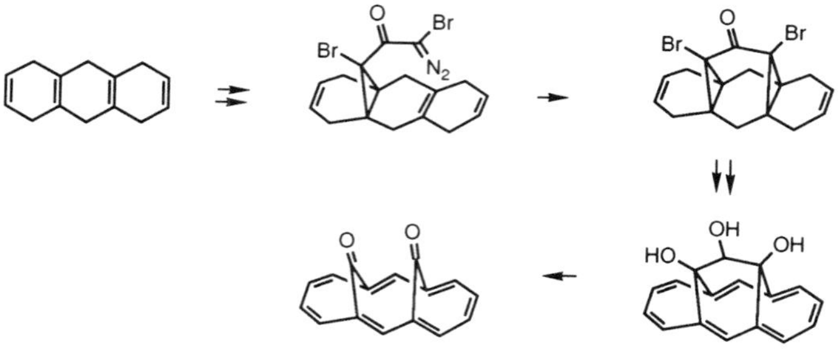

In connection with the establishment of *cis*-2,5-dialkyl-Δ^3-dihydropyran nucleus for the synthesis of the pseudomonic acids [Snider, 1983], a tandem Lewis acid-mediated ene reaction of a 1,4-diene and formaldehyde followed by a hetero-Diels–Alder reaction was extremely propitious. The latter process was directed by the coordination of the dienophile (formaldehyde) to the alkoxy-aluminium intermediate.

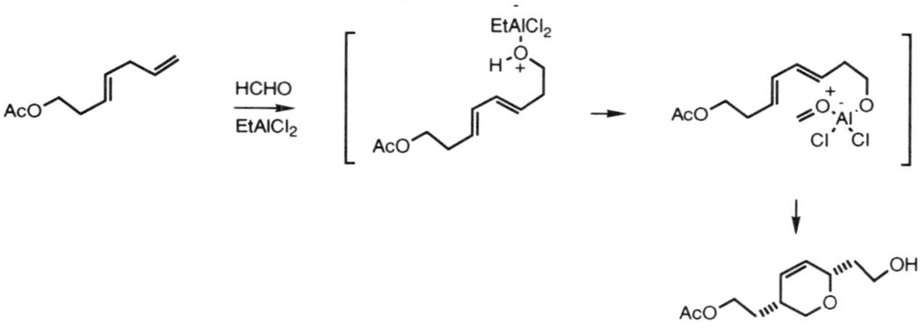

Intriguing opposite diastereofacial selectivity manifested in an intramolecular Diels–Alder reaction by changing the oxygen function at the γ-carbon of the ynoate ester segment [Trost, 1989b] has been attributed to $A^{1,3}$ strain such that the larger substituent would eventually become axial in the bicyclic product. It may be possible to explain the difference on the basis of stereoelectronic effect.

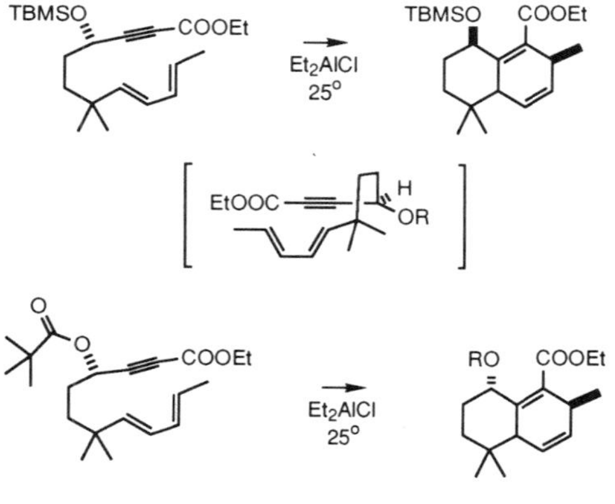

In cyclic systems the reactivity trend for an axial/equatorial pair of isomers is that the equatorial epimer is higher, owing mainly to the less steric interference to approaching reagents. However, when a monoester of 1,3-diol is sterically held together so that the *cis*-isomer exists in a diaxial conformation which permits hydrogen bonding between the two groups, this epimer would undergo hydrolysis with a faster rate [Henbest, 1957a].

An important utility of (*S*)(+)-glutamic acid is related to its stereospecific conversion into a γ-lactone acid [Cervinka, 1968] via nitrosation. The complete retention of configuration of the α-carbon atom is due to a double inversion: formation of an α-lactone intermediate and intramolecular S_N2 displacement by the γ-carboxyl function. Accordingly, two neighboring group participations are involved.

Actually the more important synthetic intermediate is the γ-hydroxymethyl γ-lactone which is the borane reduction product of the acid. (This (*S*)-lactone, also obtainable from (*R*)(+)-glyceraldehyde [Kitahara, 1984], has been used in a synthesis of (+)-brefeldin-A.) The enantiomeric (*R*)-lactone has been acquired by simple manipulations: tosylation, alcoholysis, and acid treatment [P.-T. Ho, 1983] which again exploit neighboring group participations.

COOH, COOH, NH_2 —HNO_2→ [COOH, COO^-, $^+N_2$ —$-N_2$→ COOH, H, O] → COOH (*S*)

↓ $BH_3.SMe_2$

(*R*) OH ←CF_3COOH— $COOCH_3$, H, O ←TsCl, py; CH_3ONa— OH (*S*)

HO, OH, H ⇒ TsO, MEMO, CN, H ⇒ CN, EtO, H, O ⇒ HO, O

brefeldin-A

An elegant enantioselective synthesis of α-amino acids is based on an almost exactly reverse pathway to the nitrosation. The key steps are the borane reduction of trichloromethyl ketones [Corey, 1992a] in the presence of a chiral auxiliary and the conversion of the optically active alcohols to α-azido carboxylic acid by treatment with alkaline sodium azide. Dichloro-oxirane intermediates are rapidly formed which are attacked (S_N2 reaction) by the azide ion to give the acid chlorides, and thence hydrolyzed to the acid. The key features of this method are that in the transition states of the reduction the trichloromethyl group occupies a larger void within a space created by the catalyst-borane complex (for most of the α-amino acid targets which are of interest to chemists), and the stereospecific displacement of the hydroxyl group.

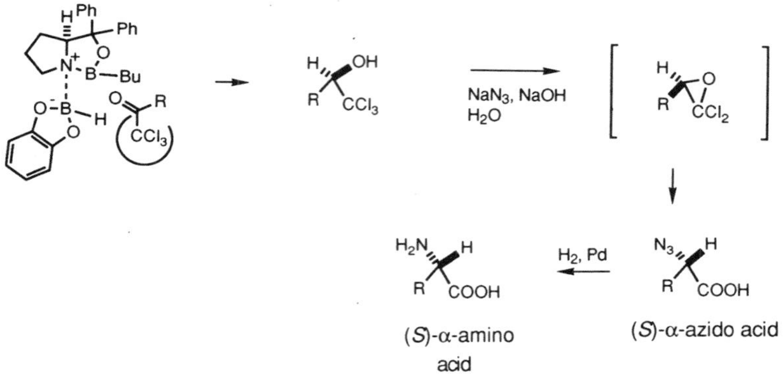

NaN_3, NaOH, H_2O

H_2, Pd

(*S*)-α-amino acid

(*S*)-α-azido acid

The formation and opening of an aziridine intermediate in a synthesis provides stereocontrol for establishment of a chiral carbon center adjacent the nitrogen atom. Necessarily the entering group becomes *trans* to the aza-substitutent. Such a stereochemical feature was exploited in a synthesis of perhydro-histrionicotoxin [Aratani, 1975].

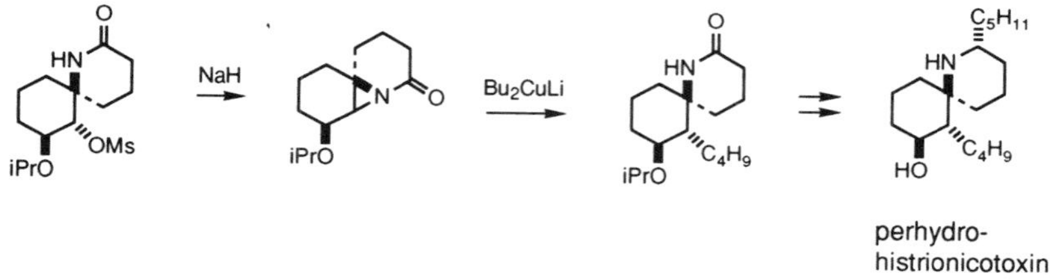

NaH

Bu_2CuLi

perhydro-histrionicotoxin

A rather unusual method for regiocontrol by changing it into a matter of chemoselection is shown in a synthesis of 13-oxoellipticine [Obaza-Nutaitis, 1986]. Formation of a tetracyclic dicarbonyl compound preceding reaction with two different organolithium reagents proved to be an important feature.

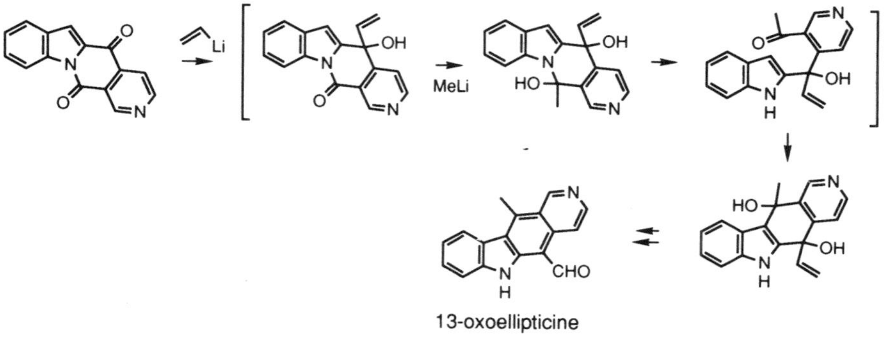

MeLi

13-oxoellipticine

A chiral AB-ring synthon for anthracycline antibiotics has been acquired via bromolactonization of a ketal derived from a tartramide [T. Suzuki, 1986].

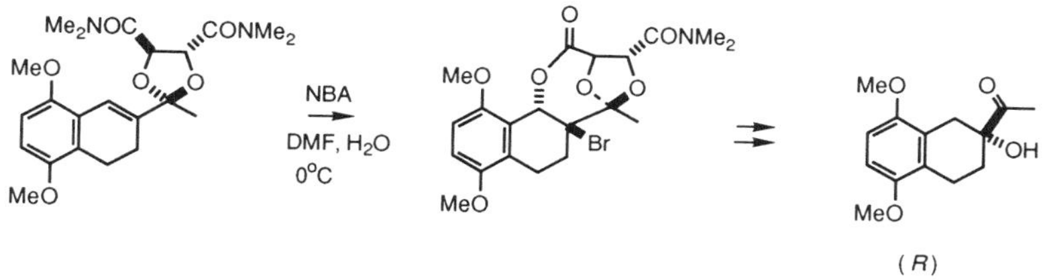

Stereocontrol of the anomeric configuration of pyranosyl derivatives upon glycosidation is well developed. Without participation of a group at C-2 the anomeric effect exerts its influence, but the configuration of C-1 is retained when C-2 contains a group such as an ester. Pivalyl is an excellent control element as attack of the nucleophile on the positively charged center of the fused dioxalanyl cation to form an orthoester is discouraged by steric effects [Nicolaou, 1991a].

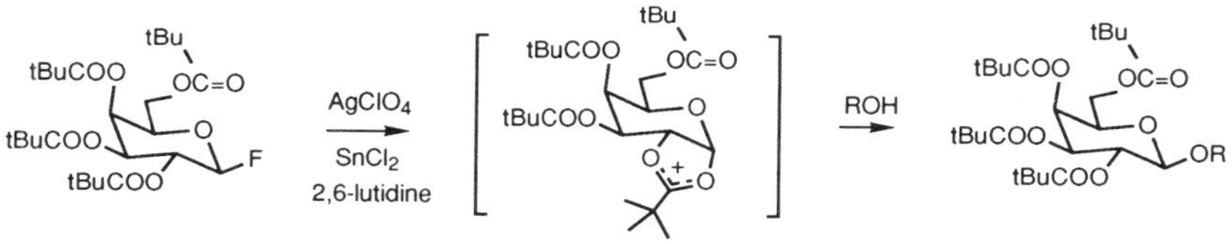

Intramolecular addition reaction can be employed to advantage in the context of both regio- and stereocontrol. This aspect is demonstrated in the stereospecific delivery of an alkoxy residue to the anomeric center of a sugar [Stork, 1992b] from a siloxane. The preparation of methyl 6-*O*-glycosyl-*β*-D-mannopyranoside hexabenzyl ether shown below is representative.

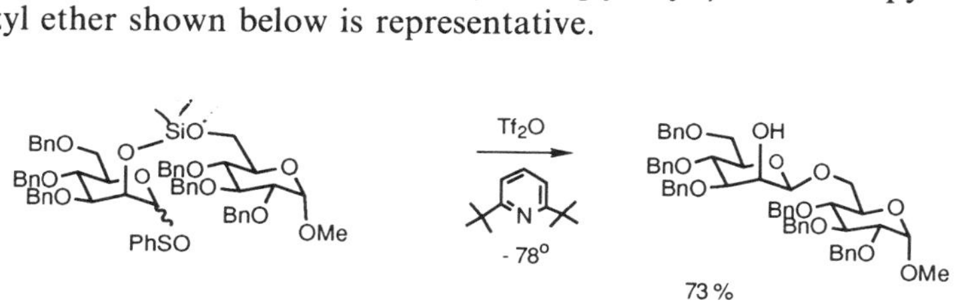

methyl 6-*O*-glycosyl-β-D-manno-pyranoside hexabenzyl ether

The attachment of an axial hydroxymethyl pendant to a tetrahydropyran subunit during synthesis of (−)-talaromycin-A [Crimmins, 1989a] by means of an oxysilylmethyl radical addition to a dihydropyran precursor solved a difficult stereochemical problem. The formation of a five-membered ring perforce established the stereocenter as desired. Oxidative cleavage of the oxasilacycle furnished the target molecule.

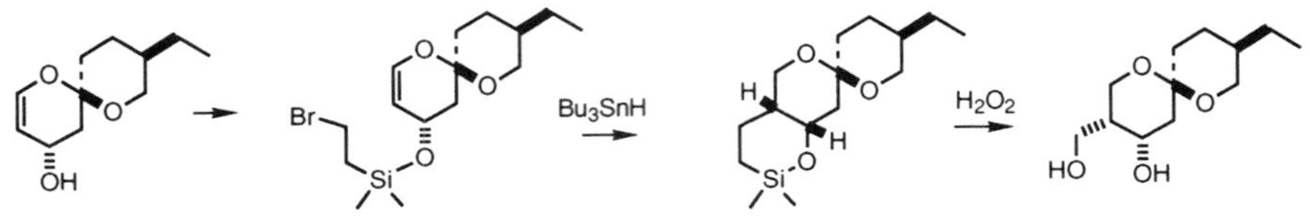

(-)-talaromycin-A

This technique can also be used to control the stereochemistry of the distal sp^2-carbon atom of an allylic alcohol [Stork, 1985; Koreeda, 1986]. In two examples concerning elaboration of steroid skeletons, the ring junction of a CD-ring synthon in one case, and the configurations of C-17 and C-20 in the other, have derived satisfactory solutions. Related tactics are represented by the indirect angular methylation of a hydrindenone system [Stork, 1989b].

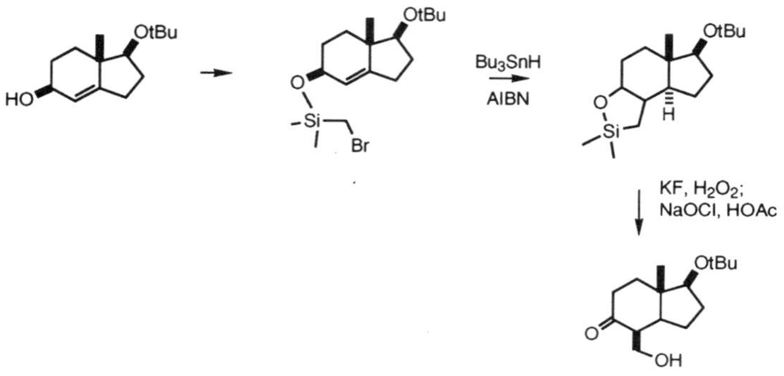

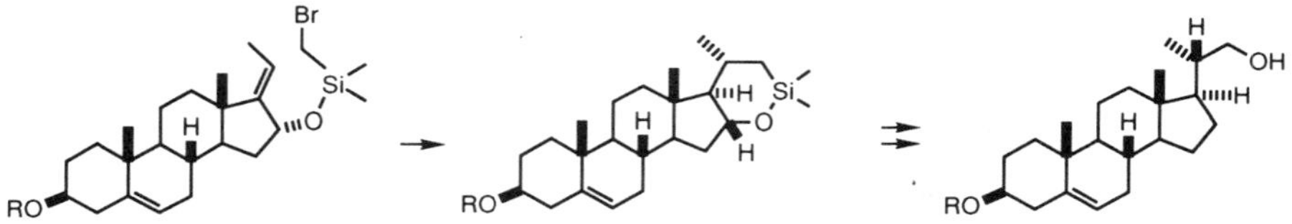

Starting from (*S*)-malic acid the synthesis of statine via pyrrolidone intermediates [Koot, 1991] resulted in 4-epistatine. However, using the oxygen functionality to direct an intramolecular delivery of the isobutyl group (in a precursorial form), a successful outcome emerged.

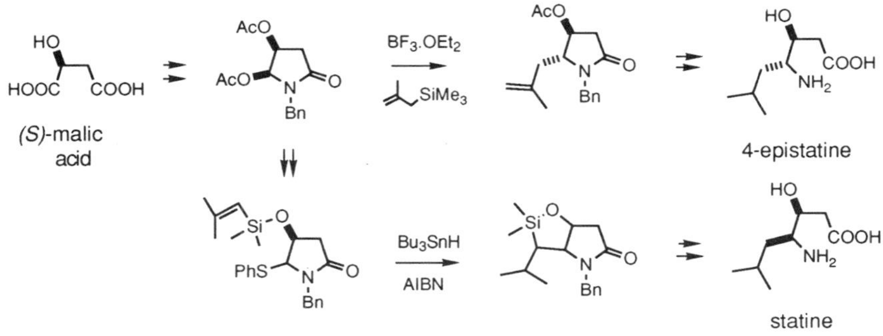

(*S*)-malic acid

4-epistatine

statine

Intramolecular radical addition to a double bond followed by trapping constitutes an efficient regio- and stereoselective method for the formation of two bonds at vicinal atoms, usually in a *trans* manner. A concise synthesis of prostaglandin-$F_{2\alpha}$ [Stork, 1986b] using this method is shown below.

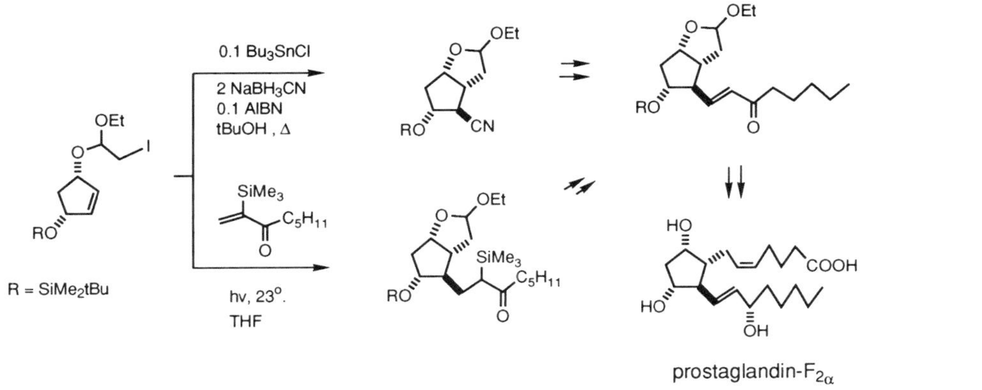

R = SiMe2tBu

0.1 Bu3SnCl
2 NaBH3CN
0.1 AIBN
tBuOH, Δ

hv, 23°, THF

prostaglandin-$F_{2\alpha}$

The reductive removal of an allylic sulfonyl group from homoallylic alcohols containing that functionality may be directed toward formation of products with a transposed double bond or those retaining the original carbon skeletons [Inomata, 1987]. The different reaction pathways consist of reduction of the π-allylpalladium species by external and internal hydride delivery, respectively.

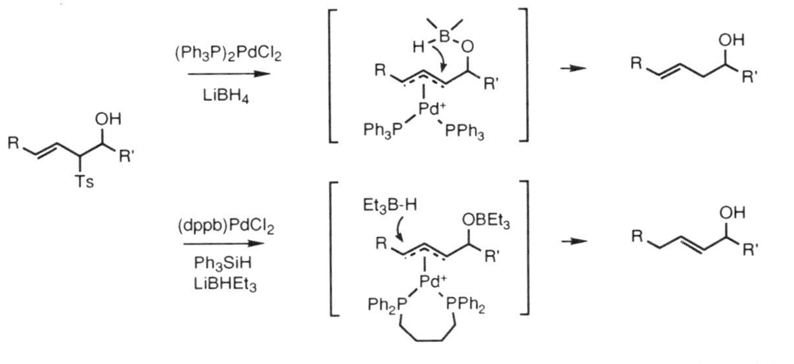

(Ph3P)2PdCl2
LiBH4

(dppb)PdCl2
Ph3SiH
LiBHEt3

Sometimes the transpositional reduction of an allylic system can be achieved via generation of diazene. The problem pertaining to an elaboration of the most intricate core of dynemicin-A [Wood, 1992] was solved using this tactic in conjunction with the intramolecular Diels–Alder strategy for assembling the molecular skeleton.

Reductive cleavage of C_2-symmetric acetals exhibits a high degree of diastereoselectivity. It is interesting that one such acetal containing a tricarbonylchromium-complexed phenyl group undergoes reductive methylation to afford a product with an unexpected absolute configuration [Davies, 1989]. However, the result can be explained by participation of the chromium atom.

$TiCl_4$ $(OC)_3Cr$ [$TiCl_4$ H $(OC)_3Cr^+$] Me_3Al → HO H $(OC)_3Cr$

6.2.3. Regiocontrol

Group introduction to a specific site of a molecule is even more important than stereocontrol in a synthetic process, because configurational changes are more readily made while it is often difficult to detach a group from a carbon skeleton. The concept also applies to functionalization reactions.

The efficacy of dehydrochlorination of a tertiary chloride precursor for the synthesis of Δ^1-tetrahydrocannabinol [Fahrenholtz, 1967] depends on the base used. The major product from the reaction with KOH in ethanol is the Δ^6-isomer, but the change to sodium hydride in THF resulted in a 3:1 mixture in favor of Δ^1-THC, presumably due to an intramolecular reaction initiated by the phenoxide ion.

	Δ^1-THC	Δ^6-THC
KOH / EtOH		major
NaH / THF	major	

Disruption of an intramolecular hydrogen bond in an intermediate by the addition of isopropanol apparently has the result of changing the electrophilic site and hence enabling the formation of a zoapatanol ring system [Trost, 1992a].

BOMO — $(dba)_3Pd_2.CHCl_3$, Ph_3P, dppp, THF → [BOMO, —Pd^+, H---O] → BOMO, OH

↓ $(dba)_3Pd_2.CHCl_3$, THF-iPrOH

BOMO, OH

Both stereoselectivity and regioselectivity are affected by hydrogen bonding in the transition states of 1,3-dipolar cycloaddition of nitrile oxides [Curran, 1990]. The hydrogen-bonded transition state causes a reversal of preferred regiochemistry in the following example.

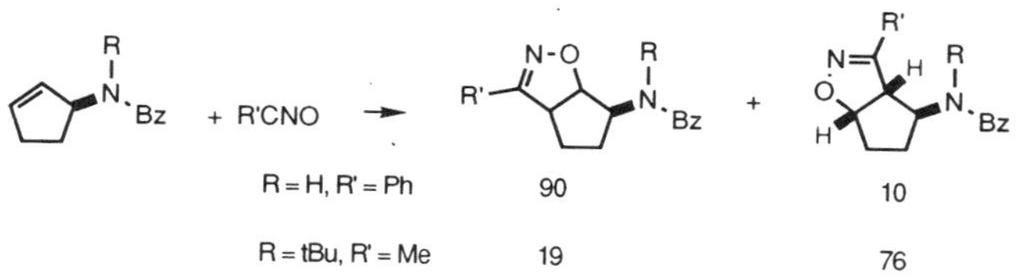

R = H, R' = Ph 90 10

R = tBu, R' = Me 19 76

It should be noted that such cycloaddition involving an allylic alcohol in the presence of a Grignard reagent proceeds via a chelated transition state in which the allylic alcohol would adopt a conformation that is relatively free from $A^{1,3}$ strain [Kanemasa, 1991]. Consequently two contiguous asymmetric centers are established.

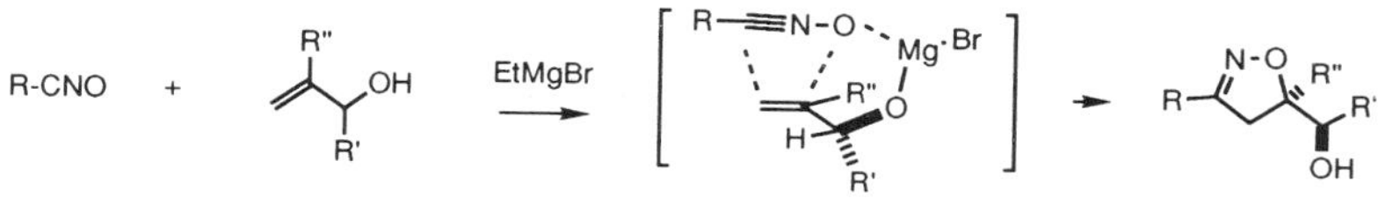

The array of five contiguous asymmetric centers in the six-membered E-ring of reserpine presents a formidable challenge to the synthetic chemist. However, the problem was solved, soon after the structural determination of this alkaloid, on the basis of an ingenious plan [Woodward, 1958] involving a Diels–Alder reaction to establish the D/E-ring juncture stereochemistry and the methyl ester pendant. The creation of the other two oxygen functions from a double bond calls for, in one scheme, a bromoetherification, and subsequent elimination–addition process. The geometry of each intermediate is such that it dictates complete regio- and stereocontrol in the reaction applied to it.

It should be noted that in a later stage the pentacyclic product obtained from Bischler–Napieralski cyclization and reduction has a 3α-H configuration. The necessary inversion is possible when the one-carbon sidechain is locked into a γ-lactone ring with the hydroxyl group at C-18. The conformational change elicited by this neighboring group participation provides a driving force for that of the configuration at a remote atom.

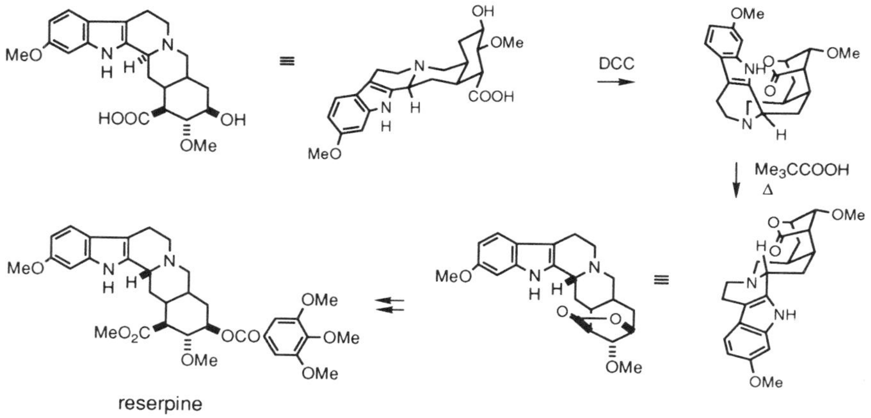

reserpine

In a synthesis of (−)-calicheamicinone [A.L. Smith, 1992] the closure of the enediyne ring produced predominantly an epimeric alcohol. However, inversion of configuration was achieved by lactonization using the ester carbonyl as the nucleophile to displace the derived mesylate.

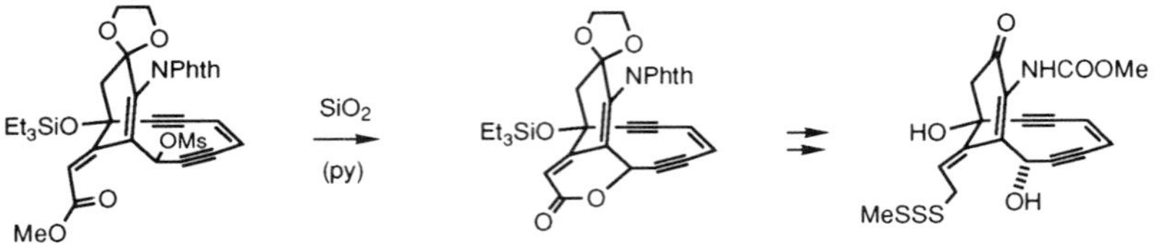

(-)-calicheamicinone

A synthesis of forsythide dimethyl ester [Furuichi, 1974] starts from *endo*-bicyclo[2.2.1]hept-5-ene-2,3-dicarboxylic acid which is readily available via a Diels–Alder reaction. Bromolactonization affixes an oxygen atom to the more remote five-membered ring, and differentiates the two carboxyl functions to permit chain extension and reduction of free carboxyl group. Conversion of the β-bromohydrin unit to a ketone is by alkali treatment (a reaction quite peculiar to bridged ring systems such as those represented by the present case). The ketone directs formation of a new cyclopentane unit, and thereafter it undergoes a photolysis to generate a diquinane compound. The two rings of this diquinane are differently substituted so that oxidative cleavage of the double bond in one of the rings leads to the synthetic target. In this synthesis an important role is played by the bromolactonization.

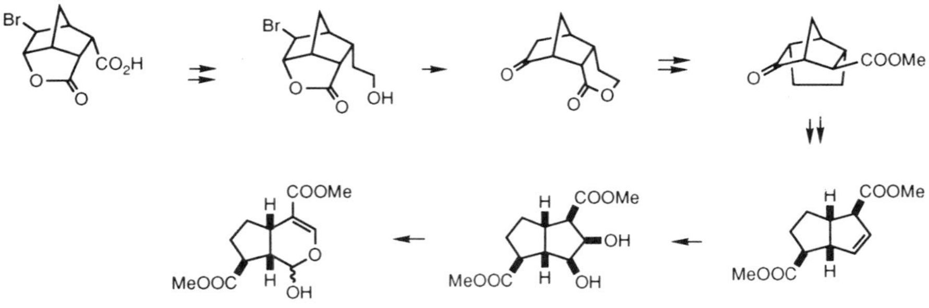

fosythide dimethyl ester

Actually, bromolactonization belongs to a large family of electrophile-induced heterocyclization [Bartlett, 1984a]. Modification and extension of the method enables the transformation of homoallylic alcohols to *syn*-1,3-diol systems which are very common in natural products. The enantiodivergent synthesis of nonactic acid [Bartlett, 1984b] from a common intermediate based on this technique shows its versatility. The reader is urged to examine closely the strategy of the assembly of nonactin by cyclodimerization of the dimer.

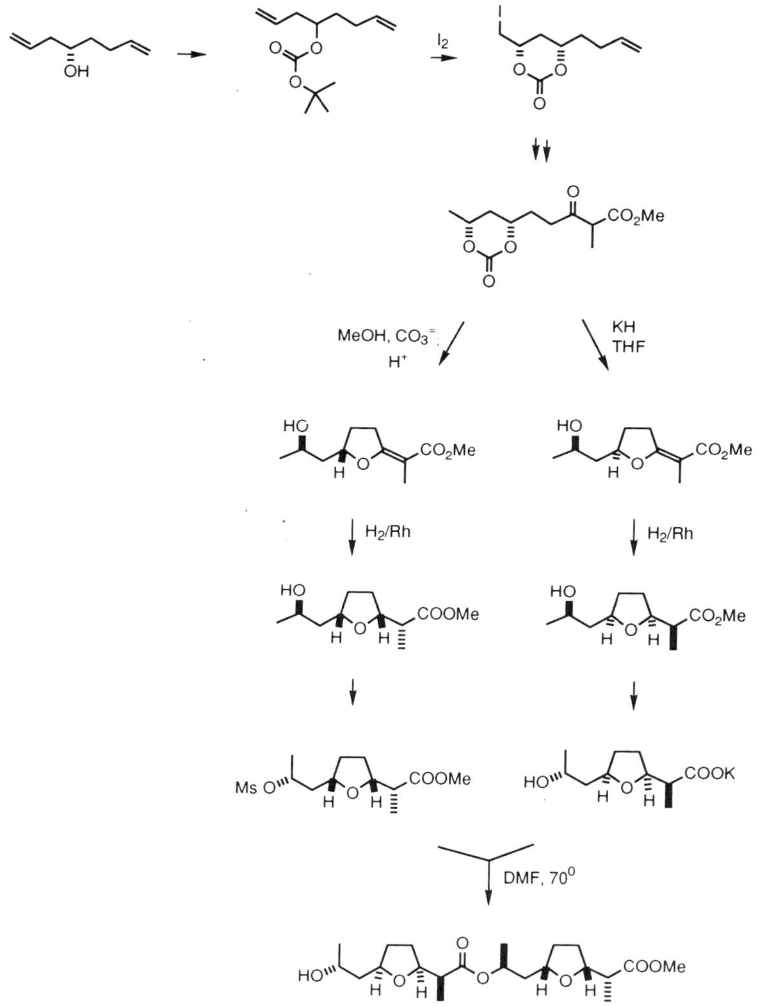

In a synthesis of chlorophyll-*a* [Woodward, 1961] the union of two dipyrrylmethanes (AD + BC) presents a problem because there are two ways available, due to the unsymmetrical nature of the reaction components. An ingenious way to circumvent this regiochemical nightmare is by a selective imine formation from the aminoethyl sidechain of the A-ring with the aldehyde group of the B-ring via the more reactive thioaldehyde of the latter subunit. The resulting imine directs cyclization which effectively links the two dipyrrylmethanes in the desired manner. As the reaction conditions are also conducive to macrocycle formation, a more advanced intermediate is produced in one step, and after tautomerization, a phlorin system results.

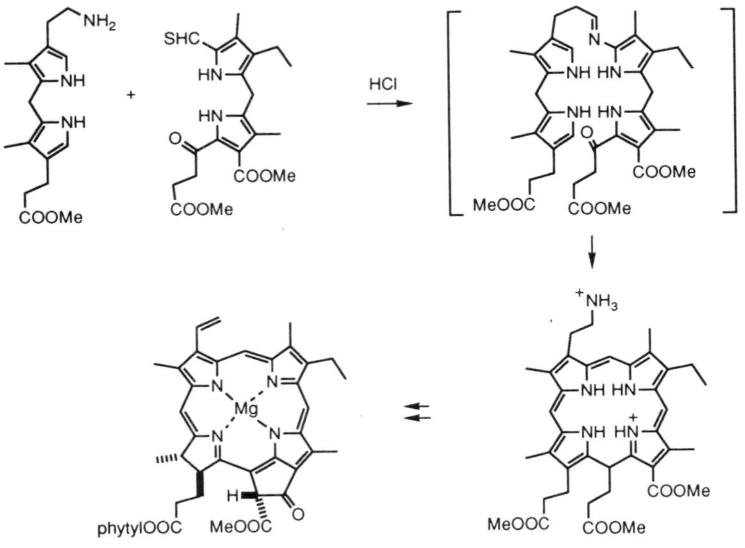

chlorophyll-*a*

The anchoring tactic also appears in a synthesis of colchicine [J. Schreiber, 1961] during the initial stage of the tropolone ring construction. Thus, the reaction of a benzosuberone derived from purpurogallin trimethyl ether with methyl propynoate leads to a fused α-pyrone. Although the benzylic position of the benzosuberone is much more reactive than the other α-methylene group, and perhaps the regiocontrol is not much of a problem, steric crowding around the benzylic position disfavors alkylation at this side. The utility of the *peri*-hydroxyl to assist the alkylation via Michael reaction is of enormous importance, as once the alkylating agent is properly attached, its transfer to the benzylic position becomes a favorable intramolecular Michael/retro-Michael reaction sequence.

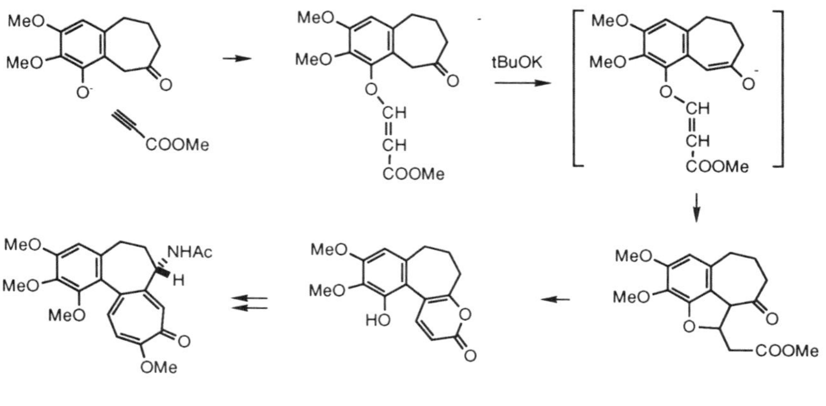

colchicine

The alkylation at C-2 of most 3-substituted cycloalkanones is problematical in terms of regiocontrol. (It is much easier to alkylate at the α'-position, such as via enamines.) With the specific aim of elaborating *cis*-jasmone [Goldsmith, 1981] from 3-oxocyclopentanecarboxylic acid, activation at C-2 was possible through the formation of a bicyclic ketolactone.

COOH COOMe tBuOK

cis-jasmone

A remarkably selective internal epoxidation of peroxyarachidonic acid to afford only the 14,15-oxidoarachidonic acid [Corey, 1979a] occurs because of very favorable stereoelectronic effects. On the other hand, the 5,6-epoxide can be obtained via iodolactonization of arachidonic acid.

CO_3H 20°C COOR

> 98%

14,15-oxido-arachidonic acid

COOH KI_3 $KHCO_3$ aq.THF, 0° LiOH; CH_2N_2 COOMe

methyl
5,6-oxido-arachidonate

Baeyer–Villiger reaction of bicyclo[2.2.1]heptane-2-ones gives mainly lactone products from migration of the bridgehead atom. However, the presence of a 7-*syn* bulky group (methyl or larger) changes the reaction course which favors migration of the methylene group. These results can be rationalized by the epimeric structures of the ketone-peracid adducts due to different directions of attack to the reagent on the ketones. From *exo*-adducts the transition states for bridgehead migration are chairlike, whereas those for methylene migration are boatlike. Reverse situations prevail in the cases of *endo*-adducts. A *syn*-7 substituent usually inhibits *exo* approach of the reagent, unless it is polar. For example, an ester group directs the addition from the *exo*-face, and its participation as relay has been speculated [T.-L. Ho, 1982b].

MeOOC AcOOH HO OMe AcO–O MeOOC COOMe

In the late stage of a luciduline synthesis [W.L. Scott, 1972] the stereospecific introduction of a methylamino group by S_N2 displacement of a tosylate prior to the Mannich cyclization is very facile, despite the requirement of nucleophile entering from the concave side of the *cis*-decalin system. This behavior is most likely due to aminal formation which enables an internal delivery of the methylamino group.

OTs MeNH2 MeNH MeNH OTs MeNH Me NH OTs H2O - MeNH2 OTs Me N HCHO MeNH O O O

luciduline

A solution to the problem of selective *N*-alkylation of a diamine is illustrated in the heptylation of *N*-heptyl-1,8-diaminonaphthalene [H. Yamamoto, 1981a]. Condensation with heptanal followed by treatment with diiso-butylaluminum hydride led to the *N*,*N'*-diheptyl derivative. On the other hand, tosylation of the animal intermediate before the reductive C—N bond cleavage changed the regioselectivity of the latter step. Detosylation then afforded *N*,*N*-diheptyl-1,8-diaminonaphthalene. In either case the animal formation was critical to controlled monoalkylation.

H_2N NHC_7H_{15} $C_6H_{13}CHO$ HN NC_7H_{15} C_6H_{13} iBu_2AlH $C_7H_{15}NH$ NHC_7H_{15}

TsN NC_7H_{15} C_6H_{13} iBu_2AlH TsNH $N(C_7H_{15})_2$ Na naphth. H_2N $N(C_7H_{15})_2$

A synthetic approach to (+)-pinidine [Dolle, 1991] contemplated the reduction of 2-methyl-6-(1-propenyl)-3,4,5,6-tetrahydropyridine which is derivable from an open-chain amino ketone. Preparation of this amino ketone can take advantage of [2.3]-sigmatropic rearrangement of a sulfonium imide with *in situ* desulfurization, and accordingly the synthesis was initiated by alkylation of methyl acetoacetate with an optically active iodide containing an allylic sulfide which can be elaborated from (−)-ethyl lactate.

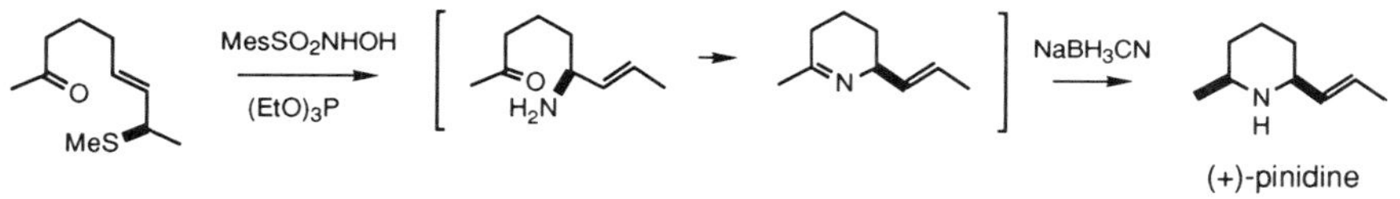

MesSO2NHOH (EtO)3P NaBH3CN (+)-pinidine

Functionalization of a thiazolidine derivative via intramolecular S → C transfer of a hydrazine function represents the first crucial step in a synthesis of cephalosporin-C [Woodward, 1966]. Inversional manipulation of the nitrogen-containing group establishes the elements of the β-lactam. The functionalization step involves formation of an aminosulfonium adduct and subsequent 1,2-rearrangement of an ylide.

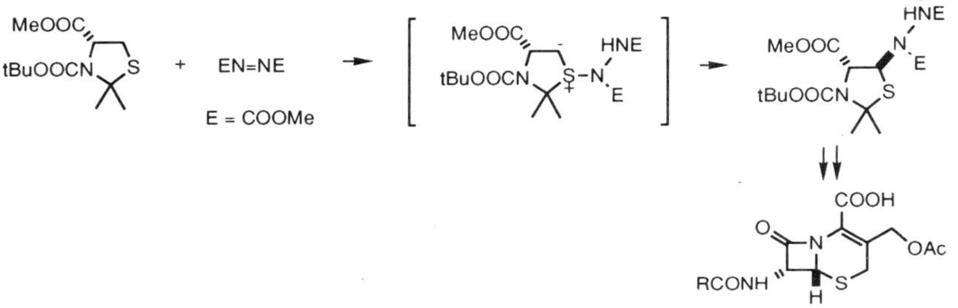

E = COOMe

γ,δ-Alkynyl and δ,ε-alkynyl ketones undergo Hg(II)-catalyzed hydration in a regioselective manner to give 1,4- and 1,5-diketones, respectively [Stork, 1964a]. Apparently the oxygen atom of the carbonyl group in these substances participates in the mercuration step. Similarly, Wacker oxidation of 2-(2-butenyl)-2-methylcylohexanone [Gratten, 1988] furnished a 91:1 mixture of 1,4- and 1,5-diketones.

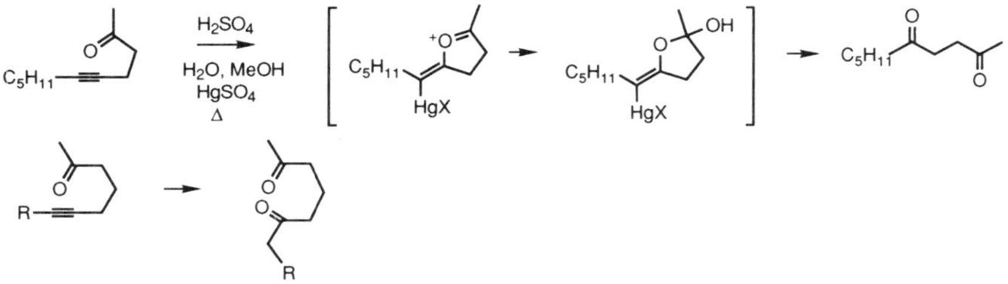

H_2SO_4 / H_2O, MeOH / $HgSO_4$ / Δ

The oxygenation of the double bond of N-allyl amides by the Wacker reaction with exclusion of water gave the aldehydes instead of the methyl ketones [Hosokawa, 1991]. The reaction pathway involves transfer of a hydroperoxy group from palladium to the terminal carbon atom of the olefin within a chelated (amido carbonyl) species. In the presence of water the chelation cannot be sustained, and the attack of water on the double bond follows the normal course.

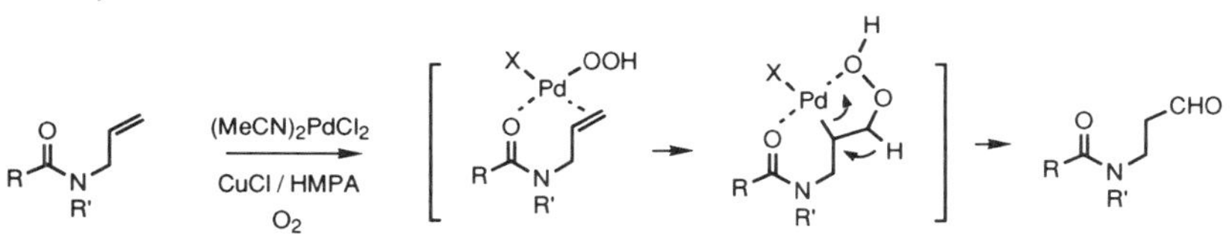

$(MeCN)_2PdCl_2$ / CuCl / HMPA / O_2

Directed hydration of the alkyne linkage of homopropargylic alcohols is possible via intramolecular hydrosilylation of the derived siloxy derivatives and subsequent oxidative-desilylation [Tamao, 1988a]. (For the hydration of allylic alcohols by the same technique, see Tamao [1988b].)

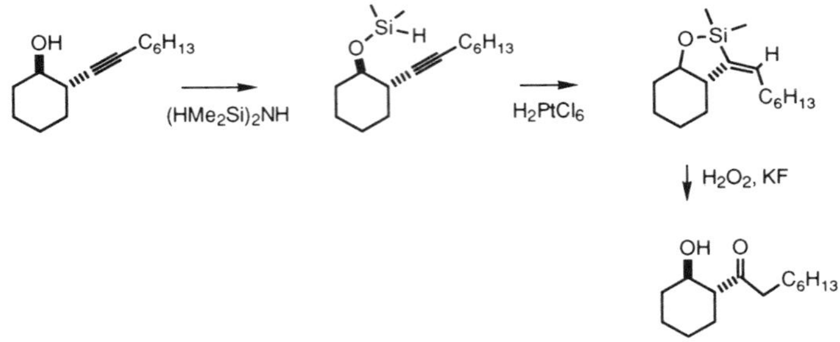

OH C_6H_{13} → (HMe$_2$Si)$_2$NH → O Si H C_6H_{13} → H$_2$PtCl$_6$ → O–Si H C_6H_{13}

↓ H$_2$O$_2$, KF

OH O C_6H_{13}

In a synthetic approach to prostaglandin-F$_{2\alpha}$ [Holton, 1977], an allylic amino group directs the regio- and stereoselective introduction of three contiguous asymmetric centers about the five-membered ring via a carbopalladation process. The only defect of this synthesis is the formation of a substantial quantity of an unsaturated side-product.

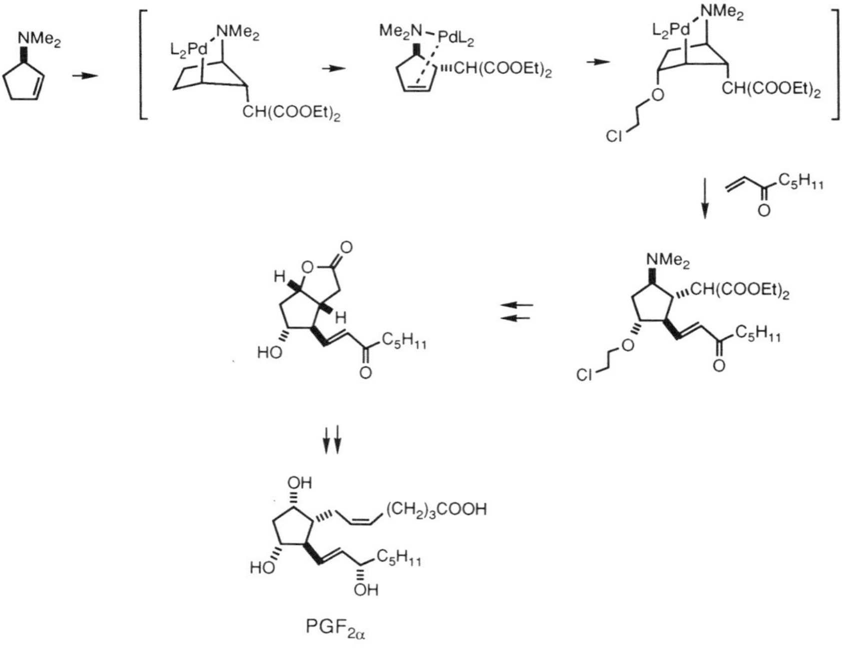

NMe$_2$ → [L$_2$Pd NMe$_2$ CH(COOEt)$_2$ → Me$_2$N–PdL$_2$ CH(COOEt)$_2$ → L$_2$Pd NMe$_2$ O Cl CH(COOEt)$_2$]

↓ C_5H_{11} O

H O O H HO C_5H_{11} O ⇇ NMe$_2$ CH(COOEt)$_2$ O Cl C_5H_{11} O

↓↓

OH (CH$_2$)$_3$COOH HO C_5H_{11} OH

PGF$_{2\alpha}$

cis-1,2-Diols can be acquired from allylic alcohols via mercuriocyclization in the presence of chloral [Overman, 1974], taking advantage of the tendency of chloral to form hemiacetals. An analogous iodocyclization has been employed to introduce the tertiary hydroxyl group at C-1 of paeonilactone-A and -B [Kadota, 1989] from an allylic alcohol derived from (−)-carvone.

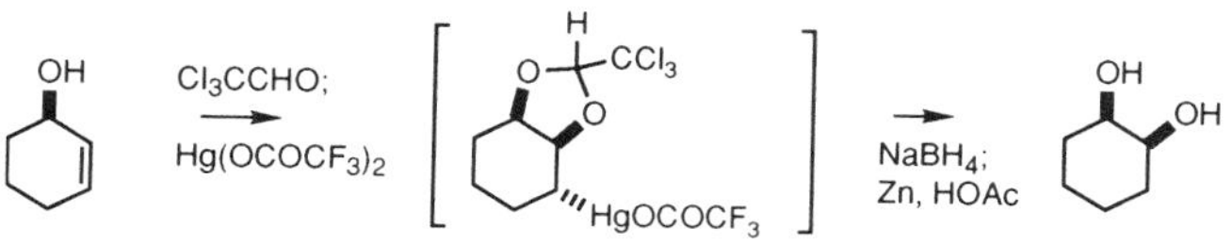

A synthesis of (+)-stoechospermol [M. Tanaka, 1985] features an intramolecular [2 + 2]-photocycloaddition. The linkage of the two addends in the form of an ester obviates the generation of regioisomer mixture.

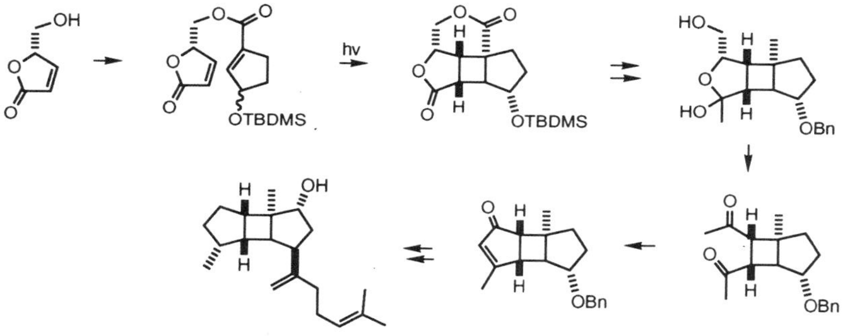

(+)-stoechospermol

The regioselectivity of several reactions of amines is the result of intervention of aziridinium species. The production of gramine derivatives by borohydride reduction of 1-(3-indolyl)-2-chloroethylamines [Julia, 1973] and the conversion of *N-t*-butylaziridinecarboxylic acids to α-chloro-β-lactams [Deyrup, 1974] by treatment with thionyl chloride, are two examples.

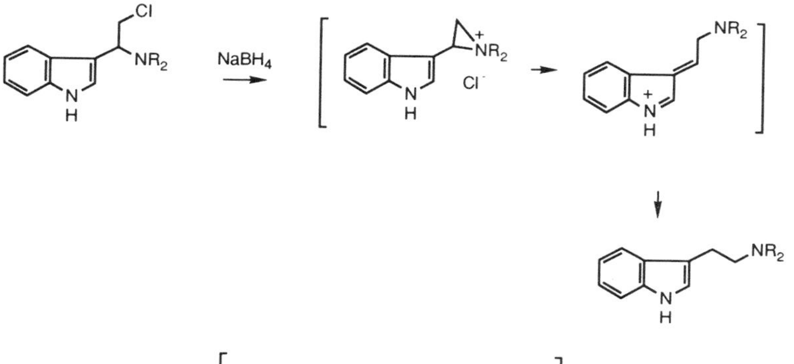

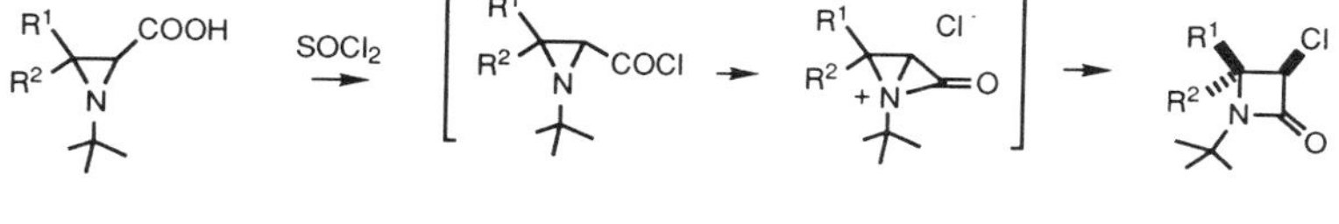

Remote functionalization is a method whereby a functional group is introduced into an unreactive position beyond the γ-atom. Regiocontrol is essential and this control is generally provided by a functional group. Further criteria for successful remote functionalization include steric accessibility of the reaction site and rigidity of the molecular framework.

The first general method for remote functionalization is the nitrite ester photolysis which gives rise to oximes at remote sites. Upon u.v. irradiation a nitrite ester undergoes homolysis and the oxy radical would abstract a hydrogen from a proximal CH_2 group before diffusion of the nitroso radical outside the solvent cage, and formation of a new C—N bond would follow. Tautomerization of the *C*-nitroso product then leads to an oxime. Molecular systems most amenable to this functionalization are those containing an axial alcohol and *cis* methyl group at the γ-carbon in a chairlike six-membered ring (1,3-diaxial relationship). Conversion of corticosterone acetate into aldosterone acetate [Barton, 1961] exemplifies the power of this method. The functionalization is at the angular methyl group between C- and D-rings.

NOCl, py; hv, PhMe

20 %

R = H corticosterone acetate

aldosterone acetate

In a synthesis of azadiradione [Corey, 1989c] based on polyene cyclization the introduction of an oxygen function to C-7 was accomplished at the tricarbocyclic stage by a Barton reaction. Although the yield was modest, the by-product, being an enal corresponding to the allylic ester, could be recycled.

hv

azadiradione

An alternative procedure for the remote functionalization is by oxidation of an alcohol with lead(IV) acetate. As with the nitrite ester photolysis, the regiochemistry is sensitive to the separation between the oxygen atom and the hydrogen to be abstracted [Burke, 1988]. Thus in a bridged ring system a 0.5Å difference in distance between the oxygen and proximal hydrogens at two δ-carbon atoms biases completely the reaction at the closer site [Beddoes, 1992]. This result has great significance in a synthetic approach to stemofoline. It must be emphasized that other structural features can influence hydrogen abstraction; geminal oxygen atom tends to activate the C—H bond [Micovic, 1969].

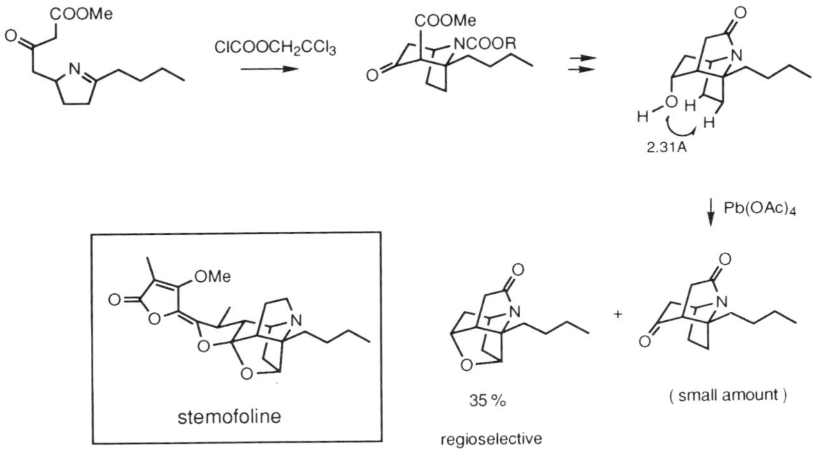

COOMe
$ClCOOCH_2CCl_3$
COOMe
NCOOR
2.31A
$Pb(OAc)_4$
OMe
stemofoline
35 %
regioselective
(small amount)

A remote functionalization tactic was capitalized to provide entry into the dodecahedrane skeleton by manipulation of a pagodane diamide [Pinkos, 1989; Melder, 1989].

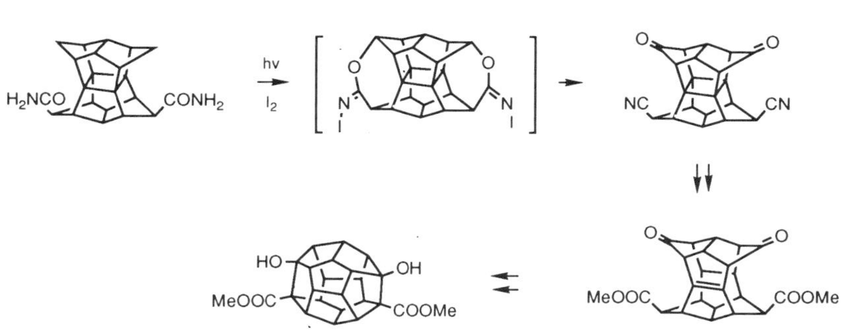

H_2NCO
$CONH_2$
hv
I_2
NC
CN
HO
OH
MeOOC
COOMe
MeOOC
COOMe

An extension of this oxidation is remote cyanation [Kalvoda, 1968] which delivers a cyano group from an α-cyanohydrin to a remote methyl substituent. The reaction involves hypoiodite formation and homolysis, $O \rightarrow C$ radical transfer, and plausibly addition of the radical to the CN triple bond, ring cleavage, and hydrogen loss.

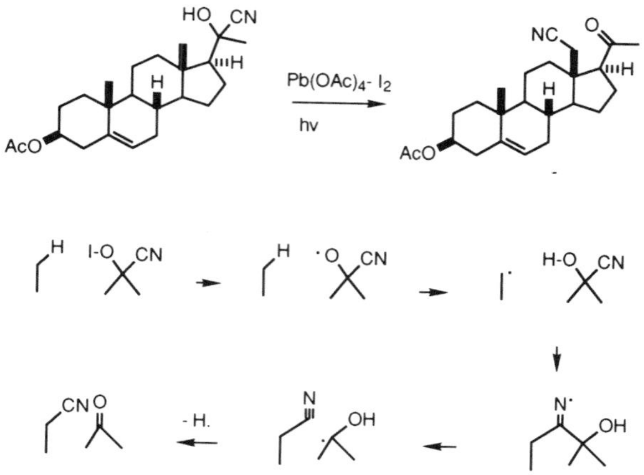

Directed chlorination of the steroid skeleton mediated by the iodine atom of iodophenylalkanoic esters [Breslow, 1977a] or *p*-(phenylthio)phenylacetic ester [Breslow, 1976] of 3α-cholestanol in free radical relay reactions has been demonstrated.

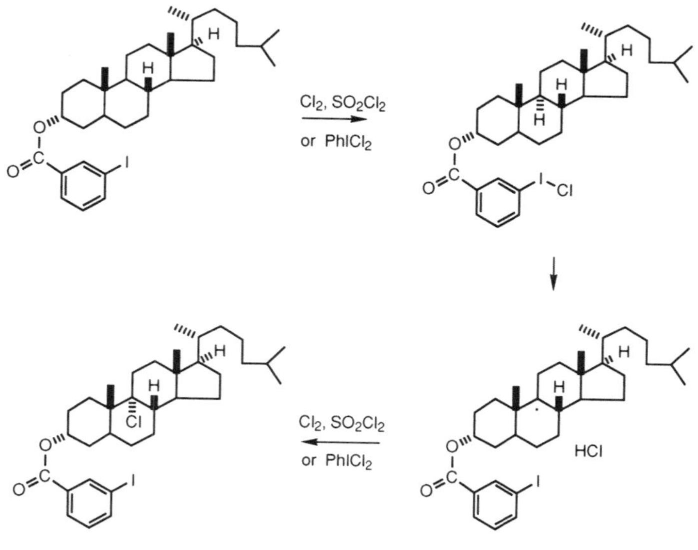

A process which involves mercury (II) ion-induced opening of cyclopropane with participation of a proximal hydroxyl group, followed by reductive scission of the C—Hg bond to generate a primary radical which rapidly removes the α-hydrogen of the cyclic ether (actually a 1,3-dioxane system) before bimolecular reaction with acrylonitrile constitutes a novel method for stereocontrolled α-alkylation of a secondary alcohol. The synthetic use of this reaction sequence has been demonstrated by a synthesis of (+)-ipomeamarone [Sugimura, 1993]. Note the C—C bond formation step is subjected to 1,3- and 1,4-asymmetric induction by steering the approach of the radicalophile.

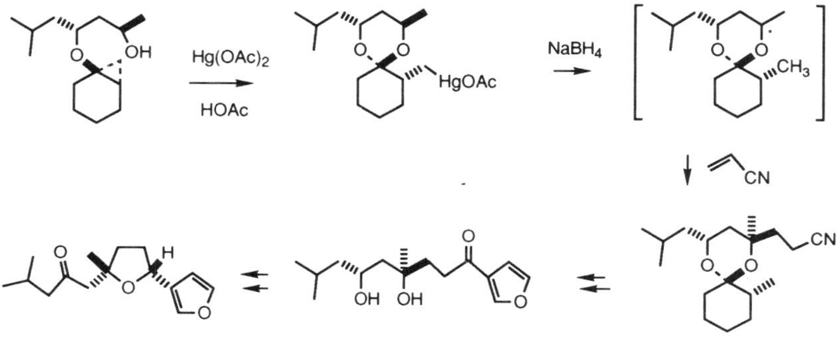

(+)-ipomeamarone

A regiocontrol problem was encountered during hydroformylation of a cyclohexene ring at the late stage of a (+)-phyllanthocin synthesis [Burke, 1986a]. A satisfactory method has been found based on providing the rhodium catalyst with an internal phosphine ligand. The dramatic control is reflected in the generation of mainly the undesirable regioisomer by using a *p*-phosphinobenzoate tether instead of the *m*-phosphinobenzoate.

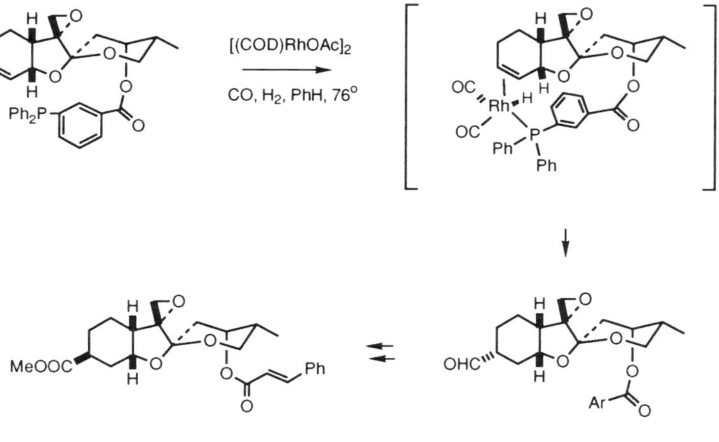

(+)-phyllanthocin

major

(vic-isomer ratio 8 : 2)

The formation of *cis*-1,2-cyclohexanediol, isolated as the diacetate, from rhodium-catalyzed hydroboration of 3-diphenylphosphinoxycyclohexane [Evans, 1992c] can be attributed to coordination effect.

Selective activation of C—H bond by palladation of oxime to form azapalladacyclopentene and subsequent cleavage of the C—Pd bond demonstrates neighboring effect. Thus it is possible to introduce an acetoxy group to C-23 (equatorial methyl) of lupanone via its oxime [J.E. Baldwin, 1985].

7

TEMPLATE AND CHELATION EFFECTS

7.1. TEMPLATE EFFECTS

Template effect is an often-encountered term in the recent chemical literature. According to definition, template is an instrument used as a guide or gauge in bringing a piece of work to a desired shape, therefore a linear block of carbocyles to which a free radical initiator and a chain terminator group are laterally linked is a chemical analogy. In the presence of this template, oligomerization of methacrylic ester is precisely controlled by the breadth of the gap between the initiator and the terminator [Feldman, 1987].

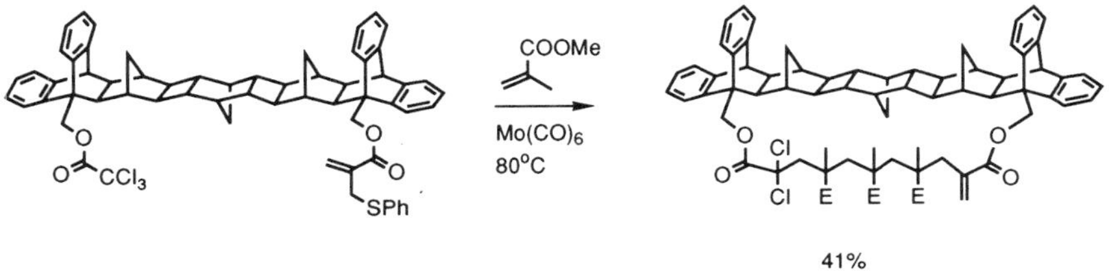

COOMe

$Mo(CO)_6$
80°C

41%

A template can also be an assembling platform for new molecules. As an illustration, it has shown that a polymer-bound hypoxanthine can act as template for the production of 1-benzyl-5-phenylimidazole [Ranganathan, 1990] by consecutive reaction with phenacyl bromide and benzylamine. Regeneration of the template was by *N*-debenzylation and treatment with formamide.

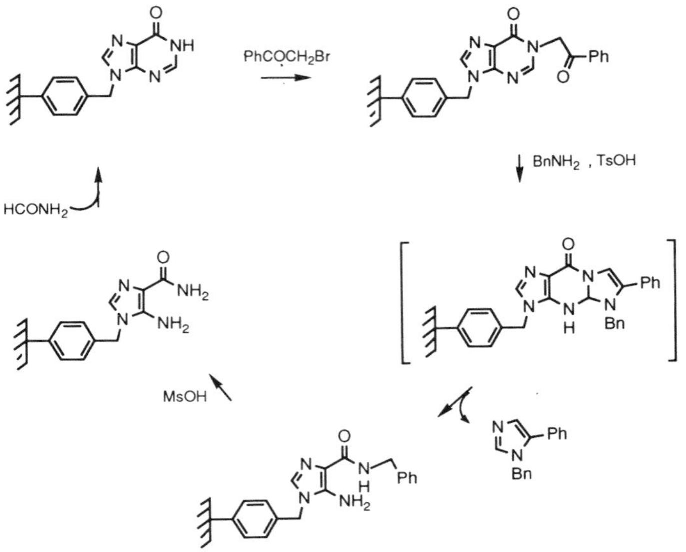

A complexing agent which exerts influence on the stereochemical course of a reaction by providing spatial bias can also be regarded as a template. Consequently, in the broad sense steric shielding by a bulky auxiliary can be regarded as template effect.

7.1.1. Stereocontrol via Facial Differentiation

By whatever reason or means, the facial blockade of a reaction substrate to an approaching reagent leads to effective diastereoselection. When the blocking instrument is readily removable, the method involving such or similar device is even more valuable for synthetic application. Thus an enantiomerically pure γ-lactone, readily obtained via nitrosation of (*S*)-glutamic acid, borane reduction and tritylation, continues to attract application in synthesis, by virtue of the 1,3-and 1,4-asymmetric induction as controlled by the bulky trityl group. Furthermore, the (*R*)-isomer is available by Mitsunobu inversion of the corresponding hydroxy acid, rendering synthesis of a chiral molecule in either enantiomeric series feasible. On the basis of proper manipulations, indole alkaloids including (−)-antirhine [Takano, 1981b], (+)-eburnamine [Takano, 1985], both enantiomers of velbanamine [Takano, 1980b, 1982b] and quebrachamine [Takano, 1980a] have been elaborated. The synthesis of several lignans and neolignans has been made possible by employing this building block: (+)-podorhizon [Tomioka, 1979], (−)-steganone [Robin, 1980], (+)-steganacin [Tomioka, 1980], (−)-megaphone [Tomioka, 1985], and a chiral intermediate for calonectrin [Tomioka, 1987]. It is noted that these syntheses involve self-immolation of the asymmetric center in the γ-lactone.

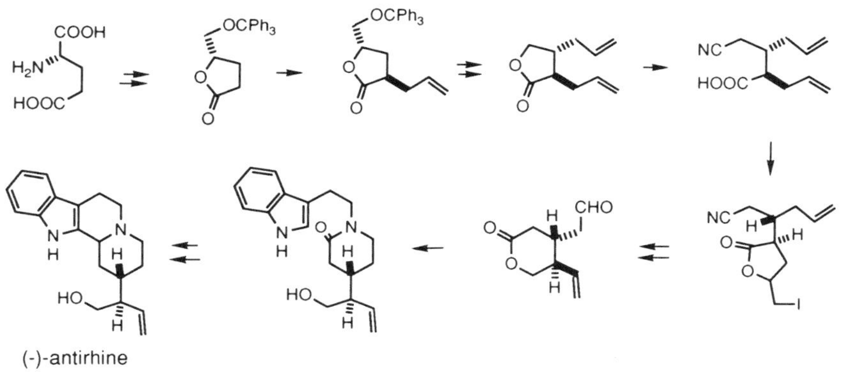

(-)-antirhine

LDA, Na_2SO_4

epimeric

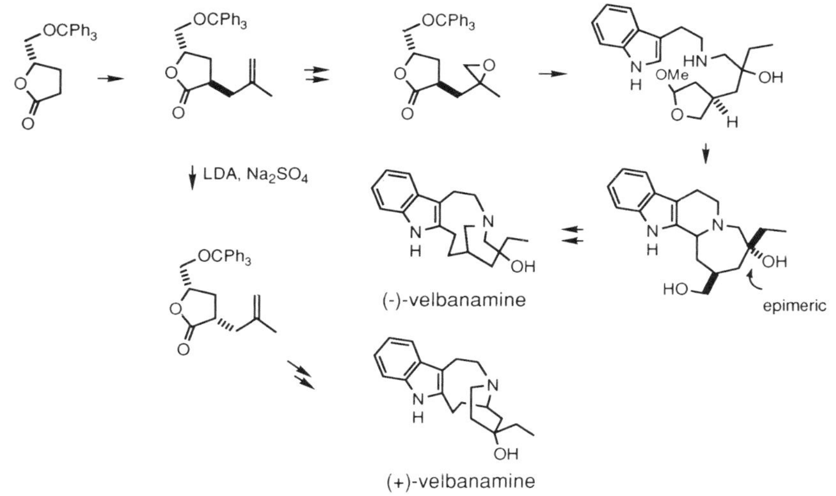

(-)-velbanamine

(+)-velbanamine

MsO^-

(+)-quebrachamine

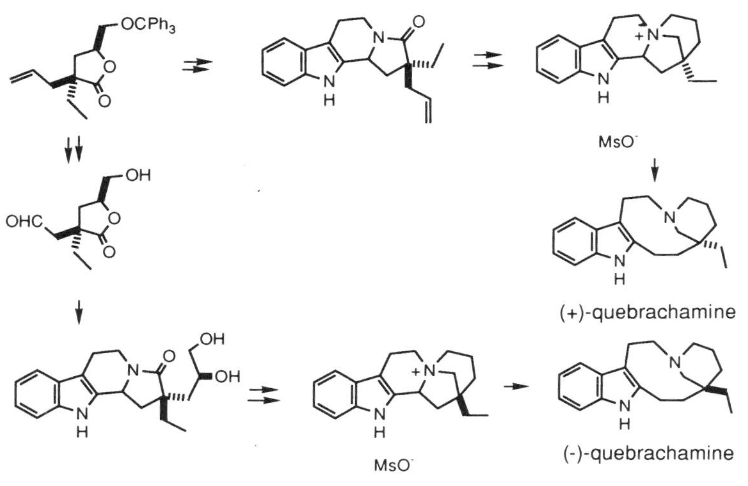

MsO^-

(-)-quebrachamine

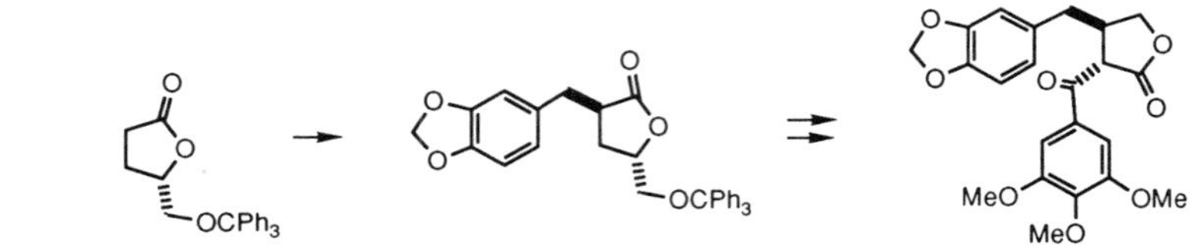

(+)-podorhizon

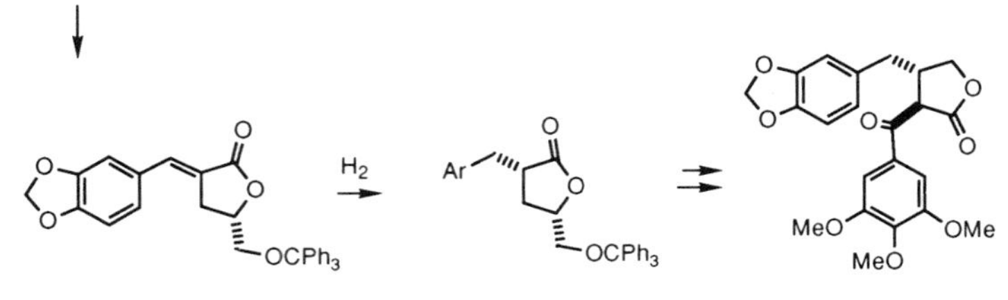

(-)-podorhizon

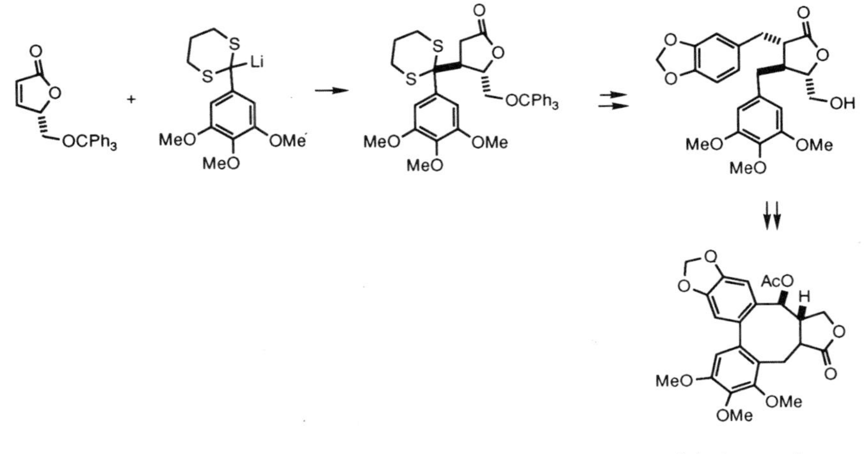

(+)-steganacin

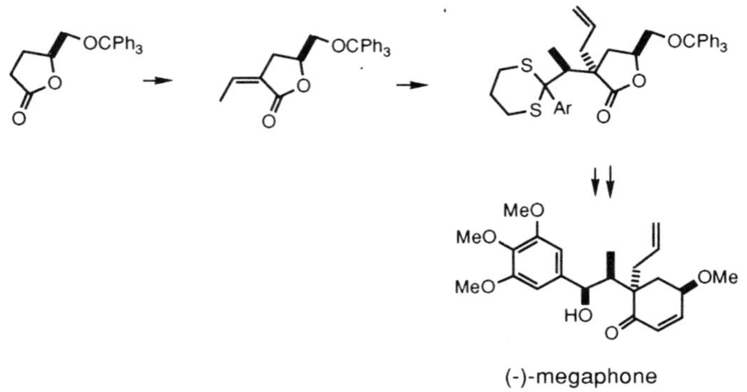

(-)-megaphone

NaH / HCOOEt ; MVK / Et_3N

(-)-calonectrin

The stereocontrol of the entering electrophile at the α-position of the butyrolactone [Hanessian, 1985a] and that of a nucleophile at the β-position of the corresponding butenolide [Hanessian, 1985b] proved extremely useful in chain building processes which have implications in the synthesis of polypropionate metabolites. The value is further increased by a demonstration of reiteratability.

The (*R*)-butenolide has also served as a building block for (−)-swainsonine [Ikota, 1987].

(*R*)/(*S*)-5-Trimethylsilyl-2-cyclohexenone is also a valuable chiron [Asaoka, 1990b] because the racemate is accessible from Birch reduction of anisole and *in situ* silylation, followed by acid hydrolysis and conjugation. Kinetic resolution by cinchonidine-catalyzed Michael addition with half-equivalent of *p*-toluenethiol which leaves the (*S*)-(+)-enone unreacted. The adduct is a source of the (*R*)-(−)-isomer (by elminination). The trimethylsilyl group of the enone directs Michael addition from the other face by virtue of its equatorial occupancy, ensuring the favorable axial attack of various reagents. Even more useful is the regiocontrol exerted by the silyl substituent on Baeyer–Villiger reaction resulting in oxygen insertion in the C—C bond closer to the silicon atom. The silyl group can also be removed oxidatively to generate a new cyclohexenone. Based on the silylcyclohexenone several natural products have been synthesized: (−)-enterolactone [Asaoka, 1988a], (+)-α-cuparenone [Asaoka, 1988d], (+)-α-curcumene

[Asaoka, 1988c], (−)-β-vetivone [Asaoka, 1988f], (+)-4-butyl-2,6-cycloheptadienone [Asaoka, 1988e], (−)-*O*-methyljoubertiamine [Asaoka, 1988b], (+)-ramulosin [Asaoka, 1989a], (+)-quebrachamine [Asaoka, 1989b], (+)-ptilocaulin [Asaoka, 1990a], (+)-magydardienediol [Asaoka, 1991].

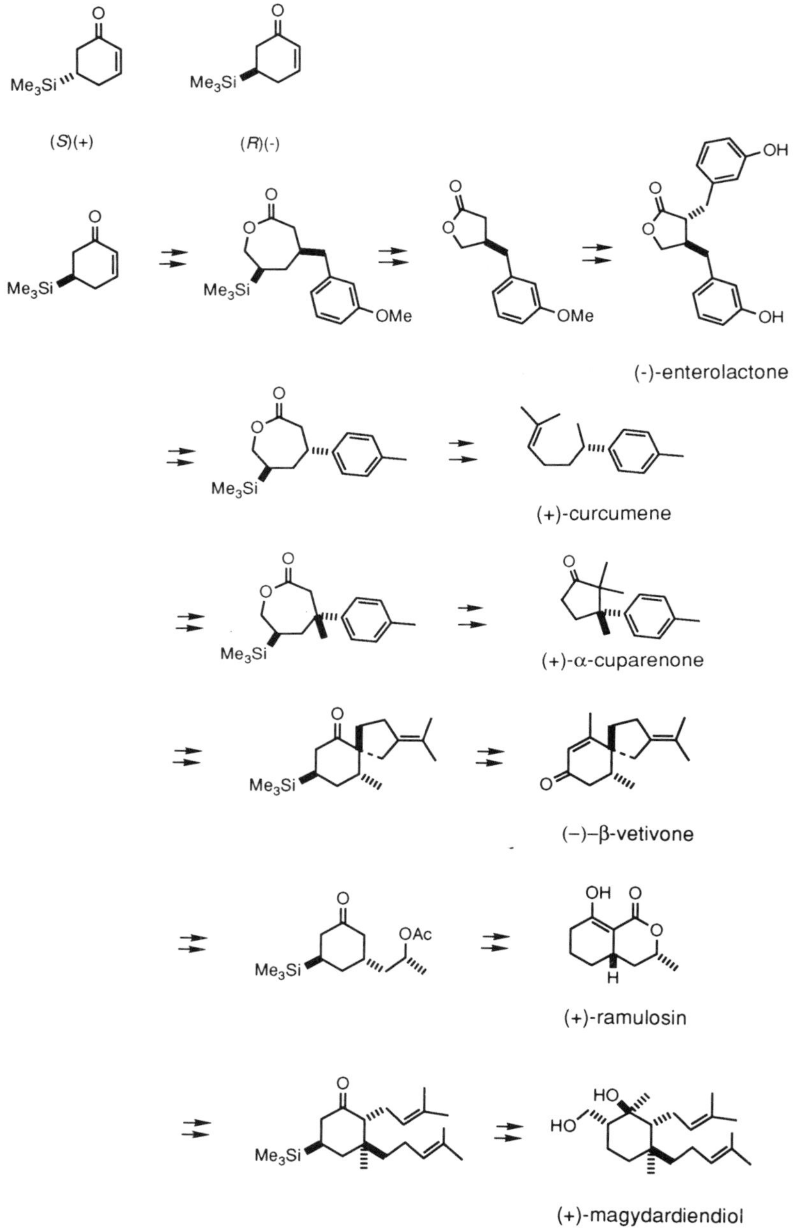

O O
Me_3Si Me_3Si

(*S*)(+) (*R*)(-)

O O O O OH
Me_3Si Me_3Si OMe OMe OH

(-)-enterolactone

O
Me_3Si

(+)-curcumene

O
Me_3Si O

(+)-α-cuparenone

O
Me_3Si O

(−)-β-vetivone

O OH O
OAc O
Me_3Si H

(+)-ramulosin

O HO
Me_3Si HO

(+)-magydardiendiol

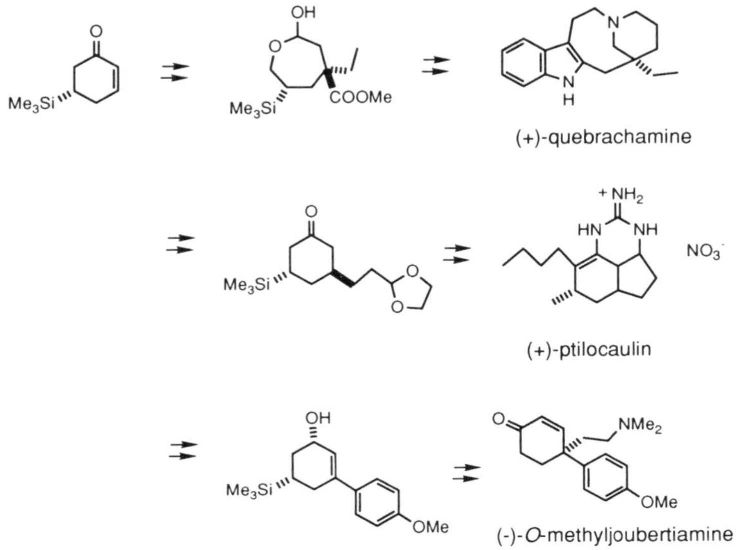

(+)-quebrachamine

(+)-ptilocaulin

(-)-*O*-methyljoubertiamine

Apparently the trimethylstannyl group can serve the same function [Piers, 1991]. Its reductive removal is effected by treatment with lithium in liquid ammonia in the presence of *t*-butanol.

An allylic silyl group can exert steric influences on functionalization of the double bond. For example, hydroboration with 9-BBN of 4-(phenyldimethyl)silyl-2-pentene gives rise to the *anti*- or *syn*-alcohol in >95:5 ratio according to whether the alkene has an (*E*)- or (*Z*)-configuration [Fleming, 1988].

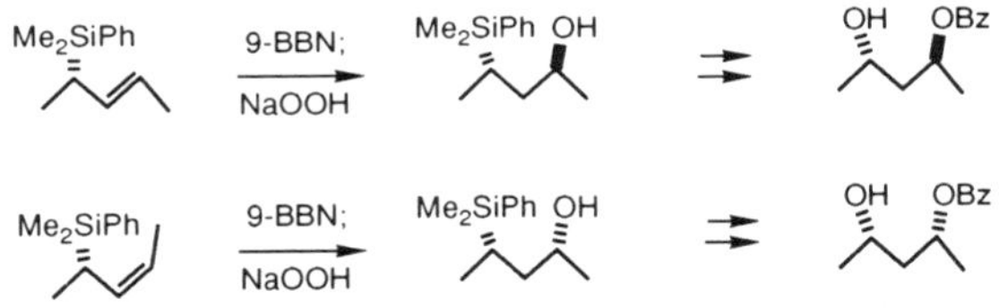

2-(*p*-Toluenesulfinyl)-2-cycloalkenones and -2-alkenolides present themselves to Michael donors with the face opposite to the aryl group [Posner, 1983, 1986]. Enantioselective access to β-alkoxycarbonylmethylcycloalkanones and lactones has become very convenient.

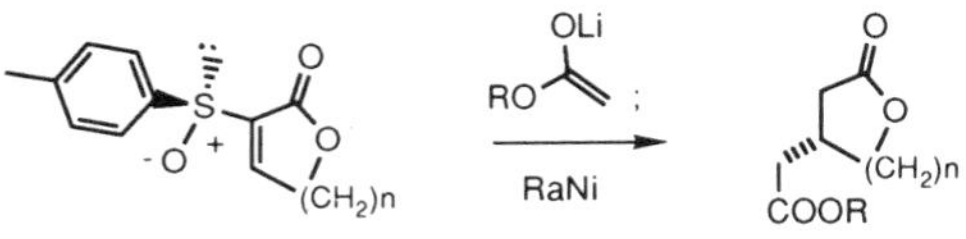

Asymmetric induction (>99% ee) has been observed in the conjugate addition of organocopper reagents to 8-phenylmenthyl crotonate [Oppolzer,

1981b]. The bulky alkoxy residue provides excellent diastereofacial guidance and enantioselectivity. Another good auxiliary is the isobornyl system in which C-10 is substituted with an *N*,*N*-dicyclohexylsulfonamide group [Oppolzer, 1983b].

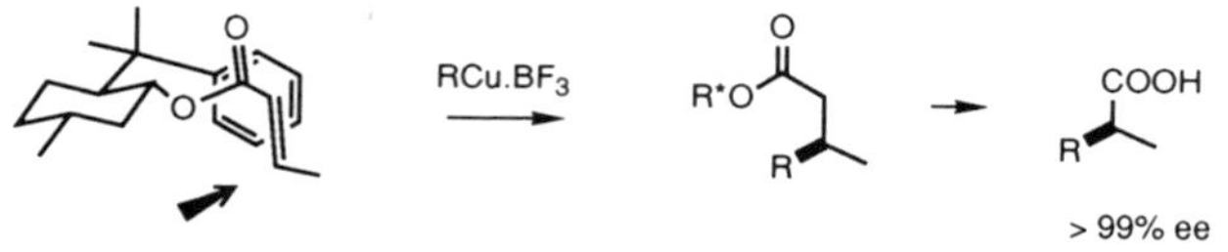

RCu.BF_3 R*O R COOH R > 99% ee

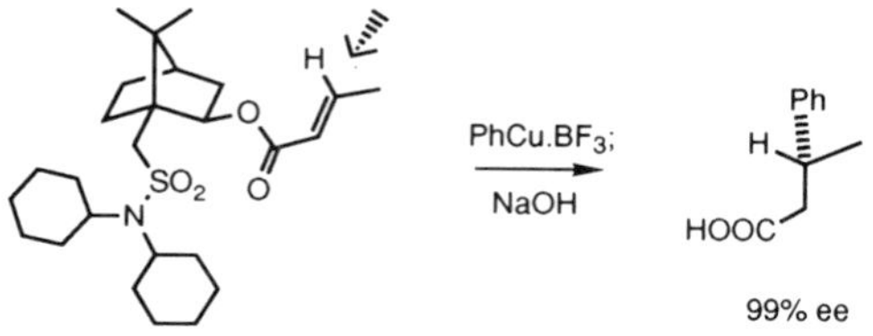

H SO_2 N O PhCu.BF_3; NaOH Ph H HOOC 99% ee

The bulky group can also play a crucial role in determining the steric course of intramolecular Michael addition [Stork, 1986a] and rearrangement [O. Takahashi, 1987]. Its effect is reflected in the excellent enantioselectivity and diastereoselectivity.

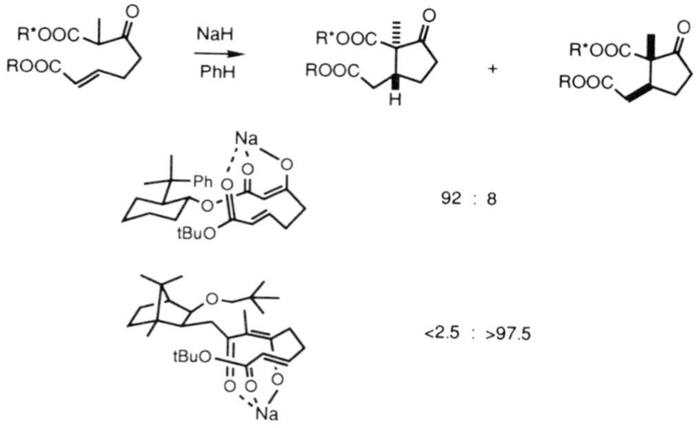

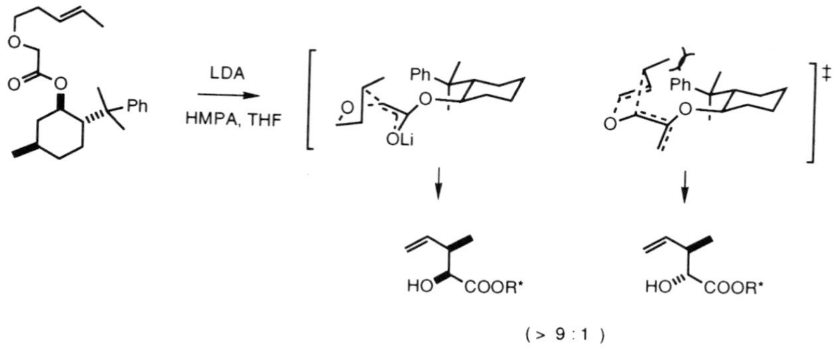

R*OOC ROOC O NaH PhH R*OOC ROOC H O + R*OOC ROOC O

Na O Ph O O tBuO 92 : 8

O tBuO O O O Na <2.5 : >97.5

O O O Ph LDA HMPA, THF [Ph O O OLi Ph O O]‡

HO COOR* HO COOR*

(> 9 : 1)

Thus it is evident that effective stereocontrol by intrinsic and auxiliary groups is possible. In the same vein the establishment of the configuration of the methyl-bearing carbon atom of annotinine by hydrogenation of a methyl-enecyclobutane predecessor is instructive [Wiesner, 1969]. The steric bias was set by ketalization.

$(CH_2OH)_2$ / H^+ ; H_2

annotinine

Sometimes the selection of conformational isomers in reactions is influenced by complexing agents. For example, the steric course of the cuprate reaction on 6-*t*-butyl-2-methyl-2-cyclohexenone is dependent on whether trimethylsilyl chloride is present initially [Horiguchi, 1989]. In such circumstances the complexation erects a template effect of the *t*-butyl group which assumes an axial orientation in order to avoid steric compression by the complexed silyl residue.

Bu_2CuLi ; Me_3SiCl / HMPA ; $OSiMe_3$; Bu

Bu_2CuLi / Me_3SiCl ; $OSiMe_3$; Bu

Substantial facial differentiation has been witnessed in reactions of *trans*-2-phenylcyclohexane derivatives. Thus peroxidation of a nonconjugated diene leads to two diastereomeric alcohols [Dussault, 1992]. The 92:8 ratio reflects the extended and the folded conformation of the diene substrate, and only one face is exposed to attack by external reagents.

folded ; extended

O_2; Ph_3P ; O_2; Ph_3P

HO ; HO

8 : 92

Heterobicyclization of 2-octenedial on exposure to a chiral oxazolidine [S.L. Schreiber, 1988] afforded a 17:1 mixture of diastereomers. The transition state in which the large substituent of the oxazolidine ring is kept away from interaction with the β-hydrogen of the enamine intermediate must be the preferred one for cyclization.

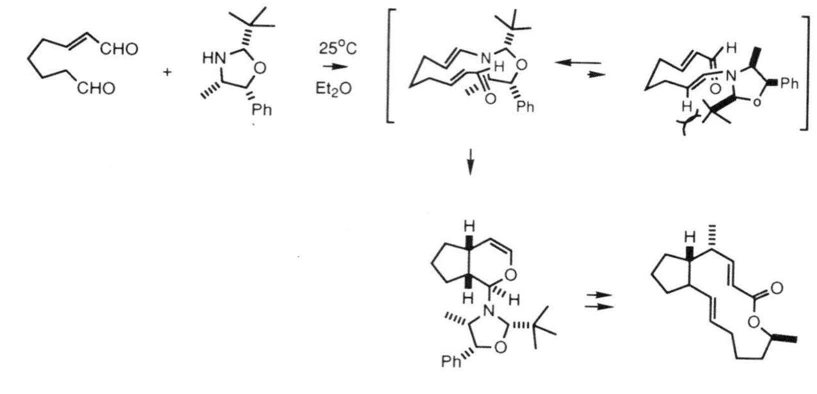

(+)-brefeldin-C

A remarkable observation is that the reduction of the C-15 ketone of a prostaglandin intermediate gives rise to the same ratio of alcohols by using lithium triorganoborohydride reagent derived from (+)- or (+/−)-limonene. It has been suggested [Corey, 1972a] that the *p*-phenylbenzoyloxy group at C-11 interacts with the enone sidechain so that a cisoid enone conformation is more favorable, and the large group also shields one face of the sidechain so that reduction by the bulky reagent is allowed from the other face predominantly.

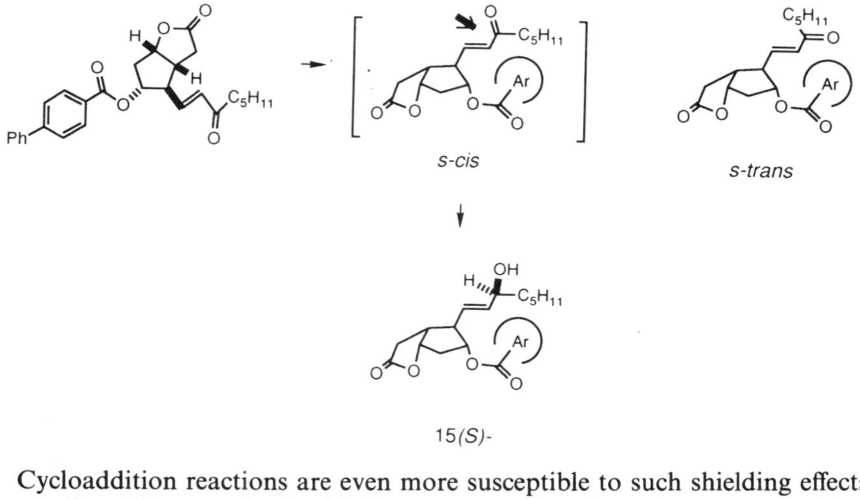

s-cis

s-trans

15*(S)*-

Cycloaddition reactions are even more susceptible to such shielding effects, as manifested by a synthesis of (−)-α-cuparenone and (+)-β-cuparenone [A.E.

Greene, 1987] in which only the *re*-face of the ketenophile is exposed. Intramolecular Pauson–Khand bicyclization is also subject to such diastero-control [Castro, 1990].

Tol Cl Cl_2=C=O Tol H Ph O Cl Cl O CH_2N_2; $Cr(ClO_4)_2$ O Cl (+) (+)-β-cuparenone O O (-)-α-cuparenone

O $Co_2(CO)_8$ Δ *RO H O *RO O 7 : 1

O $Co_2(CO)_8$ Δ *RO O *RO H O 5 : 1

Palladium-mediated trimethylenemethane cycloaddition to electron-deficient alkenes is a valuable method for creation of cyclopentane derivatives [Trost, 1986]. In a syntheis of (−)-rocaglamide [Trost, 1990b] (2*R*, 3*S*)(6*Z*)-benzyl-idene-3,4-dimethyl-2-phenylperhydro-1,4-oxazepine-5,7-dione reacts from the same face which the methine protons of the chiral auxiliary project. Note that the stereoselectivity is the same as observed in conjugate organometallic additions to this system [Mukaiyama, 1978c], but opposite to that of an intramolecular Diels–Alder reaction [Tietze, 1987].

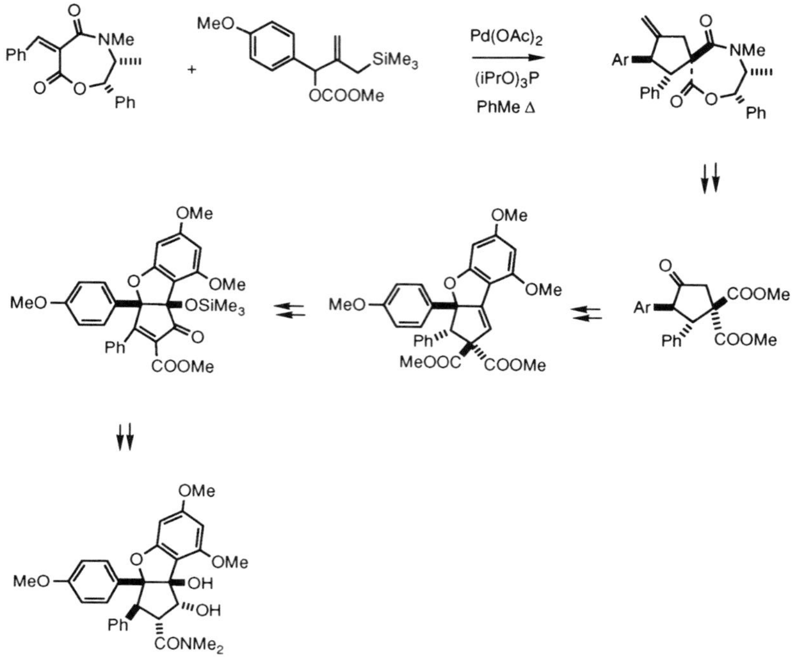

(-)-rocaglamide

The *exo*-selectivity shown by aminocarbene complexes in the Diels–Alder reaction [B.A. Anderson, 1992] is likely the result of steric hindrance by the carbonylmetal fragment to the approaching diene to attain an *endo* transition state.

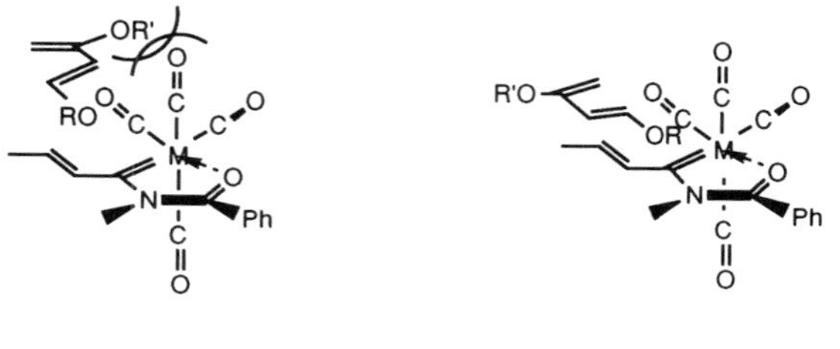

endo *exo*

α-Chloro nitroso compounds undergo hetero-Diels–Alder reactions readily and they can be used to synthesize *cis*-4-amino-2-cyclohexenols. A chloronitroso derivative based on a furanomannose diacetonide condenses with cyclohexadiene to generate an oxazabicyclo[2.2.2]octene with >96% ee [Felber, 1986], via an *endo* transition state.

A seemingly less rigid molecular system, 1-mesityltrifluoroethanol, proves to be a superb chiral controller (in one of the enantiomers) for catalyzed Diels–Alder reactions of its acrylic ester [Corey, 1991a].

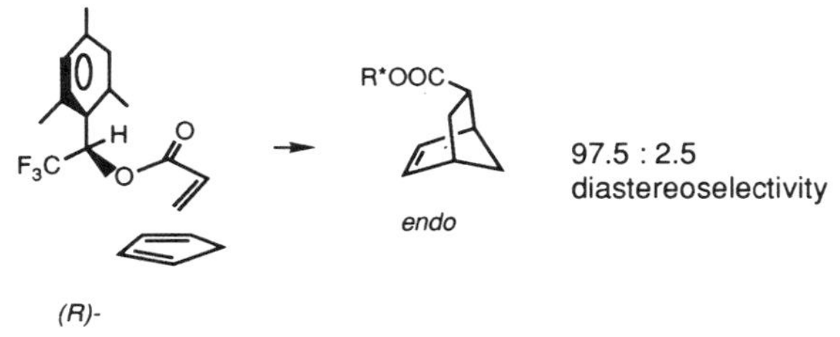

97.5 : 2.5 diastereoselectivity

A versatile chiral diazaluminolidine catalyst has been used to achieve asymmetric Diels–Alder reaction of an acrylamide [Corey, 1992b]. Nmr studies confirm the ordered structure of the dienophile-catalyst complex in which one face of the dienophile is effectively shielded from an approaching diene.

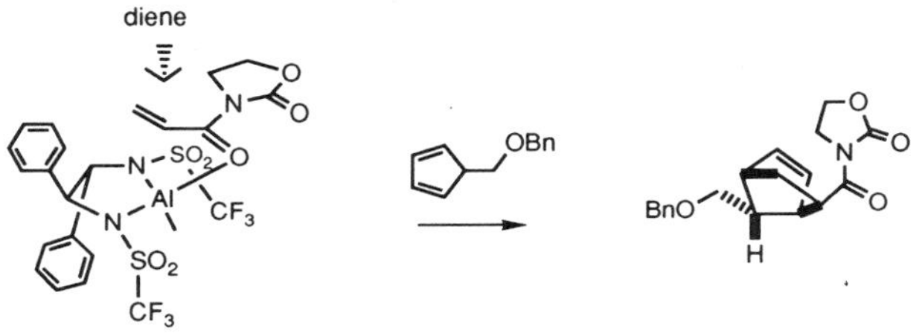

An even more elaborate catalyst designed to restrict conformation and facial availability of the dienophile is a 4-substituted 1,3,2-oxazaborolidine-5-one prepared from (*S*)-tryptophan [Corey, 1991c], and its effectiveness is shown in the reaction of cyclopentadiene with α-bromoacrolein. The preorganization involving O–B coordination and π-π interaction of the aromatic ring and the dienophile is responsible for the (*R*/S)-selectivity of 200:1.

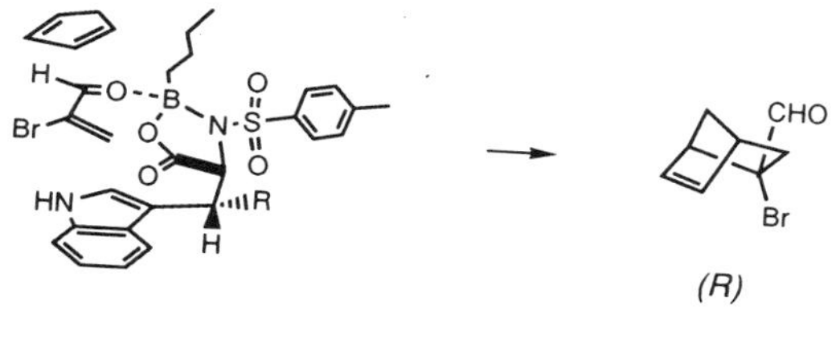

The remarkable stereobias in reactions of chiral spirocyclic dioxinones derived from (−)-menthone is due to a sofa-shape conformation of the heterocycle. A synthesis of a loganin precursor [M. Sato, 1991] exploited this structural feature.

hv

$CH(OAc)_2$

$OSiR_3$

15:2

COOMe

HO

OH

Opposite facial selection is observed in the reaction of Me_2CuLi with an *o*-(2-imidazolinyl)cinnamate ester [Alexakis, 1990b] in the presence or absence of trimethylsilyl chloride. With Me_3SiCl for entrapment of the kinetic π-complex derived from the reagent approaching from the less hindered face of the molecule, the (*S*)-isomer is observed. On the other hand, without Me_3SiCl in the reaction mixture, equilibration can be established, and within a chelated π-complex the methyl group delivery from the *re*-face is possible.

Me_2CuLi

COOEt

(*R*)

Me_2CuLi

Me_3SiCl

(*S*)

The reaction of α-(*N*,*N*-dibenzylamino)aldehydes with many carbanionoid reagents proceeds with excellent diastereoselectivity in favor of the nonchelation control products [Reetz, 1991, 1993]. An X-ray analysis of the aldehyde derived from phenylalanine clearly showed shielding of one face of the aldehyde group which is the most likely cause of the observed reaction pattern, although arguments based on ground state conformation are tenuous in most other instances. Nonchelation control also dominates in the reactions of the corresponding *N'*-tosylaldimines with Grignard reagents and with ester enolates, apparently due to depletion of electron by the tosyl group on the imino nitrogen for chelation, leaving the template effect alone to operate. The importance of the protecting group "tuning" has been further demonstrated in the reaction of the *N*,*N*,*N'*-tribenzyl analogs with organocerium compounds, cerium also being a better chelating metal.

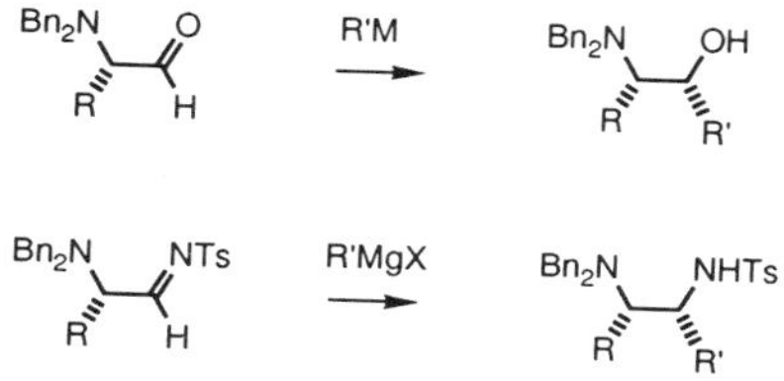

A (4Z, 8Z)-dodecadienedioic ester which is linked to a rigid template shows very high stereoselectivity in bisepoxidation [Feldman, 1992]. The incorporation of two *gem*-dimethyl groups at the α-positions further restricts conformational mobility of the carbon chain.

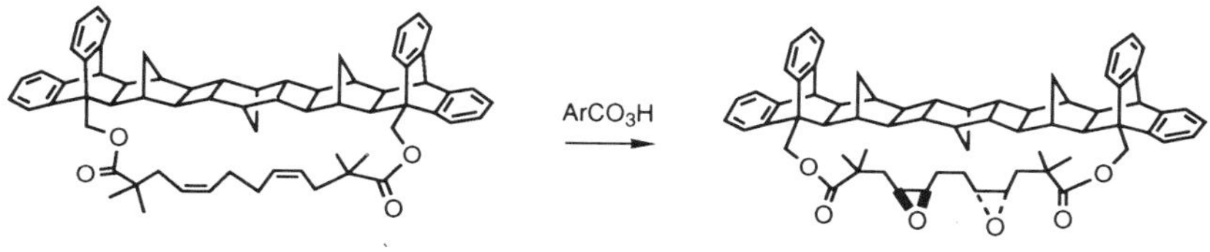

The *endo*-oriented five-membered ring of dicyclopentadiene displays very high facial discrimination in favor of *exo*-approach of reagents. This behavior persists in various derivatives and its exploitation for synthetic purposes has been abundant. Recent examples include synthesis of (−)-herbertene [Takano, 1992a] and (+)-estrone [Takano, 1992b].

In the latter synthesis the manipulation of the C/D-ring juncture is particularly significant. While *exo*-approach of the diene establishes the C-14 configuration, the angular methylation in a subsequent step also occurs from the same side to give a product possessing the *trans*-juncture. The present method is a unique solution to a long-standing problem in steroid synthesis that without the facial blockade this methylation results predominantly in the *cis*-epimer.

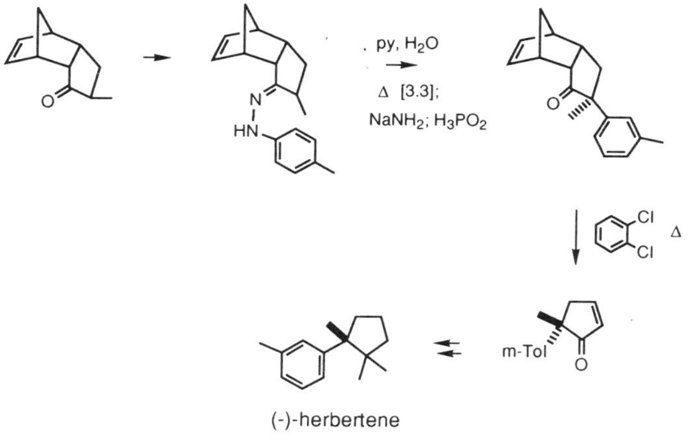

(-)-herbertene

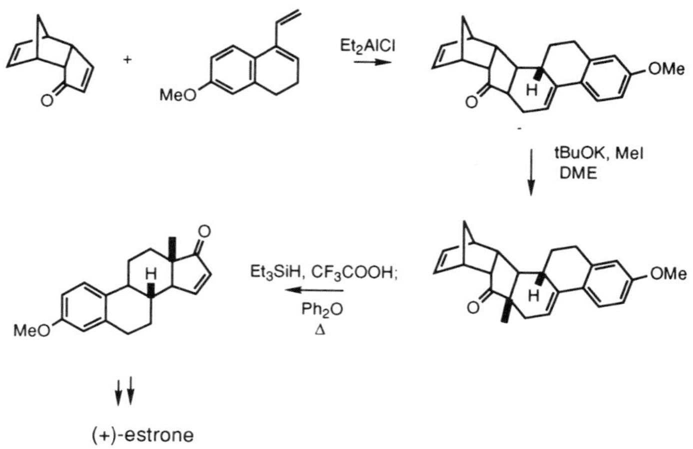

Et_2AlCl

tBuOK, MeI DME

Et_3SiH, CF_3COOH; Ph_2O Δ

(+)-estrone

A route to (+)-laburnine [Arai, 1991] via stereocontrolled intramolecular trapping of an acyliminium ion is also based on the same principle, although interestingly the protonation at the subangular position occurred from the concave side of the pyrrolizidine system.

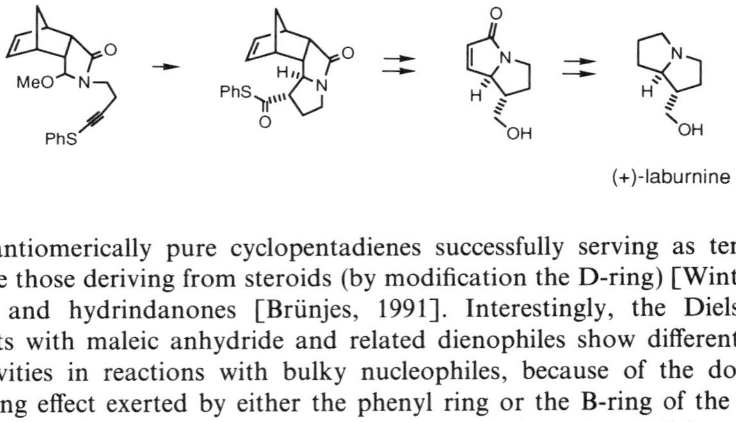

(+)-laburnine

Enantiomerically pure cyclopentadienes successfully serving as templates include those deriving from steroids (by modification the D-ring) [Winterfeldt, 1993] and hydrindanones [Brünjes, 1991]. Interestingly, the Diels–Alder adducts with maleic anhydride and related dienophiles show different regio-selectivities in reactions with bulky nucleophiles, because of the dominant shielding effect exerted by either the phenyl ring or the B-ring of the steroid skeleton, respectively, making it possible to synthesize (*R*)- or (*S*)-butenolides [Beckmann, 1990]. Furthermore, kinetic resolution of dienophiles by the Diels–Alder reaction may provide useful chiral building blocks such as that used in a synthesis of the didemnenones [Bauermeister, 1991].

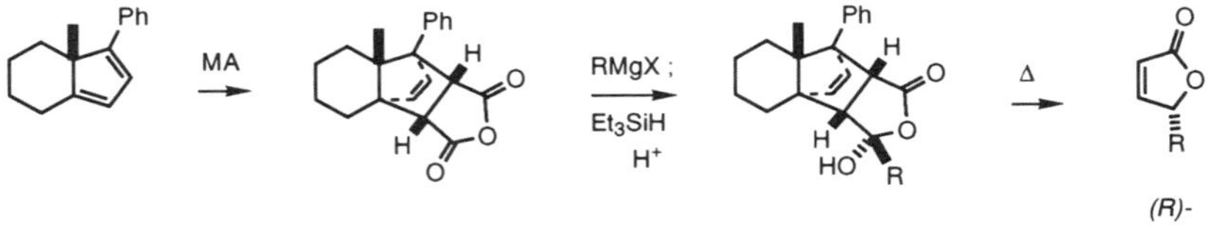

MA

RMgX; Et_3SiH H^+

Δ

(*R*)-

MA; RMgX; Et_3SiH; H^+; Δ; (S)-

The unusual structure of proline among α-amino acids makes it useful as a template for synthesis of many other optically active substances. A general synthesis of α-amino acids [Bycroft, 1975], shown in the scheme, has obvious advantages.

DCC; R'''NH_2; CF_3COOH; H_2/ Pt; H_3O^+

π-Facial selection can often be accomplished by temporary shielding with bulky metal species. For example, alkylation of indanone and α-tetralone occurs *anti* to the metal when these ketones are ligated to tricarbonylchromium [Jaouen, 1975]. The reductive methylation at the benzylic position of a phenone has also been regulated via reactions on the tricarbonylchromium complex [Uemura, 1988]. Thus it is possible to obtain either the (*R*)- or (*S*)-isomer by variation of reagents. *cis*-2,6-Dialkylation of a 3-cyclohexenone has been demonstrated in a molybdenum complex [Pearson, 1989], and *cis*-2,5-disubstituted 5,6-dihydro-2*H*-pyrans (such as those required for synthesis of monic acid derivatives) and *cis*-2,6-disubstituted tetrahydropyrans are also readily prepared [Hansson, 1990].

NaH; MeI

MeLi; Et_3SiH, $BF_3.Et_2O$

(R = H, R' = Me; R = Me, R' = H)

$NaBH_4$; Ac_2O, py; Me_3Al

A similar stereocontrol permitted the preparation of *cis*-5,7-dimethyl-1,3-cycloheptadiene which was employed in a synthesis of Prelog–Djerassi lactone [Pearson, 1988, 1990]. Actually, the synthetic pathway also involved *cis*-1,4-diacetoxylation on the basis of palladium template, that is metal complexation in opposite face to the methyl groups and acetate ion entry in the formal double inversion sense.

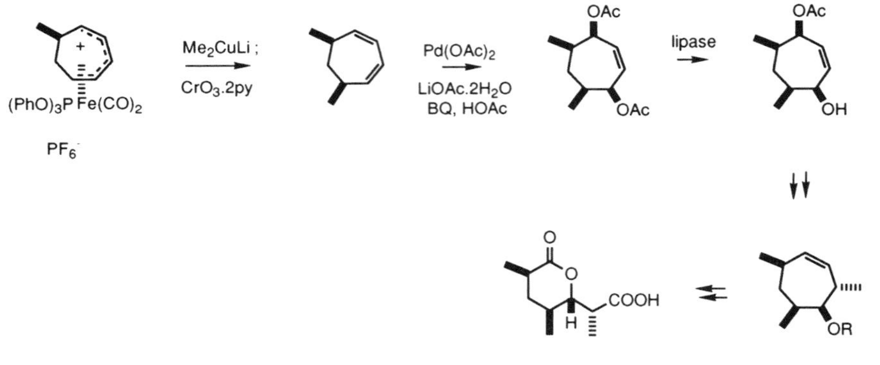

(+)-Prelog-Djerassi lactone

Most interestingly, the tricarbonyliron complex of 2-formyl-1,3-butadiene is attacked by methyllithium selectively from the exo side, but the addition occurs from the *endo*-face when the nucleophile is changed to lithium dimethylcuprate [Franck-Neumann, 1986]. The latter reaction proceeds via relay through the metal atom, probably due to the softness of the reagent.

As expected, the tricarbonyliron complex of methyl 3,5-hexadienoate undergoes methylation at C-2 from the side opposite to the metal [W.A. Donaldson, 1992].

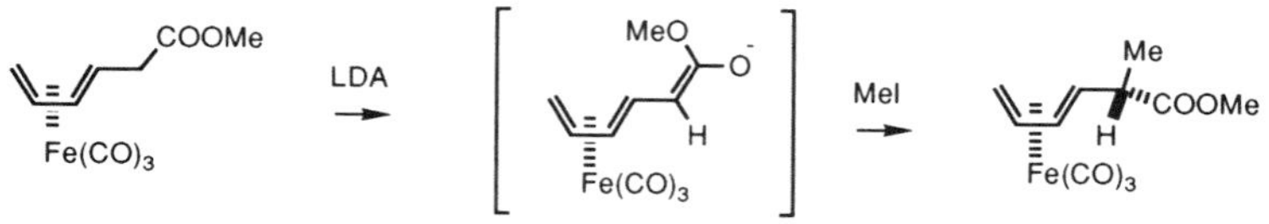

In an application of a (1*R*,4*S*)-diene irontricarbonyl complex to synthesis of a leukotriene-B_4 (LTB_4) intermediate [Nunn, 1988] the metal species serves to fix the configuration and to protect the double bonds.

A lateral template effect is witnessed in the reaction of a π-crotylmolybdenum complex with aldehydes [Faller, 1989]. The 22:1 selectivity in favor of forming the (*RR*,*SS*)-isomers over (*RS*,*SR*)-isomers indicates exposure of the *re*-face of the aldehydes to attack.

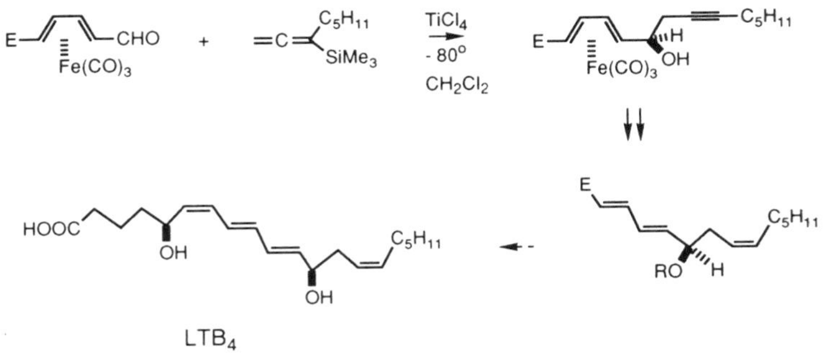

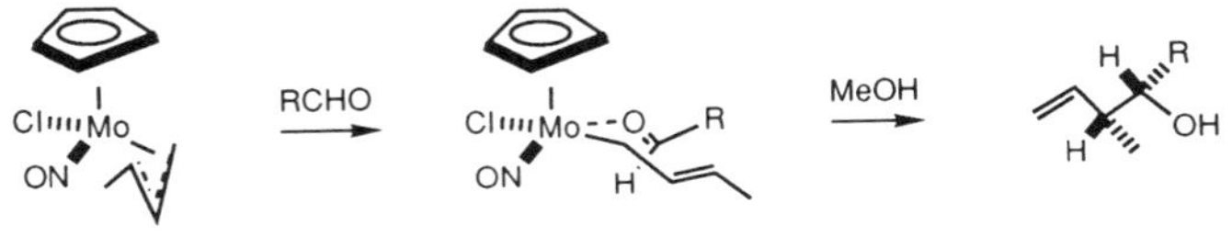

Oxidative cyclization of properly tethered alkenols has been studied with α-pinene-palladium complex [Hosokawa, 1981]. Template effect is evident.

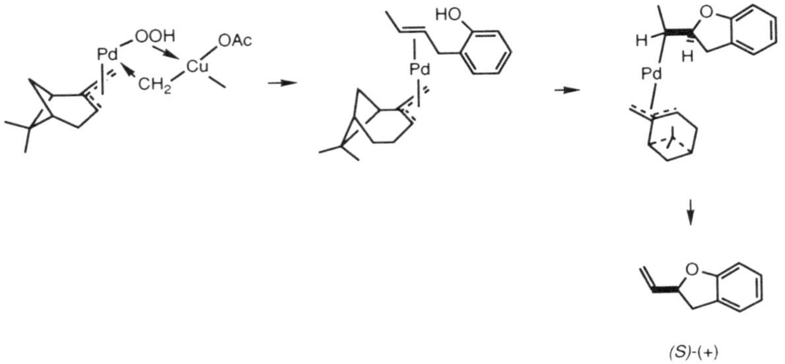

(S)-(+)

The reaction of ketene with chloral at −50°C in the presence of quinidine gives a β-lactone in virtually quantitative chemical and optical yields [Wynberg, 1985] A zwitterionic adduct of quinidine and ketene has been proposed as the nucleophilc cycloaddend [Wynberg, 1986]. The alkaloid molecule provides a template to direct the approach of chloral.

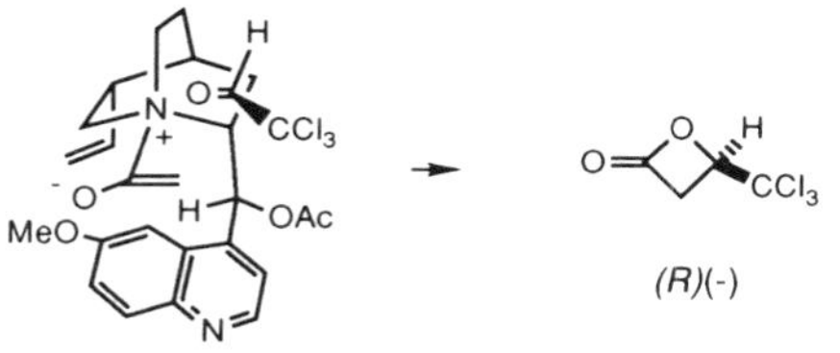

(R)(-)

The stereocontrol and regiocontrol exerted by palladium through coordination to the nitrogen atom of 3-dimethylaminocyclopentene in a synthesis of a prostaglandin intermediate [Holton, 1977] is exquisite. Palladium activation of a double bond permits *anti*-attack by a malonate anion and an alcohol in two separate steps, which is otherwise difficult (stereochemically for the second step). The resulting metallo-bridged species in undergoing C—Pd bond insertion by reaction with an enone accomplishes the objective.

NMe$_2$ L$_2$Pd NMe$_2$ CH(COOEt)$_2$ Me$_2$N-PdL$_2$ CH(COOEt)$_2$ L$_2$Pd NMe$_2$ O CH(COOEt)$_2$ Cl C$_5$H$_{11}$ O

H O O H HO C$_5$H$_{11}$ O NMe$_2$ CH(COOEt)$_2$ O C$_5$H$_{11}$ Cl O

OH (CH$_2$)$_3$COOH HO C$_5$H$_{11}$ OH

PGF$_{2\alpha}$

Stereocontrol for the formation of C-seco yohimbinoid compounds by intramolecular alkylation of an α-sulfonyl amide with a cyclohexenyl acetate was readily attained [Godleski, 1986]. With *cis*-arrangement of the amide chain and the adjacent allylic acetoxy group, the base-promoted S_N2 reaction led to a *trans*-hydroisoquinoline; on the other hand, the presence of a Pd(0) catalyst completely reversed the stereochemical outcome because a π-allylpalladium complex intervened and the CC bond formation was the second process of two consecutive S_N2 reactions. Thus the same compound may serve as intermediate for both *cis*- and *trans*-D/E yohimbine-type alkaloids.

It should be re-emphasized that the palladium metal template is so effective in the allylic substitution that a rather bulky group can be introduced. This contrasteric phenomenon is further illustrated in a synthesis of aristeromycin [Trost, 1988a] and in a preparation of *cis*-4,5-disubstituted 2-oxazolidinones [Trost, 1988b]. The origin of such outcome is the avoidance of steric compression between the ligated metal with the proximal groups(s).

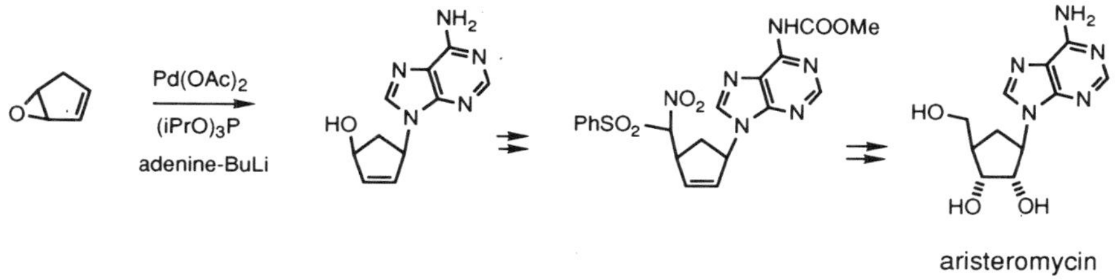

Pd(OAc)2 (iPrO)3P adenine-BuLi

aristeromycin

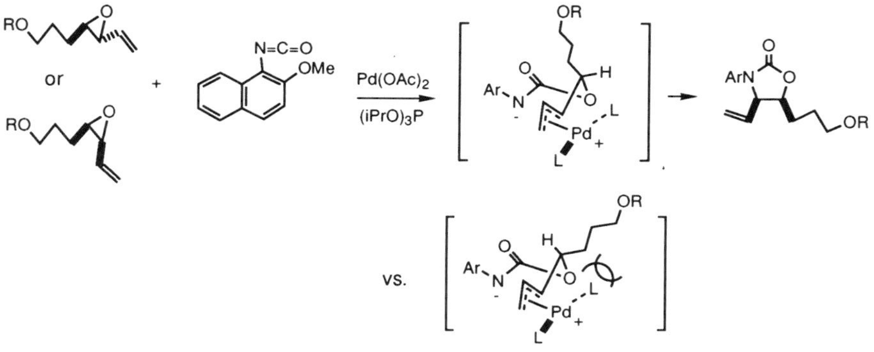

Pd(OAc)2 (iPrO)3P

vs.

Cationic metal π-complexes are diastereofacial electrophiles. The benefit of complementary template effects involving such a species is shown in a synthesis of trichodermol [O'Brien, 1989]. The nucleophile was a cyclic stannyl vinyl ether in which the allylic position carried a bulky silyl group, its reaction with a tricarbonyl-iron complex gave a product containing three contiguous asymmetric centers in the sense required for further elaboration of the terpene. The bulky silyl substituent in the nucleophile controls the approach of the metal complex, and itself was later being converted into a hydroxyl group with retention of configuration.

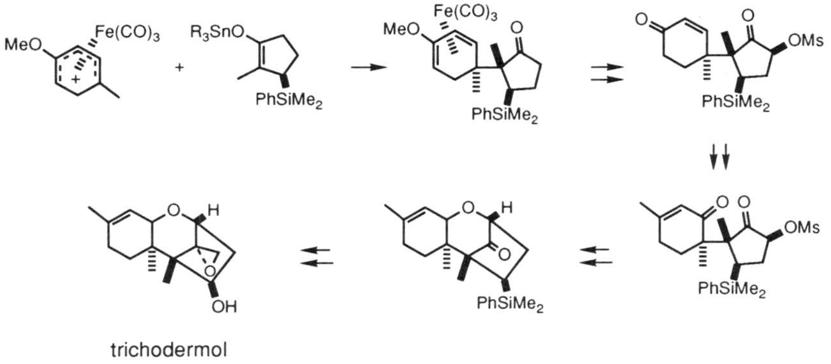

trichodermol

Osmylation of 5-trimethylstannyl-2-cyclohexenone and its ethyleneketal is stereorandom. However, it is possible to direct the reaction to occur from the *anti*-face to the trimethylstannyl group by maintaining the molecule in a conformation mimicking a bridged ring system via internal coordination between the ethyleneketal and the tin atom [Ochiai, 1988].

OsO4 ·
(PhIO)n ,BF3.Et2O;
NH4Cl
(49 : 51)
(PhIO)n
BF3.Et2O;
MeMgBr
(94 : 6)

With the formation of a chelate to fix its conformation, a carbonyl compound may exhibit facial preferences toward different reagents. For example, α-(4-oxazolinyl)acetaldehydes favor attack by relatively small organometallic species such as methylmagnesium bromide in an axial sense, while bulkier reagents (e.g. Me_2CuLi) tend to avoid 1,3-diaxial interaction with the angular hydrogen in the chelate ring [McGarvey, 1986].

R^1 = Me, R^2 = H
R^1 = H, R^2 = Me
Me2CuLi
(bulky)

MeMgBr	76	24
Me2CuLi	96	4

Me2CuLi
(bulky)

The ene reaction with glyoxylic esters is *threo*-selective and *erythro*-selective, respectively, in the presence of stannic chloride or dimethylaluminum triflate [Mikami, 1988]. With stannic chloride to chelate the two carbonyl groups the glyoxylate exists in the cisoid form, the ene reaction of which with either (*E*)- or (*Z*)-alkene can adopt a transition state leading to the *threo* product if nonbonding interactions are to be avoided. On the other hand, aluminium reagents coordinate with the aldehyde only, and the ene reaction transition states which give rise to *threo* compounds suffer from unfavorable steric compression.

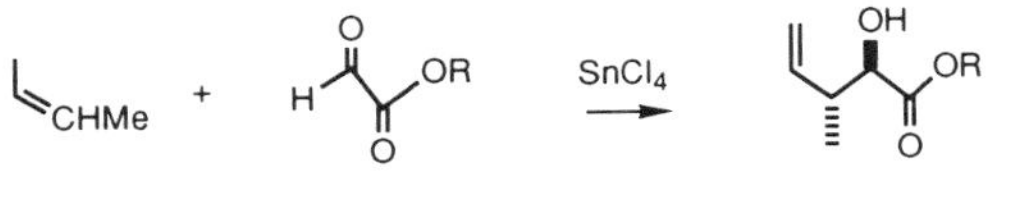

threo-

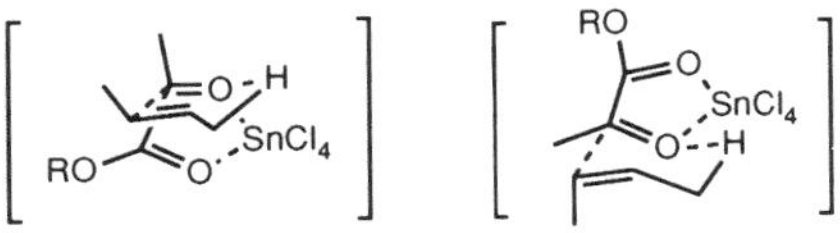

E → *threo* *Z* → *threo*

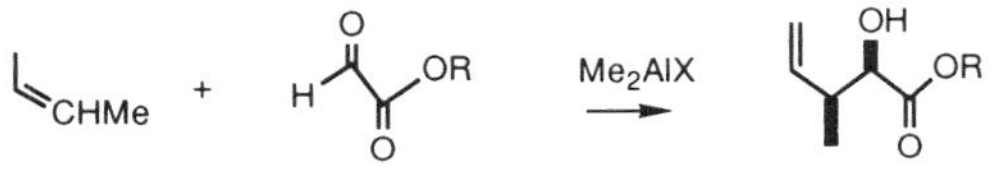

erythro-

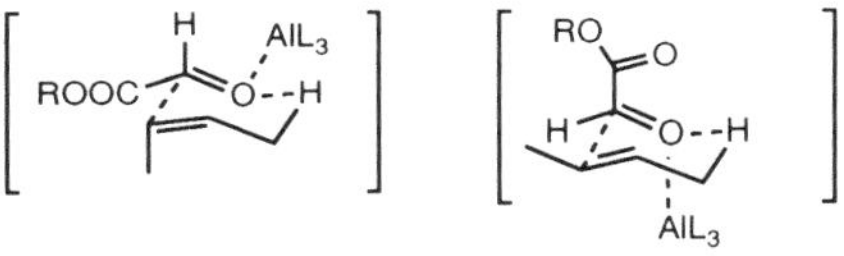

Z → *erythro* *E* → *erythro*

The bornane system provides an effective stereocontrol element. Reactions of bornane derivatives involving chelates proceed via very well defined transition states and they often exhibit excellent diastereo- and enantioselectivity. A method for α-amination of ketone (*Z*)-enolates [Oppolzer, 1992] is based on their reaction with a 2*β*-chloro-2α-nitroso derivative.

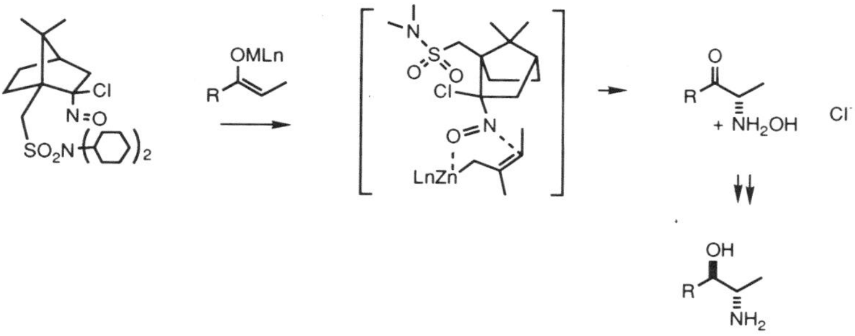

An asymmetric synthesis of the V-shaped Tröger's base unit [Maitra, 1992] has been performed on a 7-deoxycholic acid template. The condensation proceeded after the two hydroxyl groups of the steroidal substance were acylated with 4-nitrophenylacetyl chloride, and the product was reduced.

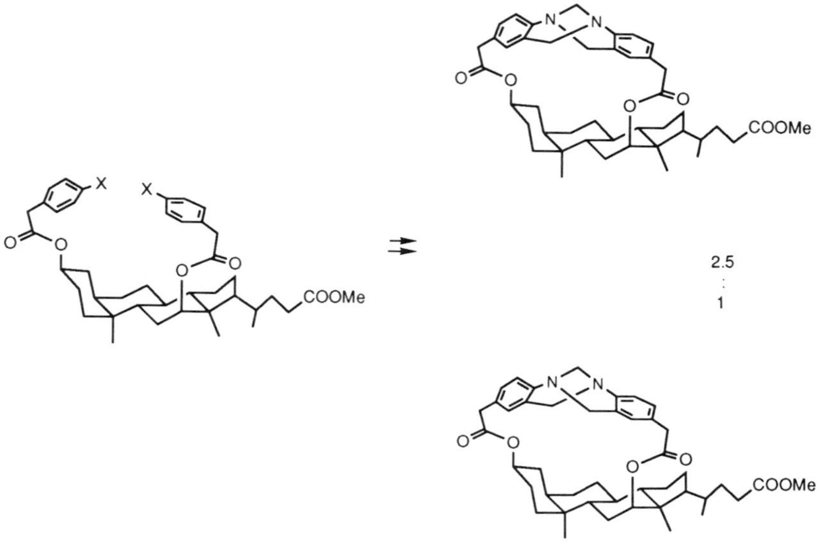

7.1.2. Organization in Ligand Sphere and Activation

In view of the often similar consequences of metal template effect and chelation, rigid subdivision in their discussion is not advisable. The common feature of the two phenomena is the organization of reactive components within bonding distances, and accordingly it facilitates reactions and exerts stereocontrol of such reactions.

The excellent catalysis of hydroxyaqua-1,4,7,10-tetraazacyclododecane-cobalt(III) complex in hydrolysis of nitriles (to amides) [Chin, 1990] is due to ligand exchange of the water molecule for the substrate, bringing together the nucleophile (OH^-) and the electrophile (CN) in a *cis*-arrangement within the coordination sphere of the metal ion.

Alkene methathesis is an interesting chemical process whereby two alkene residues are joined and then dissociated in the alternative fashion, formally transforming A=B and C=D into A=C and B=C by catalysis with certain alkylidenemetal species. A synthetically significant alkene metathesis is ring formation from the 1,(ω-1)-dienes [Fu, 1992].

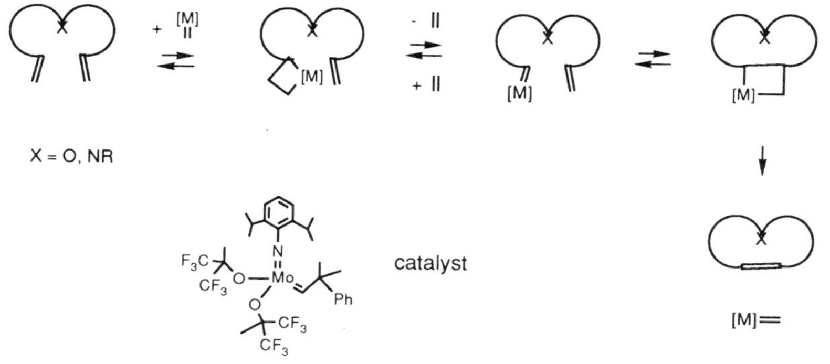

X = O, NR

catalyst

[M]=

The nickel catalyzed cyclotetramerization of ethyne to cyclooctatetraene discovered by Reppe is a typical templated process [Wilke, 1978].

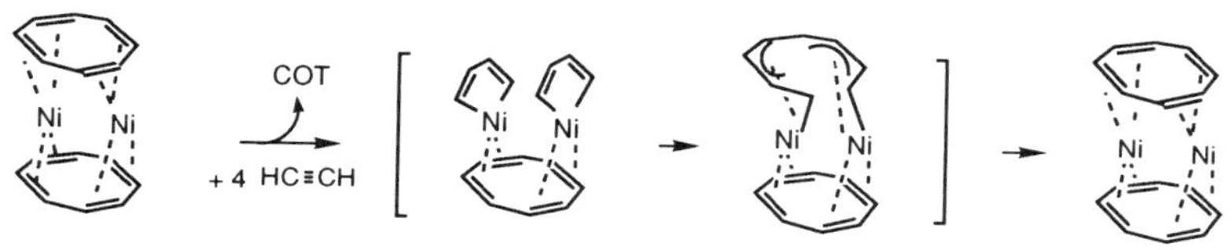

COT

+ 4 HC≡CH

In the metallo-ene reaction which involves C-M σ-bond transposition is also subject to template effect. To great advantage in a synthesis of khusimone [Oppolzer, 1982] a required, thermodynamically less stable dicarbocyclic product with an axial functionalized sidechain is formed in an intramolecular magnesium-ene reaction.

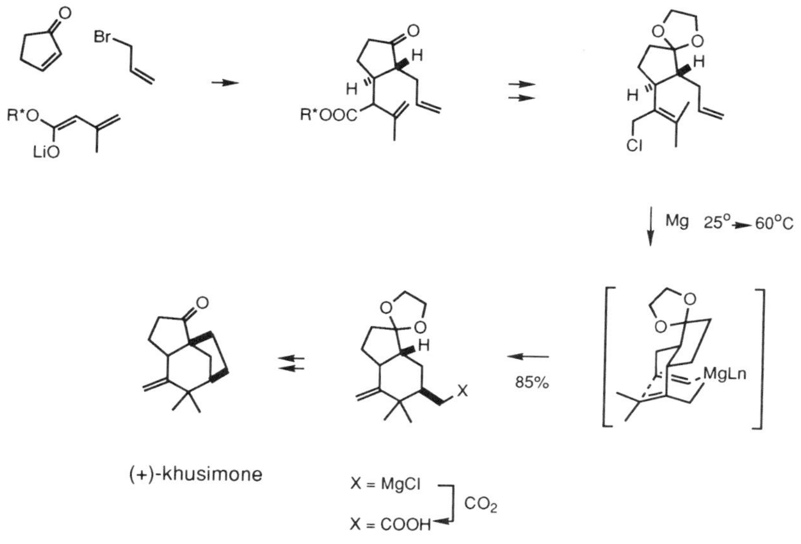

Mg 25°→60°C

85%

(+)-khusimone

X = MgCl

X = COOH

CO_2

In a synthesis of (−)-dendrobine [Trost, 1991] it was discovered that Pd(O)-catalyzed cyclization failed in the case of an allylic carbonate with more accessible double bond for complexation, yet the more hindered allylic isomer succeeded. This phenomenon was explained by invoking a precomplexation of the metal to the sidechain alkyne, therefore the intramolecular delivery of palladium to the allylic moiety was the key to the reactivity.

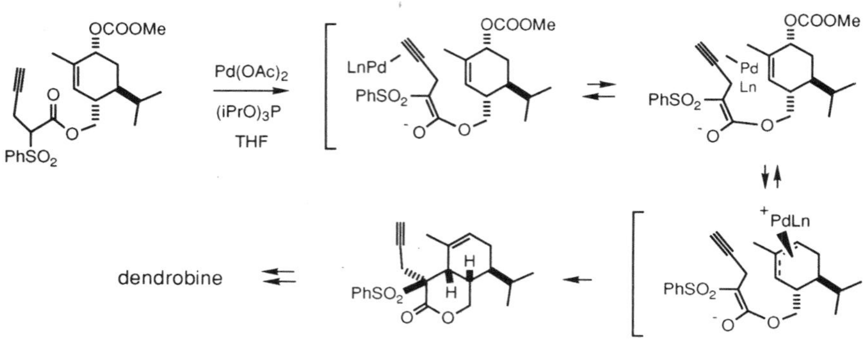

Metal templation is evidently involved in the ruthenium-catalyzed reconstitutive condensation of allylic alcohols with terminal alkynes [Trost, 1992b]. In this reaction a vinylideneruthenium complex has been postulated as intermediate in the catalytic cycle. Ligand displacement by the allylic alcohol is followed by O—C bond formation within the complex, thereby completing the transformation of the alkyne into an acyl moiety. Release of the β,γ-unsaturated ketone is by a reductive elimination process.

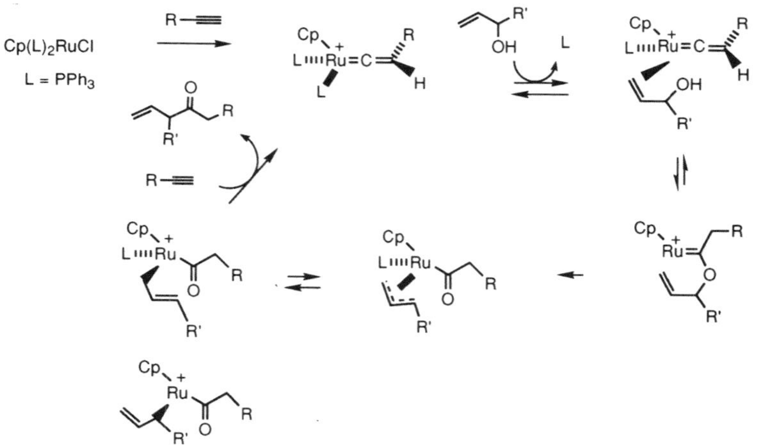

A most interesting display of stereoselection in intramolecular Heck reactions has been found during a synthetic study toward gelsemine [Madin, 1992]. Cyclization catalyzed by tris(dibenzylideneacetone)dipalladium provides a 9:1

product mixture in favor of the desired diastereomer, whereas the other diastereomer is favored when the reaction is conducted in the presence of a silver salt. In the latter reaction pathway an apparent coordination of the vinyl group with the metal atom upon dissociation of its bromide ligand directs the subsequent C—C bond formation from the *syn*-face to the vinyl substituent.

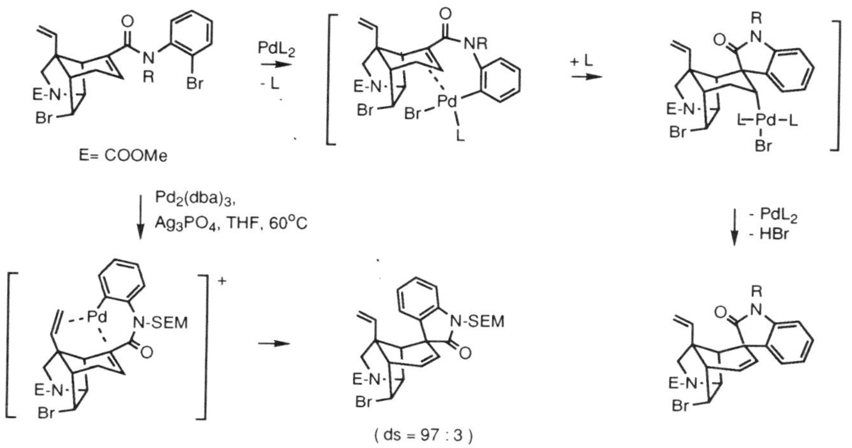

E= COOMe

PdL_2, - L; + L; - PdL_2, - HBr; $Pd_2(dba)_3$, Ag_3PO_4, THF, 60°C

(ds = 97 : 3)

A remote double bond can fix the sidechain conformation of the palladacycle derived from a 1,6-enyne by coordination to the metal. The consequence is that the 1,4-diene formation from such an intermediate is sterically prohibited [Trost, 1990]. The product is a 1,3-diene which under proper reaction conditions (sufficiently high temperature) may undergo intramolecular Diels–Alder reaction.

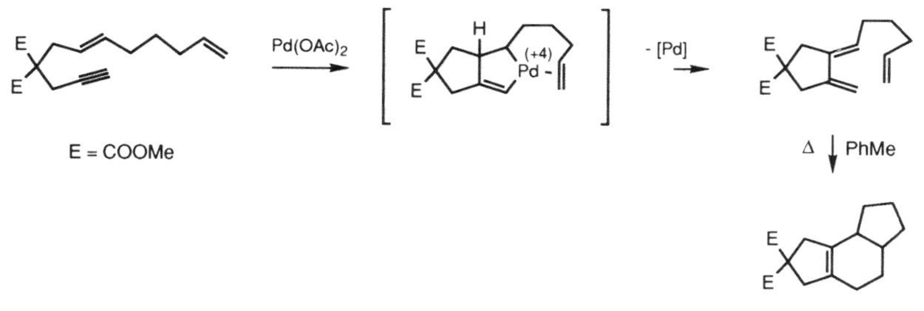

E = COOMe

$Pd(OAc)_2$; - [Pd]; Δ PhMe

Allyl halides undergo reductive coupling on treatment with nickel carbonyl to furnish 1,5-dienes [Webb, 1951]. The intramolecular version is very effective for the construction of cyclic structures, particularly terpenes containing a medium-sized ring such as humulene [Corey, 1967]. However, a caveat should be borne in mind that certain 1,5-cycloalkadiene products are liable to Cope rearrangement, as in the case of an attempted synthesis of hedycaryol which resulted in a compound of the elemane skeleton [Corey, 1969a].

From studies on simpler systems it seems that the efficiency of the cyclization leading to a medium ring substance increases greatly when the substrate contains an internal double bond. A possible role of such a double bond is coordination with nickel after reaction of one of the allylic halides, thereby reducing the number of conformations of the chain and favoring intramolecular coupling.

- $NiBr_2$ hv PhSSPh

humulene

Cyclotrimerization of alkynes to give benzene derivatives is catalyzed by transition metals. Because it is inexpensive, η^5-cyclopentadienylcobalt dicarbonyl has been used very extensively, including the preparation of estrone in two versions [Vollhardt, 1980; Sternberg, 1984].

$CpCo(CO)_2$ CF_3COOH; $(CF_3COO)_4Pb$

estrone

$CpCo(CO)_2$

$CpCo(CO)_2$; SiO_2

vitamin-B_6

$CpCo(CO)_2$

camptothecin

The [2 + 2 + 2]-cycloaddition reactions involving two alkynes and a nitrile, and two alkynes and an isocyanate, yielding a pyridine, and an α-pyridone, respectively [Vollhardt, 1984] have found applications to synthesis of vitamin-B_6 and camptothecin [Parnell, 1985; Earl, 1984]. In these studies there was noted a remarkable regiochemical domination by a silyl group of an alkyne on the cycloaddition. The intramolecular version of the process for combining two alkynes and an alkene to produce a 1,3-cyclohexadiene is the basis of synthetic routes to illudol [E.P. Johnson, 1991] and stemodin [Germanas, 1991].

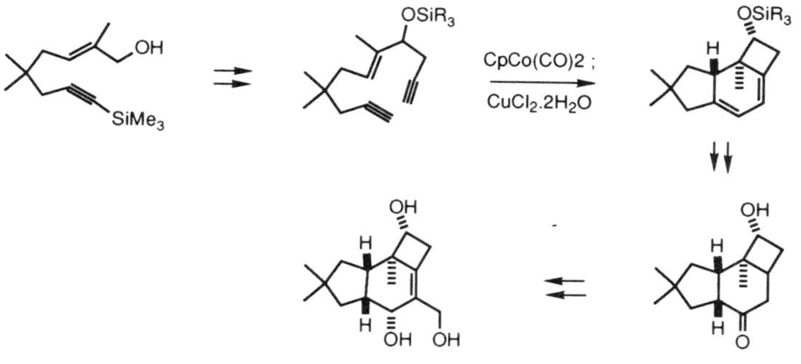

OSiR3
OH
SiMe3
CpCo(CO)2 ;
CuCl2.2H2O
H
OH
H
H
OH OH
illudol

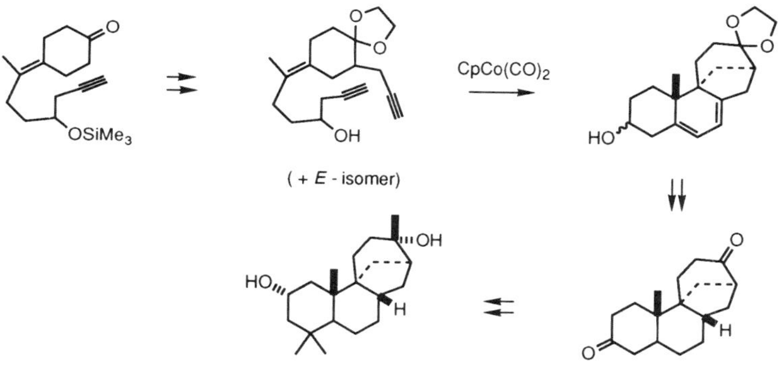

O
O
O
OSiMe3
OH
(+ *E* - isomer)
CpCo(CO)2
HO
OH
HO
H
H
O
O
stemodin

Metal-promoted formation of a cyclopentene from an alkyne, an alkene, and an insertable ligand has been explored intensely. Several transition metals can be used, for example nickel [Tamao, 1992], cobalt [Pauson, 1985; Schore, 1988], zirconium [Negishi, 1991], but the facility of each version varies. Thus the Pauson–Khand cyclization requires relatively high reaction temperatures because dicobalt octacarbonyl is a 18-electron species. Zirconium(O) entities such as "Cp_2Zr" are more effective due to simulataneous availability of one empty and one filled nonbonding metal orbital for π-complexation, and another empty orbital for the subsequent rapid carbometallation.

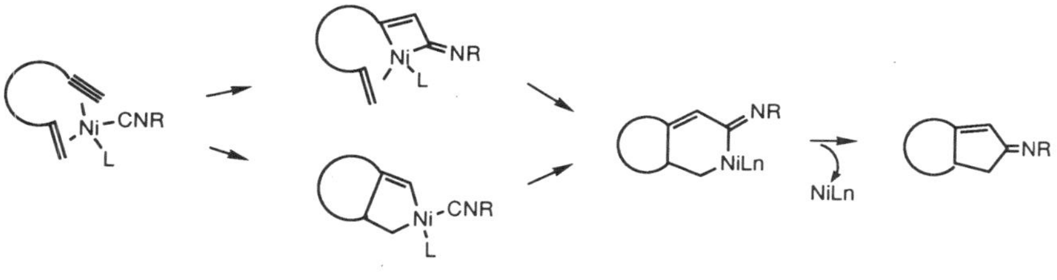

Ni-CNR
L
NR
Ni
L
Ni-CNR
L
NR
NiLn
NiLn
NR

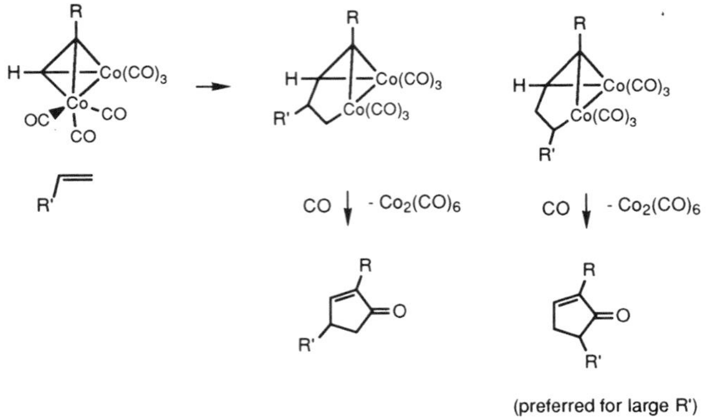

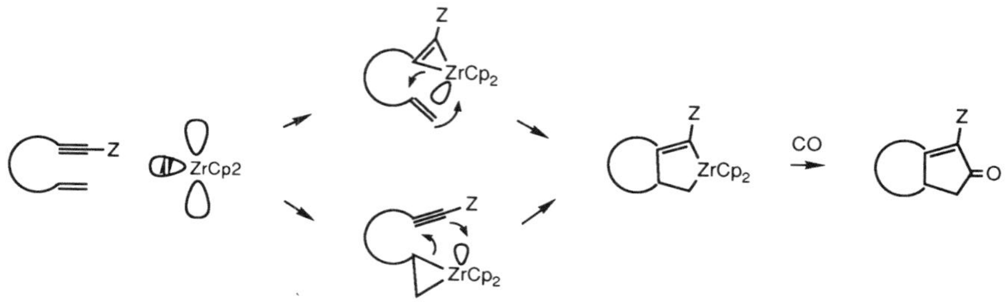

(preferred for large R')

Cylization of polyenes and polyenynes by zipper mode cascade [Negishi, 1992] is most remarkable. Such a process is initiated by the formation of (vinylic) organometallic species which interacts with a proximal unsaturated linkage, effectively transferring the active component to a new site. When another unsaturated bond is reactable by the transposed metal center the hopping of the active metal continues. At least four rings can be constructed in one step.

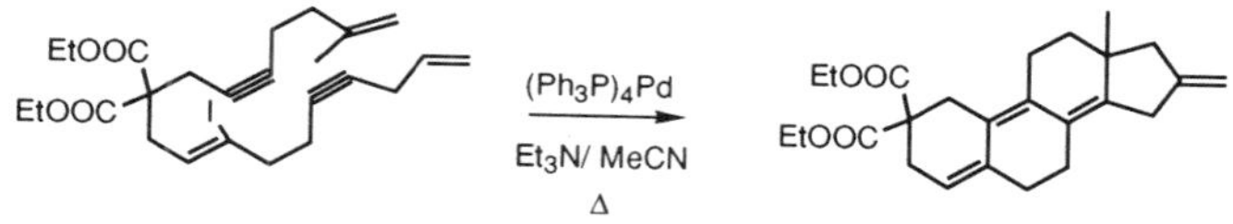

Transition metal promotion of [6 + 4]-cycloadditions has been investigated [Rigby, 1993]. While the exact mechanism is still unclear, it is quite sure that both addends are ligated to the metal atom when C—C bond formation occurs.

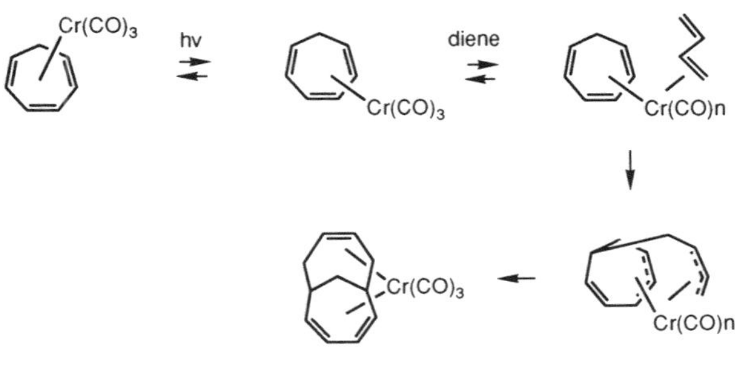

A number of transition metal-templated reactions that result in interligand bond formation are favored by a low oxidation state and an octahedral configuration of the metal. A remarkable example is the Dötz annulation [Dötz, 1984; Wulff, 1989] which is based on the Fischer carbene complexes. A low oxidation state coordinates well to π-acceptor ligands, with an octahedral configuration the coupling of the desired ligands within the coordination sphere has a high probability. In a Fischer carbene complex the carbene ligand has four *cis* CO ligands and one *trans* CO ligand, the replacement of one of the *cis* CO with an alkyne leads to a complex possessing three different ligands on the same face of the octahedron. Regioselective cyclization involving these ligands is favored (actually via formation of a chromacyclobutene). There are many elegant applications of this method, among them the skeletal assemblage of the anthracyclinones via B-ring or C-ring formation [Wulff, 1984; Dötz, 1985].

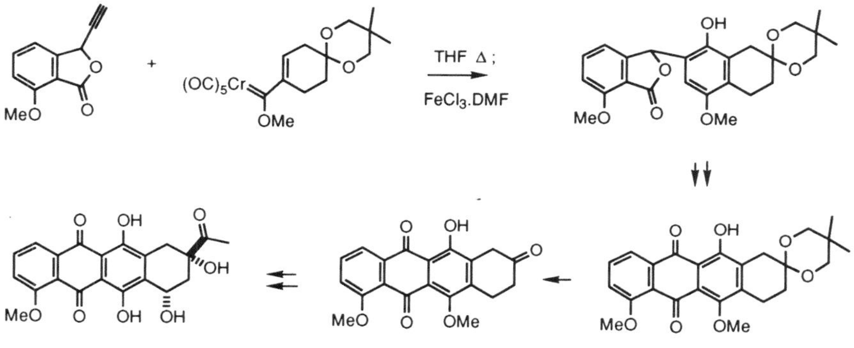

THF Δ; FeCl$_3$.DMF

daunomycinone

tBuOMe 55°C

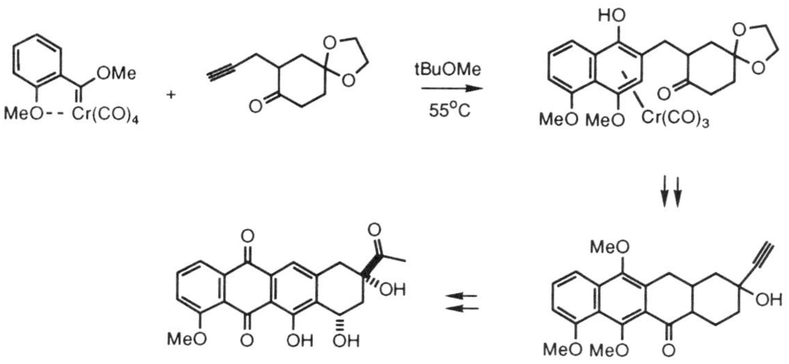

11-deoxydaunomycinone

[2.3]-Wittig rearrangement of a pair of 3,4-dialkoxyalkenes displayed varying degrees of diastereoselectivity [Wittman, 1988]. The relative stereochemistry of the two alkoxy groups was crucial because the transition states organized by the lithium atom are of different energies owing to eclipsing of two larger groups in one of them.

BuLi, - 78°C

BuLi, - 78°C

less favorable

69%

31%

The immediate precursor of spermine alkaloids such as celacinnine, celafurine, and celabenzine have been acquired [H. Yamamoto, 1981b] in about 41% overall yield over five steps from 1,4-diaminobutane. Closure of the macrocyclic diazalactam was mediated by a bicyclic triaminoborane from which the lactamization involved formation of a six-membered ring. The excellent template affect of boron is evident.

$(Me_2N)_3B$, xylene, Δ

R= Ph celabenzine
R= CH=CHPh celacinnine
R= 3-furyl celafurine

A template effect is also manifested in crown ether synthesis [Pedersen, 1972; Mandolini, 1986]. The donor–acceptor interactions between lone-pair electrons of heteroatoms and a metal cation drive the folding of a suitable chain toward ring closure. Thus with this entropic enhancement of rates the high dilution technique [K. Ziegler, 1933] for minimizing intermolecular reactions can be obviated. In the cyclization transition state the binding between a substrate and the metal ion is stronger than that of the product-metal ion complex because of proximity of the charges. The oligomerization of ethylene oxide in the presence of metal fluoroborate [Dale, 1976] to form crown ethers displays selectivities indicative of templation. Thus hexamer formation is favored (100%) by M=Rb, Cs, whereas the pentamer is the major product (90%) when M=Cu, Zn.

6 [ethylene oxide] $\xrightarrow{M(BF_4)_n}$ [18-crown-6]

Isolation of the *trans-syn-trans*-dicyclohexano-18-crown-6 but not the *trans-anti-trans* isomer has been explained [Coxon, 1978] in terms of cyclization tendency of the preorganized intermediates. The decrease in the rotation freedom component about bonds that attend adoption of the all-gauche OCH_2CH_2O conformation is an important factor to the reaction.

[Scheme: diol + ditosylate → K^+-templated intermediates (OTs); one → ? polymer; other → trans-syn-trans-dicyclohexano-18-crown-6]

In the elaboration of an A-ring building block for aklavinone [Bauman, 1984] utilizing an intramolecular Darzens condensation, a significant solvent effect was observed. In the presence of methanol the expected epoxide was formed, but in less polar solvents such as ether or THF, the enolate preferred a tight ion-pair in which the metal ion gained stabilization through chelation with the ester and the ketone carbonyl groups. Under the latter conditions a lactone instead of the epoxide was generated.

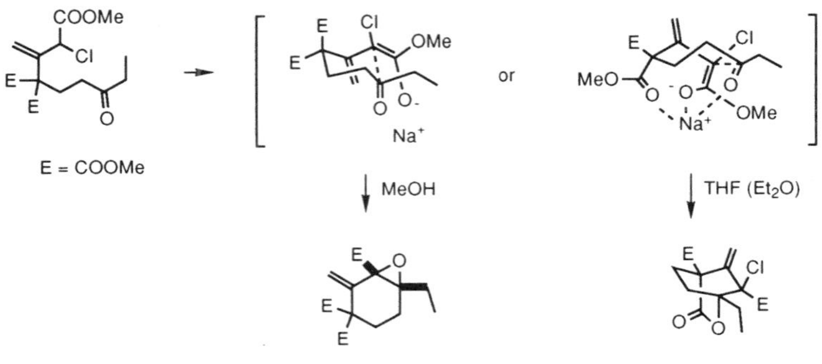

Pseudotetrahedral tetracoordinate transition metal ions such as Cu(I) provide three-dimensional templates to assemble ligands in defined geometry. An elegant application of this concept to the synthesis of catenanes and molecular knots has been reported [Dietrich-Buchecker, 1992].

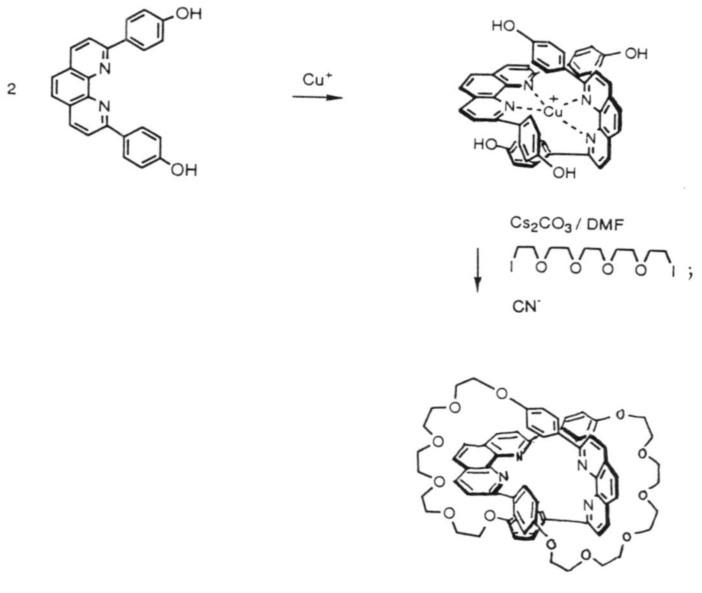

The addition of barium and strontium salts to a crown ether-spanned calix[4]arene monoacetate has the effect of enhancing methanolysis rate, due to transition state stabilization in which the ester carbonyl participates in metal ion coordination [Cacciapaglia, 1992].

Potassium ion organization of a polyether which is terminated by a benzoquinone and hydroquinone dimethyl ether moieties promotes reduction of the quinone by a dihydronicotinamide [Pierre, 1992]. This can be considered as an allosteric system. Potassium ion has no effect on the reduction of benzoquinone itself. An allosteric effect is manifested on the second-stage binding of mercury derivatives in a biphenyl to which two crown ether rings have been attached [Rebek, 1985a]. The cooperativity is due to reduction of conformational freedom of the remote binding site upon the first coordination. Metal catalysis of reactions and racemization of 3,3-disubstituted 2,2′-dipyridyl systems has also been examined [Rebek, 1985b].

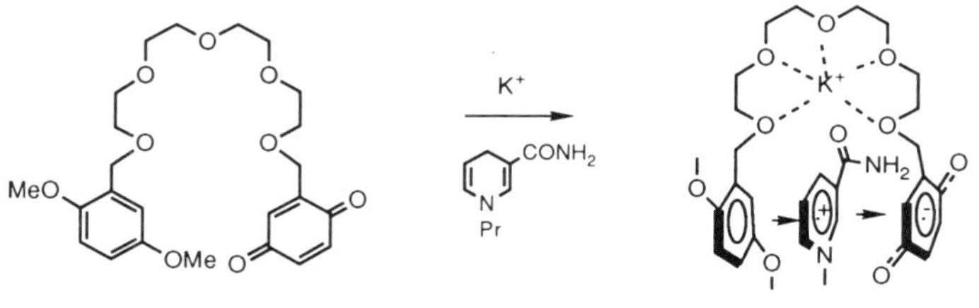

The cesium ion effect [Ostrowicki, 1991] which attributes special properties to Cs^+ in favoring cyclization has been criticized [Galli, 1992]. A template effect for the cation seems to be sufficient to cause such results.

Among complex organic syntheses the role metal template has played in promoting formation of the macrocyclic chromophore of vitamin-B_{12} from A/B- and A/D-secocorrins is unequalled. Coordination of the four nitrogen ligands brings the reactive ends within bonding distances. The brilliantly conceived and executed photochemical cyclization of the A/D-secocorrins [Eschenmoser, 1976, 1988] validates the Woodward-Hoffmann rules regarding antarafacial 1,16-hydrogen migration and conrotatory cyclization, and this process proceeds best with a cadmium complex.

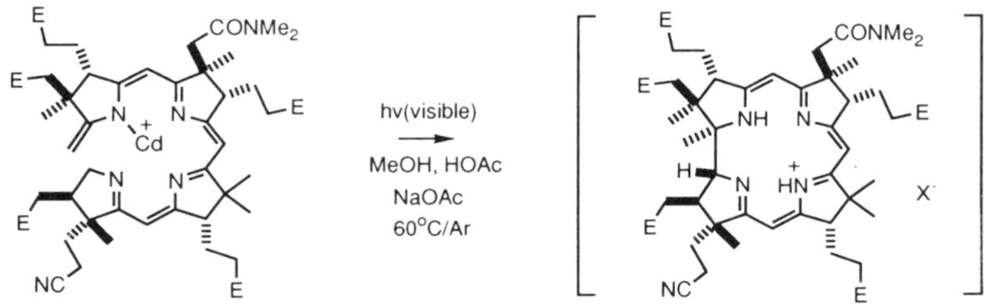

Reactions on metallic surfaces can overcome certain unfavorable factors. For example, the synthesis of medium-sized rings has been plagued by intermolecular processes, and for a long time the acyloin condensation has enjoyed an exceptional success. The reason for this success is the occurrence of reaction on the metallic surface of sodium where two ester groups at the ends of a long chain are gathered during the initial stage of the cyclization. Unfavorable entropic factors and intrachain interactions are overcome. Even if the two esters are simultaneously anchored onto the metal surface beyond bonding distances, there is little chance of interference from an ester group from another molecule, so that by a series of electron transfers likened to a hopping motion, the two ester groups of the same molecule can be brought together to start C—C bond formation. The same principle applies in the cyclization of dicarbonyl substances with low-valent titanium species [McMurry, 1989].

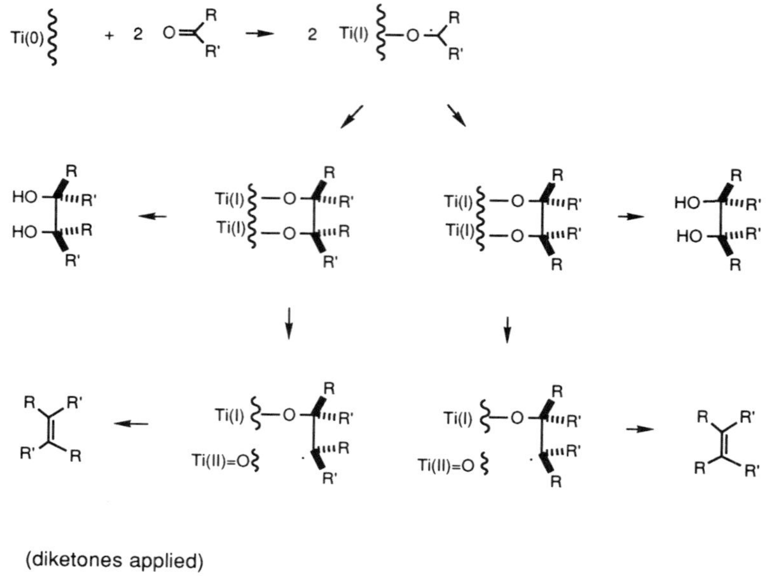

(diketones applied)

The analogy of these coupling reactions with the Kolbe electrolysis which effects decarboxylative coupling of acids should be noted.

The McMurry coupling represents exploitation of the general phenomenon that Lewis acids complex with many organic compounds containing hetero-atoms. In the classical sense such complexation accelerates attack by nucleophiles. New or improved synthetic methods can often be devised on the basis of complexation with bulky metal reagents. For example, the dramatic change of axial/equatorial alcohol ratios in the Grignard reactions of 4-*t*-butylcyclohexanone by the addition of a bulky methylaluminum bisphenolate [Maruoka, 1985] can be explained by the template effect of the latter species which, by virtue of its steric demand, occupies and shields the quatorial position, permitting approach of the reagent to approach from the axial direction. Unfortunately, this method has its limitations with regard to the nature of the Grignard reagents.

It is remarkable that the cyclohexanone carbonyl complexes with diorganocuprate and is then rapidly attacked by organolithium reagents from the equatorial direction [Still, 1976].

Organoaluminum-promoted aliphatic Claisen rearrangements of substrates in which both α and γ positions of the allyl residue bear substituents display dramatic stereochemical consequences with respect to the bulkiness of the promoter [Nonoshita, 1990]. A more bulky organoaluminum favors formation of (4*Z*)-alkenals via the chairlike transition state.

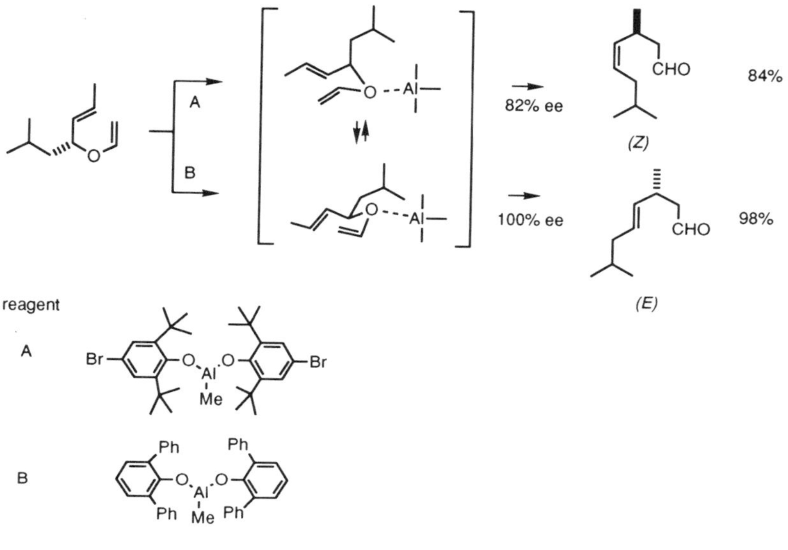

In the aromatic Claisen rearrangement the aluminum coordinates preferably from the side lacking an *ortho* substituent [Maruoka, 1990], thereby forcing the rearrangement to give a product not usually observed in the purely thermal course.

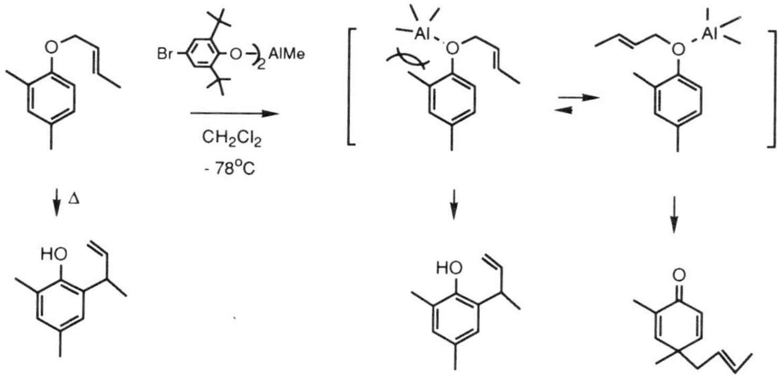

Further utility of bulky organoaluminum reagents in eliciting stereo- and regiocontrol is found in the Diels–Alder reaction of nonidentical fumaric diesters [Maruoka, 1992]. For example, the thermal condensation of the *t*-butyl/methyl esters with cyclopentadiene gives two products in essentially equal amounts, while the methylaluminum bis(2,6-di-*t*-butylphenoxide)-catalyzed reaction leads to one compound predominantly (99:1 in favor of the *endo*-carbomethoxy isomer), due to selective coordination at the smaller ester C=O and placing it in the *endo* direction.

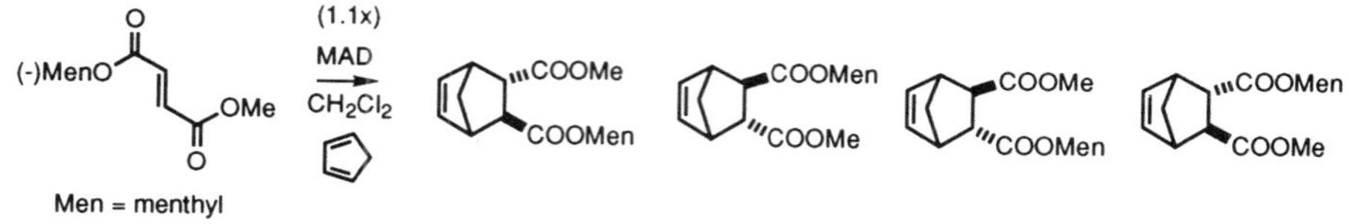

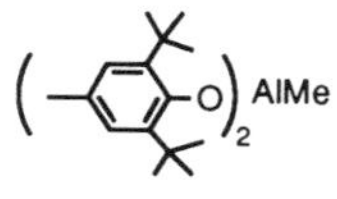

Men = menthyl

	endo (COOMe exo-side / COOMen)	endo'	exo	exo'
= MAD	91.4	7.0	0.2	1.4
Et_2AlCl	52.5	4.6	5.2	37.7
Δ	22.8	26.4	24	26.8

The intrinsic differences in chelation tendencies between Grignard reagents and diorganomagnesiums, together with the preferred maximum separation of two oxygen atoms in free α-sulfinyl ketones (*anti*-conformation) cause diverse shielding of two faces of 2-*p*-toluenesulfinyl-2-cyclopentenone, and hence the steric courses of the conjugate addition [Posner, 1982, 1984].

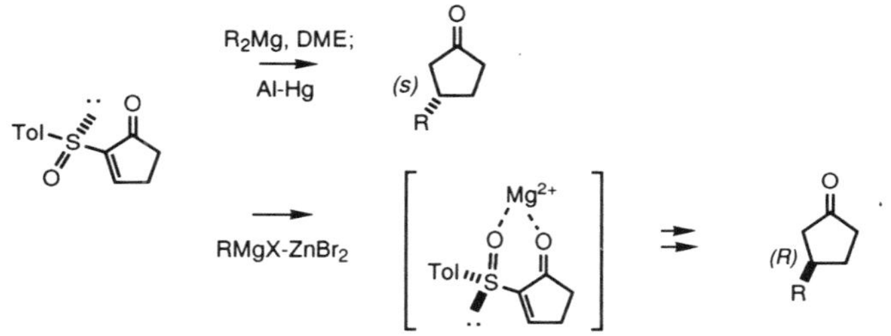

Chiral (acyloxy)borane complexes derived from tartaric acids promote synthetically significant reactions. For example, good enantioselection and diastereoselection (*erythro*:*threo* > 80:20, usually in the 90:10 range) results in the Mukaiyama condensation between enol silyl ethers and aldehydes in their presence, regardless of the configuration of the donor species [Furuta, 1991]. The catalyst in the complex covers the *si*-face of the carbonyl group so that nucleophile approach is allowed from the *re*-face only.

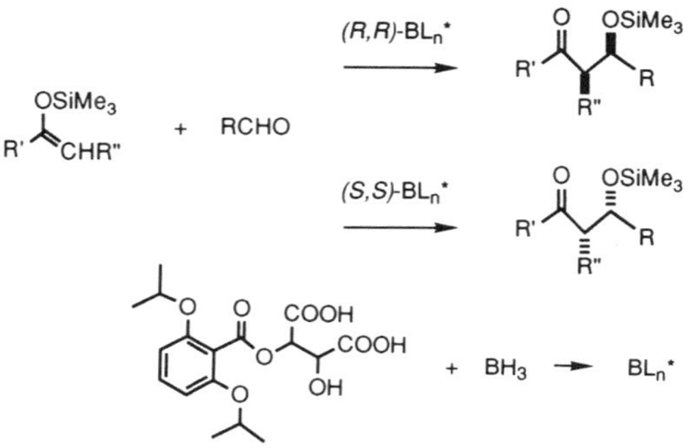

The same catalysts can be used in asymmetric Diels–Alder reactions [Furuta, 1989].

7.2. CHELATION

Chelation can change chemical reactivity profoundly. A recent example is the increase in the efficiency and chemoselectivity of alkyl transfer in the Stille coupling reaction from alkylstannanes in which the metal center is internally coordinated by an amino group [Vedejs, 1992]. In these 5-alkyl-1-aza-5-stannabicyclo[3.3.3]undecanes the Sn-R bond is exceptionally long.

Chelation of one dithioacetal unit to metal appears to be an effective means for activating another dithioacetal in proximity toward nickel-catalyzed dethiomethylenation with Grignard reagents [Wong, 1992].

The intramolecular complexation by the amidic oxygen atom to the metal of amidocarbene complexes probably was the cause of the enhanced dienophilc reactivity in comparison with complexes having one more carbon monoxide ligand on the metal [B.A. Anderson, 1992]. The chelation would permit effective delocalization of the nitrogen lone-pair electrons, and the effect might be more than compensate the back-donation of the metal elicited by the extra carbon monoxide ligand.

Ph O= N— $(OC)_4W$= > Ph— O N— $(OC)_5W$=

An unexpected catalytic role of oligo(ethyleneglycol) derivatives for nucleophilic displacement of haloanthraquinones with sodium alkoxides [T. Lu, 1990] is complexation of the cation and the nucleofuge.

4-Hydroxy-2,7-octadienyl acetates undergo cyclization in the presence of a Pd(O)-catalyst to give *cis*-2-vinylcyclopentanols [Negishi, 1989]. The palladium ene reaction is successful only when the hydroxyl group is present. A chelated transition state is indicated.

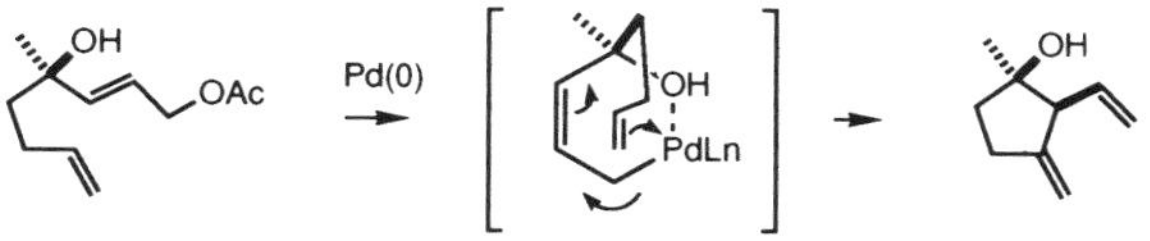

OH OAc Pd(0) → [OH PdLn] → OH

Tetrahydroisoquinolines undergo deprotonation at C-1 when the nitrogen atom forms part of a formamidine group. In the presence of a β-alkoxy substituent to the imino nitrogen, the deprotonation with butyllithium proceeds rapidly after chelation [Meyers, 1987b]. In other words, the chelation is the rate determining step. In a related work [Meyers, 1987a] it was shown that the imine prepared from 1-naphthaldehyde and (*S*)-*O*-*t*-butylvalinol underwent asymmetric addition (at C-2) with an alkyllithium, direct trapping and hydrolysis yielded chiral 1,2-dialkyl-1,2-dihydro-1-naphthaldehyde. The chelated alkyllithium reagent delivered the organic group to C-2 of the ring system to generate a metalloenamine which was alkylated.

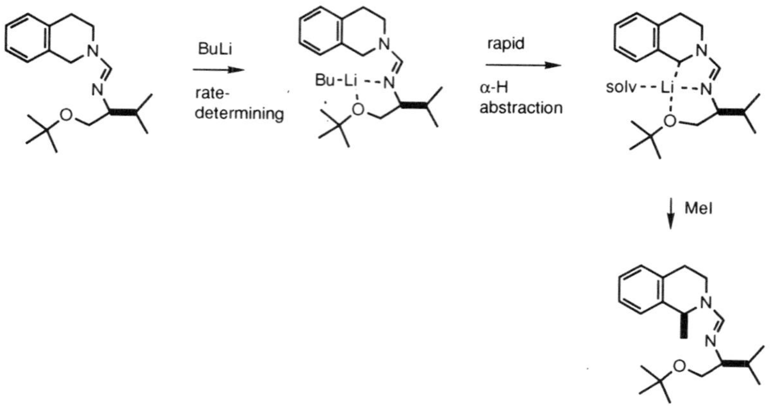

BuLi rate-determining; Bu-Li; rapid α-H abstraction; solv--Li; MeI

The Birch reduction product of anisole is readily deprotonated and alkylated, therefore it serves as a synthetic equivalent to 2-cyclohexanone α-anion. This protocol is not uniformly applicable to 3-substituted anisoles owing to difficulties in the metallation of the reduced compounds. Changing the *O*-methyl group to good chelators such as 2-diethylaminoethyl and methoxymethoxy groups to direct and assist the metallation step afforded excellent results [Amupitan, 1983; Orban, 1983].

The last but not the least important topic relating chelation to activation is hydrogen bonding. It has been hypothesized that diorganotin oxide promotes lactonization of ω-hydroxyalkanoic acids [Steliou, 1980] via adducts in which intramolecular hydrogen bond is established, the gathering of the relevant reactive groups greatly facilitating the subsequent ring formation. When lactonization is effected via (α-pyridine)thioesters [Corey, 1974b], the development of the hydrogen-bonded species into an ion pair is even more advantageous to the cyclization step. (This lactonization is also promoted by silver ion [Gerlach, 1974].)

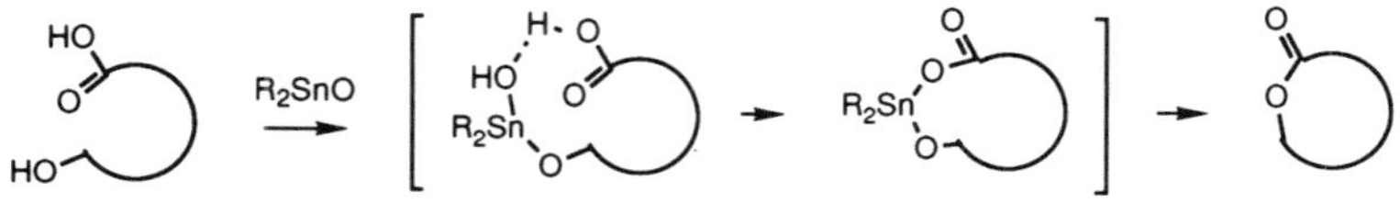

R_2SnO

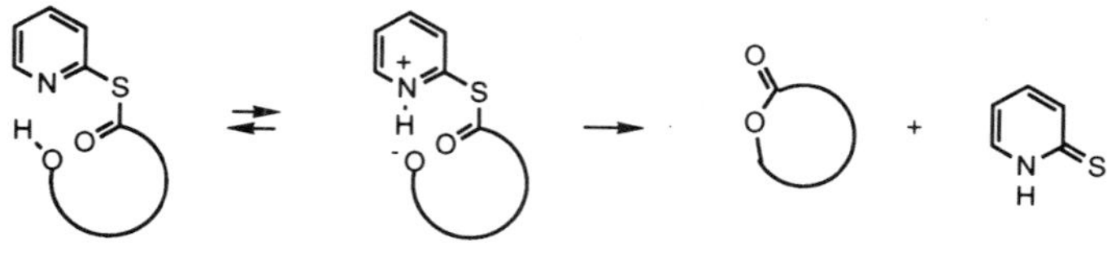

Attention must be paid to the fact that chelation sometimes has adverse effects. Thus a carbamate group attaching to the indolic nitrogen atom of a Δ^1-tetrahydrocarbazole directed an intramolecular Heck reaction in the undesired carbon site [Rawal, 1993]. Without this chelator the expected reaction leading to dehydrotubifoline proceeded.

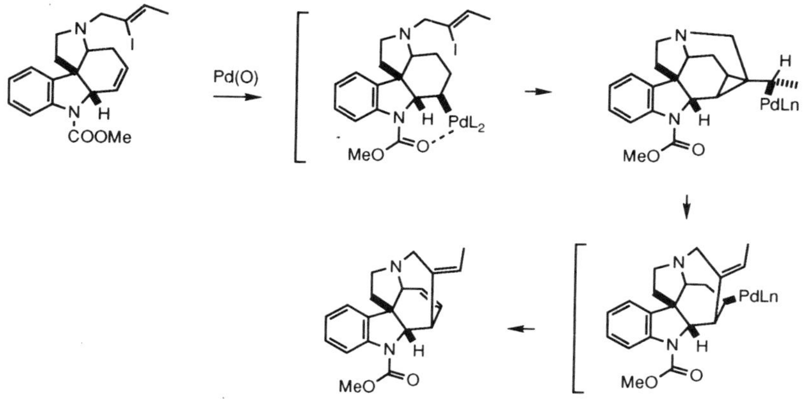

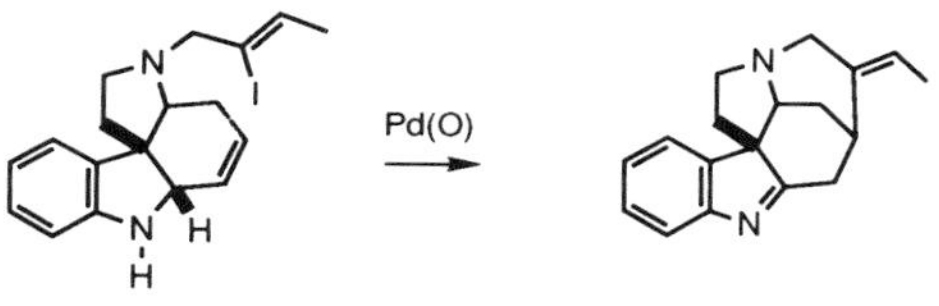

Pd(O)

dehydrotubifoline

7.2.1. Regiocontrol and Stereocontrol

Conformation fixation of a substrate is a crucial factor in many stereoselective reactions. In a precursor of ormosanine the intramolecular hydrogen bonding of the pyridinium ion with a lactam carbonyl apparently determined the course of hydrogenation [Liu, 1976]. Removal of the carbonyl group(s) rendered the hydrogenation nonstereoselective.

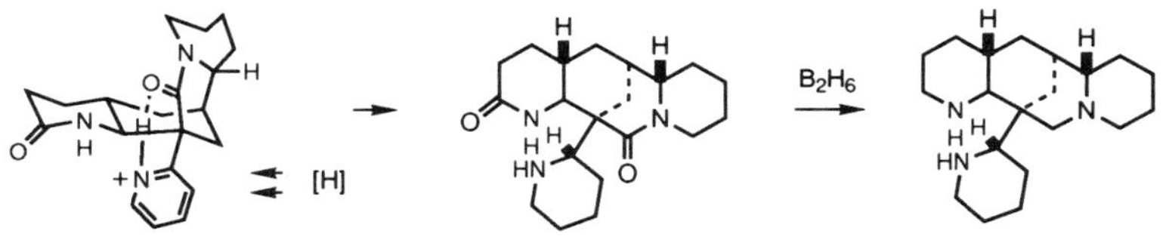

[H] B_2H_6

ormosanine

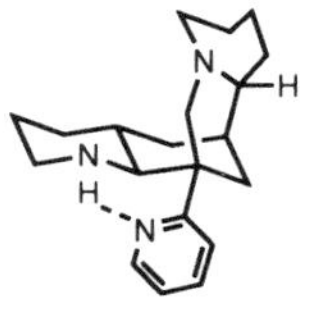

nonstereoselective hydrogenation

Isomerization of a certain compound, with respect to feasibility and completeness, depends on many factors. The equilibration of a pair of spiroketals [Kurth, 1985] on treatment with aluminum chloride in dichloromethane obviously proceeded in the direction toward bidenate chelation of the metal ion.

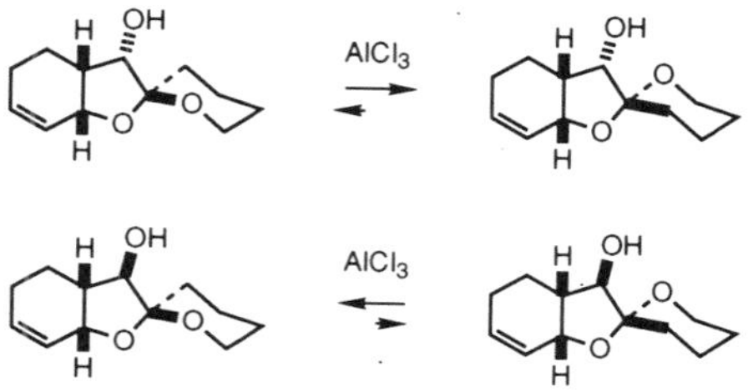

$AlCl_3$

$AlCl_3$

An intramolecular Michael reaction of a keto ester, in tandem with an alkylation to form a spirocyclic system, proceeded stereorandomly when the counterion of the enolate was sodium. However, only one isomer in which the acetic ester chain is *cis* to the ketone group was obtained when the reaction was conducted in the presence of cesium carbonate [Le Dreau, 1993]. Chelation effect is the most reasonable explanation.

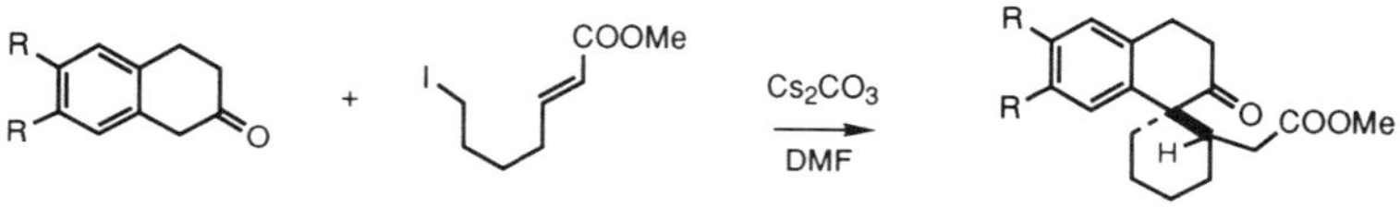

Cs_2CO_3 / DMF

The orientation of a reagent by precoordination with a donor group of the substrate necessarily restricts the side of attack. This principle is amply demonstrated in the 3-(2-oxazolinyl)benzyne system in its reaction with the nucleophilic species [Pansegrau, 1988]. Thus under kinetic control the attack of an organolithium is directed by coordination of the lithium atom with the nitrogen atom of the heterocycle, resulting in the transfer of the organo residue to the *ortho* position. On the other hand, the reaction with diorganocuprate species (thermodynamic control) a regiochemical reversal is witnessed.

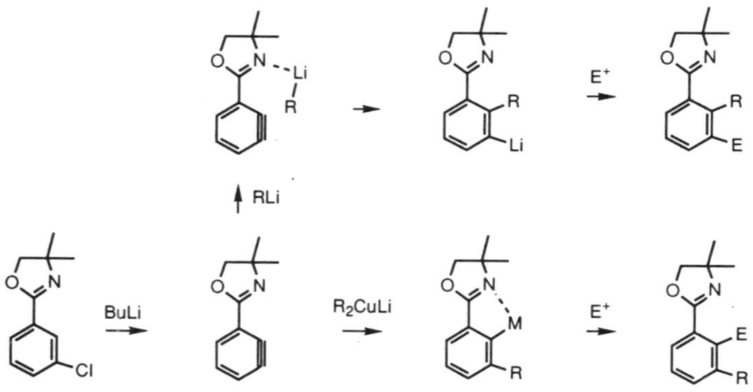

RLi; BuLi; R_2CuLi; E^+

The regioselective 6-cyanation of 1-benzyl-3,3-(ethylenedioxy)piperidine by reaction with mercury(II) acetate and cyanide ion has been attributed to oxidation from a chelate involving both the nitrogen atom and the axial oxygen of the ketal group to the mercury ion [Compernolle, 1991]. In this form the 2-axial hydrogen is more difficult to attain an antiperiplanar conformation with the metal than the 6-axial hydrogen, therefore it is more difficult to remove.

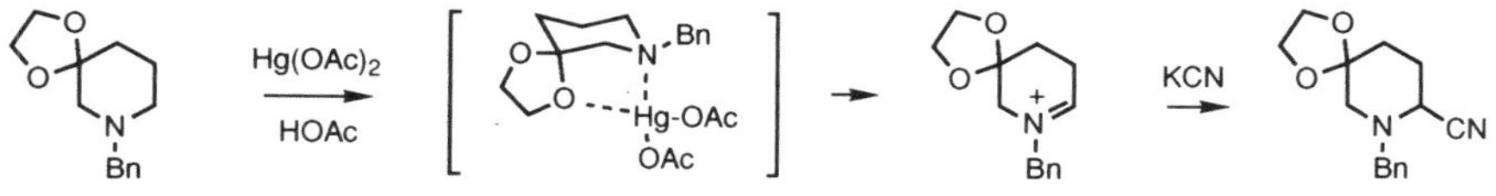

The stereochemical outcome in the second stage of 2,6-dimethylation of *N*-protected piperidines differs in the case of having a *t*-butoxycarbonyl protecting group or that having a formimidoyl group [Shave, 1991]. The α-lithiopiperidine requires chelative stabilization by the carbonyl oxygen or the imino nitrogen atom. The chelate in which the lithium is equatorial can only be accommodated in formimidoyl substrate for steric reasons, because in the carbamate derivative the *t*-butoxy group experiences severe hindrance from the 2-methyl group.

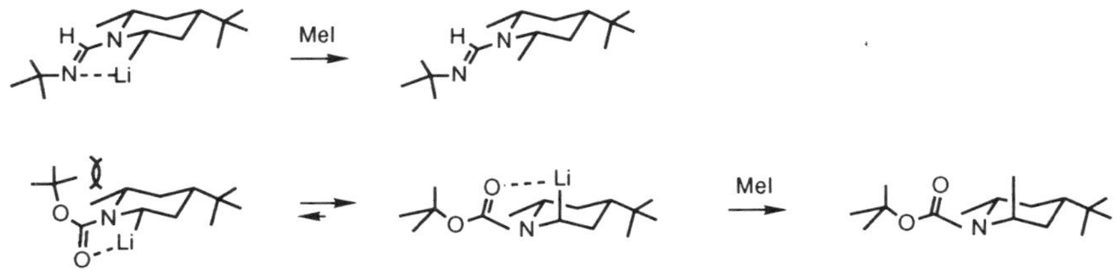

It has been found that deprotonation of α-siloxycarboxylic esters with the very bulky lithium 2,2,6,6-tetramethylpiperidide led overwhelmingly to the (*E*)-enolates, while the use of lithium hexamethyldisilazide gave the (*Z*)-enolates [Hattori, 1993]. The latter base, with a smaller size and much lower strength favors a transition state in which the lithium atom is also chelated with the siloxy oxygen atom.

As a nucleophilic glycine synthon, (2*S*,4*R*)-2-ethoxycarbonyl-4-phenyl-1,3-oxazolidine reacts with diorganozinc reagents in the presence of magnesium halide to afford ring opening products in about 98:2 diastereomeric ratio [Andres, 1992]. The presence of the Lewis acid is essential, therefore a chelate transition state is most likely involved. The ring opening of acetals with cuprate reagents in the presence of a bromoborane [Guidon, 1991] represents a similar situation.

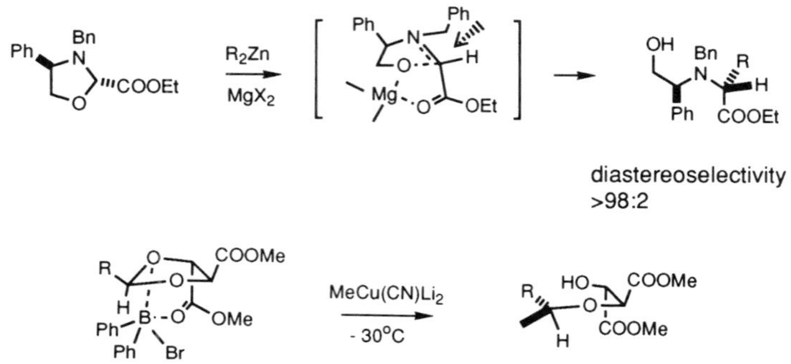

diastereoselectivity
>98:2

Regiocontrol related to chelation is apparent in the palladium-catalyzed formation of styryl ethers from aryl halides [Andersson, 1990]. Replacing butyl vinyl ether with 2-dimethylaminoethyl vinyl ether as reagent renders the nonselective arylation process virtually β-selective. The critical role played by the nitrogen atom is apparent.

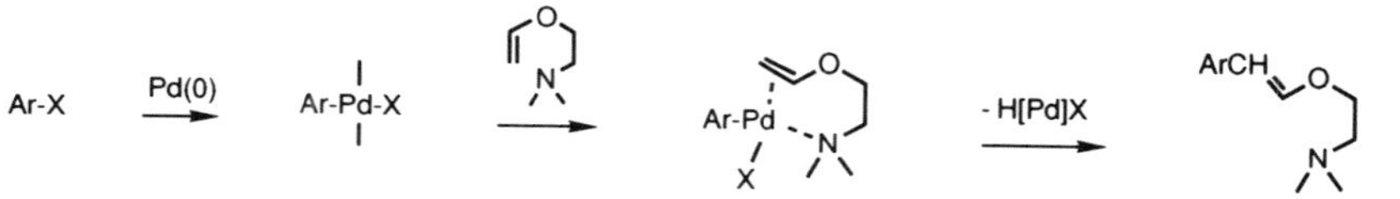

Great improvement on a synthesis of prostaglandin-$F_{2\alpha}$ which was based on epoxide ring opening with an alkynylalane was obtained by changing a dioxolane unit into a primary alcohol [Fried, 1973]. This change permits chelation of the reagent which forces the delivery of the alkynyl chain to the position proximal to the two-carbon sidechain, remedying the previously encountered lack of regioselectivity.

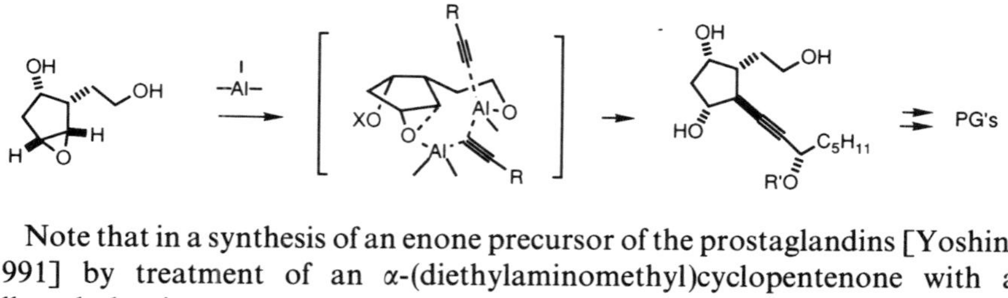

Note that in a synthesis of an enone precursor of the prostaglandins [Yoshino, 1991] by treatment of an α-(diethylaminomethyl)cyclopentenone with an alkynyl-aluminum reagent to introduce the ω-sidechain, the nucleofugal allylic amino group was an important element as its absence failed the reaction. Furthermore, the alcohol must be protected, otherwise it would exert a dominant control on the steric course for the entering alkynyl substituent (i.e., from the α-side).

Et$_2$Al—≡—R' → prostaglandins

trans-6-(2-Phenylsulfonyl-2-trimethylsilyl)ethenyltetrahydropyranyl ethers react with alkyllithiums stereoselectively [Isobe, 1981, 1984], the alkyl group transfer occurring after chelation. However, a hydroxyl group at C-5 becomes the dominant element, but in such cases high stereoselectivity is observed in reactions with Grignard reagents but not with methyllithium [Isobe, 1986a,b].

MeMgBr	95	5
MeLi	50	50

The presence of methoxymethyl and phenyl substituents at C-4 and C-5, respectively and in a *trans* arrangement, in 2-alkyloxazolines renders alkylation of the derived *N*-metalloenamines highly stereoselective [Meyers, 1976], due to the 95:5 preference for the (*Z*)-form and the alkylation from the same face of the chelator.

LDA, THF, - 78°C (95 : 5) RX

(*S*)- acid

C-Alkylation of an imine anion can be coaxed to follow an asymmetric course if a chirality center is present in close proximity to the nucleophilic site and the conformation of the metalloenamine can be fixed, for example by chelation [Whitesell, 1977b].

Chelation is undoubtedly a very important factor controlling the steric course of alkylation of a silyl carbanion [Chan, 1989].

An interesting observation pertains to the reductive methylation of *N*-(*o*-methoxybenzoyl)-2-methoxymethylpyrrolidine [Schultz, 1988]. Direct addition of iodomethane to the Birch reduction product gave virtually one isomer, whereas the other diastereomer was obtained when ammonia was removed before methylation.

X		
MeI	99.6 :	0.4
25°C; BuLi; MeI	1 :	99

Apparently different enolates were involved. In ammonia, the enolate may exist as solvated dimer. Ammonia ligands restricted rotation of the pyrrolidine ring, causing its sidechain to block approach of the alkylating agent. Evaporation of ammonia allowed configuration inversion at the nitrogen atom, and after reinsertion of ligands (presumably THF) a new aggregate resulted, which permitted alkylation from the other face.

Internally chelated (*Z*)-tin(II) enolates, when participating in aldol reactions with aldehydes, show π-face selectivity which is better than the corresponding titanium and boron enolates [Paterson, 1992]. Such an enolate is more conformationally restricted, and the transition state in which the aldehyde is arrayed by the metal is very sensitive to steric compression.

Sn(OTf)2 Et3N, CH2Cl2 RCHO 92% de (S,S)

The secondary *exo*-hydroxyl group in the bicyclic core of the esperamicin/calicheamicin class antitumor compounds is erectable during formation of the ring system by a borate-mediated synclinal aldol reaction [Magnus, 1990]. Other synthetic approaches have afforded epimeric mixtures.

Bu_2BOTf DABCO

The chromum(II)-mediated reductive condensation of 1,3-diene monoxide with aldehydes at C-2 of the former species proceeds in a highly stereoselective manner [Fujimura, 1990]. It has been proposed that the transition states involve chelated organochromium reagents in which the aldehyde molecule coordinates with the metal atom in a pro-(*E*) conformation. In other words, the oxygen lone-pair electrons participating in the chelation are *syn* to the smaller hydrogen atom.

+ R'CHO $CrCl_2$, LiI

Different diastereomers are obtainable from reduction of a chiral γ-sulfinylβ-keto ester (esp. *t*-butyl ester) with diisobutylaluminum hydride in the presence or absence of zinc chloride [Solladie, 1992]. Removal of the sulfur substituent results in enantiomeric β-hydroxy esters.

iBu_2AlH - 78°C ($ZnCl_2$)

de >95% de 90%

RaNi RaNi

Metal ion chelation of a bidentate substrate such as acrylamide has profound effects on its dienophilic character in terms of stereoselectivity as the substrate must settle in a special environment created by the stronger ligands surrounding the metal. This is the basis for the iron(III) complex which contains a bisoxazolinylmethane ligand derived from two molecules of (+)-phenyl-glycinol to show exquisite catalytic effects in the Diels–Alder reaction between 3-acrylyl-1,3-oxazolidin-2-one and cyclopentadiene [Corey, 1991b].

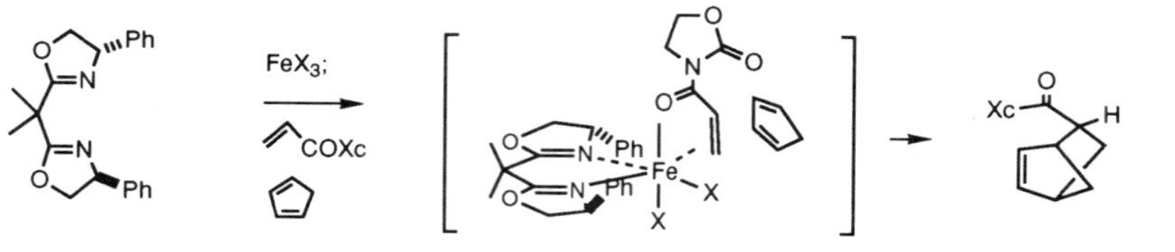

FeX_3;

COXc

enantioselectivity 91:9
endo/exo 96:4

It has been witnessed that chelation-induced conformational changes of a bidentate substrate affect the reactivity. A related observation is the acceleration of the intramolecular Diels–Alder reaction of a *N*-benzyl-*N*-(5-carboxyfurfuryl) fumaramide [Hirst, 1991] by pre-forming a complex with *N*,*N'*-bis-(6-methyl-pyrid-2-yl)isophthalamide. Hydrogen bonding of the two carboxyl groups of the substrate with the receptor brings the diene and dienophile together, therefore further entropic demand to reach the transition state is easily met. On the other hand, the analogous terephthalamide receptor tends to suppress the Diels–Alder reaction by enforcing wide separation of the reactive components through two sets of hydrogen bonds.

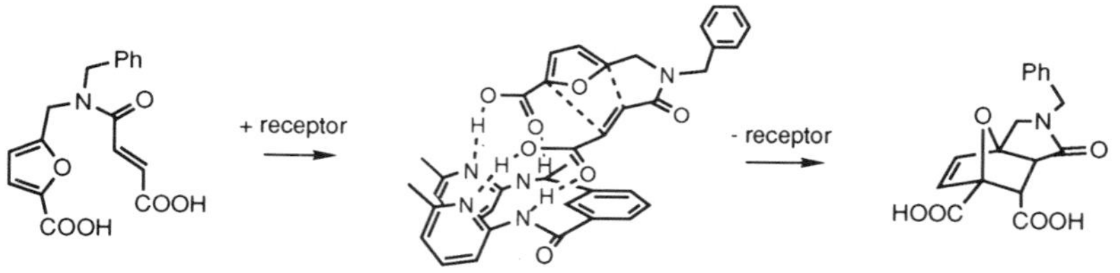

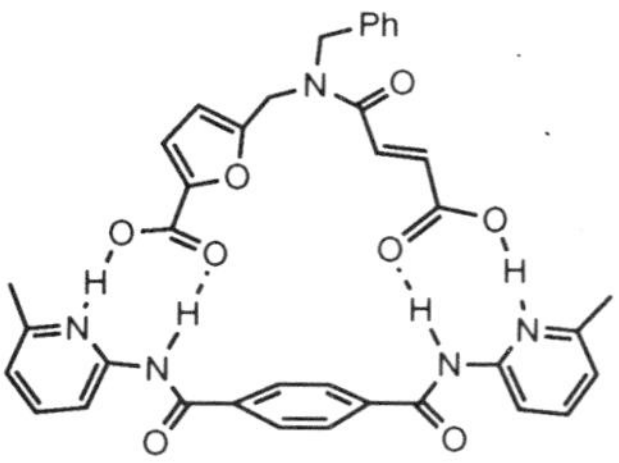

+ receptor

- receptor

unreactive for intramolecular Diels-Alder reaction

Disparate degrees of facial exposure toward an attacking nucleophile through conformational change of a substrate are illustrated by the preference of a hydrazone which bears a C_2-symmetric auxiliary for *si*-face reaction with organolithiums [Alexakis, 1991], and the complete reversal of selectivity to one occurring on the *re*-face when Grignard reagents are used [Alexakis, 1992].

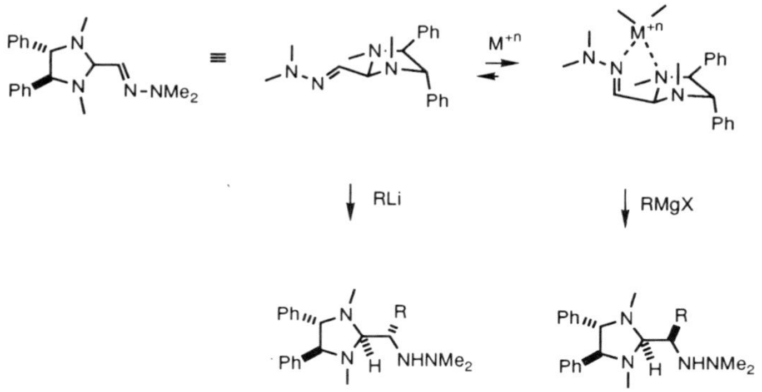

Enolates of amides derived from a C_2-pyrrolidine which contains two alkoxymethyl groups at C-2 and C-5 are chelates. Their alkylation are highly stereoselective [Kawanami, 1984]. Amides of prolinol form (*Z*)-enolates in which one lithium is coordinated with both oxygen atoms; methylation occurs predominantly from the *si*-face [Evans, 1980; Sonnett, 1980].

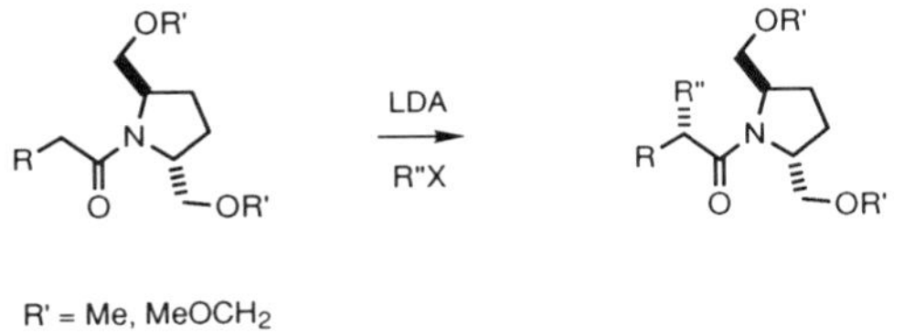

R' = Me, $MeOCH_2$

Imide enolates with high facial bias undergo electrophilation diastereoselectively. Thus, asymmetric α-electrophilation of alkanoic acids can be achieved by forming imides with an oxazolidin-2-one auxiliary derived from valine [Evans, 1982a]. Approach of the electrophile to the chelated lithium (*Z*)-enolates is directed away from the isopropyl group of the heterocyclic auxiliary.

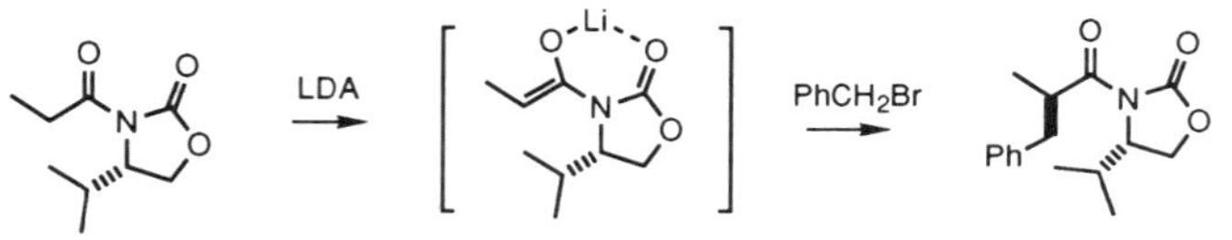

A system elaborated from Kemp's triacid provides steric environment [Stack, 1992] such that reaction occurs from the opposite face of the intentionally installed benzoxazole ring. Note that the two enantiomeric 2-methyl-4-pentenoic acids are obtained from allylation and the reaction sequence of iodination and deiodinative allylation, respectively. The contrasting results are due to involvement of chelated imide enolate in the first step, but no longer in the free radical in which the two carbonyl groups of the imide function prefer an *anti* conformation to minimize electostatic repulsion.

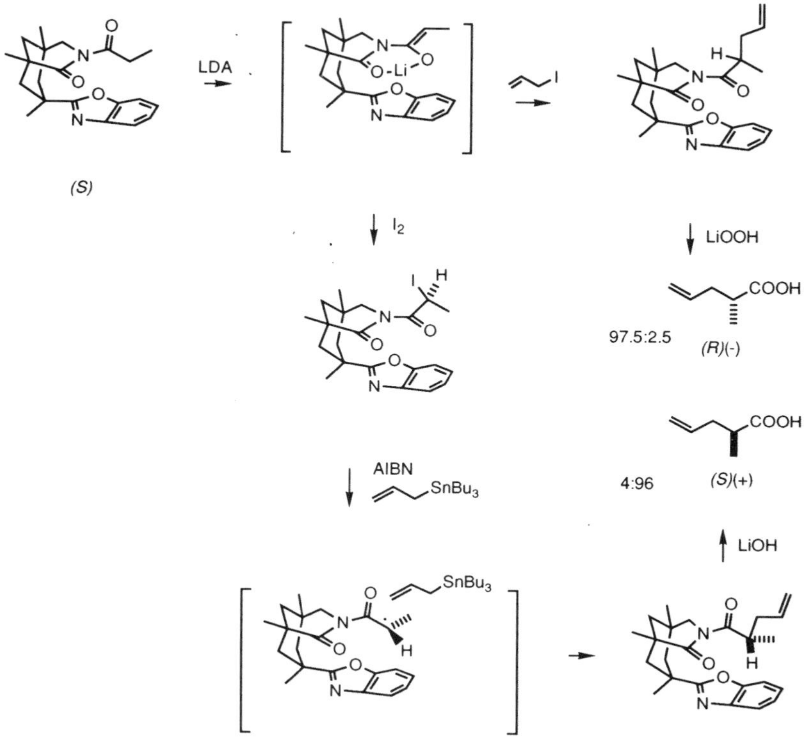

(S)

LDA

I_2

LiOOH

97.5:2.5 (R)(-)

4:96 (S)(+)

AIBN $SnBu_3$

LiOH

A tricyclic auxiliary derived from the Kemp triacid [Jeong, 1990] is also effective for diasteroselective alkylation.

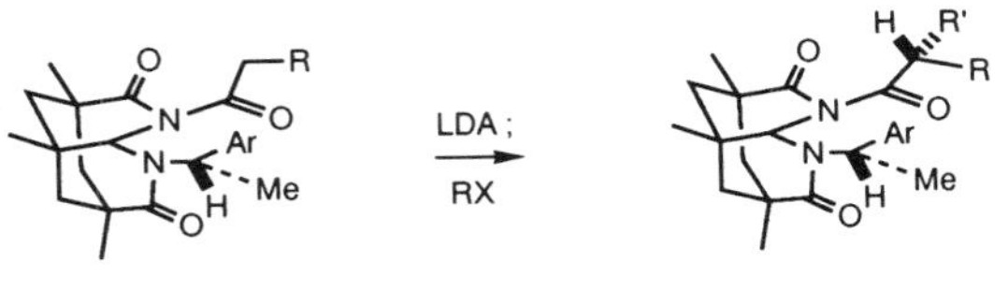

LDA ; RX

Metal chelates of imides can exhibit opposite sense of chirality transfer, as shown in the aldol-type reactions involving lithium and boron enolates [Yan, 1991].

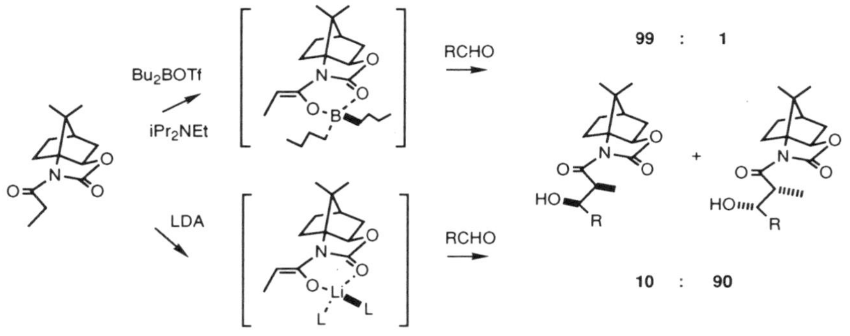

A complete reversal of selectivity in aldol reaction involving boron enolate of an imide [K. Hayashi, 1991] is attributable to the adoption of a open transition state when excess triflyloxyborate is present. Under such conditions the donor species exists as an internal chelate, as opposed to those in which the same boron atom marshals both the donor and the accepter.

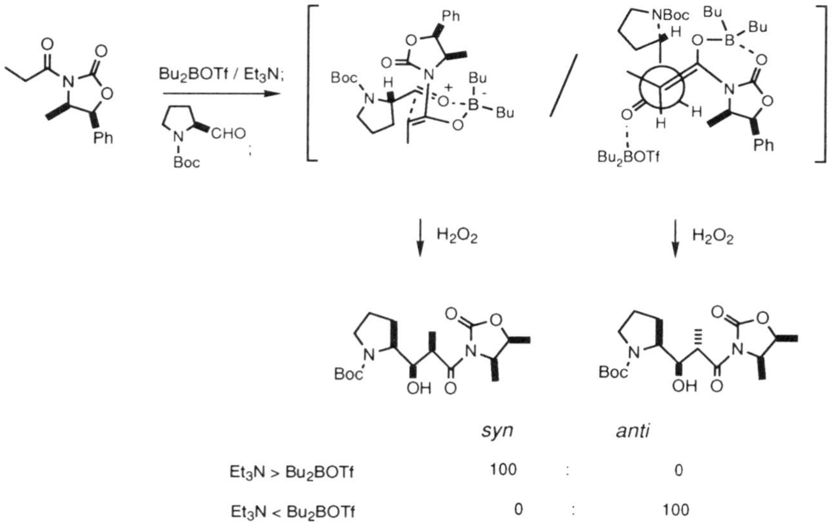

	syn		*anti*
$Et_3N > Bu_2BOTf$	100	:	0
$Et_3N < Bu_2BOTf$	0	:	100

Equally remarkable is the observation that aldol reactions of enolates of (*S*)-*N*-propionyl-2-oxazolidinone with aldehydes can display totally different diastereofacial selectivities, by choosing either titanium or boron counterion [Nerz–Stormes, 1991]. Thus it is possible to prepare either enantiomer of a β-hydroxy-α-methylcarboxylic acid. Chelation involving two oxygen atoms from the donor species as well as the aldehyde is very likely a factor in the case of titanium-mediated reactions.

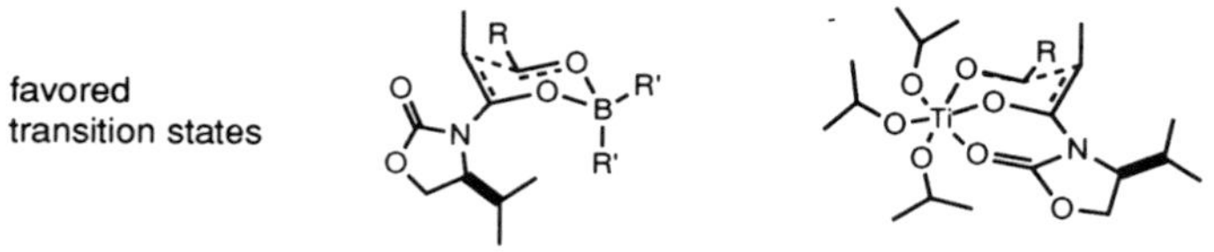

favored
transition states

Reactions such as the above have invoked various transition state models to account for different sets of results, their generalization is not always valid. For example, the aldol reactions of the kinetic enolates of (*S*)-4-siloxyl-5,5-dimethyl-3-hexanone were inferred to proceed via two-point or three-point coordination with the reaction partners [Van Draanen, 1991]. Lithium and magnesium enolated appear to undergo reaction through transition states with three-point contact, whereas boron and titanium enolates adopt the two-point coordination states. (Note (*E*)- and (*Z*)-enolate structures).

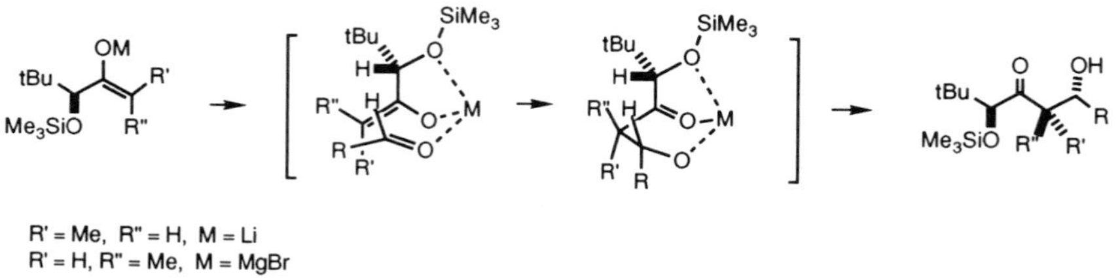

R' = Me, R" = H, M = Li
R' = H, R" = Me, M = MgBr

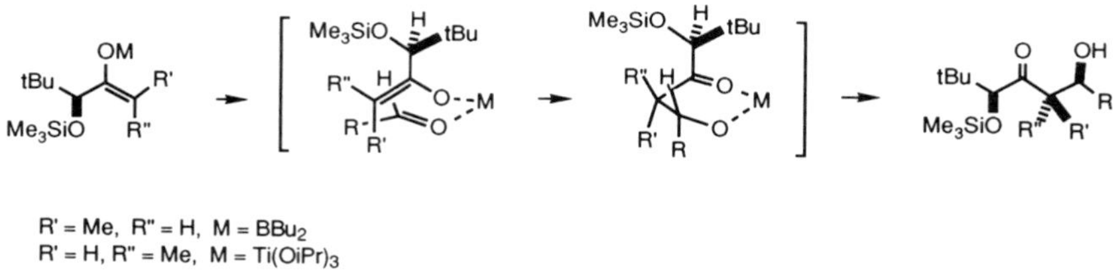

R' = Me, R" = H, M = BBu_2
R' = H, R" = Me, M = $Ti(OiPr)_3$

Synthetically significant is the use of 2-(trimethylsiloxy)-3-pivaloxy-1-butene as aldol donor [Trost, 1990c] because the stereocontrol does not require a substituent on the enol carbon. The organization of ligands around the titanium atom under the influence of steric effects and dipole–dipole interactions is sufficient to render the reaction highly stereoselective.

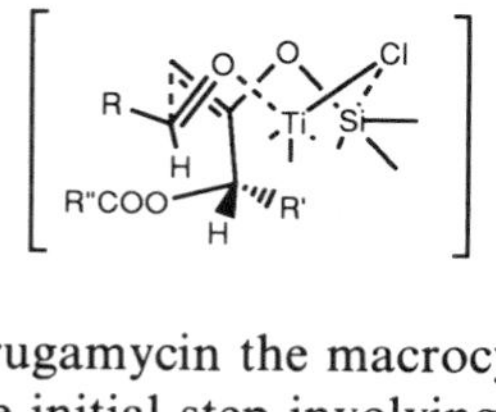

In a synthesis of (+)-ikarugamycin the macrocyclic portion was built from a tricarbocyclic scaffold. The initial step involving vinyl group addition to an unsaturated aldehyde mediated by an imine derivative [Paquette, 1990b] is subject to chelation control. By using a bulky chiral amine as auxiliary to exploit the diastereochemistry, a moderate success of kinetic resolution was materialized. The unwanted, minor diastereomer arose from a "mismatched pair" of reactants, the delivery of the vinyl group occurring from the congested *endo*-face of the chelate.

"matched" "mismatched"

(83 : 17)

By employing a 2,2′-dihydroxy-1,1′-binaphthyl monoether to form γ-keto esters, and conducting Grignard reactions on the latter compounds in the presence of magnesium bromide, chiral butenolides have been obtained [Tamai, 1992]. The 1,7-asymmetric induction becomes possible because chelation of the ethereal and carbonyl oxygen atoms dictates a highly ordered transition state.

$MgBr_2$-OEt_2

R′MgX

n = 0, 2 good
n = 1 poor

(R)

The nitrogen atom of 2-iodo-3,4,5-trimethoxybenzaldimine played an important role in the nickel(O)-mediated coupling with arylzinc chlorides [Larson, 1982], in view of the corresponding ester (instead of the imine) failing to react. Chelation to the nickel fixed the conformation of the *o*-substituent and reduced steric hindrance around the reaction center. Most importantly, chelation has a stabilizing effect on the arylnickel species.

A much simpler ligand-assisted nucleophilic addition is involved in the reaction of 4-hydroxy-2,5-cyclohexadienones with Grignard reagents [Swiss, 1990]. A slow formation of diorganomagnesium species which is coordinated to the carbinol oxygen is followed by rapid transfer of the organic residue. The relative configuration of C-4 and C-5 of the products is preordained.

Contrarily, cuprate reagents deliver the organic group from the opposite face of the hydroxyl [Swiss, 1992].

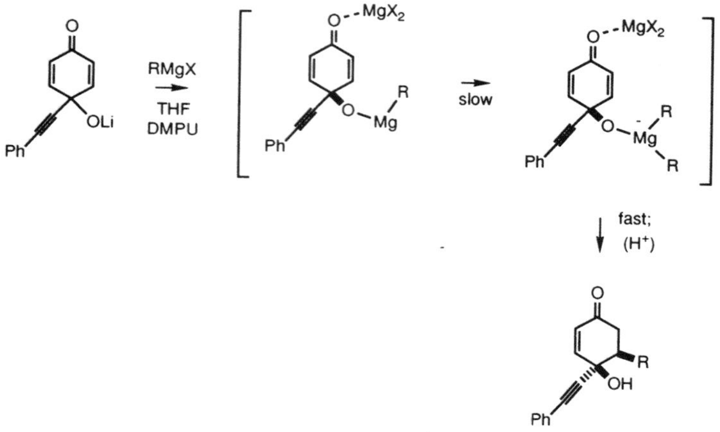

Conjugate addition of allylmagnesium bromide to a cyclohexenyl sulfoxide was the crucial step for a synthesis of (−)-sibirine [Imanishi, 1991]. Participation of a methoxy group of the acetal pendant as well as the sulfoxide in chelating the magnesium atom ensured the stereochemical outcome. Note that the alternative transition state was disfavored because of $A^{1,3}$-strain.

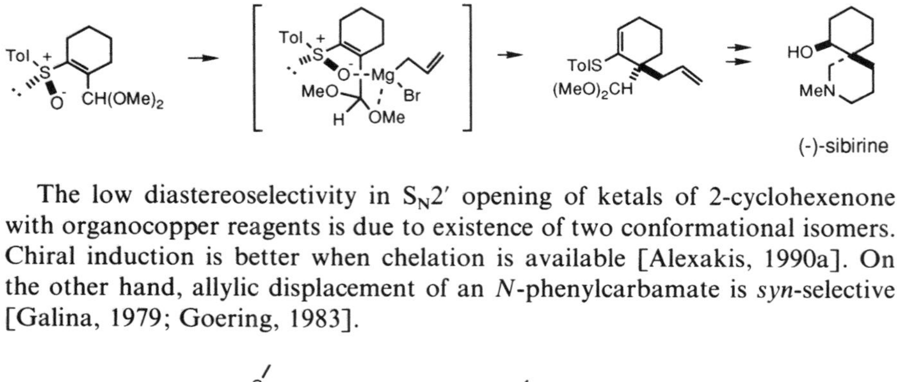

(-)-sibirine

The low diastereoselectivity in S_N2' opening of ketals of 2-cyclohexenone with organocopper reagents is due to existence of two conformational isomers. Chiral induction is better when chelation is available [Alexakis, 1990a]. On the other hand, allylic displacement of an *N*-phenylcarbamate is *syn*-selective [Galina, 1979; Goering, 1983].

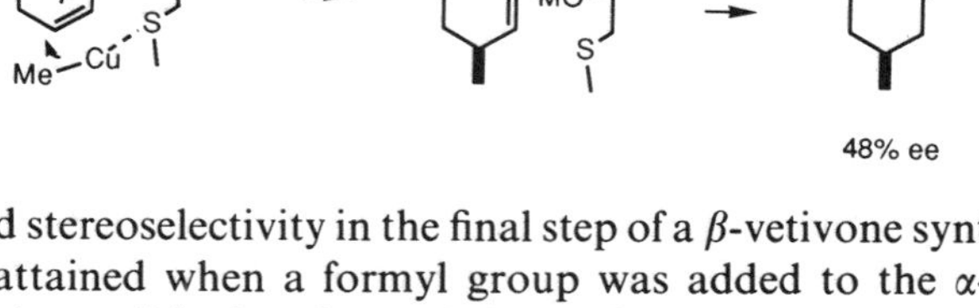

48% ee

Improved stereoselectivity in the final step of a β-vetivone synthesis [Bozzato, 1974] was attained when a formyl group was added to the α-position of the substrate. It is possible that the methyl transfer occurred within a copper chelate involving the aldehyde oxygen atom and the tetrasubstituted double bond which is exocyclic to another ring.

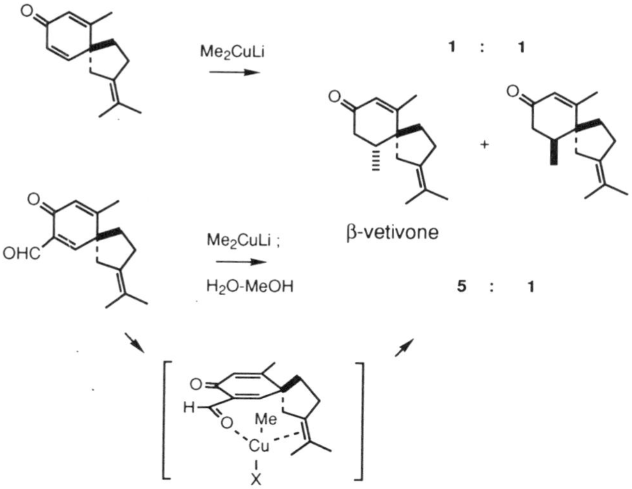

Me_2CuLi → 1 : 1

β-vetivone

Me_2CuLi ; H_2O-MeOH → 5 : 1

An interesting aspect of a perhydrohistrionicotoxin synthesis [Evans, 1982] pertained to internal chelation of the alkoxide by the imino nitrogen during reduction of the C=N bond. In this particular conformation the attack by hydride ion would give the new Sp^3-carbon a correct configuration.

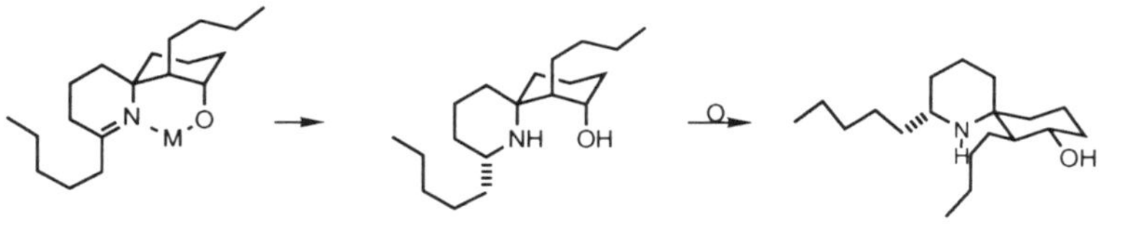

perhydro-histrionicotoxin

A carbon-bound metalloid atom complexed internally to a ketone can be exploited as conformational control for diastereoselective addition to the carbonyl group [Molander, 1992]. Because the organometallic moiety is of further synthetic utility, such systems are valuable.

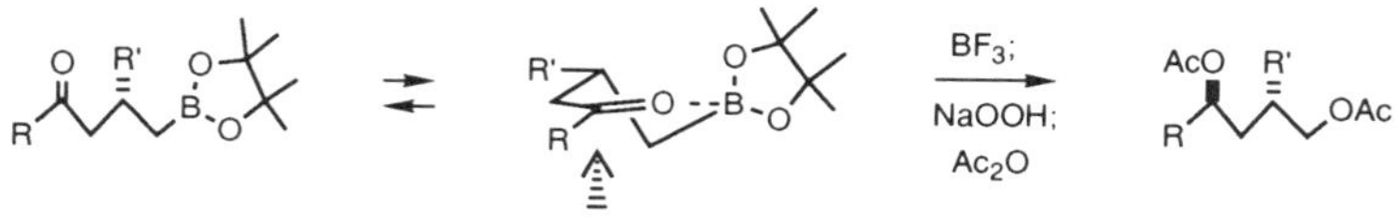

BF_3; NaOOH; Ac_2O

Dialkylzinc compounds in the presence of various catalysts, such as those based on aminoisoborneol [Kitamura, 1986; Itsuno, 1987; Oppolzer, 1988], prolinol [Soai, 1987], prolineamine [Corey, 1987b], ephedrine and pseudoephedrine [Corey, 1987b] effectively add to aldehydes with good enantioselectivity. Bimetallic complexes seem to be involved in such organozinc reactions.

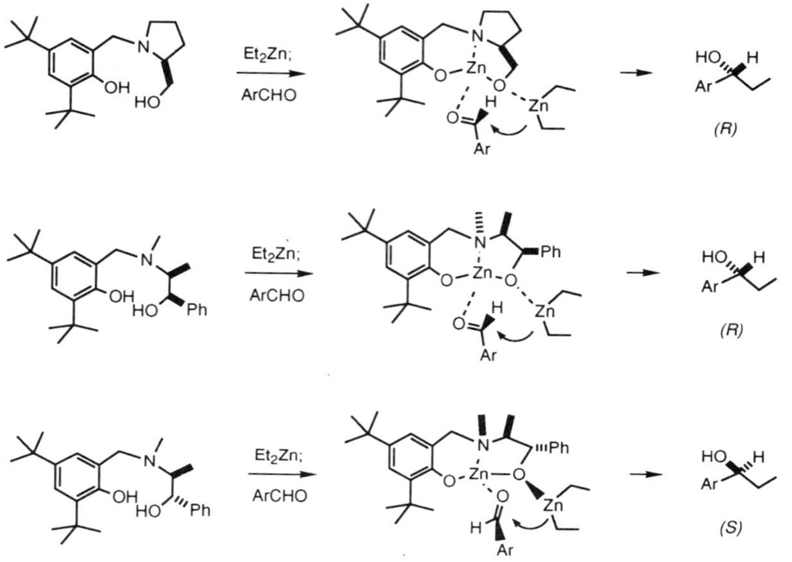

The diastereoselectivity for addition to imines by organometals [Ukaji, 1991] and ester enolates [M. Shimizu, 1992] can exhibit divergence depending on different metals. This phenomenon reflects varying coordinating ability of the metals, the oligomeric states of reactive species, and their steric bulk.

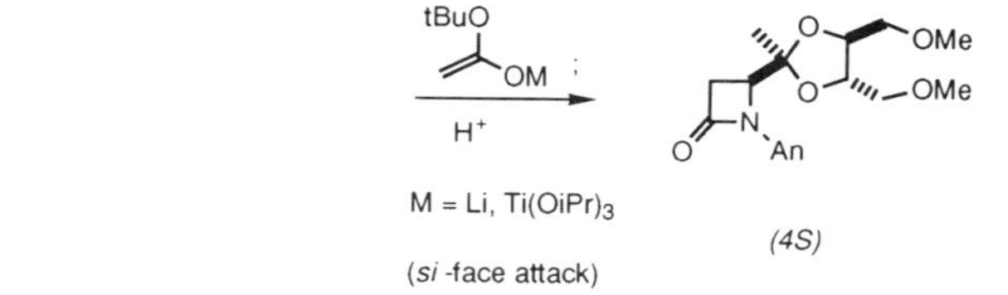

The coordinating Lewis acid affects the stereochemical course of the allylmetal reaction of an α,β-epoxy imine [Beresford, 1992].

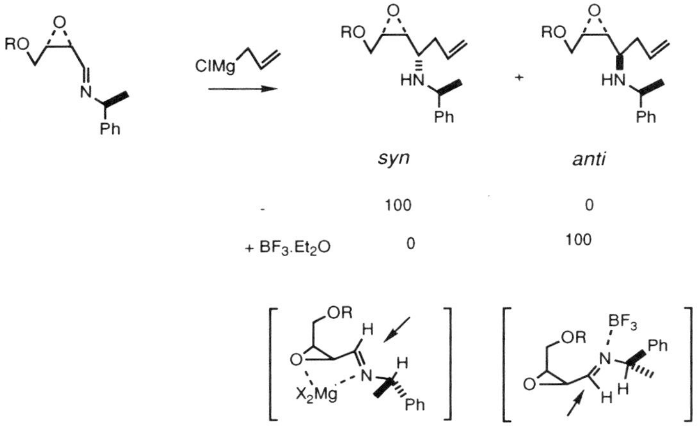

	syn	*anti*
-	100	0
+ $BF_3.Et_2O$	0	100

Worthy of note is the behavior of certain cyclic hemiaminals toward Grignard reagents [Nagai, 1992]. Chelation-assisted heterolysis of C—O bond, when it is enabled by a proximal alkoxy moiety, must be responsible for the high reactivity as compared with the deoxy analog.

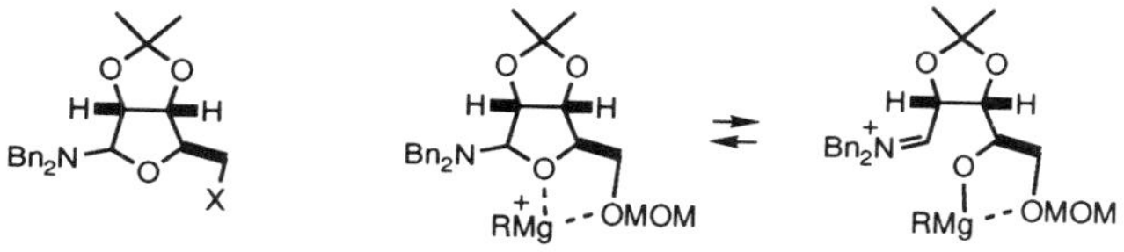

Sometimes it is very difficult to predict the onset of chelation control. For example, Grignard reactions of a 3-*exo*-7-oxabicyclo[2.2.1]hept-5-ene-2-*exo*-carbaldehyde are apparently under chelation control in THF, but not in THF [Bloch, 1987].

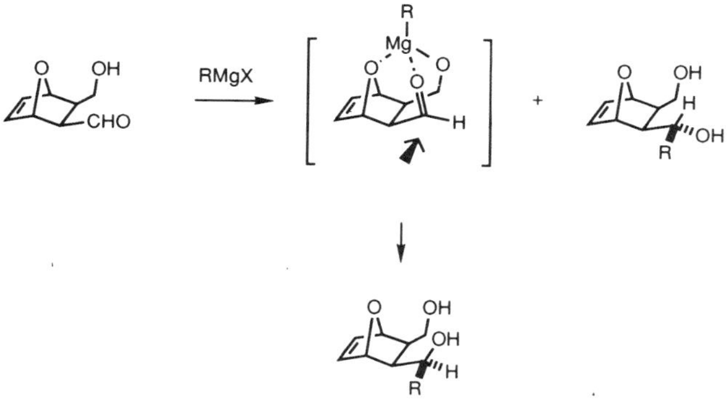

Treatment of a γ-benzyloxy α,β-unsaturated ester which contains a methylenecyclopropane unit with Pd(O) reagent led to a fused methylenecyclopentane possessing three well-defined asymmetric centers [Motherwell, 1991]. The benzyloxy group proved to be of utmost importance and its role was to act as a ligand for the palladium ion of the triethylenemethane complex. In the debenzyloxy substrate the intramolecular cycloaddition pathyway was superseded by simple isomerization by ring opening.

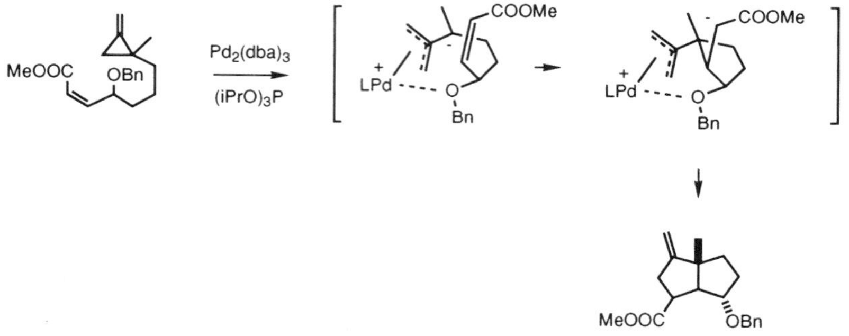
MeOOC
OBn
Pd2(dba)3
(iPrO)3P
COOMe
LPd
O
Bn
COOMe
LPd
O
Bn
MeOOC
OBn

The regiochemistry of the Pauson–Khand reaction can be directed, particularly by a homoallylc substituent (*N*, *S*-based) [Krafft, 1991]. The results may be rationalized by considering a bidenate mononuclear transition state.

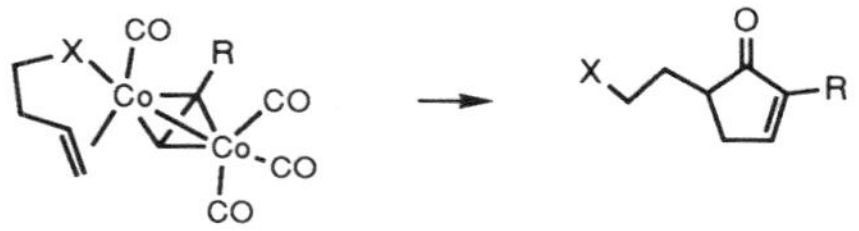
X
CO
R
Co
CO
Co
CO
CO
X
O
R

A high level of asymmetric induction has been observed in the $SnCl_4$-catalyzed ene reaction of 8-phenylmenthyl glyoxylate with alkenes [Whitesell, 1982]. It is likely that the alkene attacks the chelated carbonyl group in a chair transition state.

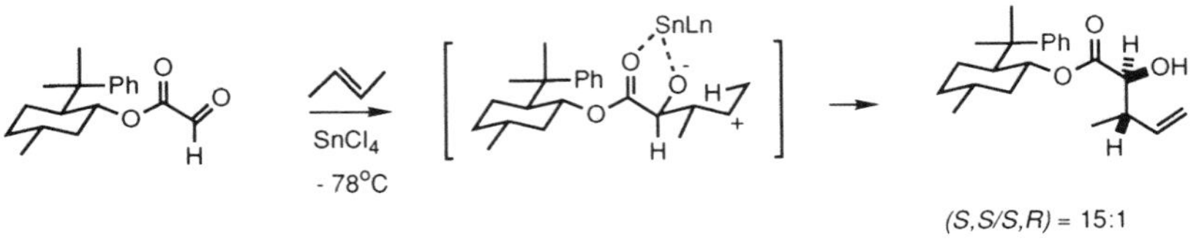
Ph
O
O
O
H
SnCl4
- 78°C
SnLn
Ph
O
O
H
H
Ph
O
H
OH
O
H
(S,S/S,R) = 15:1

Strong chelation of a homoallylic oxygen atom to the metal in the transition state must be responsible for the much enhanced stereoselectivity of [2.3]-Wittig rearrangement shown below when it is mediated by a titanium enolate (vs. lithium enolate) [Yamaguchi, 1989].

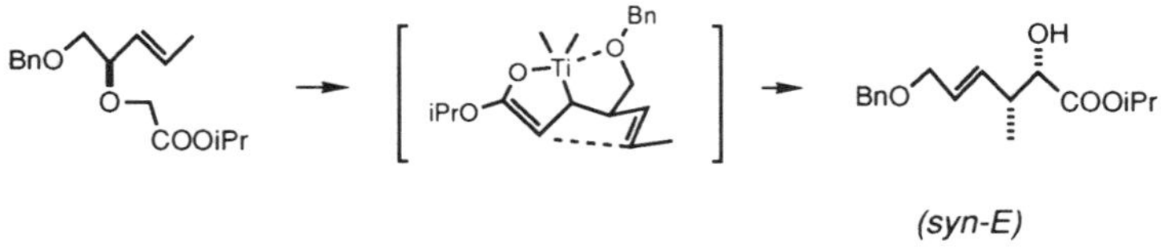
BnO
O
COOiPr
Bn
O
O
Ti
iPrO
OH
BnO
COOiPr
(syn-E)

Dramatic improvement of regioselectivity by Lewis acid catalysis in the Diels–Alder reaction has been discovered in recent decades. The nature of the catalyst is of the utmost importance in relation to the structure of an addend (mostly the dienophile). Thus, the condensation of 2-methoxy-5-methyl-1,4-benzoquinone with isoprene gives one major adduct in the presence of BF_3 etherate, whereas chelation of tin(IV) to the quinone leads to the formation of a regioisomer [Tou, 1980]. The site of principal coordination differs as the result of steric effect and chelation. A *peri*-hydroxy of a 1,4-naphthoquinone participates in chelation readily, and accordingly, polarization gives rise to regiocontrol [Kelly, 1978]. In fact, by virtue of intramolecular hydrogen bonding, juglone itself also exhibits regioselectivity in Diels–Alder reactions [Kelly, 1977].

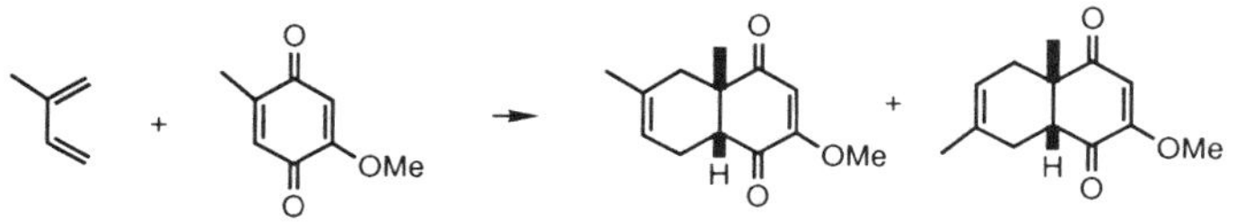

Δ	1	1
BF_3	1	24
$SnCl_4$	20	1

Both regio- and diastereofacial control of the Diels–Alder reaction are rendered possible when juglone is chelated to a boron atom which simultaneously coordinates with 1,1′-bi(3-phenyl-2-naphthol) [Kelly, 1986].

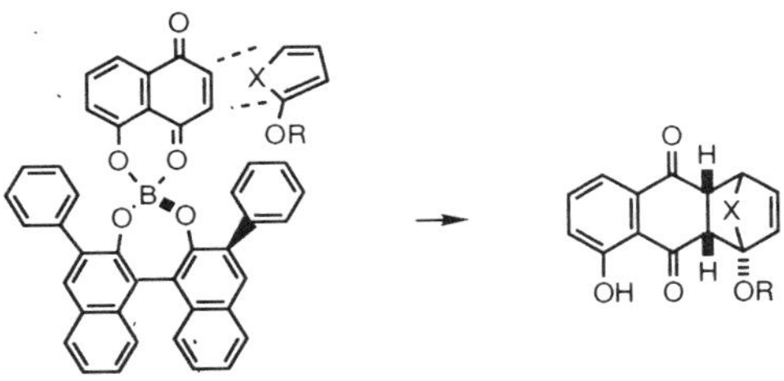

A synthesis of isomitomycin-A [Fukuyama, 1987] started from reaction of a chalcone and 2-ethylthio-5-trimethylsiloxyfuran in the presence of tin(IV) chloride. The major role of the Lewis acid was to coordinate the oxygen atom of the enone and that of the siloxy substituent, thereby ensuring the required regiochemistry of the product.

isomitomycin-A

Bicyclic bisacetal auxiliaries derived from pentitols (e.g. ribose) can be used to form acrylic esters which in the presence of a Lewis acid undergo highly stereoselective Diels–Alder reactions [Gras, 1992].

(R)

The intramolecular Diels–Alder reaction of furan and acrylyl components is rendered stereoselective by providing chelation possibility in the intervening chain, and it is possible to acquire an adduct of high optical purity [Mukaiyama, 1981].

By providing a macrocyclic cavity lined with three zinc-porphyrin units, 3-(4-pyridyl)furan and *N*-(4-pyridylmethyl)maleimide can be trapped within by their individual chelation to the metal atoms. The effect is rate enhancement of Diels–Alder reaction and geometry fixation of the transition state for the *exo*-adduct [Walter, 1993].

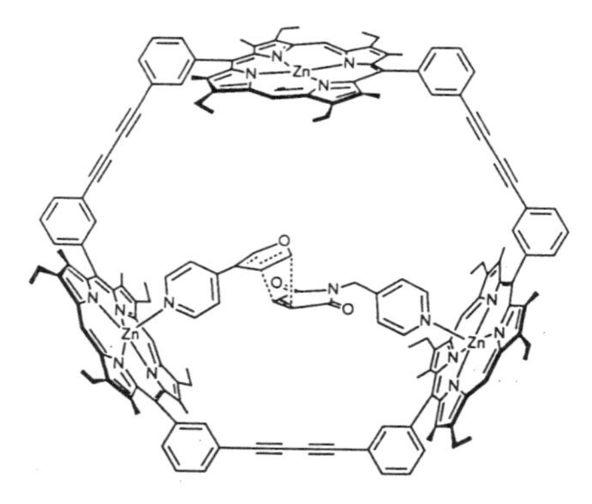

It is noted that in the synthesis of a linear oligomer comprising an even number of units, each of the two identical end groups must be monoprotected before coupling. If the unit containing two reactive termini arising from double deprotection can be directed to cyclize by a proper template, it can be removed and by such tactic the separation of products would be greatly simplified [S. Anderson, 1993]. A prototypical example of this approach is the scavenging of porphyrin-substituted phenylacetylene by 4,4′-bipyridyl [S. Anderson, 1992]. The concept is shown below schematically.

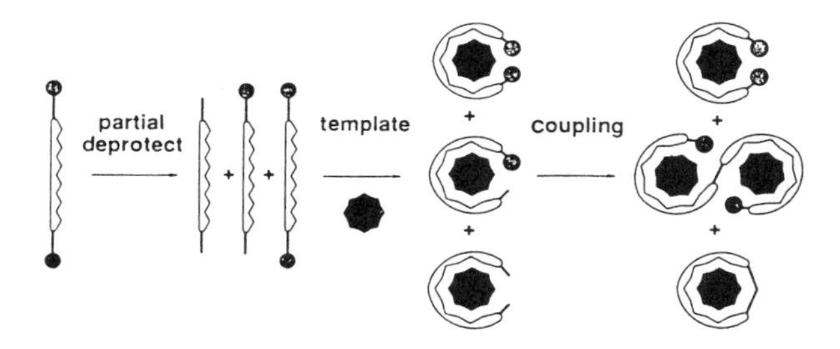

On the other hand, directed synthesis of cyclic porphyrin dimer or trimer can be achieved via intramolecular Glaser reaction of the 4,4′-bipyridyl or 2,4,6-*tris*(4-pyridyl)-1,3,5-triazine complexes of the zinc-porphyrin molecules, respectively [H.L. Anderson, 1990].

A tandem free radical cyclization of an *o*-bromoaniline derivative to form a piperidinohydrocarbazole [Jenkinson, 1992] proceeded only with the *erythro* isomer. The failure of the *threo* isomer to form the tetracyclic product has been attributed to *N* ... *Si* chelation which placed the enamine acceptor farther away from the vinyl radical.

Bu_3SnH

AIBN

does not cyclize

A chelation control for ring expansion rearrangement of spiro epoxides [Tobe, 1985a] has been exploited to achieve the synthesis of several terpenes including modhephene [Tobe, 1984] and isocomene [Tobe, 1985b]. The chelation locks up conformation to allow migration of the methylene group but not the angular carbon atom.

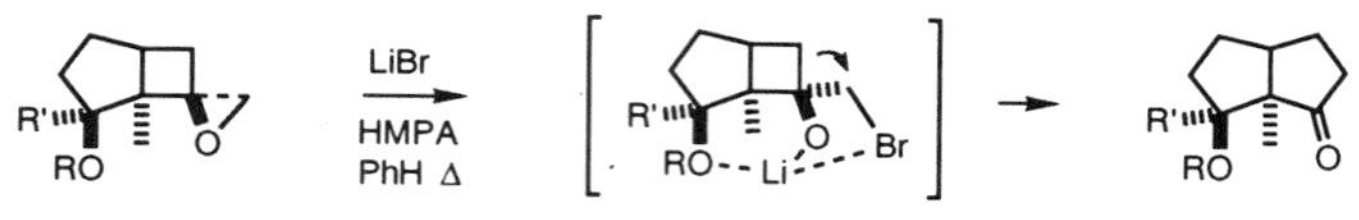
LiBr
HMPA
PhH Δ
R'
RO
O
RO--Li
Br

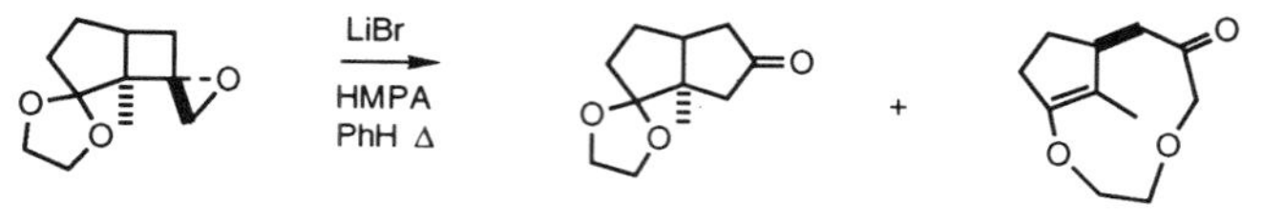
LiBr
HMPA
PhH Δ
+

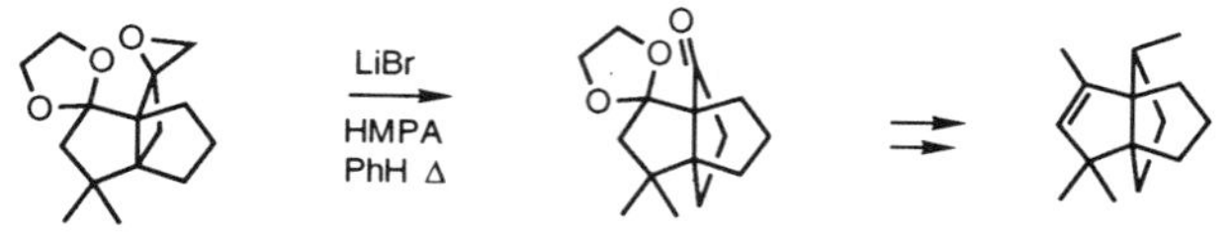
LiBr
HMPA
PhH Δ

modhephene

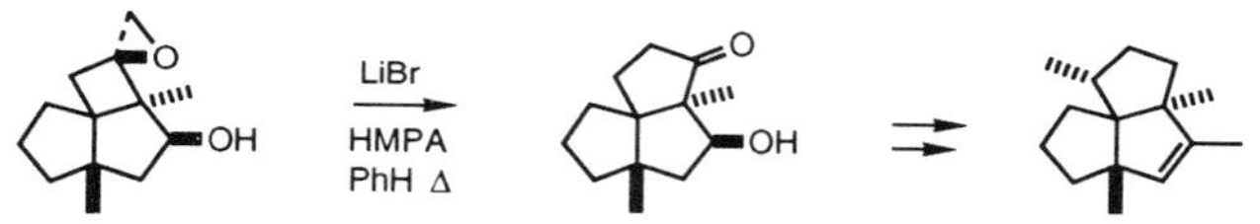
LiBr
HMPA
PhH Δ
OH

isocomene

The 1,2-Wittig rearrangement of glyceryl allyl ethers is subject to chelation control [S.L. Schreiber, 1987b]. Although the reaction proceeded in low chemical yield, it showed excellent *syn*:*anti* ratio (>90:10).

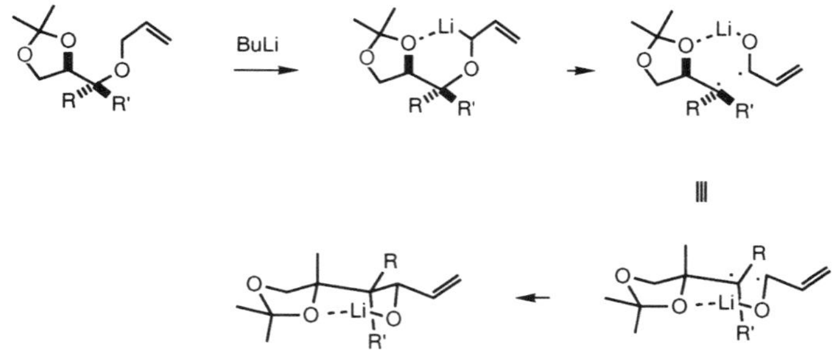
BuLi
Li
R
R'

A word of caution pertains to the necessity of chelation control. In certain situations, such as formation of homopropargylic alcohols through $S_{E'}$ addition of chiral allenylstannanes to α- and β-alkoxy aldehydes [Marshall, 1992], chelation can cause steric interactions between reactants in the transition state.

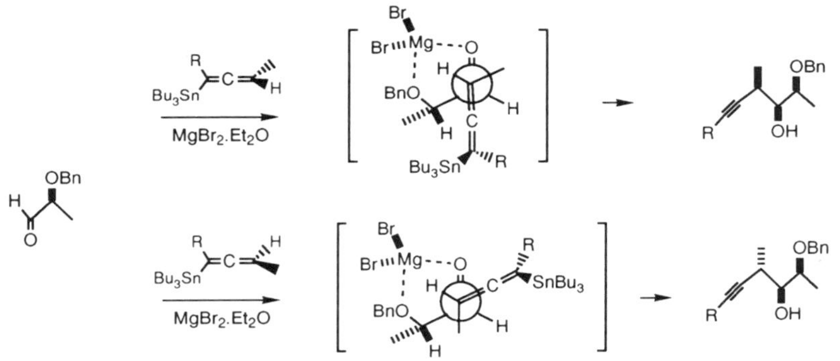

Furthermore, chelation is not always a beneficial factor in synthesis. Good judgements must be made in some cases to avoid chelation. For example, in a synthesis of (−)-brevianamide-B [R.M. Williams, 1988] the intramolecular S_N2' cyclization effected by sodium hydride in benzene gave largely an undesired isomer, due to chelation of the metal ion by both the enolate and the nucleofugal chlorine atom. Completion of the synthesis thus relied on changing the reaction conditions of this step by addition of 18-crown-6 to permit the formation of the epimer in a higher ration (3.85:1).

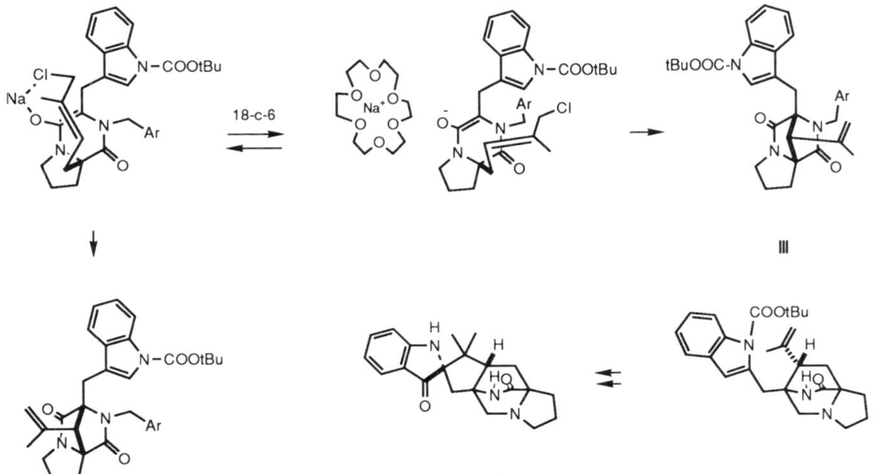

(-)-brevianamide

7.2.2. Cram Rule; Addition to the Carbonyl Group

Diastereoselectivity was recognised by Emil Fischer in 1894 as occurring when the ratio of diastereomers is related to bias exerted by existing asymmetric center(s) during formation of a new one. However, the prediction of diastereoselectivity arising from chemical reactions of prochiral groups has been quite recent [Cram, 1952; Prelog, 1953]. According to Cram, a carbonyl group adjacent to an asymmetric center bearing small (R^S), medium (R^M), and large (R^L) substituents is attacked by a nucleophile via a conformation in which the angle $<R^SCR^M$ is bisected by the bulky C=O (due to complexation) and it proceeds from the side of the small substituent. The correct conclusion of asymmetric induction was actually reached from incorrect premises since the eclipsing of the large substituent and the α' group is not preferred, and the Curtin–Hammett principle is applicable (small free energy difference between ground state conformations compared with the free energy of activation) and the ground state conformations are rather irrelevant to the reaction selectivity. Furthermore, the approach of the nucleophile is not only influenced by steric factors, as electronic interactions with existing substituents also play a role.

A more satisfactory rationale of the Cram rule is based on a transition state model [Cherest, 1968] in which the R^L group is perpendicular to the carbonyl group, and the nucleophile attacks from the opposite side of the large substituent. It was proposed that there are two reactive conformations, the favorable one with the oxygen end of the carbonyl nearest to the R^M substituent. The predilection of this particular conformation remained unexplained until the nonperpendicular trajectory of the attacking species was taken into consideration [Anh, 1980]. The nucleophile approaches the carbonyl group at an angle of *ca.* 100° in order to avoid electostatic repulsion by the oxygen atom. Note that this repulsion is nonexistent in the addition to C=C [Paddon-Row, 1982].

Cram model

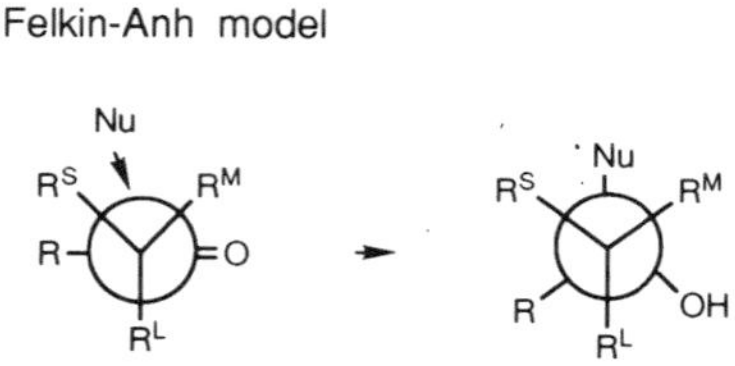

Felkin-Anh model

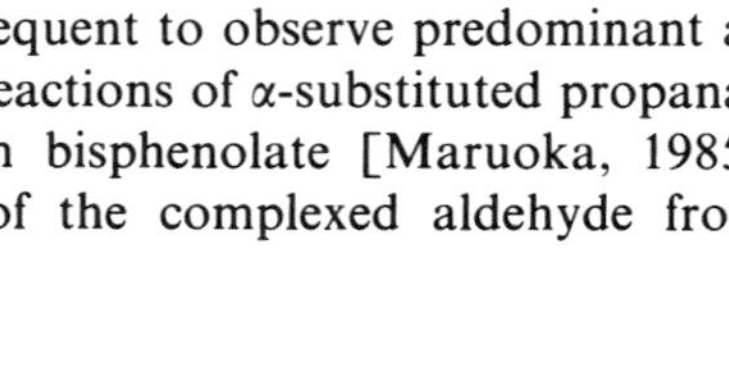

However, it is not infrequent to observe predominant anti-Cram additions, such as in the Grignard reactions of α-substituted propanals in the presence of a bulky methylaluminum bisphenolate [Maruoka, 1985], which is due to conformational change of the complexed aldehyde from the conventional disposition.

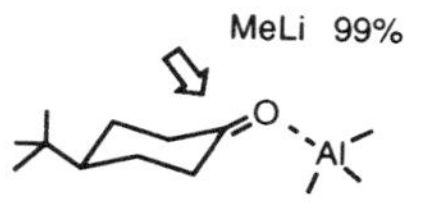

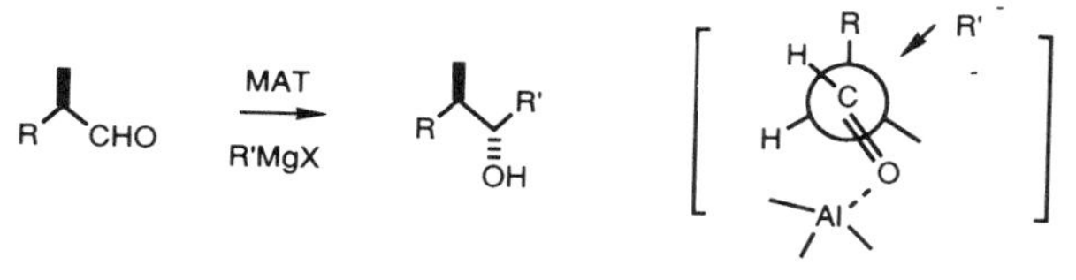

The Cram rule may also become invalid when one of the substituents at the asymmetric center is capable of functioning as a ligand for the metal ion during reaction with an organometallic reagent. In such cases the cyclic model, better known as the chelate Cram model applies (but unfortunately not always: apparently other factors such as the Lewis acidity of the metal ion also contribute to the choice of reaction pathyway). Bond formation at the carbon atom from the side of the small substituent ensues.

In the reaction of several α-alkoxy ketones with dimethylmagnesium, the rates of fast reacting compounds and the proportion of products predicted by Cram's chelate rule can be correlated. But the major product of the slowest reacting ketone does not arise from analogous transition state. In other words, there is a simultaneous effect of chelation on reactivity and stereoselectivity [Chen, 1992].

Traditionally, the diastereoselectivity from reactions with organolithium and Grignard reagents has been quite disappointing. However, reactions with other organometallic reagents such as alkyltriorganotitanates [Reetz, 1982a; Weidmann, 1983], diallylzinc [Fronza, 1982] and diallyltin [Mukaiyama, 1983] reagents show superior results.

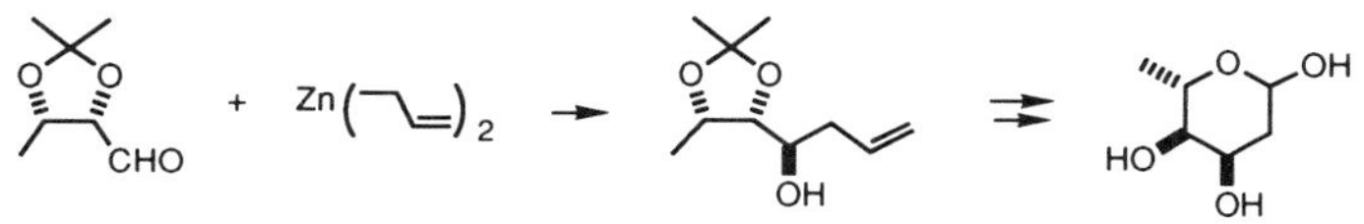

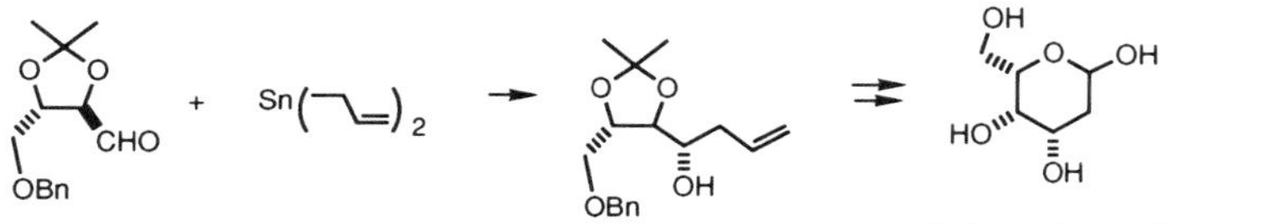

Grignard reactions of certain chiral α-alkoxy- and α-amino carbonyl compounds [Meric, 1973; Mukaiyama, 1978a, 1979] afford carbinols diastereoselectively. The 2-methoxymethylpyrrolidinyl group is a bidentate *N*,*O*-ligand which can be used to direct Grignard, organozinc, and organolithium reactions when it is placed in an α-position of a ketone [Fujisawa, 1988]. Unfortunately, organolithium reagents which can be used to provide high level of diastereoselection are very limited.

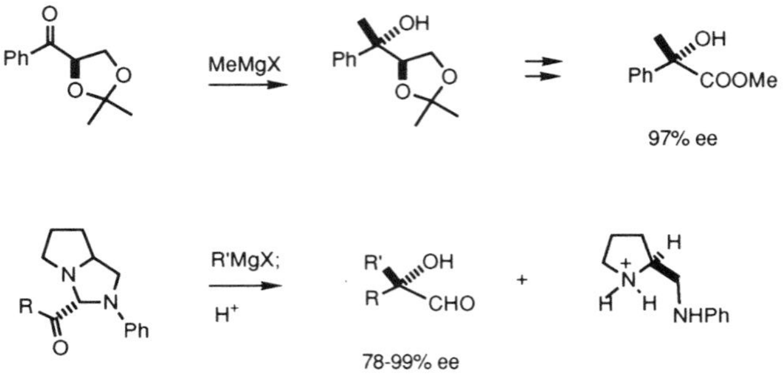

When the chelate Cram model is operative, a high degree of stereoselection is observed. An application is in a synthesis of muscarine [Still, 1980d]. The effects of an α-alkoxy group have been systemically investigated [Still, 1980b]. Intriguingly, a β-alkoxy group efficiently chelates with cuprate reagents but not with RMgX or RLi [Still, 1980c].

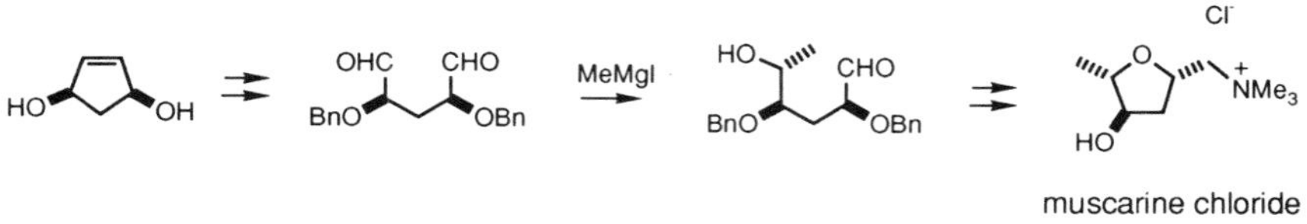

Chelation-controlled 1,2-asymmetric induction is not uniformly successful with α-alkoxy aldehydes except in aldol reactions involving silyl enol ethers as donors [Reetz, 1984a]. The preferred formation of the *syn*-isomer, regardless of the configuration of the silyl enol ether simplifies the operation.

2-Acyl-1,3-oxathianes derived from 8-hydrothiomenthol are attacked by Grignard reagents via Cram-type chelates in which the metal coordinates with both of the oxygen atoms [Eliel, 1982]. Note that the soft sulfur atom is less prone to complex with the hard metal ion.

R'MgI
- 78°C

(better R = Ph)

ee > 90%

(+)-pulegone

An exploitation of stereocontrol by chelate is an intramolecular allyl addition to chiral carbonyl compounds [Reetz, 1988]. Aldols in which the hydroxyl group is protected with an allyldimethylsilyl function undergo intramolecular reaction on treatment with a Lewis acid. With titanium(IV) chloride the *syn*-1,3-diols are favored (>90% de), due to their formation from a transition state in which the carbinolic substituent occupies a pseudoequatorial position. The selectivity is opposite to the intermolecular reaction of the corresponding benzyl ethers with allyltrimethyl silane.

$TiCl_4$

>90% <10%

Complexation by titanium probably also affords the best 1,4-induction [Reetz, 1983b].

Remote asymmetric induction is observed in α-selective addition of allystannanes to aldehydes when the reagents contain an allylic alkoxy group [McNeill, 1990, 1992] or homoallylic alkoxy functionality [J.S. Carey, 1992]. It is noted that internal complexation of the metal center by the alkoxy group in the reagent controls the transition states, and the reactions with chiral aldehydes conform to theFelkin–Anh model when the pair is matched. For an unmatched pair the *re* face approach to the aldehyde is not as favorable, and alternative processes are allowed to compete, leading to lower diastereoselectivity.

$SnCl_4$ RCHO

R = Ph *anti* 96
syn 4

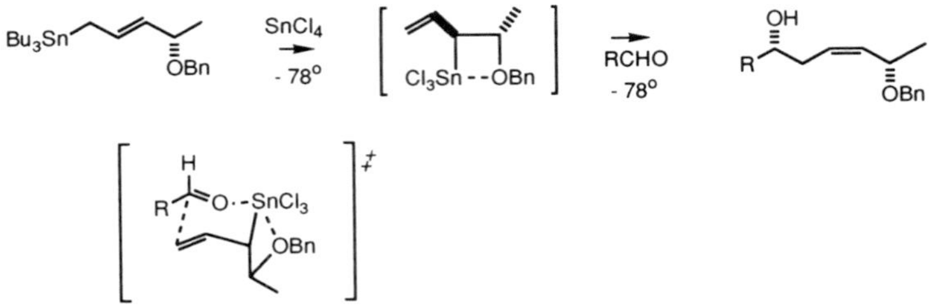

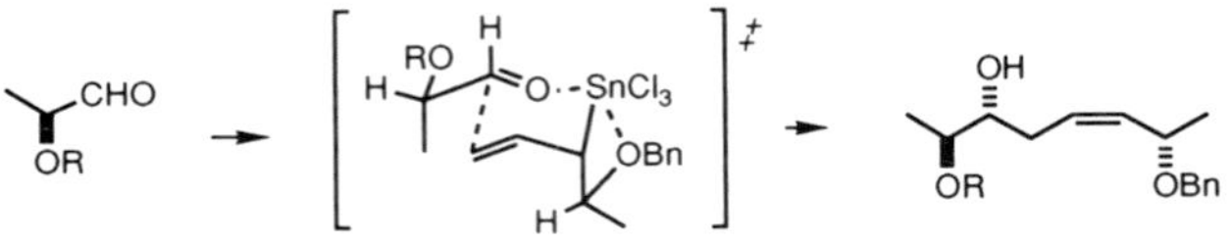

matched Felkin-Ahn

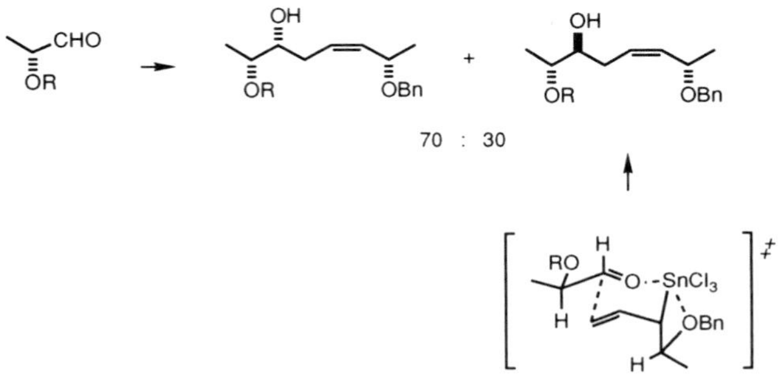

70 : 30

There is enantiofacial differentiation in the reaction of allenyl boronic esters with aldehydes in the presence of a tartaric ester [Ikeda, 1986]. The results indicate a transition state in which the R group of aldehyde RCHO is rotated along the C=O bond in order to avoid steric interaction with the ester moiety within the complex, thus the antiplanar arrangement between the R group and the Lewis acid center no longer prevails.

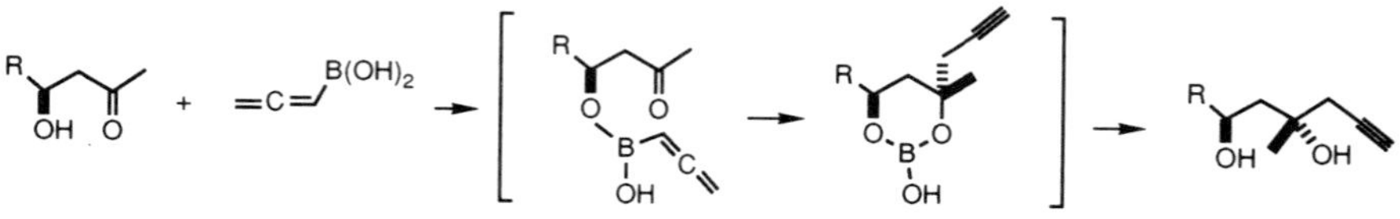

The condensation of cyclohexanone enolate with nascent cyclopropanone furnishes a tricyclic keto diol as the major product [J.T. Carey, 1986]. This reaction is strongly dependent on the presence of magnesium ion, apparently its chelation effect.

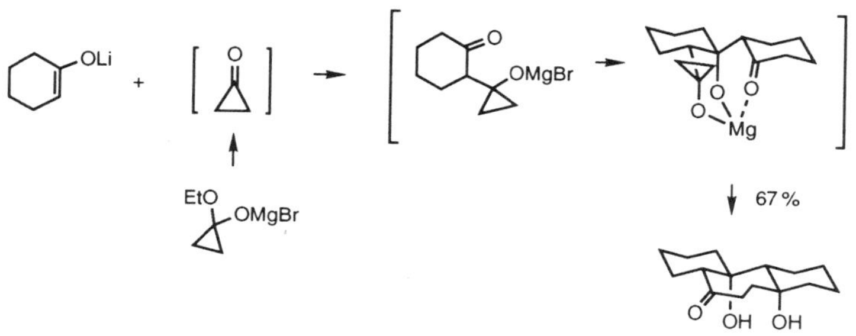

Six-membered ring chelates have been exploited very extensively in various chemical processes. Complex metal hydride reductions can be directed by chelation and it is the reason that zinc borohydride is a superior reducing agent for the reduction of a variety of β-heterosubstituted ketones in terms of diastereoselectivity. The combination of enolate acylation of chiral amides and subsequent reduction constitutes a very efficient method for the synthesis of 2-alkyl-3-hydroxycarboxylic acids [Y. Ito, 1984].

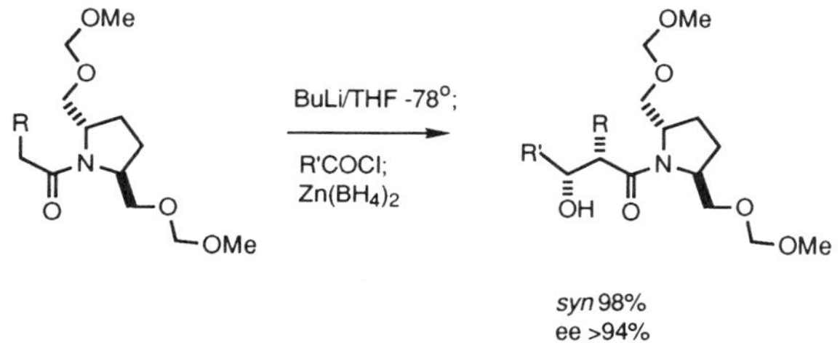

BuLi/THF -78°; R'COCl; $Zn(BH_4)_2$

syn 98%
ee >94%

On successive treatment of a β-hydroxy ketone with a borane and sodium borohydride the reduction proceeds via the chelate to furnish almost exclusively the *syn*-1,3-diol [Narasaka, 1984b]. On the other hand, *anti*-diastereoselective reduction of the same class of compounds with tetramethyl triacetoxyborohydride is preferred [Evans, 1988b]. This latter reaction involves an acid-promoted ligand exchange of acetate for substrate alcohol by the triacetoxyborohydride anion, and a fast internal hydride transfer. The reduction of a hydroxydiketo ester to give predominantly the *anti,anti*-triol indicates an initial reduction diastereoselectivity of 47:1 and the second stage selectivity of 7:1 from the *anti*-diol.

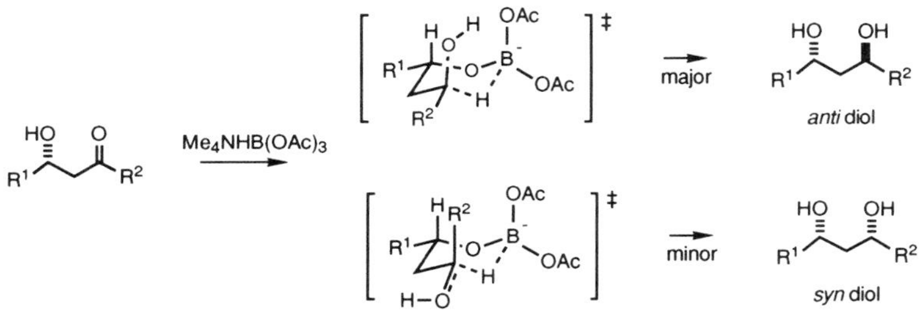

$Me_4NHB(OAc)_3$

major — *anti* diol

minor — *syn* diol

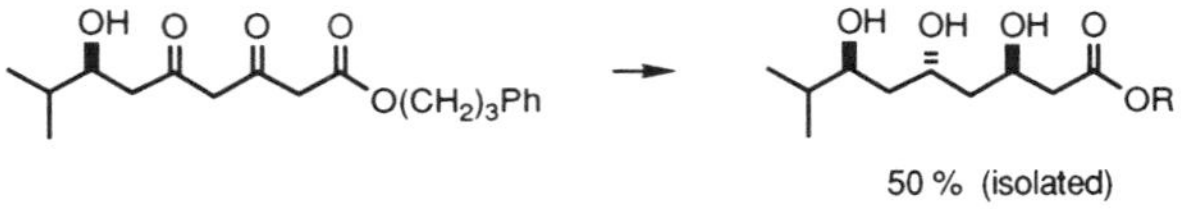

50 % (isolated)

diastereomer	ratio
anti-anti	85 %
anti-syn	13 %
syn-anti	2 %
syn-syn	0 %

A stereoselective cyanohydrin formation from β-hydroxy ketones in the presence of zinc iodide [Brunet, 1991] is attributable to the axial attack by the cyanide ion on a six-membered chelate.

Coordination plays an important role in the reductive ring opening of 2-substituted *trans*-4,6-dimethyl-1,3-dioxanes [Mori, 1987] as the binding of the Lewis acid to the oxygen atom geminal to the axial methyl group initiates the reduction. Hydride transfer within the complex occurs by an exchange of a C—H bond for the C–O bond when an aluminum hydride functions as both the Lewis acid and the hydride donor. On the other hand, reagent combinations such as $TiCl_4$-Et_3SiH act formally in an S_N2 fashion. Necessarily, the epimeric ether is obtained. The two different reactions represent, respectively, chelate and template control. The ring opening is stereoselective when the 1,3-dionxane is prepared from an aldehyde or from a ketone in which the two organic groups differ greatly in steric bulk so that there is a clear conformational preference (with the large group occupying an equatorial position).

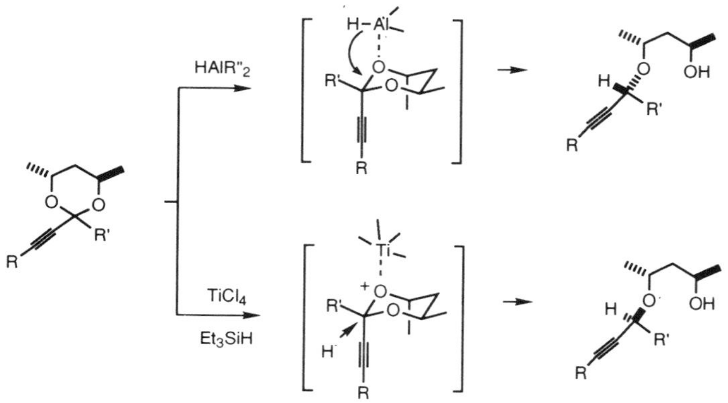

The same kind of selective ring opening is observable in the generation of 6-substituted 7,9-dioxabicyclo[4.2.1.]nonanes [Kotsuki, 1989].

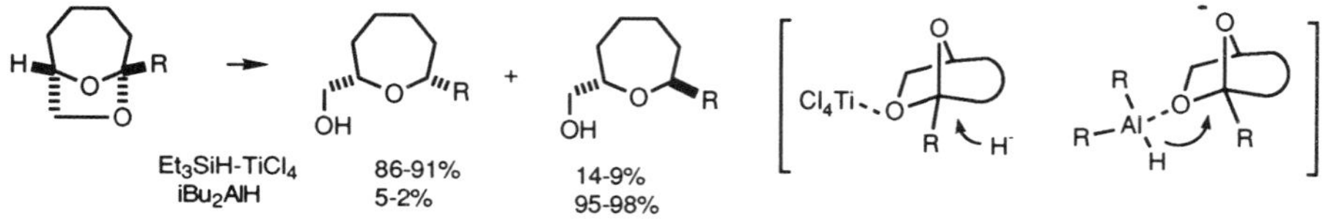

Et_3SiH-$TiCl_4$	86-91%	14-9%
iBu_2AlH	5-2%	95-98%

Underlying the stereoselective carbonyl reduction of (1*S*,2*S*)-2-acylthiane *S*-oxides with diisobutylaluminum hydride [Armer, 1993] is the same principle. The formation of the (*S*)-alcohols is due to internal delivery of the hydride ion within a complex in which the α-sulfinyl ketone system assumes preferably a cisoid conformation to avoid steric interaction of the (bulky) sidechain with one of the isobutyl group of reagents coordinating to the sulfoxide. In the presence of $ZnCl_2$, both oxygen atoms of the ketone and the sulfoxide participate in chelation, and an external hydride approaches from the *si*-face.

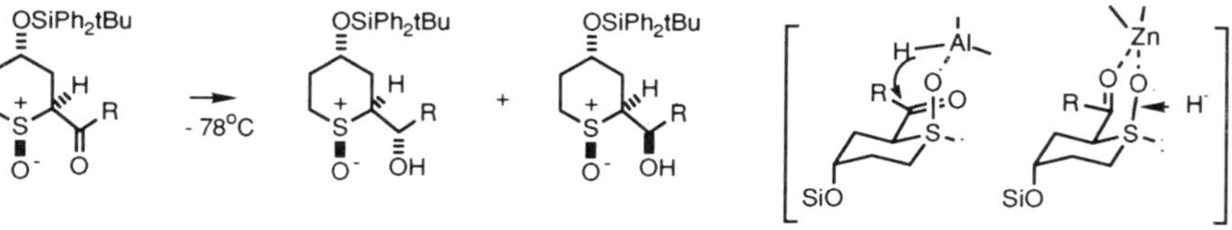

iBu_2AlH 100 : 0

iBu_2AlH / $ZnCl_2$ 0 : 100

The enantioselective reduction of an alkynyl ketone with *B*-3-pinanyl-9-borabicyclo[3.3.1]nonane [Midland, 1980] proceeds via an ordered complex. Transfer of the 9-BBN molecule from α-pinene to the ketone accomplishes the reaction.

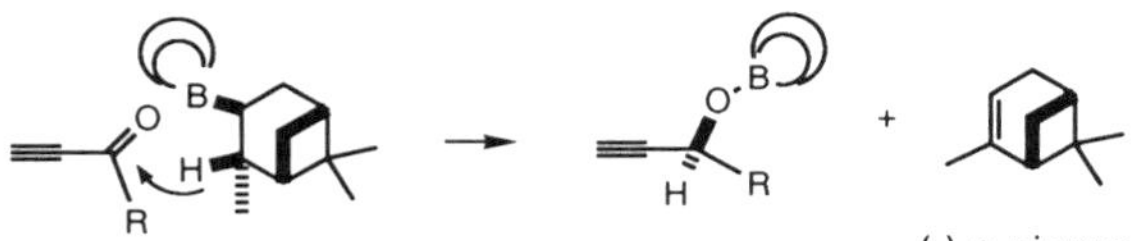

(-)-α-pinene

Formation of angular hydroxydecalins by SmI_2-mediated reductive addition of cyclohexanone to an unsaturated ester sidechain at the α-position is subjected to chelation control when the sidechain contains a hydroxyl group [Kito, 1993]. Two such diastereomers have been converted into *cis*- and *trans*-decalindiols, respectively, in which the hydroxyl groups are *cis*.

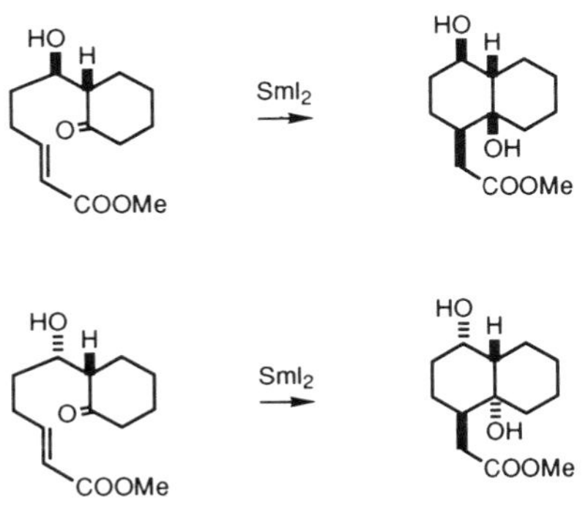

β-Hydroxy ketones undergo intramolecular Tishchenko reaction with an aldehyde in the presence of samarium(II) iodide [Evans, 1990e]. The products are mono esters of *anti*-diols. Notably both *syn*- and *anti*-α-methyl-β-hydroxy ketones follow the same stereochemical course with equally high levels of asymmetric induction. The distal hydroxylated stereocenter dominates the outcome.

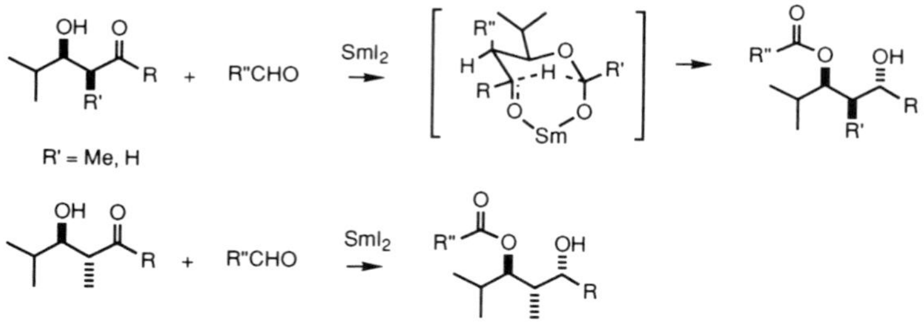

R' = Me, H

The reductive coupling of aromatic aldehydes with Cp_2TiCl_2 and iPrMgI leads to *syn*-diols predominantly [Handa, 1987]. It has been postulated that C—C bond formation occurs within a trimetallic complex in which the aromatic rings are disposed in an *anti* arrangement to minimize steric repulsions.

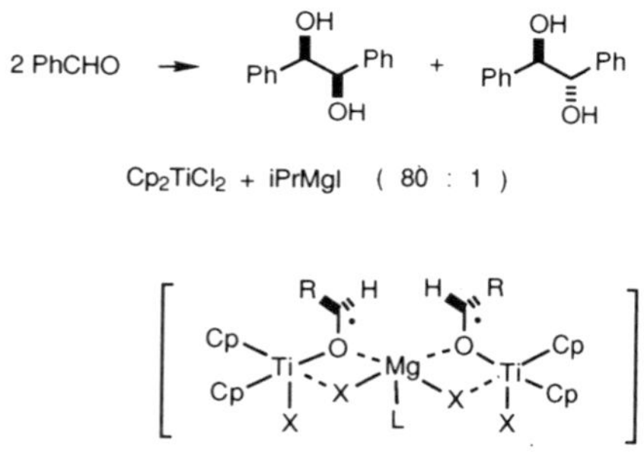

Cp_2TiCl_2 + iPrMgI (80 : 1)

The S_N2' displacement of the chiral amine residue from 2-aminomethyl-2-cycloalkenones with an organo group from an organcopper reagent to give 3-substituted 2-methylenecycloalkanones [Tamura, 1992] occurs within a metal ion chelate. Lithium ion directs the nucleophile transfer through further complexation with the copper reagent, and the diastereoselectivity is high when the cycloalkenone ring is not flat (as in the case of the cyclopentenone). Much worse selectivity is shown when the chelate metal is zinc instead of lithium because it does not complex with the copper reagent.

90% ee

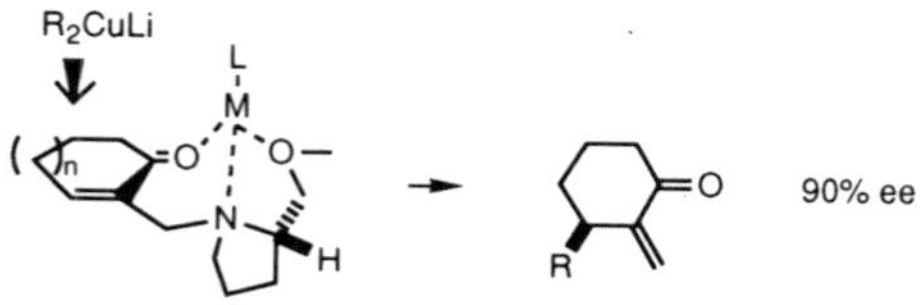

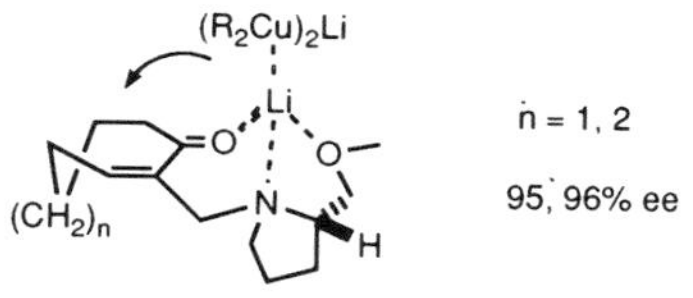

n = 1, 2

95, 96% ee

An aromatic ring coordinated to a tricarbonylchromium group is susceptible to attack by organolithiums. Such attack is also facilitated and directed toward the *o*-position by an oxazoline substituent, apparently due to chelation. Diastereocontrol in the addition of nucleophiles to the aromatic nucleus is enabled by the introduction of a bulky pendant at C-4 of the oxazoline moiety [Kündig, 1992] which sterically inhibits reaction from the rotamer in which the bulky substituent and the $Cr(CO)_3$ group are *anti*.

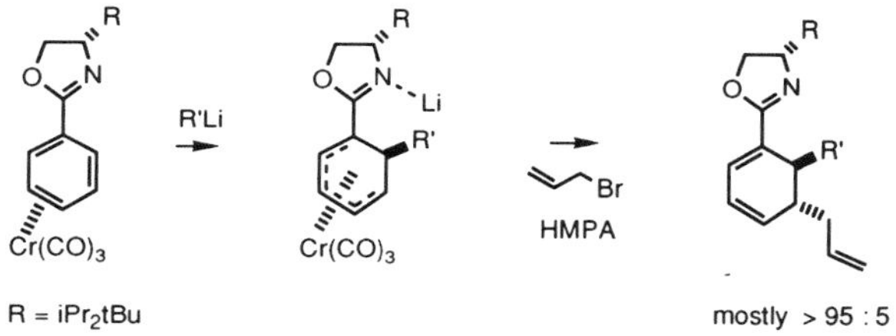

R = iPr$_2$tBu

mostly > 95 : 5

The highly diastereoselective alkylation of β-hydroxy esters to give *anti* products has been attributed to chelate formation [Frater, 1984]. Enantioselective alkylation of ketones can be performed on chiral hydrazone derivatives, and the most extensively used hydrazones for this purpose are those derived from 1-amino-2-methoxymethylpyrrolidine [Enders, 1984], in which the oxygen serves as a ligand for the lithium ion. One advantage of this method arises from the availability of both enantiomers of the chiral hydrazine (RAMP and SAMP) from (*R*)-glutamic acid and (*S*)-proline, respectively).

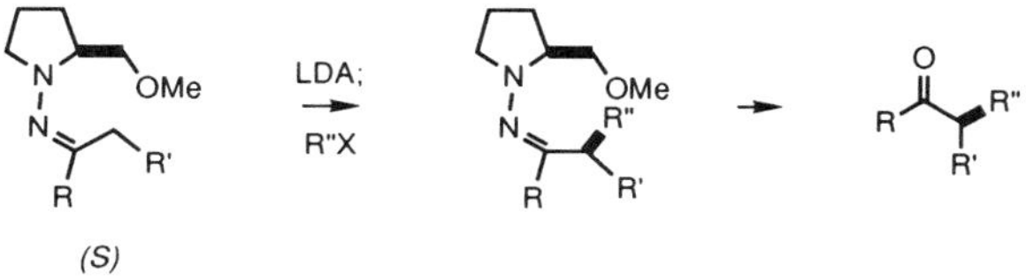

(S)

Asymmetric electrophilation [Evans, 1984a] has become an active area for research. Chiral lactams derived from α-amino acids can be transformed into imides which undergo electrophilation via chelated enolates. Newer chiral auxiliaries include a dihydropyridinone prepared from (*S*)-asparagine [Chu, 1992] and a C_2-symmetrical 2,3-bisbenzyloxyaziridine derived tartaric acid [Tanner, 1991].

Chelate-enforced intra-annular chirality transfer is particularly effective, and it is possible to synthesize α-amino acids by this method [Evans, 1986; Gennari, 1986; Trimble, 1986].

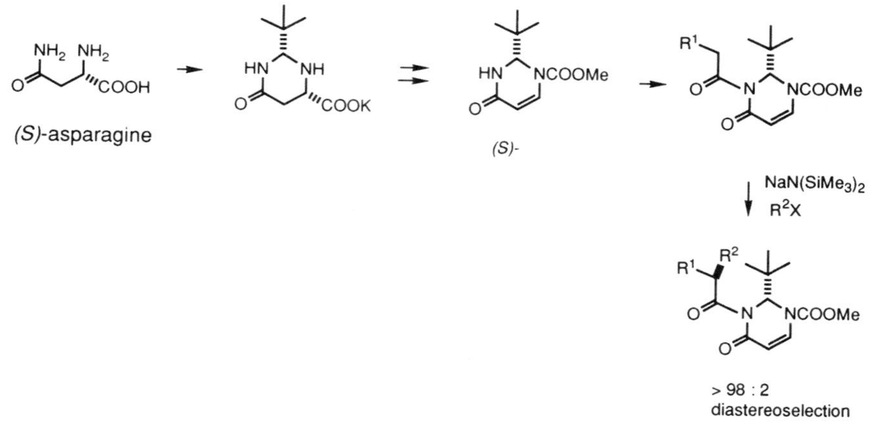

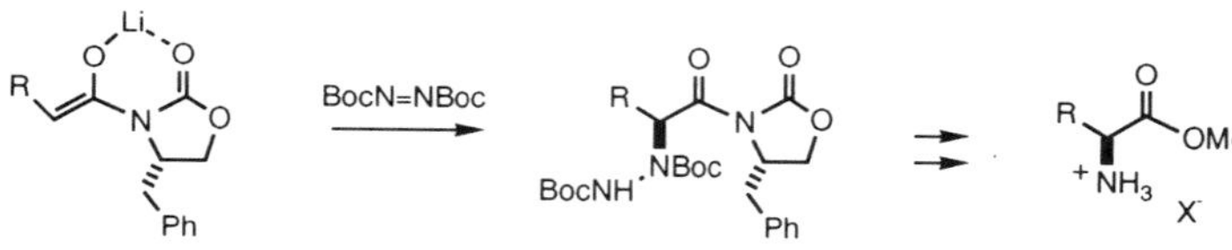

(S)-asparagine

(S)-

$NaN(SiMe_3)_2$
R^2X

> 98 : 2
diastereoselection

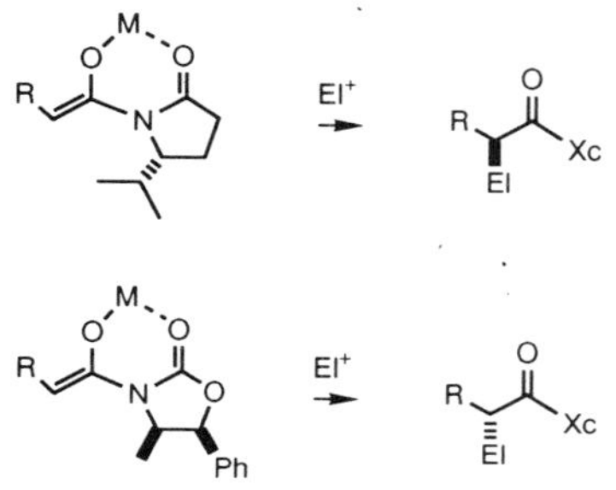

In the α-hydroxylation of a chiral amide the oxygen atom enters the molecule from the side opposite to the hydroxymethyl chain of prolinol [Davis, 1985]. The reaction via the conformationally locked lithium enolate gives a product with (*S*)-configuration, as a result of frontside attack, whereas the less covalent nature of the Na—O bond in the corresponding Na enolate of the chiral amide permits the two oxygen atoms to adopt a conformation devoid of mutual dipole–dipole interactions, leading to a (*R*)-product.

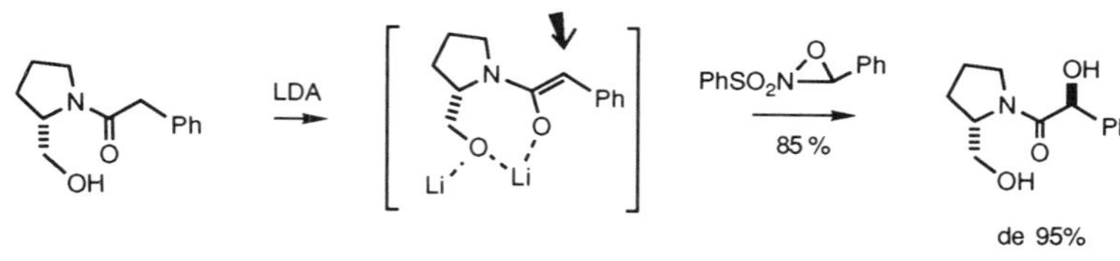

LDA

85 %

de 95%

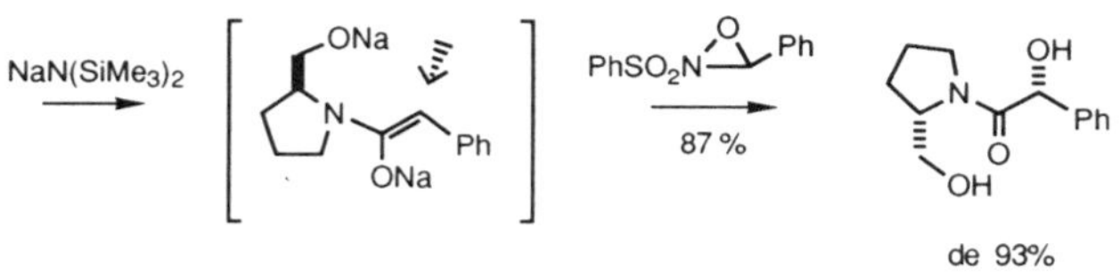

Certain carbohydrate frameworks are suitable chiral templates to which stereoselectivity of reactions may accrue by chelation effects. Thus the lithium enolates of 1,2;5,6-di-*O*-isopropylidene-α-D-allofuranose esters undergo methylation at very low temperatures to give predominantly the (*R*)-esters [Kunz, 1988].

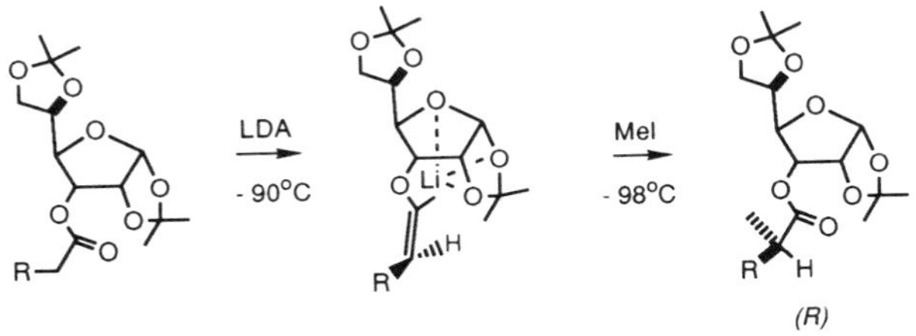

The steric course of the aldol condensation can be predicted with the aid of cyclic transition state models involving metal ion coordinating to the carbonyl acceptor. Three factors governing the product configuration are: (1) ring geometry (chair or boat), (2) enolate configuration (*E* or *Z*), and (3) mode of approach (*lk* or *ul*). In most circumstances the chair transition states are preferred. Acyclic enolates of defined geometry can be generated by specific methods and analyzed spectroscopically.

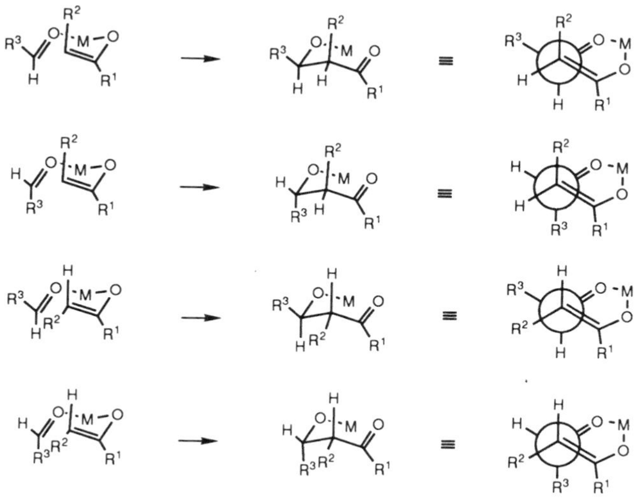

Diastereoselective aldolization from β-keto imide derived enolates [Evans, 1990b] constitutes a convenient synthetic approach to the polypropionate systems. Thus, tin(II) and titanium(IV) catalysts mediate two different reaction pathways.

$Sn(OTf)_2$, Et_3N; iBu_2AlH; $NaBH(OAc)_3$; > 20 : 1; > 20 : 1

$TiCl_4$, iPr_2NEt; $Zn(BH_4)_2$; > 20 : 1

By coordinating the vacant *d*-orbitals of tin(II) enolates with bidentate ligands it is possible to achieve stereoselective aldol condensation [Iwasawa, 1982]. Furthermore, β-hydroxy carbonyl compounds of high optical purity are available from such reactions of 3-acylthiazolidine-2-thiones in the presence of a chiral amine [Mukaiyama, 1984]. Interestingly, the *anti*/*syn* ratio of the products can be reversed by excluding the amine ligands.

The reaction of chiral *N*-alkenyl-1,3,2-oxazastannoles with aldehydes to afford β-hydroxy ketones proceeds via a highly ordered transition state [Narasaka, 1984a].

LDA; $SnCl_2$; R'CHO; H_3O^+

Reversal of aldehyde diastereofacial selectivity in the aldol condensation with methyl ketones can be achieved [Evans, 1990d] by changing from a lithium enolate donor to a silyl enol ether under Lewis acid catalysis. The former reaction is under chelation control while the latter proceeds according to the Felkin–Anh model.

OLi
- 78°C
O OR
H
OSi
BF$_3$
O OH OR
> 80 : <20
2 : 98

Asymmetric Diels–Alder reactions have been accomplished using chiral imides [Evans, 1988a] and amides [Waldmann, 1988] which bear a conjugated acyl group. An additional carbonyl group present in proximity to the amide C═O provides a second ligand to a complexing Lewis acid thus enabling approach of a diene to the activated, conformationally fixed dienophile from a predetermined face.

Et$_2$AlCl
Al Et$_2$AlCl$_2$
Ph
80% de
(+)-α-terpineol

The π-face selectivity of $TiCl_4$-catalyzed Diels–Alder reactions of the benzyl ester of *N*-acrylylproline is reversed by replacing the Lewis acid with $ZnCl_2$, $EtAlCl_2$, or BF_3 which do not have six coordination sites. In the complexes with these other Lewis acids the *s-cis* conformation of the acrylyl group is less favorable than the *s-trans* conformation. The consequences are that in the $TiCl_4$ reaction the diene comes from the *si*-face, and in the other reactions, from the *re*-face.

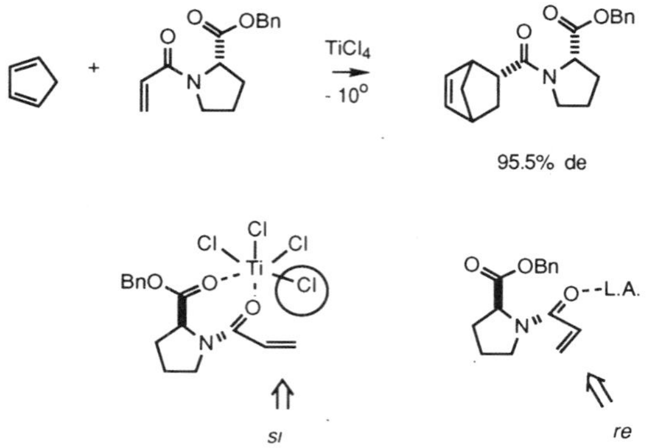

A titanium complex formed from 3-*O*-acrylylglucofuranose monoacetonide reacts with cyclopentadiene on its *si*-face, giving an adduct mixture of 93:7 (*R*:*S*) [Kunz, 1987].

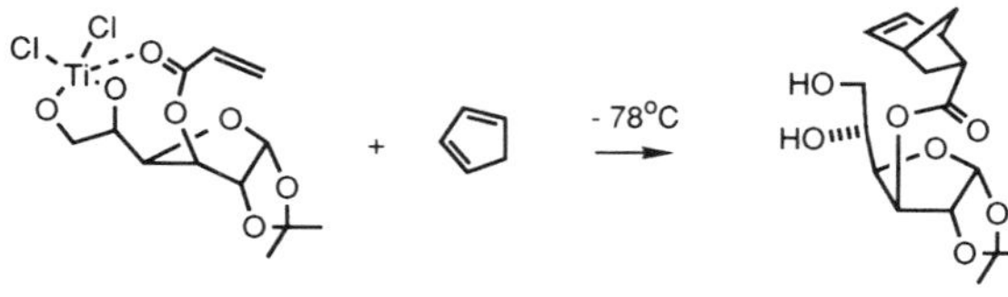

Strecker synthesis on tetra-*O*-pivalyl-*β*-D-galatosylimines with trimethylsilyl cyanide and zinc chloride proceeds differently in isopropanol than in chloroform [Kunz, 1991]. The rationale is that in the attack on the chelate involving the nitrogen atom and the C-3 pivalyloxy carbonyl, the silyl cyanide must slip inside for the zinc-bound chloride to disengage the silyl group from the nucleophile. As the Si—C bond is loosened, the cyanide immediately adds to the imine in its vicinity. On the other hand, isopropanol hydrolyzes the reagent to release cyanide ion which approaches the imine from the less hindered side.

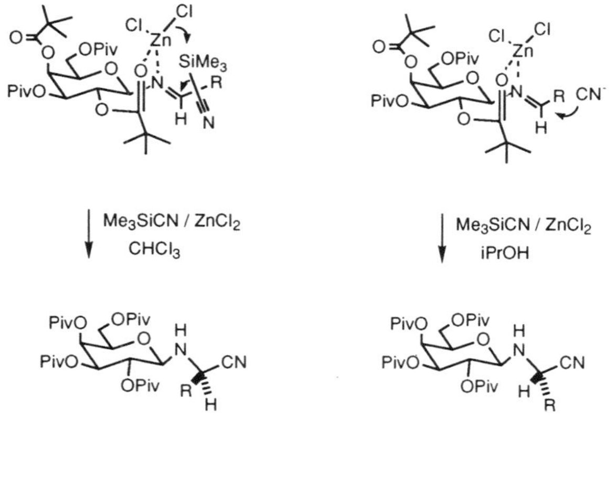

The formation of β-lactams from aldimines and ketene silyl ethers in the presence of $TiCl_4$ [Ojima, 1980] proceeds via chelated titanate species. When the substituent on the nitrogen atom is chiral, asymmetric induction is achieved upon rehybridization of the prochiral carbon atom.

Dianionic dioxy-Cope rearrangement of 1,1′-bisallylic alcohols leads to *trans*-4,5-disubstituted 1-cyclopentenecarbaldehydes because the ensuing bis-enolates undergo aldolization immediately on aqueous quenching [Saito, 1992]. The remarkable selectivity of the (*S*,*S*)(*E*,*E*)-alcohols to afford levorotatory products, the (*S*,*S*)(*Z*,*Z*)-isomers to give the dextrorotatory enantiomers, and the (*S*,*S*)(*E*,*Z*)-isomers to provide the racemates is due to chelate formation which fixes the transition state in each case.

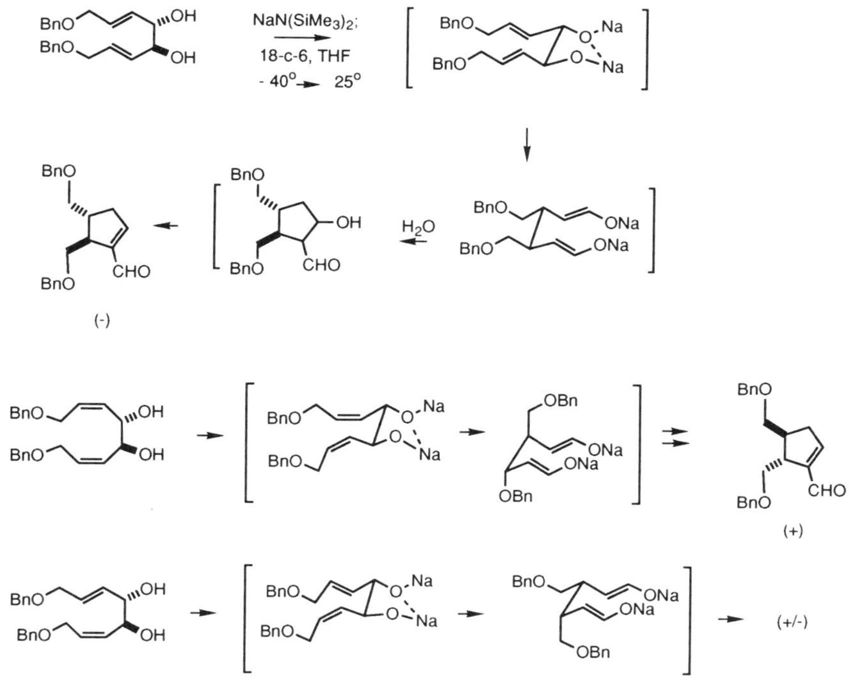

Direction by sulfoxide [Hauser, 1984] and sulfoximine [C.R. Johnson, 1984] in osmylation has been demonstrated. The sulfoxide group is oxidized under the reaction conditions, but only after the osmylation. A sulfone group does not exert diastereoselection.

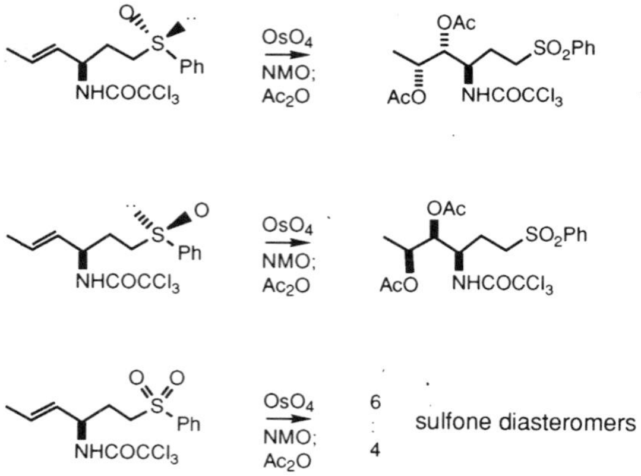

The effect of chelation or its absence accounts for different products upon protonation of hydroxyalkylnitronate anions [Seebach, 1982]. Protonation of adducts from doubly deprotonated nitroalkanes and aldehydes gives mainly the *syn*-alcohols. When epimeric mixtures of the carbinols are subjected to silylation, remetallation, and protonation, the *anti*-alcohols are produced.

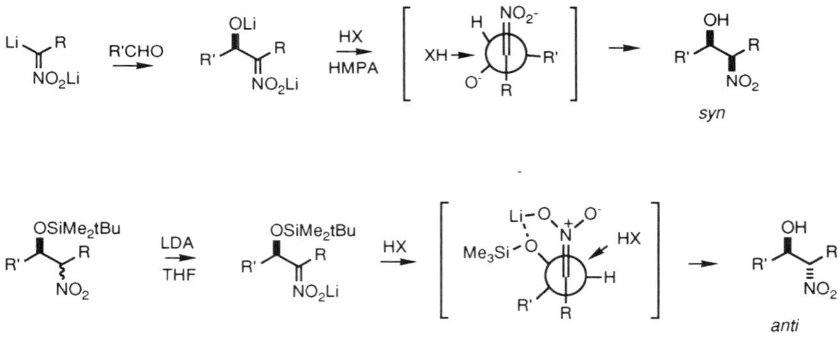

7.2.3. Directed Metallations

Alkylation and functionalization of alkenes and aromatic compounds containing heteroatom substituents or heteroalkyl substituents has been exploited very industriously in synthesis [Gschwend, 1979; Beak, 1982; Snieckus, 1990]. This branch of chemistry was independently initiated in the 1939–1940 period by Gilman and Wittig who observed the selective lithiation of alkyl aryl ethers.

The metallation is most frequently performed with an alkyllithium or a lithium dialkylamide as the base. Alkyllithiums exist in polymeric forms and the addition of coordinating solvents can break down such aggregates, thereby increasing the basicity of such reagents. *N,N,N′,N′*-Tetramethylethylenediamine

(TMEDA) is particularly effective, and it is known that TMEDA depolymerizes hexameric *n*-butyllithium to the kinetically more reactive monomer by virtue of coordination of the bidentate ligand with the lithium atom. Intramolecular coordination occurs prior to *o*-lithiation of *N*,*N*-dialkylbenzylamines, whereas both coordinative and inductive effects are operative in the cases of alkyl aryl ethers. The following observations [Slocum, 1976; Sundberg, 1973, 1976] concerning competition between two substituents are noteworthy.

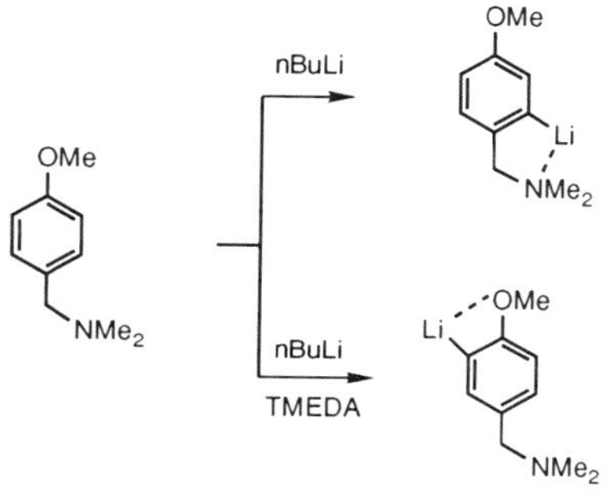

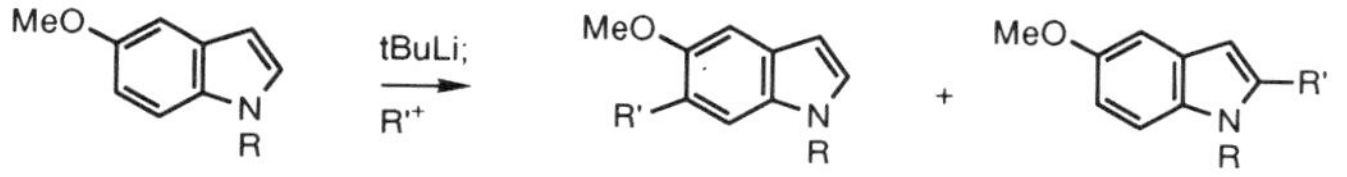

R= Me	+	+
R= CH_2OMe, SO_2Ph	-	+

In each case the competitive β- and α-lithiation of 2-(2-pyridyl)thiophene [Kauffmann, 1973] and that of *N*-phenylsulfonylindole [Saulnier, 1982] represents kinetic and thermodynamic control processes, respectively.

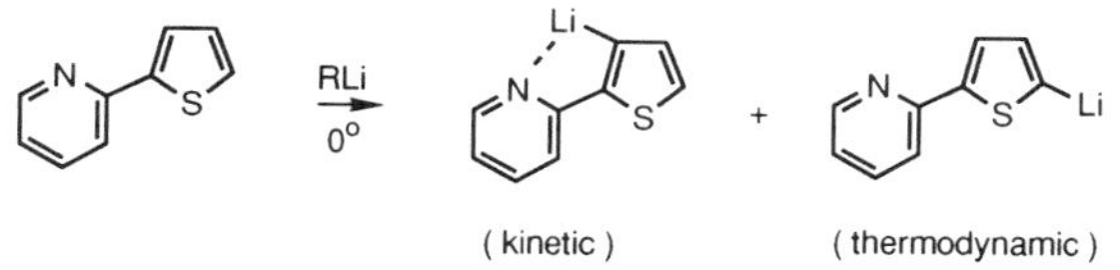

	(kinetic)	(thermodynamic)
n-BuLi / THF	4	93
n-BuLi / Et_2O	62	13

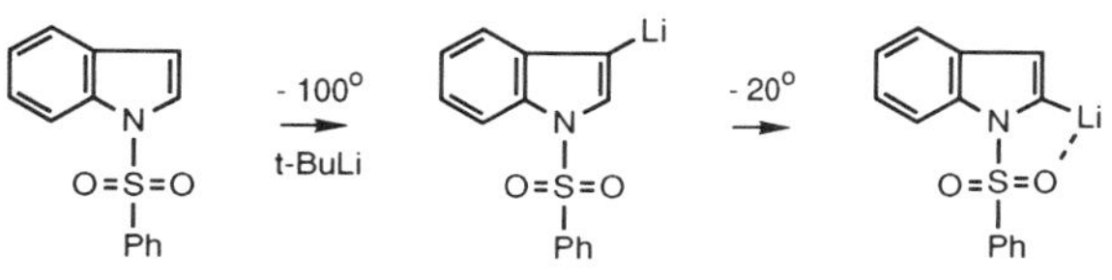

o-Lithiation of substituted aromatic compounds is easiest when the directing group is coordinative and electron-withdrawing. A heteroatom is an obligatory component of such a group. From many studies the ability of the directing group roughly falls in the series: (tBuLi) $OCONEt_2$ > SO_2tBu > CON^-R > $CONR_2$, OMOM > 2-oxazoline > SO_2NR_2, SO_2N^-R, CH_2NR_2, OMe, Cl > $CH_2CH_2NR_2$, NR_2, CF_3, F. Anilines have poor *o*-directing abilities for lithiation, but pivalamides [Fuhrer, 1979] and phenyl isocyanide [Walborsky, 1978] can be used to achieve such reactions indirectly.

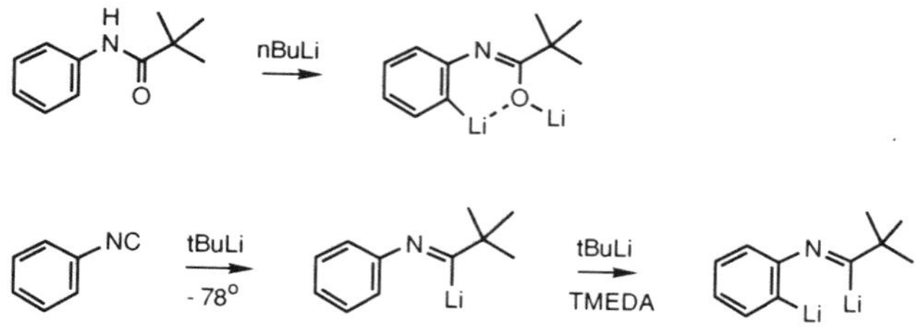

Since the carbamates and amides are most effective *o*-directors of lithiation of aromatic systems, they have been employed in syntheses of various types of target molecules. Iterative directed metalation processes serve to introduce nuclear substituents systematically.

The following routes leading to pancratistatin [Danishefsky,1989b], ellipticine [Watanabe, 1980], imeluteine [Zhao, 1991] and oxynitidine [R.D. Clark, 1988] are only a few examples demonstrating the synthetic utility of the method. In the imeluteine synthesis, activation is at an *o*′-position of a biphenyl which is five bonds apart from the amide carbonyl, and intramolecular reaction leads to a fluorenone. The oxynitidine synthesis involves activation at the sidechain, using an amide base.

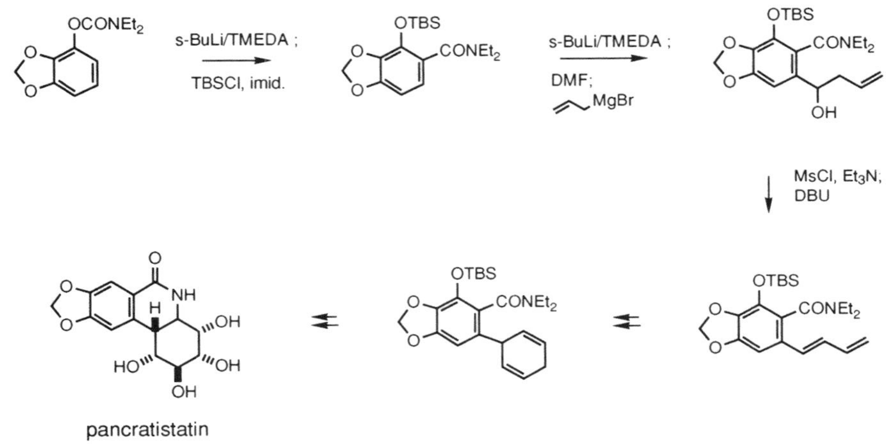

pancratistatin

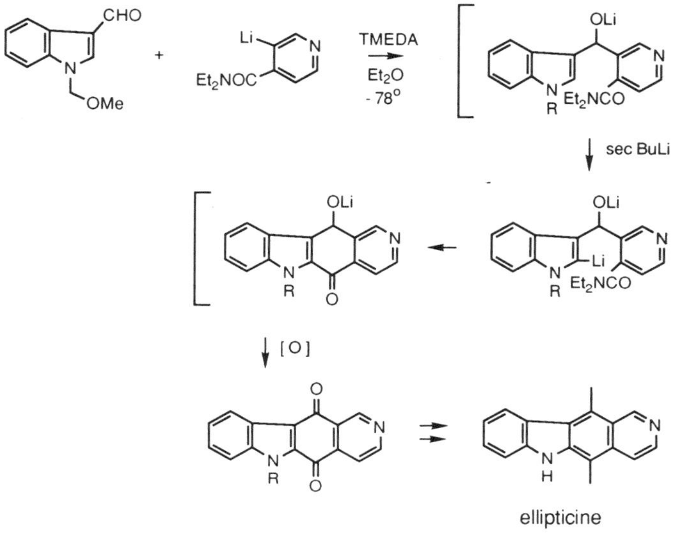

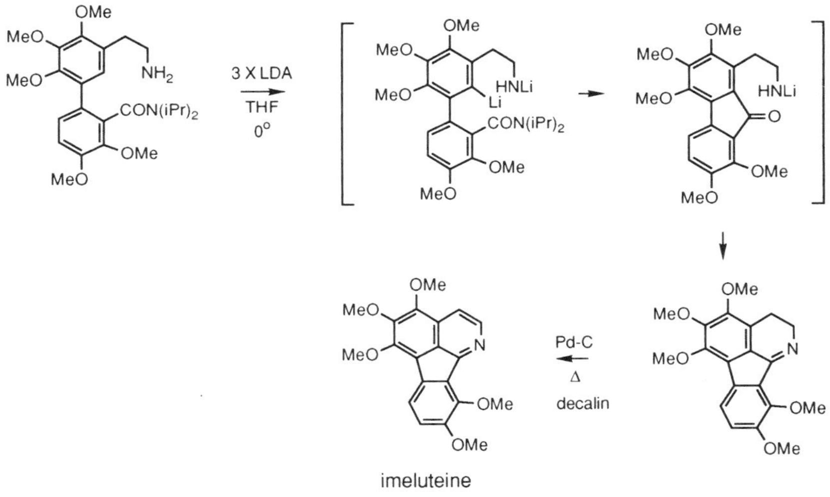

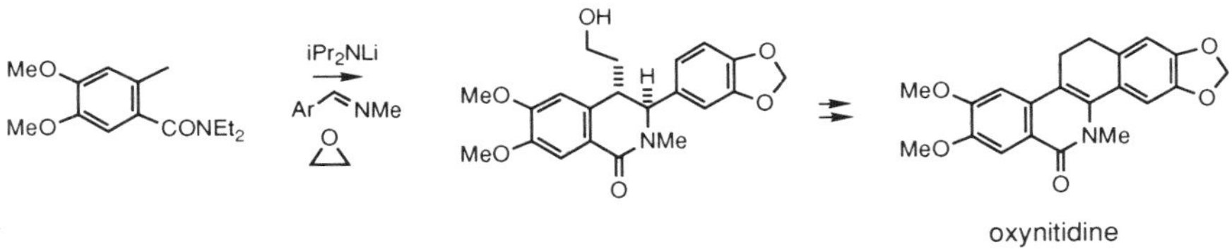

ellipticine

imeluteine

oxynitidine

The heteroratoms of various heterocycles are capable of directing metalation, thus enabling chain extension at the α-positions. Hetero-substituted alkenes undergo α-lithiation accordingly. These systems include enamines, vinyl isocyanides, formamides, thioformamides, alkyl alkenyl ethers and allenic ethers, alkenyl sulfides and sulfoxides. Haloalkenes after α-lithiation are susceptible to α-elimination to form vinylcarbenes, therefore they must be generated at temperatures below −100°C.

The presence of two hetero-substituents in an alkene can lead to interesting situations. Generally, the organolithium compound is stabilized. Examples shown below indicate the chelated structures of lithiated THP ether [Hartmann, 1974], ethoxyvinyl sulfide [Vlattas, 1976], 3-methoxypropenyl sulfoxide and propenyl 2-pyridyl sulfoxide [Okamura, 1978].

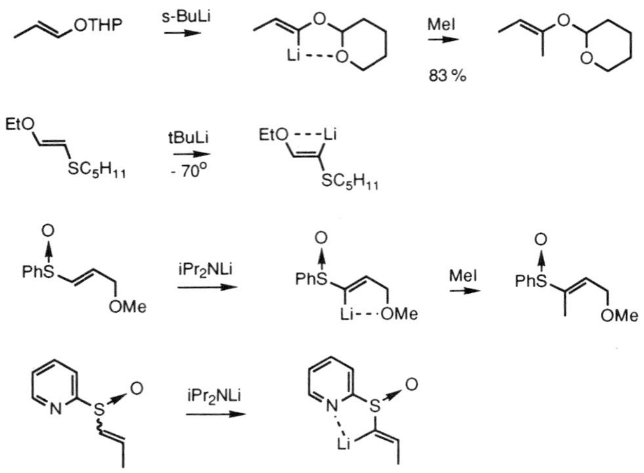

OTHP —s-BuLi→ Li---O —MeI (83 %)→

EtO, SC_5H_{11} —tBuLi, -70°→ EtO---Li, SC_5H_{11}

PhS(O), OMe —iPr_2NLi→ PhS(O), Li---OMe —MeI→ PhS(O), OMe

N, S→O —iPr_2NLi→ N, S→O, Li

m-Sidechains in aromatic systems containing ligating donors can direct metalation at the position ortho to both chains. An exploitation of this regiochemistry to palladize an aromatic ring is related to the elaboration of narwedine [Holton, 1988]. Bond rotations within the resulting complex were decreased, therefore the oxidative coupling was much more efficient. Interestingly, the palladium species served as a leaving group in a subsequent rearrangement step.

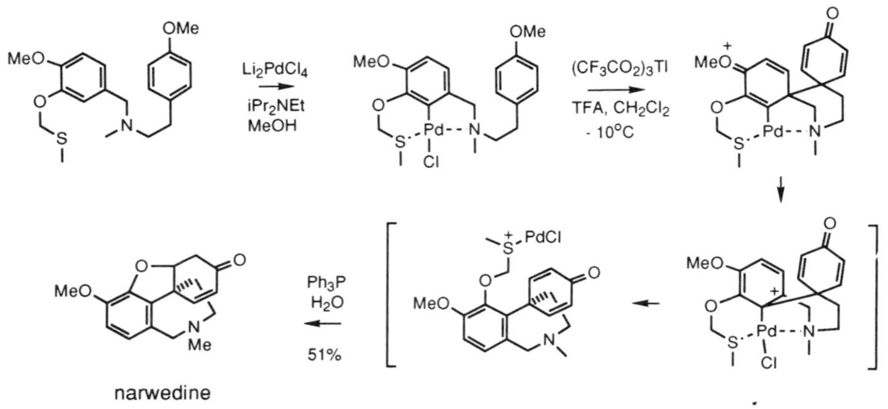

MeO, O, S, N, OMe —Li_2PdCl_4, iPr_2NEt, MeOH→ MeO, O, S, Pd, N, Cl, OMe —$(CF_3CO_2)_3Tl$, TFA, CH_2Cl_2, -10°C→ MeO, O, S, Pd, N, O

MeO, O, S, Pd, N, Cl, O → [S, PdCl, O, MeO, N, O] —Ph_3P, H_2O, 51%→ MeO, O, N, Me, O

narwedine

8

SYMMETRY CONSIDERATIONS

There are molecules possessing some sort of symmetry elements such as plane (σ), center (i), and alternating axis (Sp), and other molecules totally devoid of them. In synthesis design a consideration of these geometrical properties can be of paramount importance [T.-L. Ho, mip]. A cardinal rule concerning efficient synthesis of symmetrical molecules is to select symmetrical precursors and employ reactions which do not inordinately disturb the symmetry elements. This is also true for approaching target compounds having local symmetry. Certain unsymmetrical synthetic targets may benefit from using symmetrical starting materials, provided that a desymmetrization process can be incorporated along the pathway. Actually, desymmetrization is often performed in an earlier part of a synthesis.

In recent years several methods for desymmetrization have been developed. *Meso*-ketones can be regioselectively deprotonated with a chiral base [Cox, 1991], while achiral acids can be induced to undergo asymmetric reactions (e.g. alkylation) at the α-carbon atom by forming amides with C_2-symmetric amines. Perhaps the most intensively studied method is the enzymatic hydrolysis and transesterification of diesters having a σ-symmetry [J.B. Jones, 1986; M. Ohno, 1986]. The commercial availability of many lipases, esterases and baker's yeast contributes to the popularity of this approach. Sometimes a reversal of stereospecificity arises [Sabbioni, 1984], but as long as recognition is made there is no detraction from synthetic applicability of the enzymes. Of rarer use is selective oxidation such as that achieved with the aid of horse liver alcohol dehydrogenase (HLADH).

Astute manipulation of functionalities in molecules often overcomes problems concerning chirality such as the acquisition of an enantiomeric isomer from a more readily available compound. For example, the versatility of chiral glycidols as building blocks for synthesis [Hanson, 1991] derives from their accessibility by relatively straightforward maneuvers. Both enantiomers of the parent glycidol are obtainable from D-mannitol via its acetonide by oxidative cleavage, reduction, and ramified operations in which the exposed hydroxyl function is either activated as a leaving group to induce epoxide formation when the acetonide is hydrolyzed, or protected to allow manipulation of the acetonide into the epoxide.

The chemist's understanding of natural product biosynthesis, despite its incompleteness and occasional misconception, has provided a potent impetus for synthesis design [van Tamelen, 1961]. Processes frequently used by nature are those involving unsymmetrical dimerization and further desymmetrization of a symmetrical dimer and many of these have been emulated in the laboratory. A striking example is the synthesis of carpanone [Chapman, 1971] by a one-step oxidation of an *o*(1-propenyl)phenol. The dimer underwent a hetero-Diels–Alder reaction spontaneously, establishing five contiguous asymmetric centers in one operation.

carpanone

8.1. SYNTHESIS OF SYMMETRICAL MOLECULES

8.1.1. Concoctive Molecules

Molecules of nature and those imagined by the human mind may possess a symmetry element. The following are examples selected for illustrating the principle of symmetry exploitation and the advisability of symmetry maintenance in order to minimize the number of steps.

Dodecahedrane is a hydrocarbon analog of the most complex of the Platonic solids. The supreme I_h symmetry of dodecahedrane necessarily demands utmost attention in synthetic design, including choice of starting material and execution of individual steps. Intermediates which have been desymmetrized temporarily must be readily remedied in the next few steps. But with a meticulous plan, rapid ascent to the structural summit can be achieved by elaboration of a polyfunctional precursor. A corollary to this principle is that a synthesis would go hopelessly astray if the symmetry feature were not strictly observed. The triumph of a scaffold synthesis of dedecahedrane [Ternansky, 1982; Paquette, 1984] was in fact founded on symmetry considerations in all phases.

The synthesis started from a domino–Diels–Alder reaction of 9,10-dihydrofulvalene and dimethyl acetylenedicarboxylate. One of the adducts has a C_{2v} symmetry and it contains four interlocking five-membered rings of the target molecule. Moreover, the high symmetry of the diene diester and the pairwise proximity of the functionalities render its conversion to a symmetrical diketone very facile. Annulation of the two cyclopentanone moieties of the resulting compound can then be undertaken to give an intermediate encompassing all the skeletal carbon atoms of dodecahedrane.

It should be noted that the annulation via a bis(spirocyclobutanone), a dilactone, and a dicyclopentenone was followed by catalytic hydrogenation, and the extra C—C bond between the α-carbon atoms of the two ester groups was left intact during the manipulation. This C—C bond served as a locking device for the polyquinane system and as an effective shield against invasion of external reagents into the core of the various intermediates. Thus the hydrogenation delivered a useful precursor of dodecahedrane.

Upon conversion of the diketone to a dichloro compound a reductive cleavage of the extra C—C bond also caused ring closure in the desired manner. Addition of an alkylating agent after the reduction forced the unreacted ester group inside the molecular surface as defined by the incomplete sphere, enabling involvement of this ester group in subsequent skeletal C—C bond formation. Of course the alkylating agent must be equipped with a removable mechanism in order that the dodecahedrane itself can be unraveled.

From the heptacyclic stage onward thermodynamic factors became an ally to coordinate closure of the remaining carbocycles. The rigidity of the ring system permitted formation of only one isomer in each of the C—C formation processes. In fact, photochemical cyclization was used three times, each time to create a new ring. The final step was a palladium-catalyzed transannular dehydrogenation.

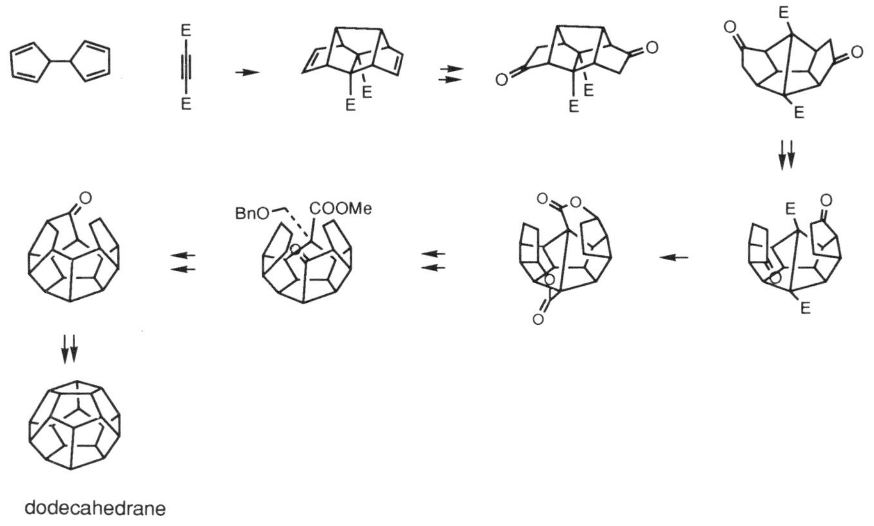

E E E E E E O O E O O E E O E O O O O BnO COOMe O O

dodecahedrane

The cube is a more familiar and simpler Platonic solid. The corresponding hydrocarbon, cubane, yielded to synthesis at an earlier date [Eaton, 1964]. Note the compromise of symmetry loss in the middle part of the synthesis due to methodological restriction to design. Key reactions of this approach include intramolecular [2 + 2]-photocycloaddition, and Favorskii rearrangement of the cage compound contracted the five-membered rings. The necessary decarboxylation steps are conceptually trivial.

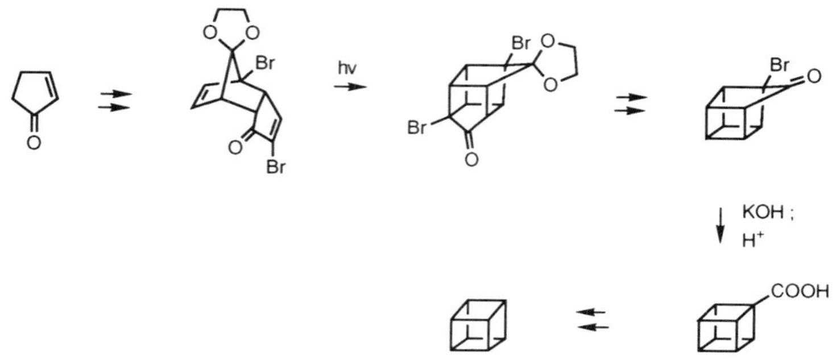

hν

KOH ; H^+

COOH

An improved version [Barborak, 1966] involved trapping cyclobutadiene with 2,5-dibromo-*p*-benzoquinone. Photocycloaddition of the adduct gave a symmetrical diketone which was subjected to Favorskii rearrangement and decarboxylation. Note that the return to a C_2-symmetric molecule upon photocycloaddition enabled twofold ring contraction in one operation.

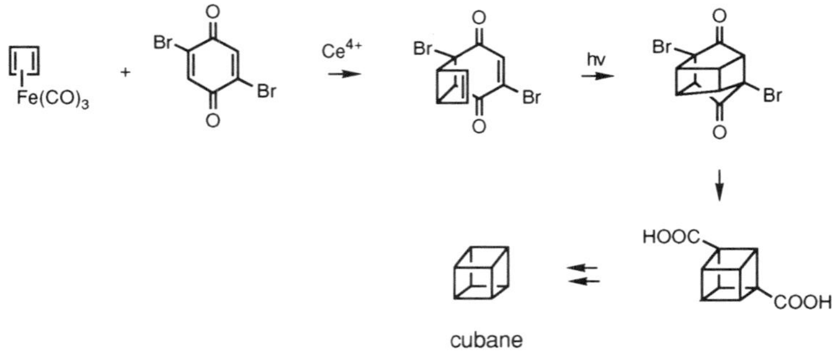

$Fe(CO)_3$ + ... Ce^{4+} ... hν

HOOC ... COOH

cubane

Barrelene commands theoretical centering on the three unconjugated double bonds, among which through-space interactions are expected. It is noted that the three double bonds cannot overlap simultaneously without causing destabilization of the molecule.

The presence of symmetry axes and planes in barrelene indicates that synthetic approaches based on construction of symmetrical intermediates are favored, particularly when intermediates already containing the bicyclic skeleton can be prepared in a single step. Accordingly, a route was developed from the reaction of α-pyrone with methyl acrylate [Zimmerman, 1969]. The 1:1 cycloadduct lost carbon dioxide to regenerate a conjugated diene unit which can undergo the Diels–Alder reaction for the second time. The final product was a diester from which the introduction of the two double bonds via proper degradation methods was readily conceived.

COOMe Δ $-CO_2$ COOMe COOMe MeOOC COOMe

$Me_3\overset{+}{N}$ 2I $\overset{+}{N}Me_3$ Δ

barrelene

A dimerization approach to dodecahedrane from triquinacene was entertained in the 1960s. The first synthesis of triquinacene [Woodward, 1964] is a paradigm of symmetry exploitation in terms of starting material selection and operation. From the pentacyclic diketone onward all the intermediates possess a plane of symmetry, allowing degradation at both enantiotopic positions.

O HO OH O HOOC COOH Me_2N^+ $^+NMe_2$ ^-O O^- Δ

triquinacene

Adamantane and its aggregates are strain-free hydrocarbons whose skeletons resemble the diamond lattice. One of the most efficient rational syntheses of the adamantane skeleton involved α,α′-dialkylation and Dieckmann cyclization [Stetter, 1968]. All the intermediates maintained a plane of symmetry.

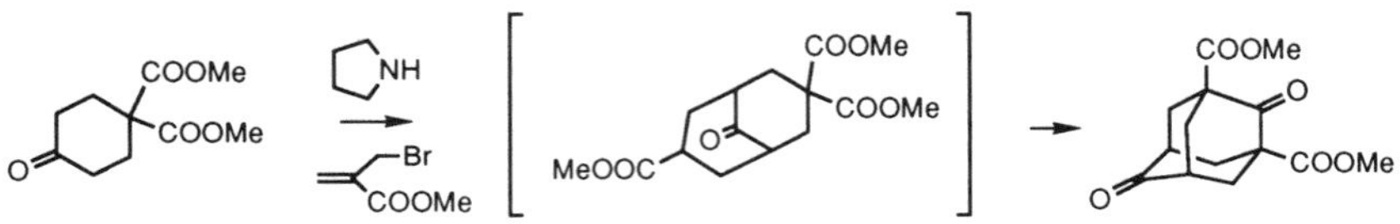

Access to the adamantanoid hydrocarbons based on thermodynamic properties is relatively facile. Polycyclic hydrocarbons which are isomeric to these compounds undergo rearrangements to the latter species under extreme conditions. Thus, tetrahydrodicyclopentadiene was converted into adamantane [Schleyer, 1960] and norbornene dimer into diamantane (congressane) [Cupas, 1965]. After addition of two CH_2 groups a dimer of cyclooctatetraene underwent isomerization to afford triamantane [V.Z. Williams, 1966]. Symmetry is the ultimate arbiter of these transformations.

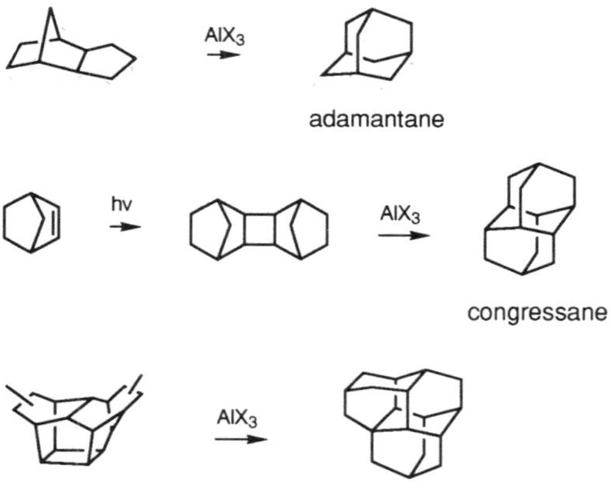

adamantane

congressane

8.1.2. Natural Products

Next we examine the synthesis of some symmetrical natural products. The carotenoids are pigments of certain plants and animals, they are characterized by the presence of a conjugated polyene chain. The majority of these oligoterpenes are symmetrical, that is, they have two identical end groups. Consequently, strategies for carotenoid synthesis are well defined, with main variations lying in the assemblage of the components or building blocks [Isler, 1963]. In the case of C_{40}-carotenoids the most efficient coupling strategies are those involving symmetrical fragments: 20 + 20, 19 + 2 + 19, 18 + 4 + 18, 16 + 8 + 16, 15 + 10 + 15, 14 + 12 + 14, 13 + 14 + 13, 10 + 20 + 10, and 5 + 30 + 5. Except for the 5 + 30 + 5 approach, all of them start from building blocks with performed trimethylcyclohexene end groups. As expected the Wittig reaction plays a very prominent role in a great number of the pathways.

A classical route to *β*-carotene [Isler, 1956] relying on Darzens condensation ($C_{13} + C_1$), aldol condensations via acetals ($C_{14} + C_2$ and $C_{16} + C_3$), and coupling with the di-Grignard reagent of acetylene ($C_{19} + C_2 + C_{19}$) may be mentioned. The need for selective semihydrogenation of the symmetrical monodehydro-*β*-carotene intermediate in this synthesis led directly to the development of the Lindlar catalyst. Finally, isomerizaton of the (*Z*)- double bond to the (*E*)-configuration was achieved simply by heat.

Darzens

CHO

$HC(OEt)_3$, H^+ ; VnOEt, $ZnCl_2$; HOAc, NaOAc H_2O

CHO

$HC(OEt)_3$, H^+ ; OEt ; $ZnCl_2$; HOAc, NaOAc H_2O

CHO

OH

OH

β-carotene

Since the two halves of *β*-carotene (and many other carotenoids) are joined by a double bond, a most direct approach would be to dimerize two C_{20} building blocks with trigonal terminus, for example by reductive coupling of the aldehyde [McMurry, 1974]. This route represents the reversal of the biological oxidation of *β*-carotene.

CHO

Ti^{II}

β-carotene

The constitution of squalene is different in that the C—C bond separating the molecule in two identical halves is bisallylic. There are many more methods available for accomplishing the union of allylic substrates.

A bidirectional, stereoselective chain extension based on two sets of organolithium reaction and Claisen rearrangement to construct a C_{24} intermediate is a very efficient approach to squalene [W.S. Johnson, 1970b]. The $C_{11} + C_8 + C_{11}$ assembly mode that features coupling of homogeranyllithium with 3,6-dichloro-2,7-octanedione and dechlorohydroxylation represents an even shorter route. [Cornforth, 1959].

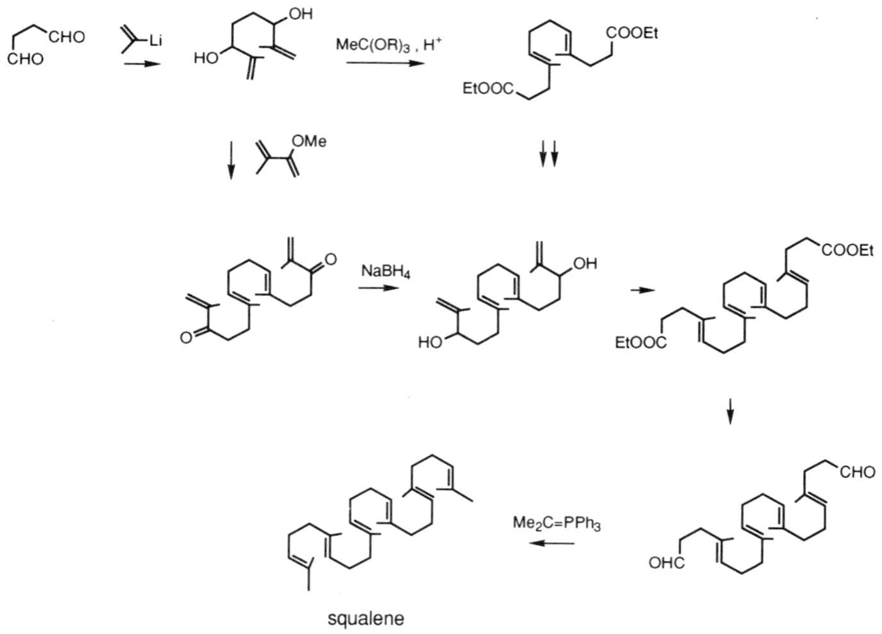

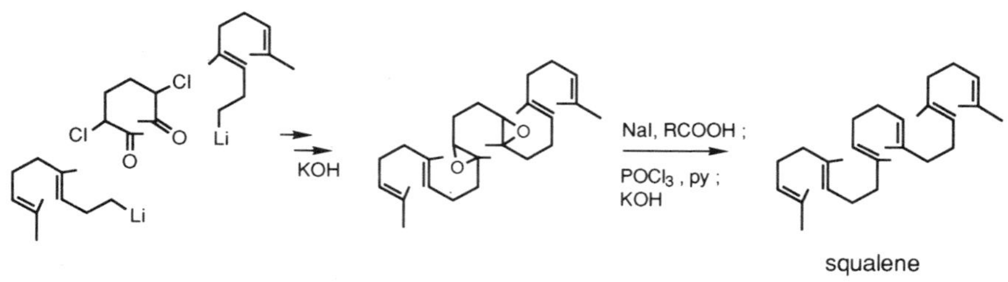

CHO CHO; Li; HO, OH; $MeC(OR)_3$, H^+; COOEt, EtOOC; OMe; O, O; $NaBH_4$; HO, OH; COOEt, EtOOC; CHO, OHC; $Me_2C{=}PPh_3$

squalene

Cl, Cl, O, O; Li; Li; KOH; O, O; NaI, RCOOH ; $POCl_3$, py ; KOH

squalene

Arcyriaflavin-A is a symmetrical bisindolobenzene, and one of its synthetic approaches [Bergman, 1989] took advantage of this structural feature. Accordingly, a double Fischer indolization was performed on the substituted 1,2-cyclohexanedione bisphenylhydrazone. The latter substance was in turn obtained from a Diels–Alder reaction, followed by *in situ* hydrolysis of the bistrimethylsilyloxycyclohexene, and reaction with phenylhydrazine in a manner analogous to osazone formation.

H N O O + RO OR Δ H N O O RO OR H N O O PhNHN NNHPh H N O O N H N H

R = SiMe3

arcyriaflavin-A

The discovery of an efficient [4 + 3]-cycloaddition involving conjugate dienes and oxyallyl cations has had important ramifications in the synthesis of cycloheptanones, tropones, and related substances. Entry into 8-oxa- and 8-azabicyclo[3.2.1]oct-6-en-3-ones is possible by employing furans and pyrrole derivatives as the 4-electron addends [Noyori, 1978]. The alkaloid scopine is thus readily assembled.

NCOOMe + Br Br Br Br O Fe2(CO)9 ; Zn-Cu NE O O NMe OH

scopine

A better known construction of the tropinone ring system is that based on a biogenetic conjecture. Thus, Mannich reaction of methylamine, succindialdehyde, and acetonedicarboxylic acid indeed afford tropinone [Robinson, 1917]. A yield of >90% can be reached when the reaction is carried out under *physiological* conditions [Schöpf, 1937].

CHO CHO + MeNH2 + COOH O COOH NMe O

tropinone

HOOC COOEt O e COOEt O O COOEt MeNH2 COOEt NMe COOEt NMe O

tropinone

MeNH2

COOEt COOEt

Tropinone has also been obtained by a Dieckmann cyclization of dimethyl *cis*-pyrrolidine-2,5-diacetate, followed by removal of the ester group [Willstätter, 1920; W. Parker, 1959]. Another route to tropinone (and cocaine) consists of reductive cleavage of a 7-azabicyclo[2.2.1]heptane-2,3-dicarboxylic ester [Krapcho, 1985]. Note that all these syntheses take advantage of the symmetry elements of the target molecule.

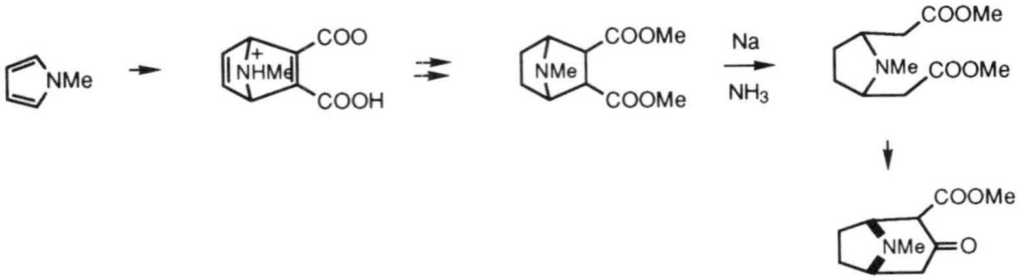

Precoccinelline is a σ-symmetrical compound elaborated by ladybird beetles for defence. One synthesis [Stevens, 1979] paid great attention to the symmetry features and was rewarded by high efficiency. The double Mannich reaction of 5-aminononanedialdehyde, which was derived from the ketodialdehyde bisdimethylacetal, led directly to the pehydro[9b]azaphenalene framework.

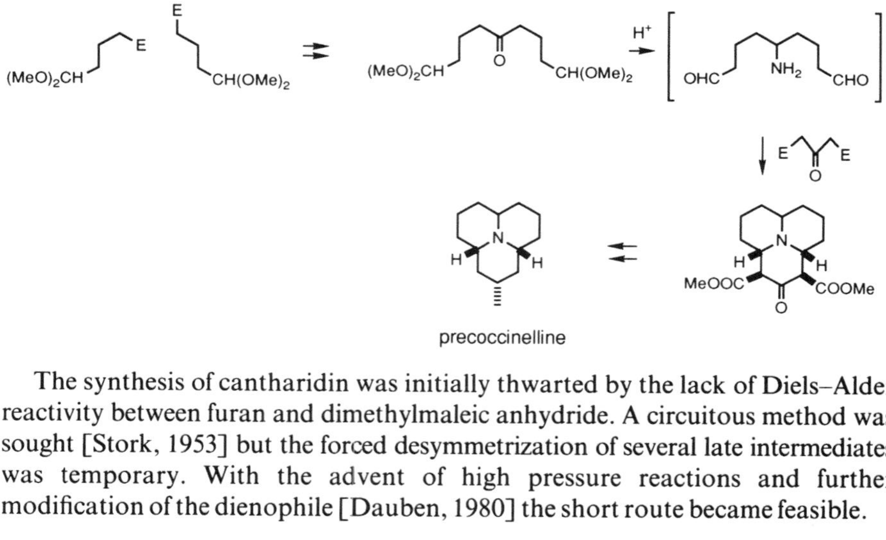

precoccinelline

The synthesis of cantharidin was initially thwarted by the lack of Diels–Alder reactivity between furan and dimethylmaleic anhydride. A circuitous method was sought [Stork, 1953] but the forced desymmetrization of several late intermediates was temporary. With the advent of high pressure reactions and further modification of the dienophile [Dauben, 1980] the short route became feasible.

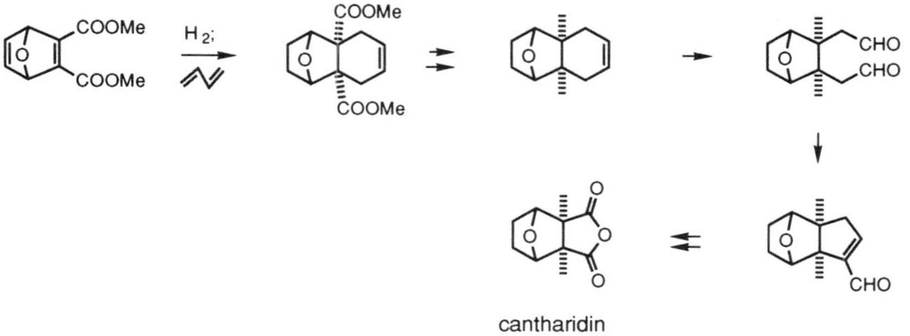

cantharidin

15 kbar

RaNi
H_2

(major)

cantharidin

α-Carophyllene alcohol, an acid-catalyzed rearrangement product of the monocyclic sesquiterpene humulene, has a plane of symmetry. Considering that the hydroxyl group can be derived from a carbocation precursor, a synthetic scheme involving cationic rearrangement would have great appeal. A more highly strained substrate would facilitate the rearrangement and render it unidirectional. The rationale for a 1,2-rearrangement devolves a 5:4:6-fused tertiary alcohol as a proper precursor. Migration of a cyclobutane bond relieves a large amount of ring strain.

It is now a matter of devising a route to such an unsymmetrical precursor [Corey, 1965b]. One involving photocycloaddition gained focus. Furthermore, the stereochemistry of the target compound is supportive of the approach in which an *exo* transition state for the cycloaddition is favored.

Because of reactivity requirements an enone must be used as one of the photocycloaddends. Naturally the other addend is the symmetrical 4,4-dimethylcyclopentene. One of the striking features of the synthesis is the molecular symmetrization by rearrangement.

hν

H^+

(major)

caryophyllene alcohol

Sesamin is C_2-symmetric. A route to this lignan starts from dibenzyl bicyclo[3.3.0]ocane-3,7-dione-2,6-dicarboxylate [Orito, 1991] and includes steps of arylation, decarbalkoxylation, Baeyer–Villiger oxidation, and degradation of the extra carbon atom, all simultaneously at the two halves of the substrate. Symmetry confers expediency!

$ArCO_3H$;
iBu_2AlH

sesamin

Ar =

By viewing such a C_2-symmetric lignan as anhydro derivative of the tetraol, the possible assemblage of the latter species from a *trans*-2,3-disubstituted butanolide becomes evident. Such a butanolide is derivable from conjugate addition of a 2-lithiodithiane to 2-butenolide, and trapping the enolate with an aldehyde [Jansen, 1991].

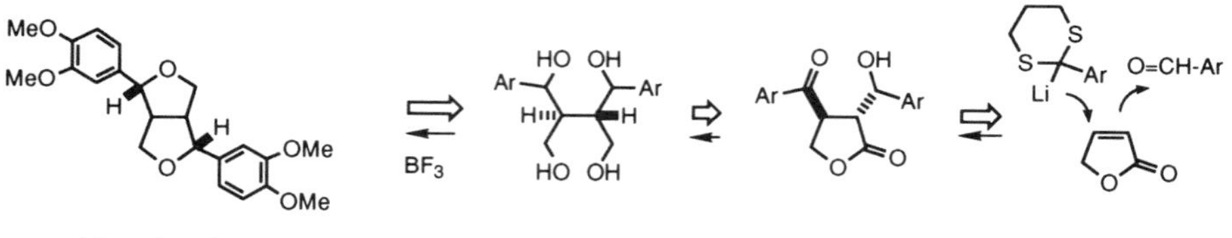

(-)-eudesmin

The calycanthaceous alkaloids are modified dimers of N_b-methyltryptamine, consequently the most expeditious synthetic pathway is biomimetic [Hall, 1967]. Treatment of the indolylmagnesium halide with ferric chloride led to several compounds including *rac*-chimonanthine and *meso*-chimonanthine. *rac*-Chimonanthine was converted into a 1:4 equilibrium mixture with calycanthine as the major component on heating with dilute acetic acid for 30 hr at 90°C.

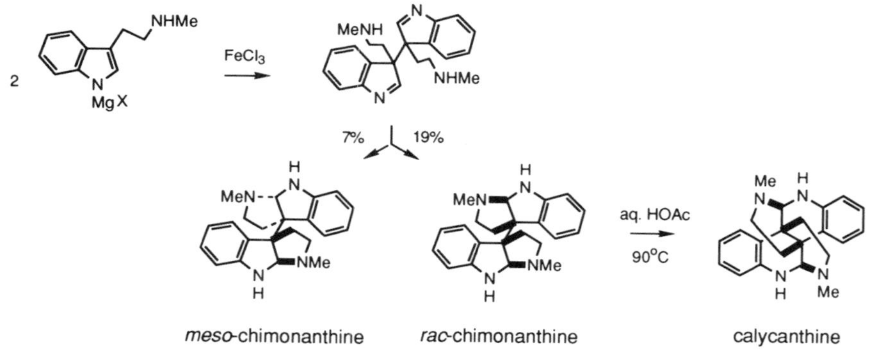

meso-chimonanthine *rac*-chimonanthine calycanthine

8.2. SYNTHESIS OF UNSYMMETRICAL MOLECULES FROM SYMMETRICAL PRECURSORS

8.2.1. From Five-Membered Carbocycles

The asymmetry of albene is due to the disubstituted double bond. Apparently racemic albene is accessible by manipulation of a symmetrical tricyclic intermediate. Indeed, the [3 + 2]-cycloaddition involving a synthetic equivalent of trimethylenemethane and a norbornene derivative served adequately as the key reaction [Trost, 1982]. In view of reactivity problems methyl norbornene-2,3-dicarboxylate was used as an addend. The desymmetrization was at the penultimate step.

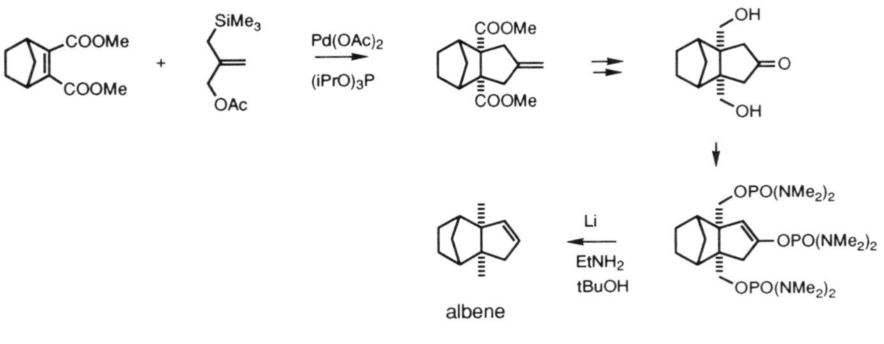

albene

Among the reported syntheses of gymnomitrol the most concise route is that involving cycloaddition of 1,2-dimethylcyclopentene with a *p*-benzoquinone monoketal [Büchi, 1979]. In one step a tricyclic intermediate was formed. This intermediate contains all the functionalities for elaboration of the sesquiterpene.

The cycloaddition combining symmetrical and unsymmetrical moieties necessarily resulted in an unsymmetrical product. The critical aspect of the synthesis is the recognition of local symmetry in the cyclopentane portion of the molecule. Exploitation of this structural feature proved very beneficial.

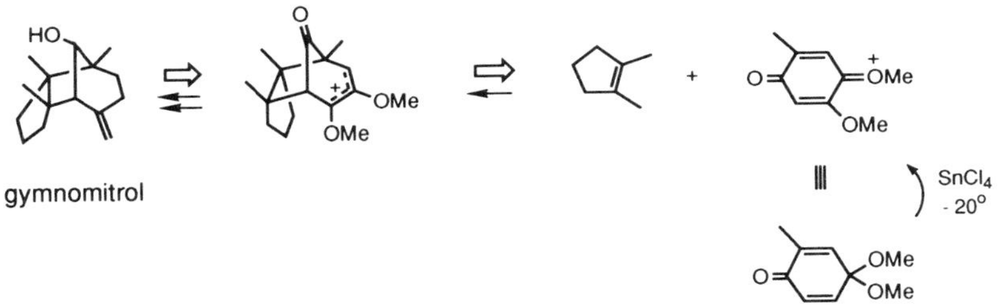

gymnomitrol

An early application of the deMayo reaction in natural product synthesis dealt with loganin [Büchi, 1973]. Delaying the introduction of the *C*-methyl group to the molecule simplified the initial stages of the synthesis (but cf. Partridge, [1973]), and fortunately the attachment of the methyl group via hydroxymethylenation of the cyclopentanone intermediate led mainly to the desired regioisomer.

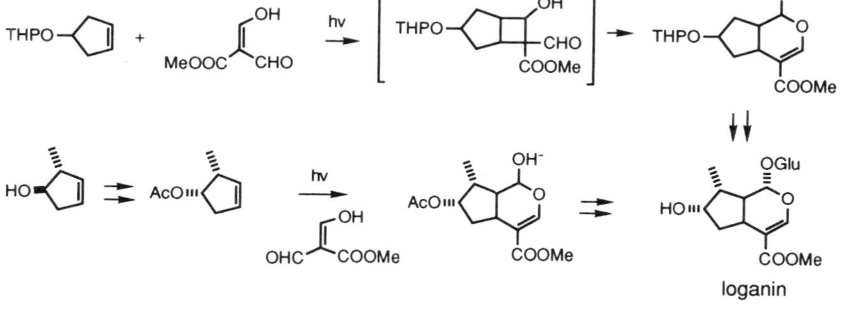

loganin

Within the tricyclic framework of protoillud-7-ene the cyclopentane ring bearing a *gem*-dimethyl group is locally symmetrical. Consequently, synthetic schemes based on reductive 1,2-dialkylation of 4,4-dimethylcyclopentene should hold certain advantages over other routes. Particularly well suited for the pursuit seems to be the DeMayo reaction with 2,4-pentanedione [Takeshita, 1979] from which a diketone is easily obtained. Intramolecular aldolization and a photocycloaddition with ethylene are then needed to elaborate the molecular skeleton of the sesquiterpene.

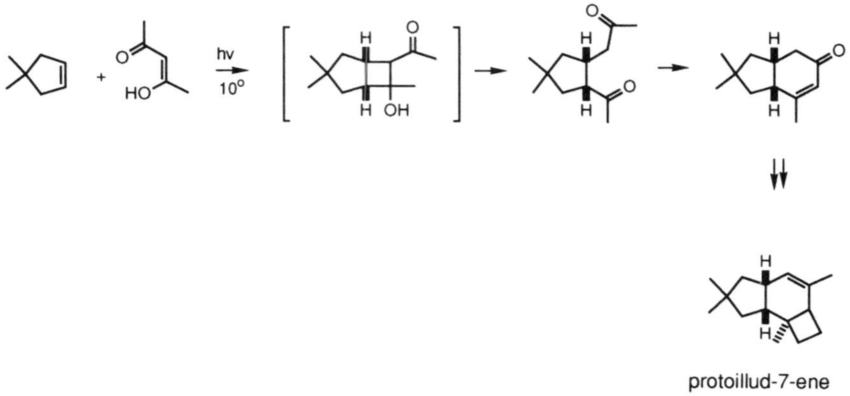

hv 10°

protoillud-7-ene

By virtue of the same reasoning on symmetry grounds synthetic approaches to hirsutene based on annulation of 4,4-dimethylcyclopentene (cf. α-caryophyllene alcohol synthesis [Corey, 1965b]) are worthy of consideration. Indeed, a convenient route to hirsutene [Iyoda, 1986] consisting of photocycloaddition and cyclomutation of the 5:4:6-fused ring system by reaction of the symmetrical diketone with iodotrimethylsilane has been developed.

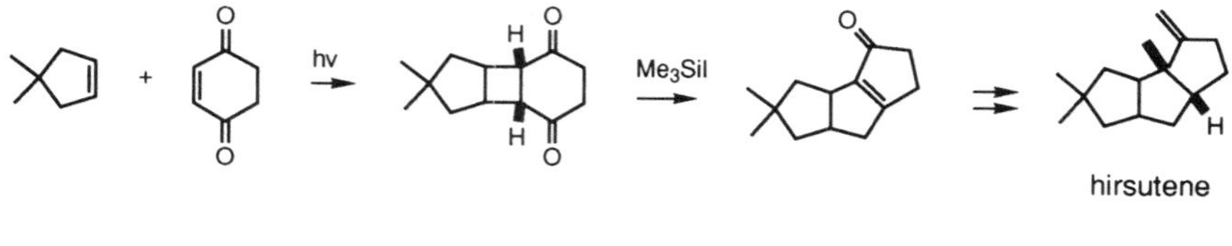

hv Me_3SiI

hirsutene

A method for α-chlorocyclopentanone synthesis via chloroketene cycloaddition and ring expansion on reaction with diazomethane is endowed with good regioselectivity. An iterative application of this reaction sequence on 4,4-dimethylcyclopentene, with proper modification of various intermediates, resulted in hirsutene [A.E. Greene, 1980]. Hirsutic acid C has also been synthesized in an analogous manner [A.E. Greene, 1985], although there was only moderate stereoselectivity (3:1) in the first annulation. It did not matter that the plane of molecular symmetry vanished upon the first cycloaddition step.

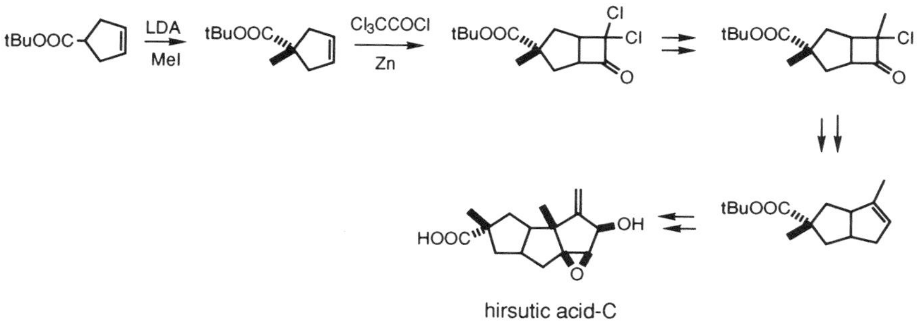

hirsutic acid-C

cis-4-Cyclopentene-1,3-diol is produced by reaction of cyclopentadiene with singlet oxygen followed by reductive cleavage of the peroxy linkage. It is of interest to note that this *meso* diol has been transformed into muscarine [Still, 1980d].

The glutaraldehyde derived from the cyclic diol is still symmetrical. However, reaction of its diacetate (a tetrahydropyran) with excess methylmagnesium bromide is *threo*-selective as a result of chelation control. Formation of the tetrahydropyranoxide anion prevented over-reaction. Internal etherification, after regeneration of the original alcohol groups, with inversion at C-5 delivered a precursor of muscarine having the correct stereochemistry.

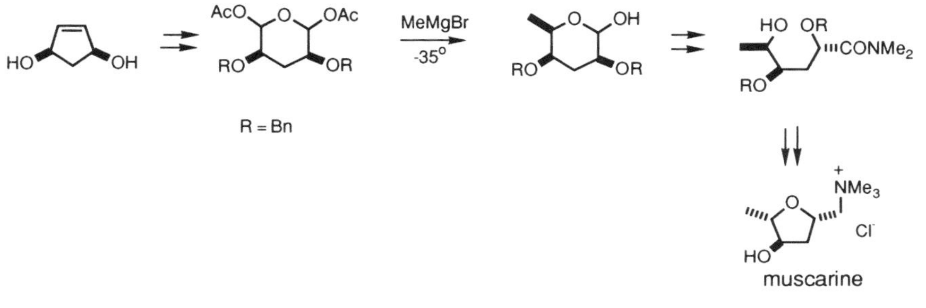

R = Bn

muscarine

From the molecular symmetry viewpoint the conversion of the all-*cis* 2,3-epoxycyclopentane-1,4-diol to terrein [Auerbach, 1974] via hydrolytic cleavage and oxidation–dehydration steps is of great interest. Reactions involving the 1,3-diol segment did not disturb the symmetry prerequisite of terrein.

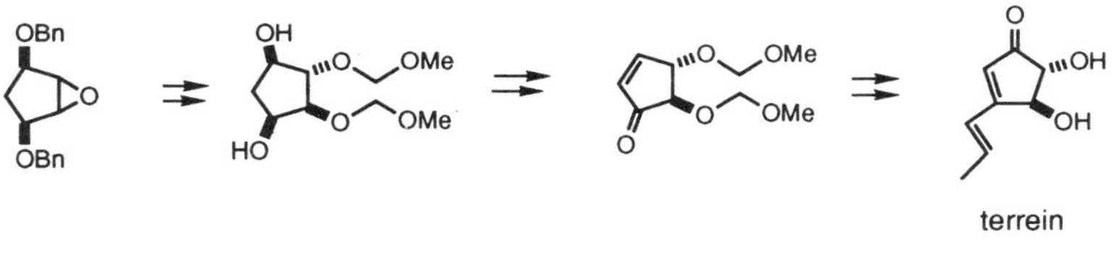

terrein

The use of *cis*-4-cyclopentene-1,3-diol in prostanoid (PGs) synthesis is logical owing to the excellent attributes of the diol in terms of its availability and functional group correlation with the prostaglandins. Desymmetrization of the cyclopentenediol by monoetherification permitted selective functionalization of the double bond. Cyclopropanation via intramolecular carbenoid interception followed by an organocopper reaction led to useful intermediates for the synthesis of various subclasses of prostaglandins [Corey, 1972].

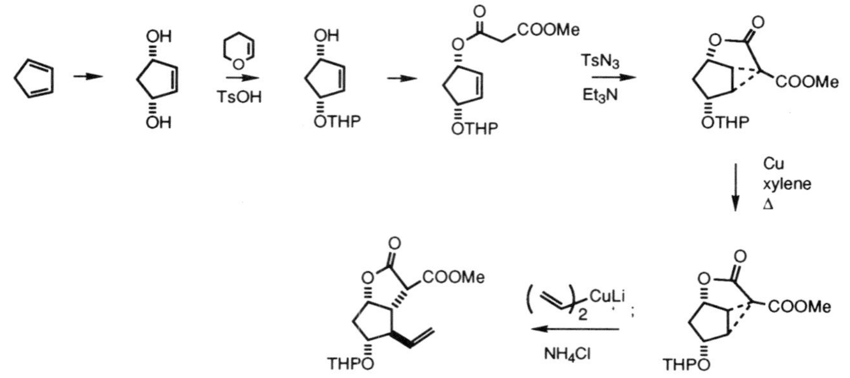

Both enantiomers of the cyclopentenediol monoacetate are now available in high purity by enzymatic hydrolysis of the diacetate [Laumen, 1984]. Thus, pig liver esterase and cholinestrase have complementary selectivities.

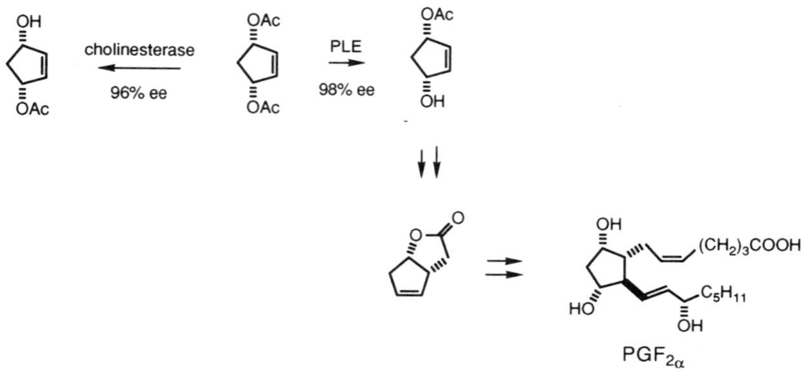

A very efficient synthesis of $PGF_{2\alpha}$ based on chain attachment via free radical reactions has been achieved [Stork, 1986b]. The starting material is the (−)-monoacetate of the cyclopentenediol. Magnificent stereoselectivity of the tandem cyclization–alkylation process was observed.

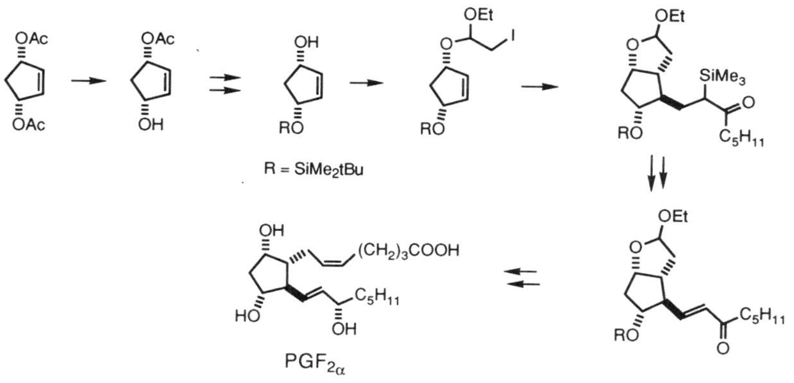

R = $SiMe_2tBu$

$PGF_{2\alpha}$

In the context of prostanoid synthesis, strategies involving desymmetrization of the cyclopentenediol system by oxidation has been very successful. An adequately protected cyclopentenolone is susceptible to conjugate addition by use of organocopper reagents. Trapping the enolate species with alkylating agents furnishes precursors of prostaglandins [Stork, 1975].

Many synthetic applications of 1,3-cyclopentanediones, particularly the 2-methyl analog, have been witnessed. The access to didehydroestrone [Ananchenko, 1963] via alkylation of the diketone with 1-hydroxy-6-methoxy-1-vinyltetralin and acid catalyzed cyclization is still one of the most efficient methods.

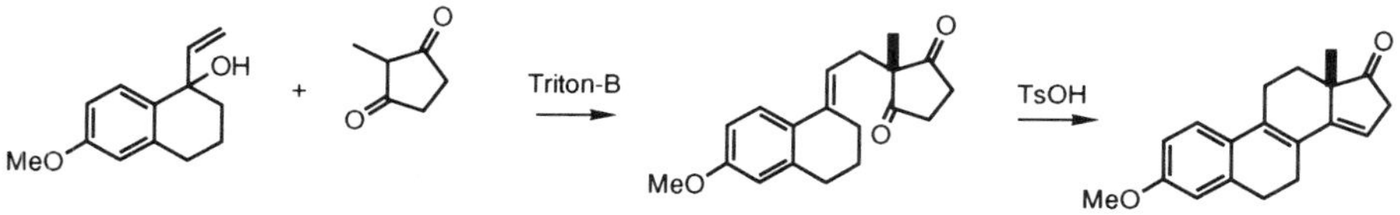

In cephalotaxine there is a monoprotected α-diketone subunit. By disconnection at the strategic C—N bond and the C—C bond joining the aromatic ring to the other carbocycle, two building blocks emerge. The azaspirocyclic portion, if put in a diketone form, has a plane of symmetry. Since common-sized cyclic diketones are very readily prepared from diester via acyloin condensation and oxidation a synthesis based on such considerations is advantageous [Semmelhack, 1975].

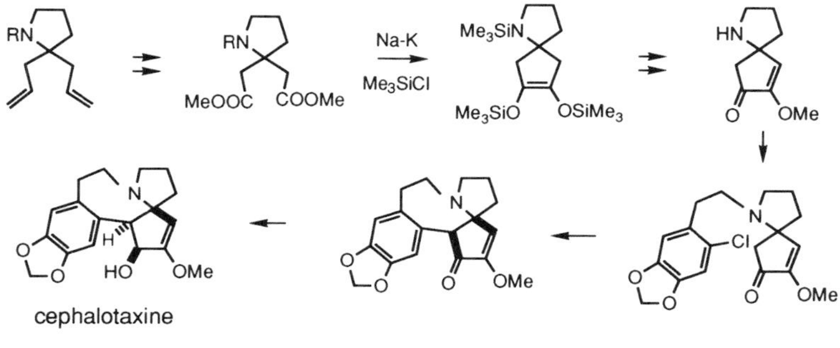

cephalotaxine

Modhephene is an unsymmetrical propellane sesquiterpene, yet a synthesis of modhephene [Wrobel, 1983] was based on the Weiss reaction to construct a symmetrical triquinanedione. The introduction of the various methyl substituents after removal of one of the ketone groups was facilitated by its high symmetry. Desymmetrization began with carbenoid insertion to block one side of the ketone allowing *gem*-dimethylation, and using the cyclopropane ring to induce attack by lithium dimethylcuprate, which stereoselectively placed a methyl group in a second cyclopentane ring.

N_2

modhephene

8.2.2. From Six-Membered Carbocycles

Many symmetrical aromatic compounds have been utilized as building blocks for synthesis of unsymmetrical molecules. The structure of oxogambirtannine evidently suggests a construction plan comprising the union of tryptamine with 2,6-dicarboxyphenylacetic acid [Merlini, 1967].

HOOC COOH HOOC

COOH HOOC

MeOOC

oxogambirtannine

A stereocontrolled synthesis of nootkatone [Dastur, 1974] from 3,4,5-trimethylanisole involved desymmetrization in the first step. The Birch reduction product underwent a Diels–Alder reaction (via isomerization to the conjugated diene) to set up two adjacent asymmetric centers, leaving one allylic methyl group to be extended into a vinyl chain. Devolution of the bridged ring system by way of fragmentation was followed by the formation of an octalone.

OMe OMe OMe COOMe OMe OH

HCOOH

OCHO

nootkatone

The elaboration of morphine [Gates, 1950, 1956] from 2,6-naphthalenediol is interesting in that both rings went through an *o*-quinone stage. The second quinone played the activating role for introducing the potential ethanamine bridge and the last carbocycle was formed by a Diels–Alder reaction.

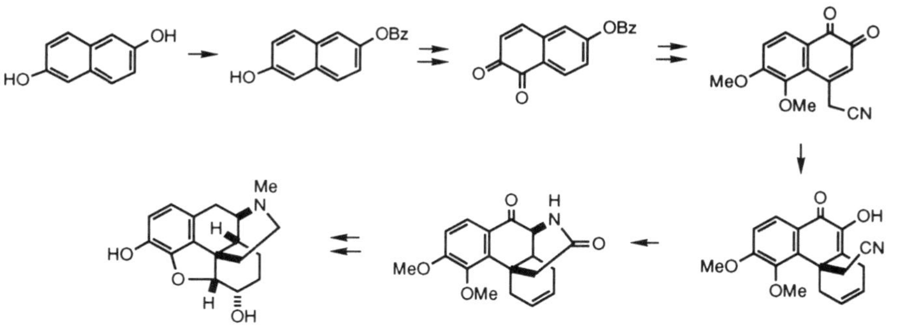

morphine

The exploitation of the enzymatic *cis*-dihydroxylation of arenes is only a recent event. This methodology has already shown great promise in the synthesis of chiral substances in very few steps [Hudlicky 1992; C.R. Johnson, 1992a]. Often both enantiomers of a particular compound are accessible through desymmetrization protocols, such as shown in approaches to conduramine-C [C.R. Johnson, 1992b].

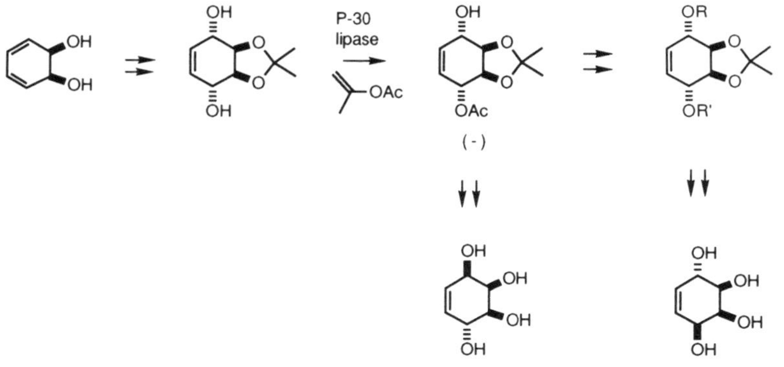

P-30 lipase

(-)

(-)-conduritol-C (+)-conduritol-C

A prostaglandin-$F_{2\alpha}$ synthesis [Woodward, 1973b] based on ring contraction started from *cis*-1,3,5-cyclohexanetriol (derived from phloroglucinol) exploited the protection of two hydroxyl groups while eliminating the third one to set up a solvolytic C—C bond formation in creating a lactol ether. Translocating the double bond and functionalizing it made the molecule susceptible to the ring contraction maneuver.

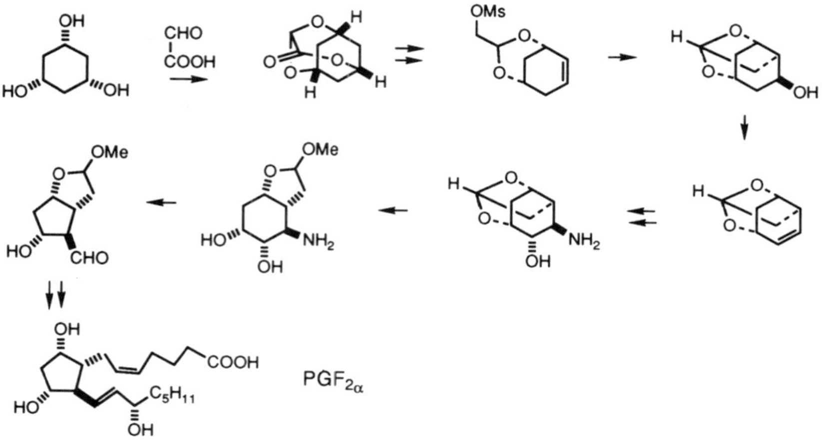

In one particular synthetic design hexahydrogallic acid contributed a six-carbon unit to emetine [Burgstahler, 1959]. Upon sidechain extension furnishing a diazoketone and reacting the latter with homoveratrylamine, the carbocycle was degraded by deleting the central carbon atom of the contiguous triol. Two aldehyde chains thus resulted, in a locally symmetrical situation, was automatically selected in the subsequent Pictet–Spengler cyclization so as to give a more stable product, i.e. *trans*-quinolizidine containing all substituents in equatorial positions.

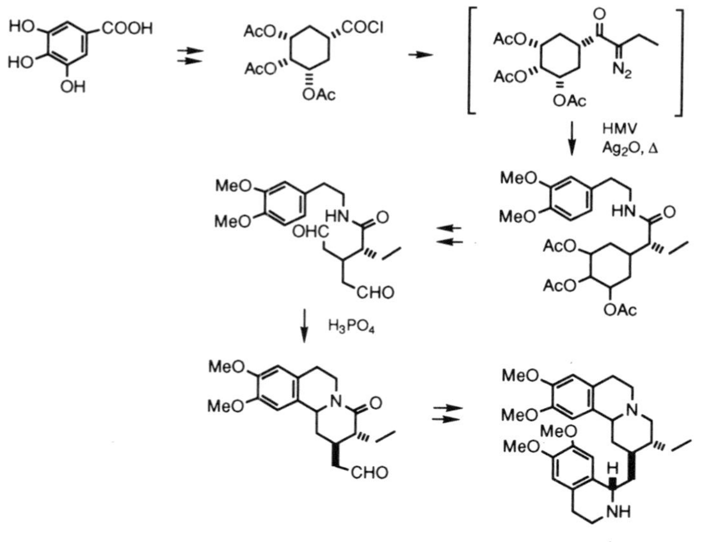

emetine

The formal Michael addition of di(+)-menthyl *trans*-4-cyclohexene-1,2-dicarboxylate to an alkynoic ester followed by cyclization led to a bicyclic compound containing all the skeletal carbon atoms of bilobalide [Corey, 1988].

Only through shrewd analysis of the molecular characteristics could this synthetic route by formulated. The C_2 symmetry of the diester was also critical.

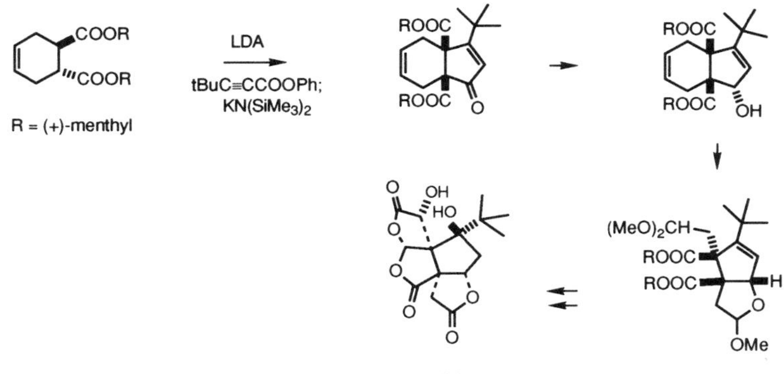

R = (+)-menthyl

LDA

tBuC≡CCOOPh; KN(SiMe3)2

bilobalide

The preparation of a bridged azabicyclic compound by a double Mannich reaction on dimethyl cyclohexanone-2,6-dicarboxylate greatly facilitated the synthesis of atisine [Ihara, 1990]. The two ester groups were subsequently reduced and differentiated.

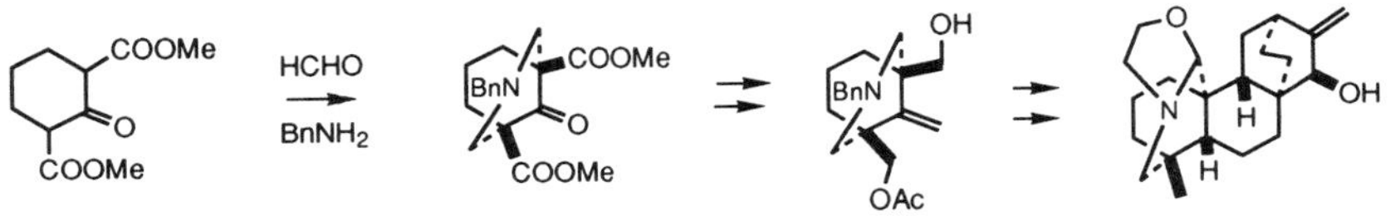

HCHO

BnNH2

atisine

An interesting approach to the alkaloid cystodytin-A [Ciufolini, 1991] involved preparation of a symmetrical 2,6-bis(2-azidobenzylidene)cyclohexanone and formation of a fused pyridine from it. One of the benzylidene groups was obliterated along with the ketone in the annulation, while the other benylidene persisted as a latent carbonyl in a position corresponding to that present in the target molecule.

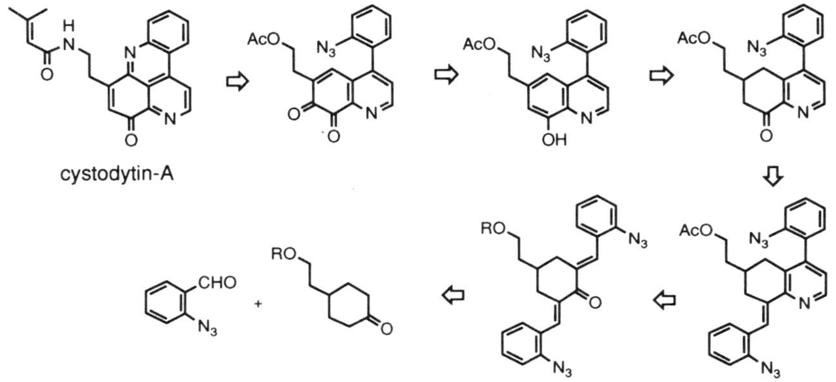

cystodytin-A

Hundreds of syntheses are known to have begun from symmetrical 1,3-cyclohexanedione and various substituted analogs. Thus, the Wieland–Miescher ketone is one of the derivatives that frequently serve as starting materials. Also, the facile monoprotection of such diketones enables C—C bond formation via alkylation or attack by organometallic reagents, resulting in unsymmetrical intermediates.

It is apparent that cannabichromene can be prepared from a symmetrical precursor. An alternative to a *C*-alkylation route involving olivetol, consisting of aldol condensation of 5-*n*-pentyl-1,3-cyclohxanedione with citral [Tietze, 1982], is equally tenable.

CHO + C_5H_{11} $RNH_3^+ OAc^-$ C_5H_{11}

OH C_5H_{11} O C_5H_{11}

cannabichromene

2,2-Dimethyl-1,3-cyclohexanedione undergoes microbial reduction (e.g. with baker's yeast) to a hydroxyketone, therefore a valuable chiral substance is readily available. Using this compound a synthesis of (−)-polygodial and its enantiomer has been achieved [K. Mori, 1986].

HO RO

R = $SiMe_2tBu$

CHO CHO COOMe COOMe RO COOMe COOMe RO CHO CHO H

(-)-polygodial

(+)-polygodial

2,3,5,6-Tetramethylidene-7-oxanorbornane is a bisdiene which has been employed in the elaboration of 4-demethoxydaunomycinone [Bessiere, 1980]. The initial steps are two Diels–Alder reactions, with methyl vinyl ketone and benzyne, respectively.

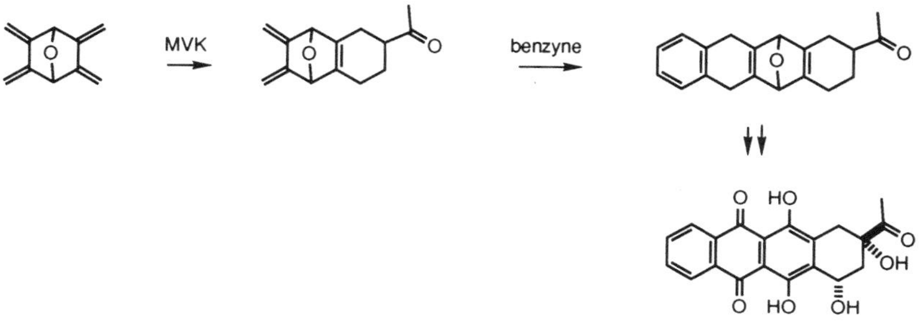

4-demethoxydaunomycinone

8.2.3. From Small and Large Carbocycles

The synthesis of the sex pheromone of comestock mealybug is in principle approachable by transpositional allylic oxidation of 2,6-dimethyl-2,5-heptadiene. Actually the C_2-symmetric *trans*-cyclopropane-$\alpha,\alpha,\alpha',\alpha'$-tetramethyl-1,2-dimethanol proved to be an excellent precursor [Skattebol, 1989].

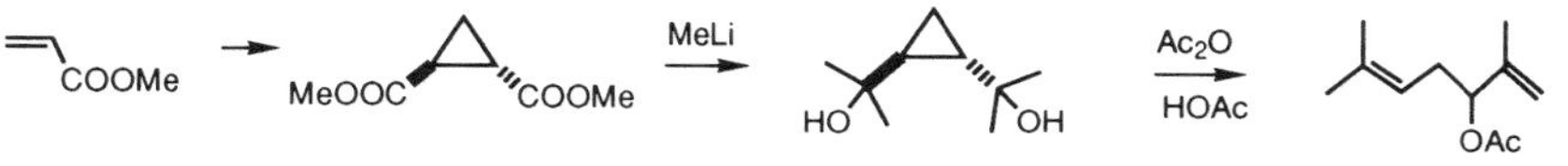

comestock mealybug pheromone

cis-1,2-Cyclobutanedimethanol undergoes oxidation in the presence of horse liver alcohol dehydrogenase to a lactone which is suitable for elaboration of (+)-grandisol [J.B. Jones, 1982]. Thus the derived methyl ester was methylated and the two sidechains were individually lengthened and modified.

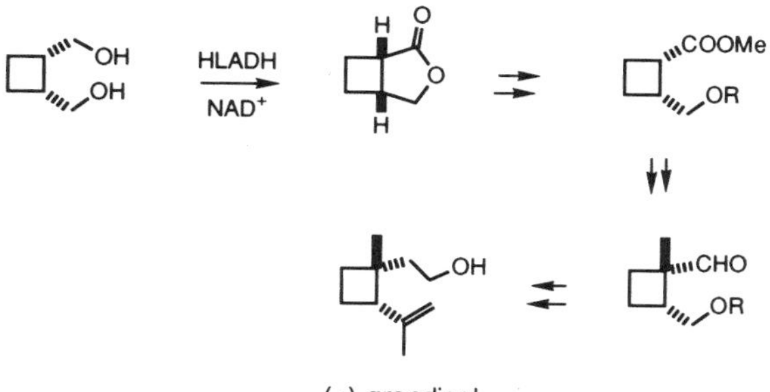

(+)-grandisol

The *meso*-diacetate of all-*cis* 4,6-dimethyl-1-cycloheptene-3,7-diol is susceptible to lipase-catalyzed hydrolysis to afford the (*S*)-alcohol which has been transformed into Prelog–Djerassi lactone [Pearson, 1988].

Prelog-Djerassi lactone

Anatoxin-*a* does not possess an element of symmetry, therefore a casual analysis of the structure of anatoxin-*a* would not reveal synthetic pathways that can take advantage of symmetry. However, when disconnection at the N—C bond joining the nitrogen atom to the allylic bridgehead carbon atom is made, the emerging synthon is cyclooctenyl cation pairing with an amide ion. Importantly, the structure is symmetrical. From synthetic perspective the replacement of the acetyl group with a halogen atom is most propitious because the intermediate is expected to be generated by solvolytic cleavage of a condensed ring system comprising a *gem*-dihalocyclopropane subunit [Danheiser, 1985].

anatoxin-*a*

It is instructive to compare this synthesis with the one starting from 1,5-cyclooctanediol [Wiseman, 1986] which proceeded via the symmetrical 5-methylaminocyclooctanone. An alternative intermediate was obtained upon bromination and internal alkylation.

anatoxin-*a*

The adduct from cyclooctatetraene and ethyl diazoacetate underwent thermal rearrangement to a dihydroindenecarboxylic ester. The desymmetrization is due to the occurrence of a second electrocyclic reaction involving 6e which was followed by the 8e ring opening. The product is useful for a synthesis of genipin [Büchi, 1967b].

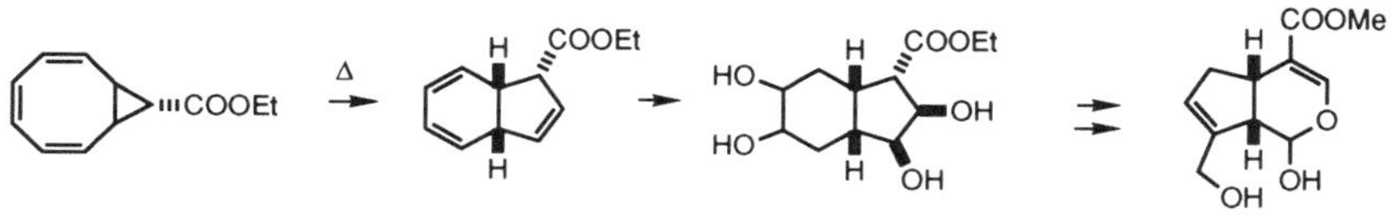

genipin

Acyloin condensation is one of the few cyclization methods which are relatively immune to ring size effect (due to entropy, torsional strain, transannular interactions, etc.). A synthesis of muscone [Y. Ito, 1977] starting from the preparation of bistrimethylsilyloxycyclotetradecene involved cyclopropanation, oxidative ring cleavage to give 1,3-cyclopentadecanedione, conversion into the monoenol acetate, methylcuprate reaction, and hydrogenation. Desymmetrization occurred at the enolacetylation step.

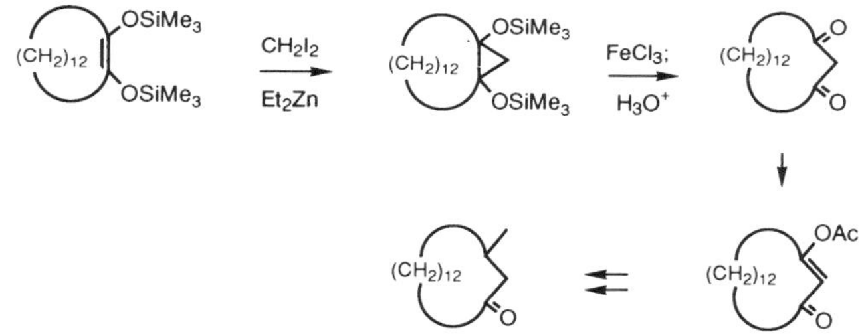

muscone

The availability of cyclododecanone was the major reason for its utility in a synthesis of muscopyridine [Biemann, 1957]. The rationale was to prepare the symmetrical [10](2,6)-pyridinophane and introduce the methyl group at the last stage. This was possible due to the location of the methyl group as oxidation of the picolinyl position is facile. Methylation of the resulting ketone followed.

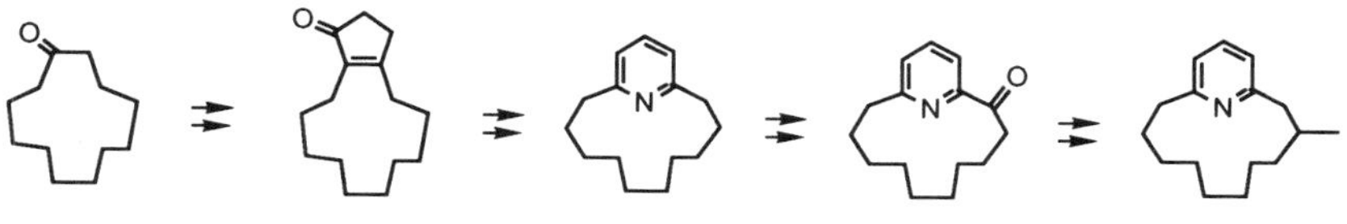

muscopyridine

8.2.4. From Heterocycles

The structural correlation of ascididemin with 1,10-phenanthroline pointed to a synthetic pathway [Moody, 1990] which involved the monoanil derivative of the 9,10-quinone. Photochemically induced electrocyclization (and *in situ* aromatization by dehydroiodination) completed the synthesis.

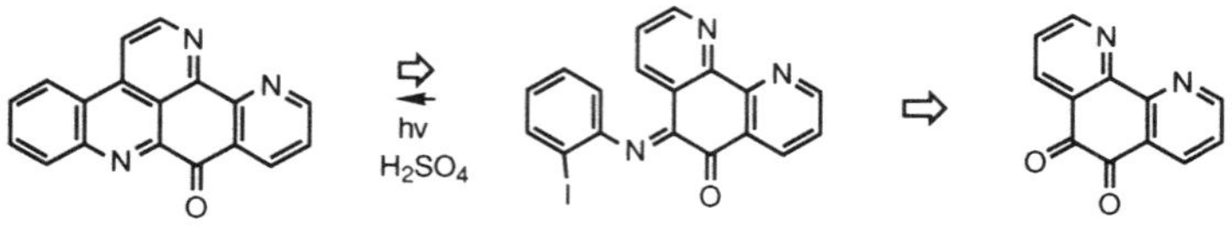

ascididemin

Very extensive use of *N*-alkyl-4-piperidinones has been witnessed in alkaloid synthesis. Usually the molecule is desymmetrized on alkylation at C-3 or annulation, an example of the latter maneuver is in a synthesis of yohimbine [Stork, 1972a].

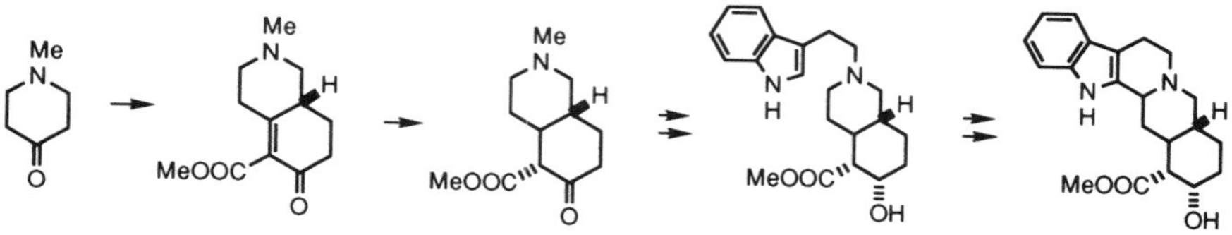

yohimbine

Synthesis of complex molecules (e.g. alkaloids) from simple azacycles generally involves desymmetrization. For those azacycles lacking functional groups in the vicinity of the nitrogen atom, activation by electrochemical oxidation [Shono, 1988] is quite efficient. On the other hand, cyclic imides can be selectively reduced to the ω-hydroxylactams which are precursors of acyliminum ions [Speckamp, 1985].

A biogenetically patterned synthesis of sparteine [van Tamelen, 1960] started from 1,5-bis(piperidin-1-yl)-3-pentanone which is a Mannich reaction product of acetone, formaldehyde, and piperidine. Mercury(II) acetate dehydrogenation served to desymmetrize the molecule, and *in situ* cyclization led to the alkaloid skeleton. Wolff-Kishner reduction completed the synthesis.

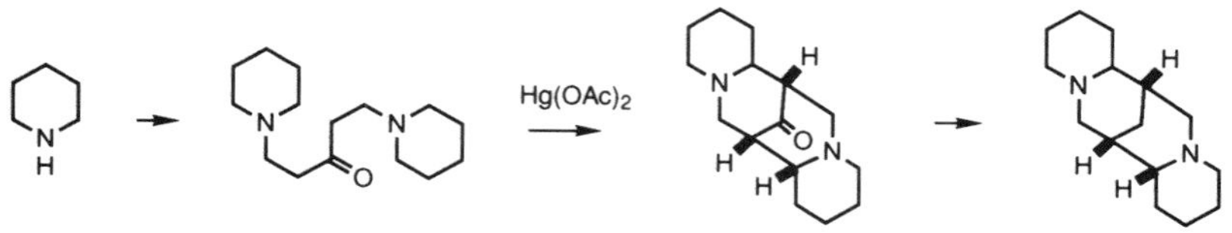

sparteine

The recognition of pseudosymmetry in emetine led to the formulation of a synthesis [D.E. Clark, 1962] which was also based on the Mannich reaction. The *meso*-bis(tetrahydroisoquinolin-1-yl)acetone underwent Michael addition to methyl vinyl ketone, but the subsequent cyclization can occur with only one of the sidechains. The remaining tasks involved adjustment of functional groups.

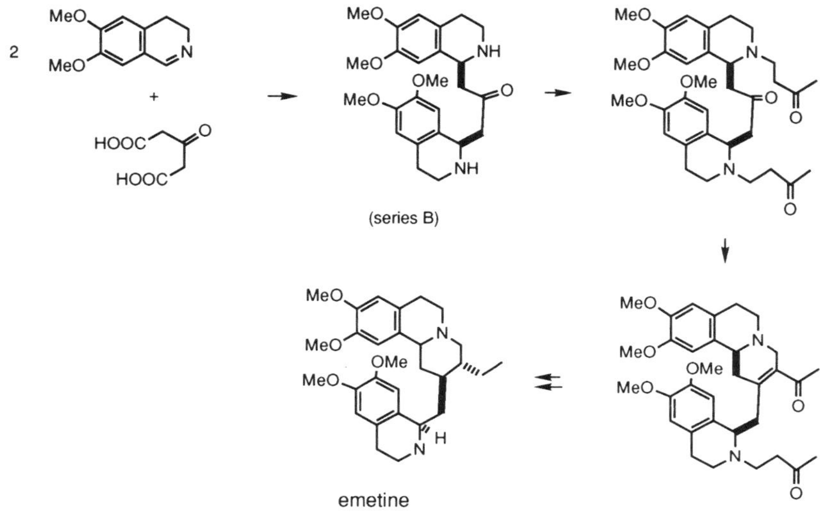

(series B)

emetine

(3*S*,4*S*)-1-Benzylpyrrolidine-3,4-diol derived from L-tartaric acid can be activated in the nitrone form toward reaction with Grignard reagents. Although two diastereomeric hydroxylamines were obtained by reaction with *p*-anisylmethylmagnesium chloride, one of them proved to be a precursor of (−)-anisomycin [Ballini, 1992].

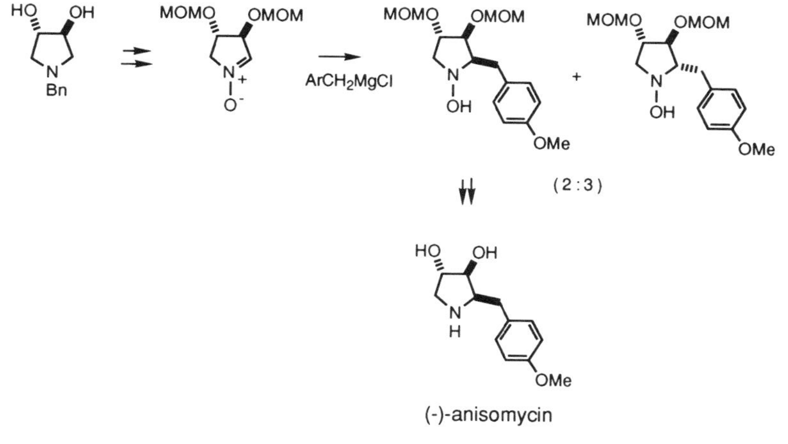

(−)-anisomycin

The partial structure of gephyrotoxin around the nitrogen atom is a *cis*-2,5-disubstituted pyrrolidine. This local symmetry, especially pertaining to the sidechains being two-carbon units terminating at a hydroxyl group and a ring junction atom, respectively, would suggest a synthetic route using *cis*-2,5-bishydroxyethylpyrrolidine as starting material. Accordingly, mono-protection of the diol and linkage of the nitrogen to a six-membered ring which is equipped with an activating group for the formation of the central piperidine constituted the major steps [Fujimoto, 1980]. 1,3-Cyclohexanedione as the six-membered ring component has the attributes that its condensation with the pyrrolidine is facile, the product resists retro-Michael elimination, and the activator is present in the correct position. The ketone also provides a pedestal for chain extension.

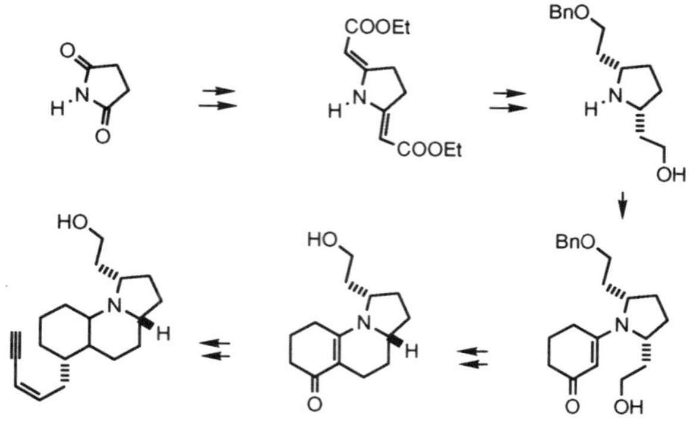

gephyrotoxin

The structure of biotin is such that the removal of the sidechain reduces it to a symmetrical bicyclic heterocycle. In synthesis it would be expedient to construct such a heterocycle with proper protecting groups on the nitrogen atoms and alkylate at the carbon adjacent to the sulfur atom. The major problem of this scheme is the stereocontrol of the alkylation step. When in fact the alkylation was achieved via the sulfoxide it proceeded to give a product with relative configuration corresponding to biotin [Lavielle, 1978].

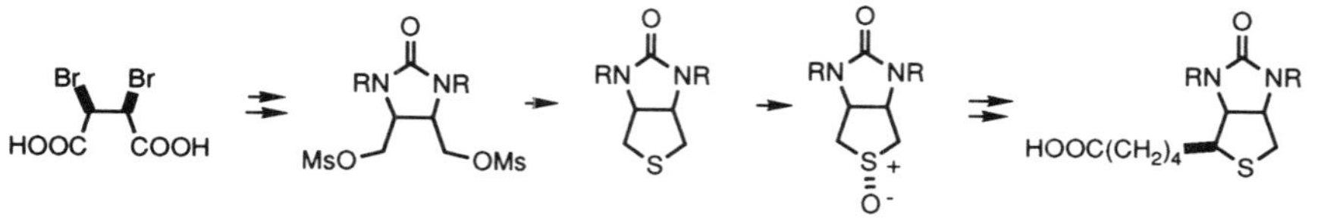

major (9:1)

biotin

Desymmetrization always ensues when an atom is inserted next to a functional group of a symmetrical molecule. In a synthesis of a Lolium alkaloid skeleton [Glass, 1978; S.R. Wilson, 1978b] the Beckmann rearrangement product of a bicyclic ketone was reduced and induced to undergo a transannular cyclization.

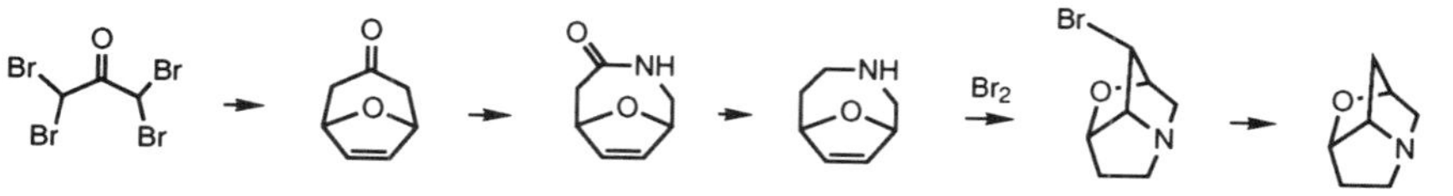

A Baeyer–Villiger-type reaction was used to degrade an oxabicyclic diketone *en route* to (−)-muricatacin [Scholz, 1991]. The symmetrical diketone was prepared by a Diels–Alder reaction of furan with cyclooctyne, followed by selective hydrogenation and ozonolysis.

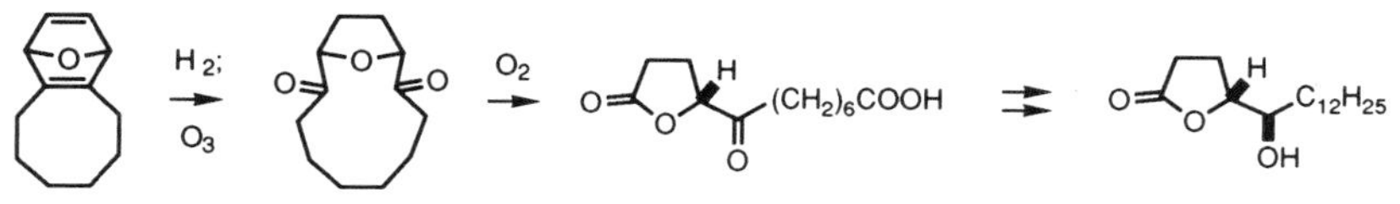

(-)-muricatacin

With the aid of a chiral base enolization of a symmetrical ketone becomes regioselective. Thus a synthesis of (+)-monomorine-I [Momose, 1990] made use of this phenomenon to effect cleavage of an azabicyclo[3,2,1]nonanone to obtain a chiral pyrrolidine hydroxy ester.

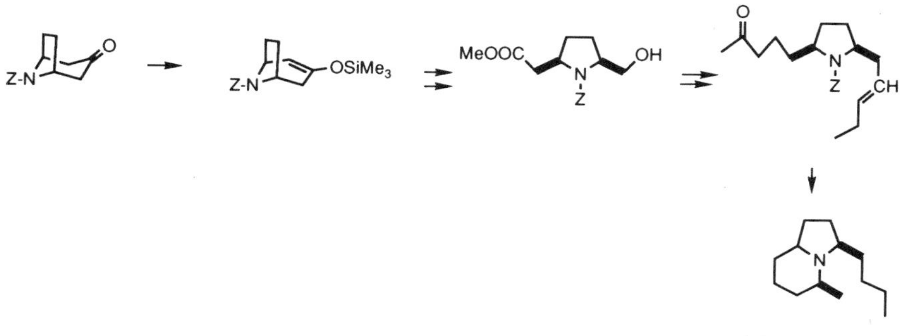

(+)-monomorine-I

8.2.5. From Aliphatic Compounds

The preeminent requirement for a chemist's choice of a symmetrical difunctional precursor to synthesize an unsymmetrical target is the ability to differentiate the two functional groups effectively at one stage of the operation. The differentiation is much easier to achieve when rendered intramolecular. For example, a disconnection at the S—C (allylic) bond of cephalosporin-C suggests a locally symmetrical synthon. In the synthetic scheme an alkylidenemalonaldehyde served as the required fragment to unite with the bicyclic lactam by a Michael reaction [Woodward, 1966]. Upon cleavage of the dihydrothiazine ring the thiol immediately reacted with one of the aldehyde groups.

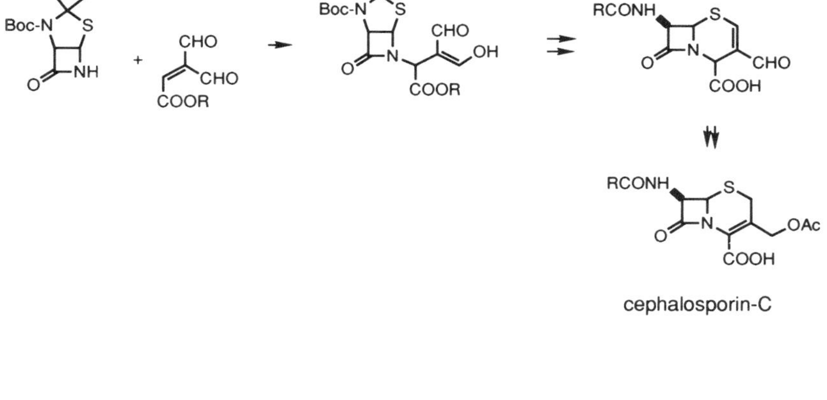

cephalosporin-C

The desymmetrization of the two primary alcohols in 2-substituted 1,3-propanediols is crucial to many of their synthetic applications. One method is via ketal formation with chiral ketones, as demonstrated in the preparation of two building blocks, a functionalized chroman and a C_{14} alcohol, for the synthesis of α-tocopherol [Harada, 1987].

HO OH OH d-menthone HO O O HO OTHP OH HO OH

OHC HO HO l-menthone H O O HO PhS HO

The popularity of C_2-symmetric compounds in the synthesis of chiral substances is well deserved because, all groups being pairwise homotopic, only one product is generated from a symmetry-breaking reaction. Among them (*R*,*R*)- or (*S*,*S*)-tartaric acid fulfills many of the qualifications for an ideal chiral building block [Seebach, 1980] in view of availability, economy, and numerous possibilities of reactions in effecting deoxygenation (at different sites), inversion of one asymmetric center to change it into a *meso*-compound, chain shortening, chain elongation, and branching.

An interesting synthesis of *exo*-brevicomin [Masaki, 1982] from tartaric acid is via formation of a dioxolane in which one of the two substituents at C-2 is a phenylsulfonylethyl chain. When the two ester units were transformed into tosyloxymethyl groups to induce cyclization, stereoselection for the latter reaction was automatic.

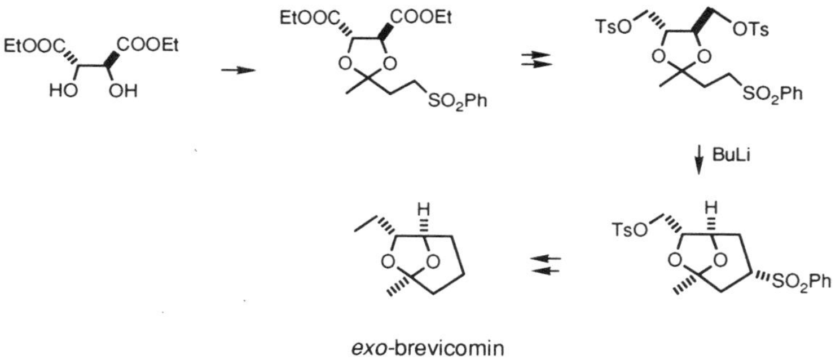

exo-brevicomin

Also related to symmetry-breaking maneuver is the formal inversion of configuration of each of the asymmetric sites in an array containing multiple centers. Thus D-glucose has been transformed into a carbon chain characteristic of the D-xylo configuration and further to the L-xylo series [Fürstner, 1990], and D-allose through the D-ribo array to L-ribo array.

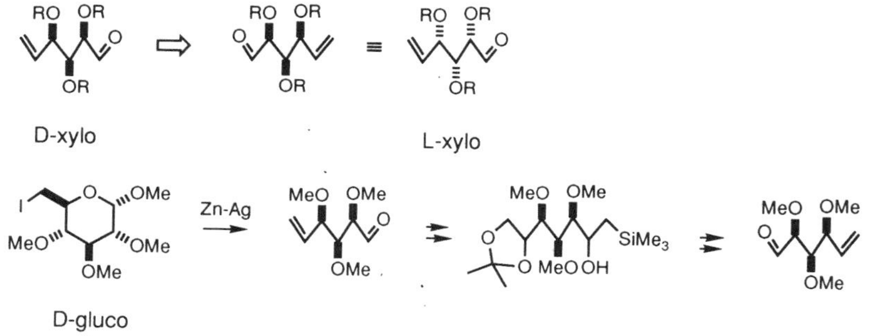

D-xylo

L-xylo

D-gluco

Selective reaction of a difunctional *meso*-compound is desymmetrizing. Cyclic anhydrides such as *cis*-2,5-dimethylglutaric anhydride and 3-substituted glutaric anhydrides are readily transformed into the monoesters and used for synthesis. Among the many applications is a synthesis of (+)-conglobactin [Schregenberger, 1984] which showed the absolute configuration of the natural (−)-isomer to have the (5*R*,7*S*,8*S*,13*R*,15*S*,16*S*)-configuration.

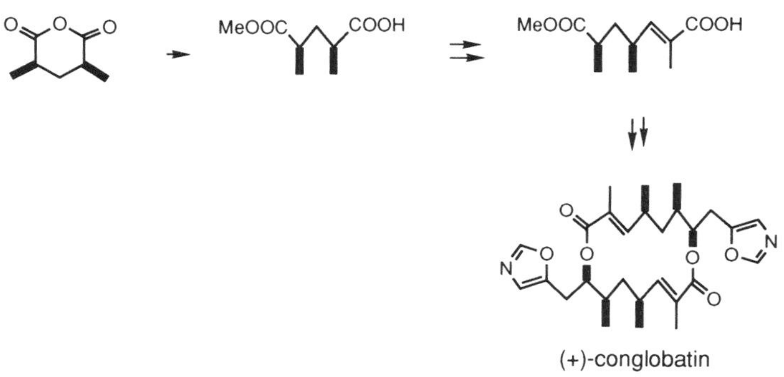

(+)-conglobatin

The alcoholysis of cyclic anhydrides is perhaps exceptionally easy in terms of functional group differentiability. However, the more difficult monoprotection of two identical functional groups is frequently required in synthetic processes. A synthesis of periplanone-J [Harada, 1992] used two symmetrical aliphatic diols, each of which was desymmetrized and coupled by a Wittig reaction to give a triene which contains all the framework atoms. The cyclization was achieved by intramolecular alkylation.

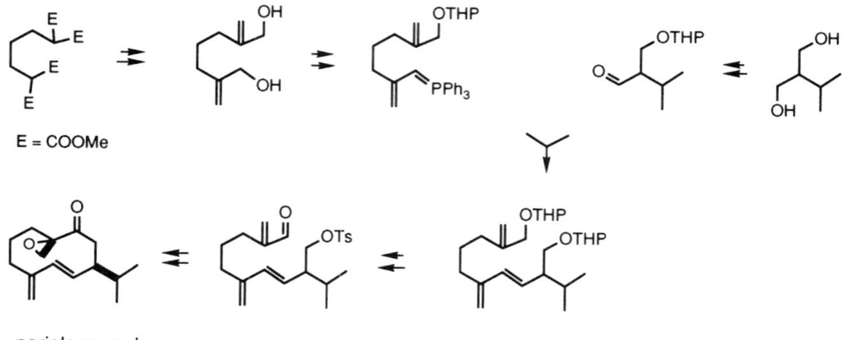

E = COOMe

periplanone-J

Two-directional chain synthesis [S.L. Schreiber, 1987a] involving simultaneous homologation of a nascent chain in two directions significantly decreases the number of total reaction steps. One example appears in the assembly of the C-10/C-19 segment of the immunosuppresant compound FK-506 [Nakatsuka, 1990].

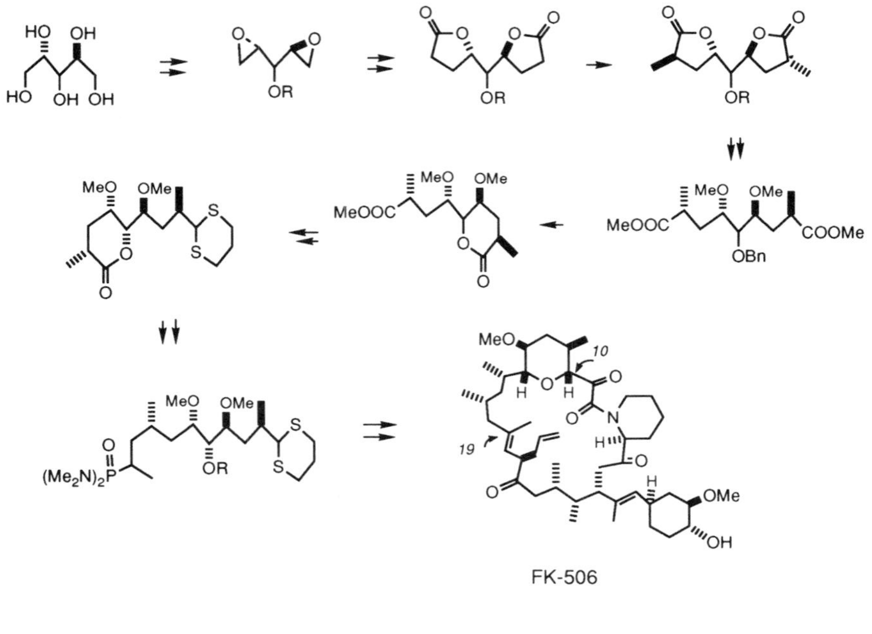

FK-506

The enormously important method of desymmetrization of bisallylic alcohols is via the Sharpless asymmetric epoxidation. This kinetic resolution method would provide the desired enantiomer by proper choice of the tartrate ester catalyst. In this manner the epoxytetrahydrofuran unit of (+)-verrucosidin was acquired [Hatakeyama, 1988].

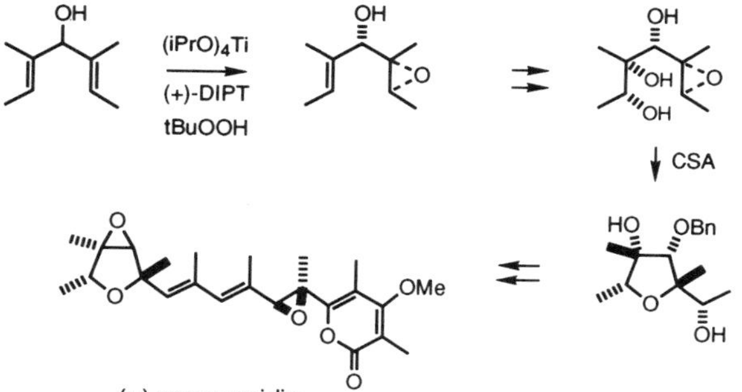

(iPrO)$_4$Ti, (+)-DIPT, tBuOOH; CSA

(+)-verrucosidin

Tricarballylic acid has been used in a synthesis of avenaciolide [W.L. Parker, 1969]. On removing the exocyclic methylene group and disconnecting the ethereal O—C bonds of the natural product a symmetrical structure is revealed. A synthesis from tricarballylic acid would require replacement of the hydroxyl group of the central carboxyl function with a carbon chain and establishment of the two lactone rings.

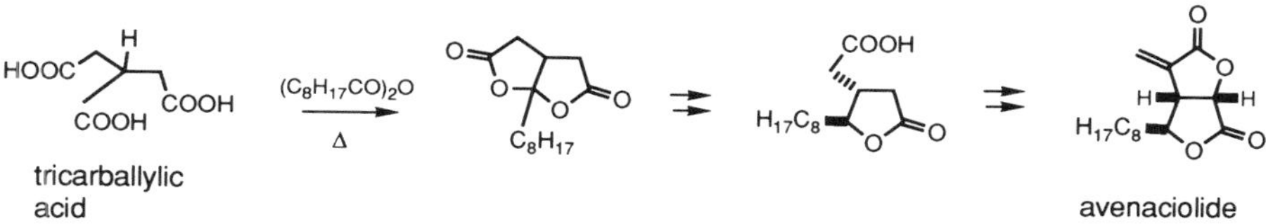

tricarballylic acid; $(C_8H_{17}CO)_2O$, Δ; avenaciolide

An enantioselective synthesis of (−)-gloeosporone [Takano, 1988b] from (*S*)-*O*-benzylglycidol proceeded via desymmetrization at the stage of a 1,2;8,9-diepoxynonane by an organocuprate reaction. Similarly, a synthesis of (−)-pyrenophorin [Machinaga, 1993] from (*R*,*R*)-1,2;5,6-diepoxyhexane owes its efficiency to the same factors governing the reaction sequence.

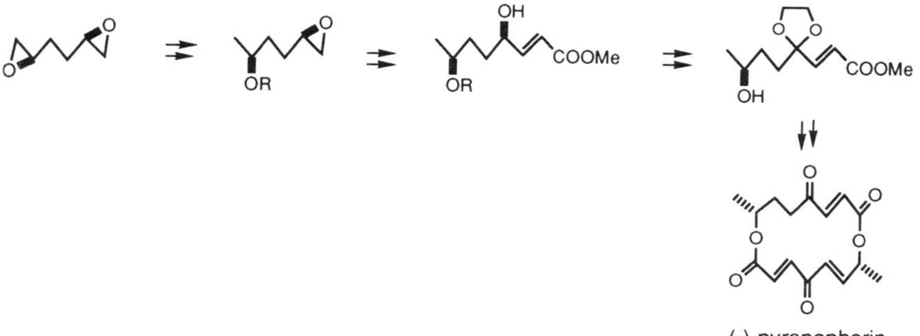

(-)-pyrenophorin

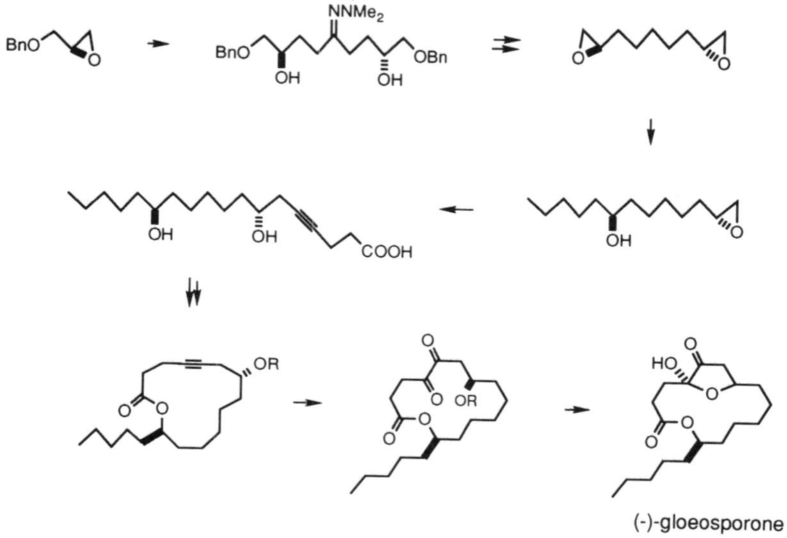

(-)-gloeosporone

A report on a synthesis of the tetracyclic alkaloid porantherine [Corey, 1974a] was the first to demonstrate explicitly the power of retrosynthetic analysis which revealed in this case symmetric intermediates. The desymmetrization occurred when one of the two exposed ketone groups underwent an intramolecular condensation with the secondary amine. A bridged azabicyclic system was formed by a Mannich reaction, and modification of the pentenyl sidechain initiated the assembly of the remaining rings.

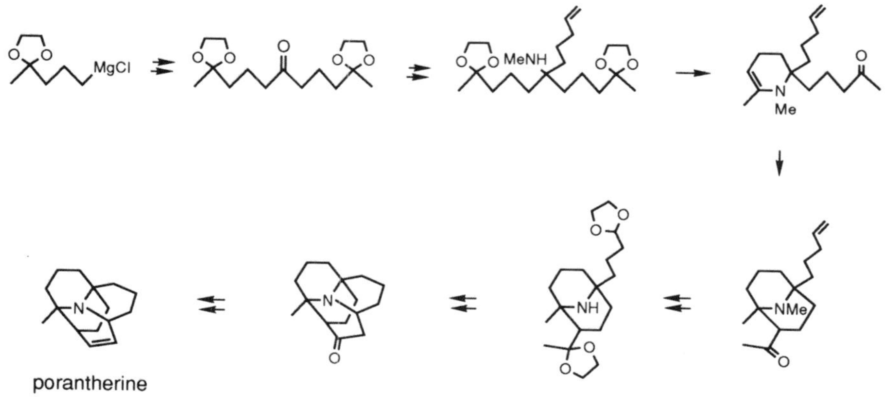

porantherine

The condensation of 3-(2-aminoethyl)oxindole with dimethyl 4-ethyl-4-formylheptanedioate led to a tetracyclic intermediate for the aspidosperma alkaloids such as aspidospermidine [Castedo, 1969] and vincadifformine [Laronze, 1974]. Advantages of the symmetrical building block are evident.

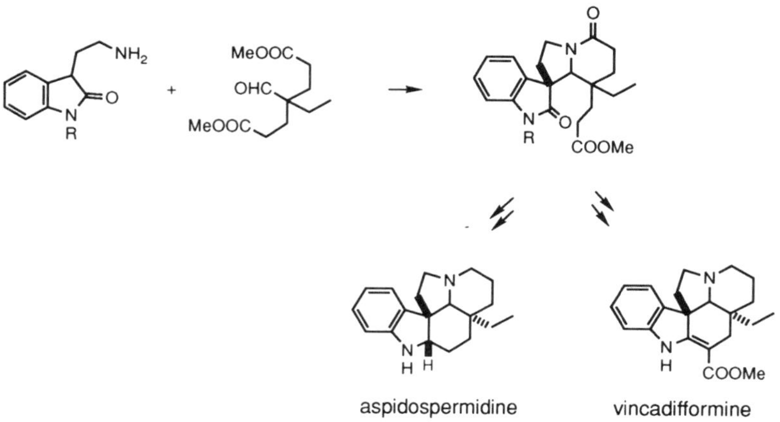

aspidospermidine

vincadifformine

The Fischer indolization of 5-oxononanedioic acid is a tricyclic indole. Upon degradation of the carboxyl group to a primary amine the product was rearranged to a nine-membered lactam which could be transformed into condyfoline and tubifoline [Ban, 1983].

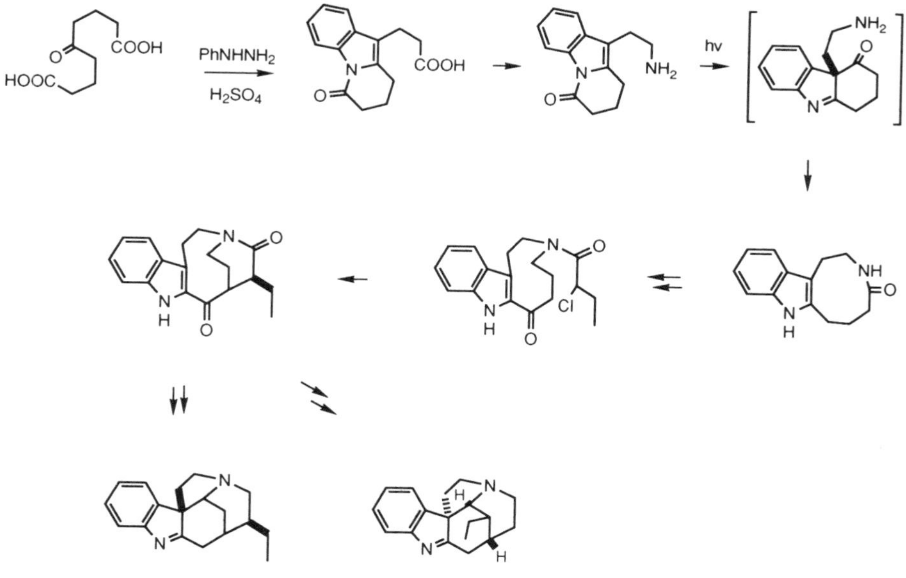

tubifoline

condyfoline

Isoalchorneine is a pseudosymmetric bicyclic guanidine, therefore it can be elaborated from symmetrical intermediates [Büchi, 1989]. Desymmetrization occurred during formation of the trisubstituted guanidine in which one of the nitrogen atoms bears a methoxy group. This product was induced to cyclize.

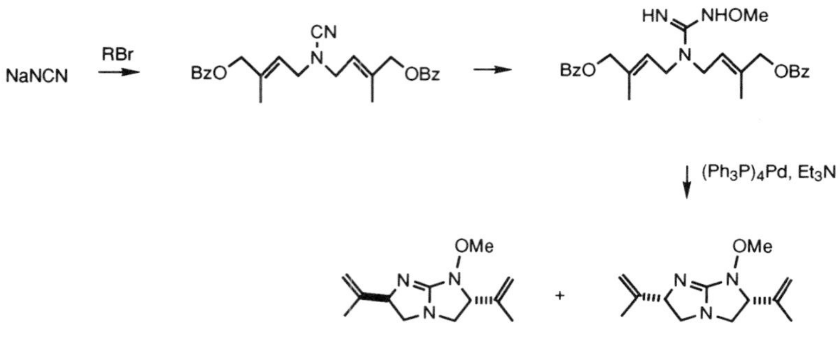

isoalchorneine

The Dieckmann cyclization desymmetrizes a diester such as diethyl *N*-benzyl-5-azanonanedioate. In this particular product the amino group and the ketone interacted strongly and on hydrogenation a pyrrolizidine system was formed. Hydride reduction of the ester led to isoretronecanol [Leonard, 1969].

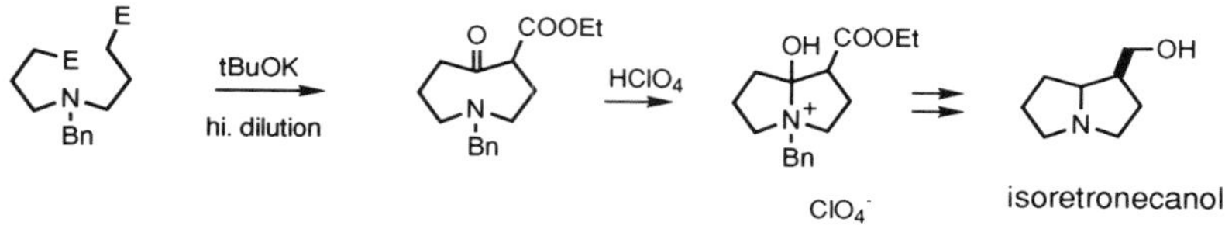

isoretronecanol

The Dieckmann cyclization furnishes bifunctional molecules. The ester pendant can be retained as indicated above, or it may serve as an activator for further skeletal modification. Thus in a synthesis of nootkatone [Marshall, 1970] a Dieckmann cyclization of dimethyl 4-oxo-heptanedioate was effected under the Wittig reaction conditions. The ester group was used to direct and facilitate a Robinson annulation before being transformed into the angular methyl substituent.

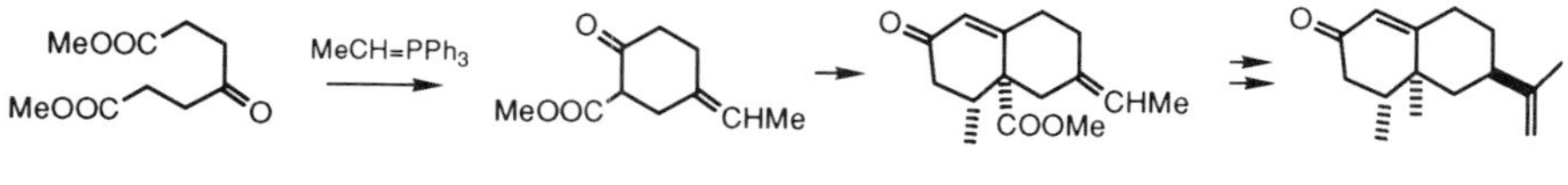

nootkatone

A valuable reminder to conclude this section is that even the synthesis of certain chiral targets or intermediates does not require desymmetrization, the task is greatly benefited by elaboration of C_2-symmetric molecules into the proper precursors of such synthetic targets/intermediates followed by cleavage of the central bond. Redundancy in terms of examples notwithstanding, it seems worthwhile to witness one more in the preparation of an aldehyde intermediate for the leukotriene LTB_4 [Le Merrer, 1985].

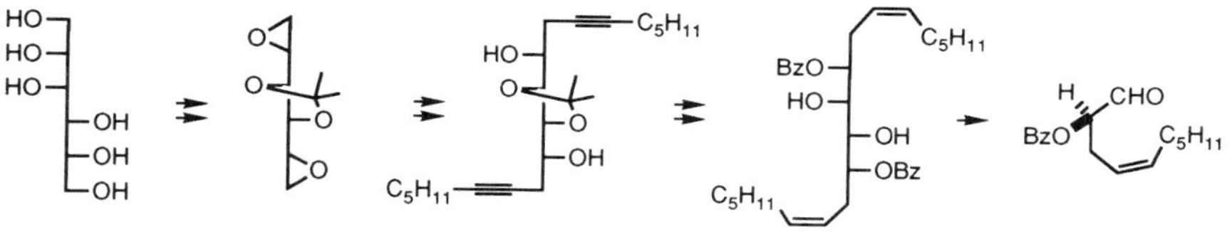

D-mannitol

9

MISCELLANEOUS TACTICS

9.1. EQUILIBRATION AND ISOMERIZATION

Even at the design stage the course of a reaction employed in a synthesis may be recognized as leading to a product other than that desired. However, if remedial action requires only equilibration and isomerization, and such processes are available the synthesis may be prosecuted as planned. An unexpected reaction product arising during a synthetic endeavor would still be usable if it is an isomer amenable to similar correction. Such a situation arose during synthesis of reserpine [Woodward, 1958] that inversion of the C-3 configuration was achieved by equilibration of a bridged lactone. Lactonization changed the molecular conformation which also destabilized the compound to allow the necessary epimerization.

A longifolene synthesis [Corey, 1964c] starting from Wieland–Miescher ketone proceeded through ring expansion and intramolecular Michael addition. While only an enedione with a *cis* ring fusion would undergo the latter reaction, there was no need to maintain such stereochemistry as the two possible stereoisomers interconverted under the reaction conditions.

Et_3N, $(CH_2OH)_2$, 225°C

longifolene

By using a Diels–Alder approach in a synthesis of cholesterol [Woodward, 1952] the C/D-ring juncture must be rectified. A triggering device was provided for the isomerization in the form of the subangular ketone group. There was a similar requirement when the adduct of butadiene and *p*-benzoquinone was used in a yohimbine synthesis [van Tamelen, 1958].

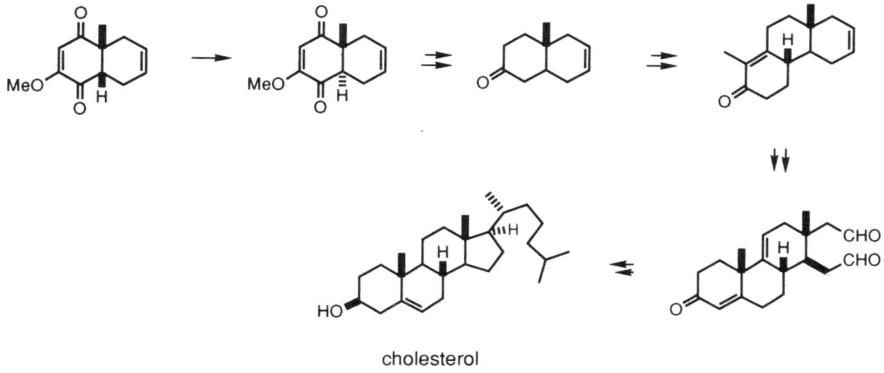

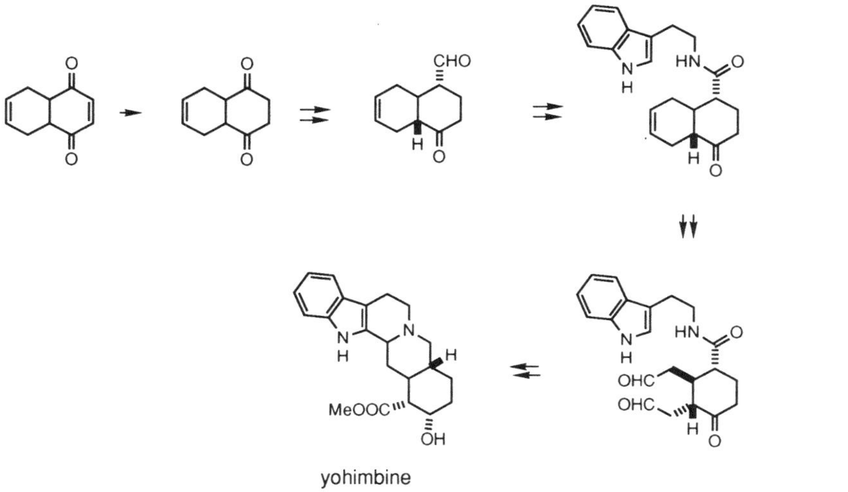

cholesterol

yohimbine

The regiocontrol and stereocontrol element during annexation of the A-ring to a precursor of conessine [Stork, 1962] was consigned to the *cis* B/C-ring juncture. This feature was inherent to the phenolic ancestry of the B-ring and the method (hydrogenation) of its generation. Later, inversion of configuration at C-8 was accomplished after making it vinylogous to the ketone group. In a synthesis of morphine [Gates, 1956] the *cis*-juncture between the two hydroaromatic rings, established during a Diels–Alder reaction, was changed to the *trans*-state upon treatment of the α-bromoketone with dinitrophenylhydrazine (to effect dehydrobromination).

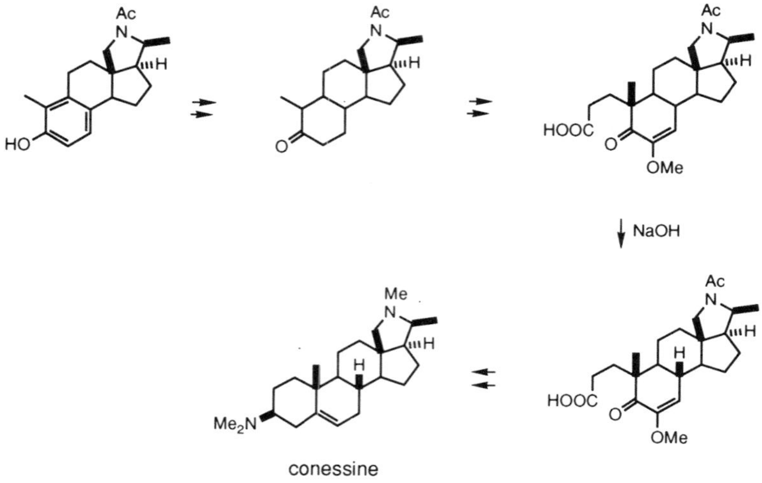

NaOH

conessine

Ecdysone is an AB-*cis* steroid, therefore its synthesis from a *trans* intermediate would require ring juncture inversion [Siddall, 1966]. This was enabled by the presence of a 2β-hydroxyl group which interacts diaxially with the 10β-methyl substituent. The equilibration was actually favored by forming an acetonide from the 2β,3β-diol subunit.

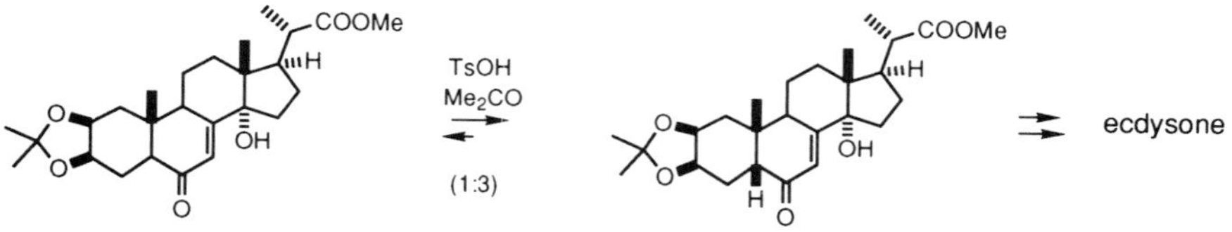

TsOH, Me_2CO (1:3)

ecdysone

The lactone carbonyl of annotinine interacts sterically with the secondary methyl group projecting from the cyclobutane ring. Accordingly during synthesis [Wiesner, 1969] a precursorial ester underwent epimerization upon saponification. Fortunately an equilibrium could be set up between the two epimeric acids and lactonization was induced from the *cis*-isomer. Completion of the synthesis depended on this trapping of the less stable compound.

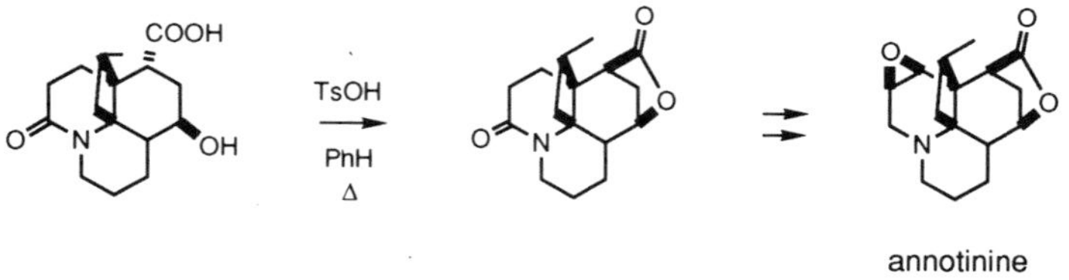

TsOH, PhH, Δ

annotinine

In a synthesis of (+)-lepicidin aglycone [Evans, 1992a] the five-membered ring was formed by an intramolecular aldol reaction. The failure of the major product to undergo dehydration could have meant disaster, but fortunately it could be converted into a diastereomer (ratio 2.5:1) in favor of the latter compound) on treatment with sodium methoxide. This diastereomer behaved agreeably.

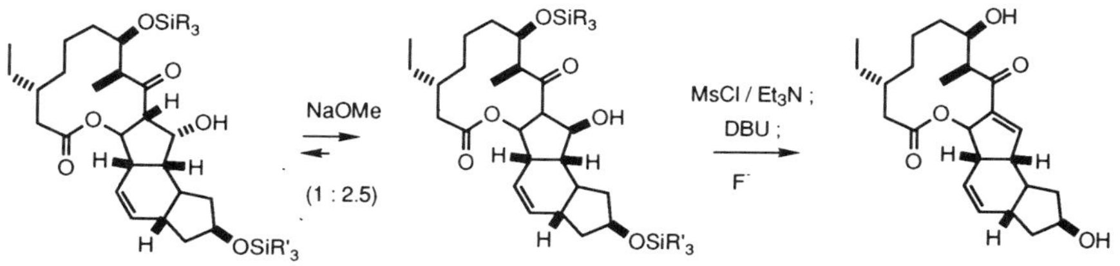

NaOMe (1 : 2.5)

MsCl / Et_3N ; DBU ; F^-

(+)-lepicidin aglycon

If a chiral synthetic target can be approached by manipulation of both enantiomers, such an approach is enantioconvergent. A route to a potential precursor of the prostaglandins [Trost, 1978] relied on 1,3-hydroxyl shift of a cyclohexenol via its chiral urethane derivative by means of mercury(II) trifluoroacetate catalysis. Diastereomer separation and resubjecting the undesirable isomer to equilibration conditions realized the utility of the racemate.

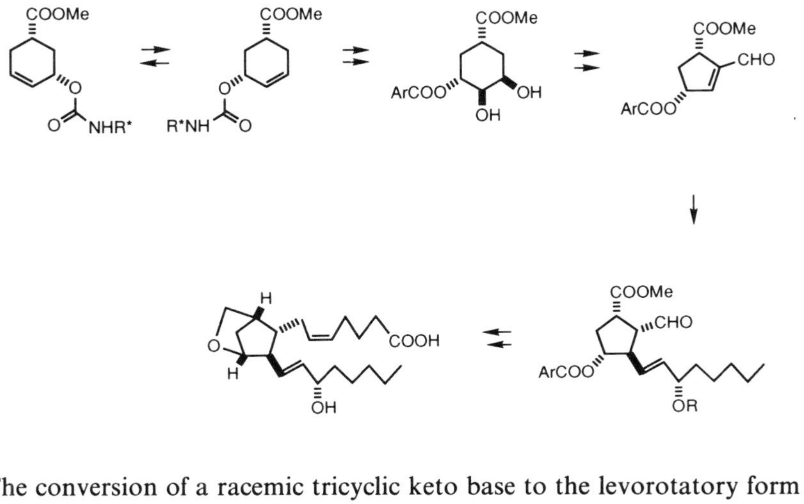

The conversion of a racemic tricyclic keto base to the levorotatory form for the synthesis of (−)-emetine [Openshaw, 1963] by refluxing with equimolar (−)-10-camphorsulfonic acid in ethyl acetate involved stereoselective equilibration by way of a reversible Mannich reaction. In a route to vincamine [Oppolzer,

1977] the acquisition of a "*cis*" racemate aldehyde whose *p*-toluenesulfonic acid salt is more soluble in dioxane was accomplished by heating the salt of the undesired "*trans*" isomer separating the solution containing the "*cis*" compound from the unchanged salt, and recycling the latter. The "unnatural" enantiomer obtained from resolution of the "*cis*" racemate with (+)-malic acid could also be put back into the equilibration.

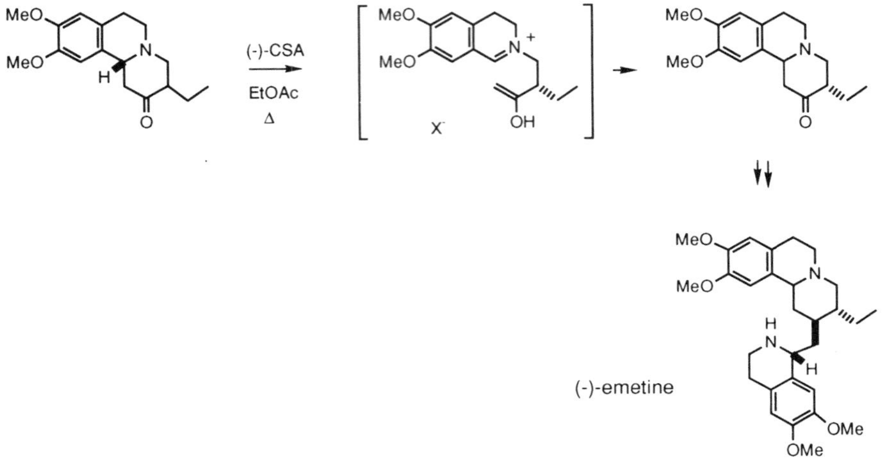

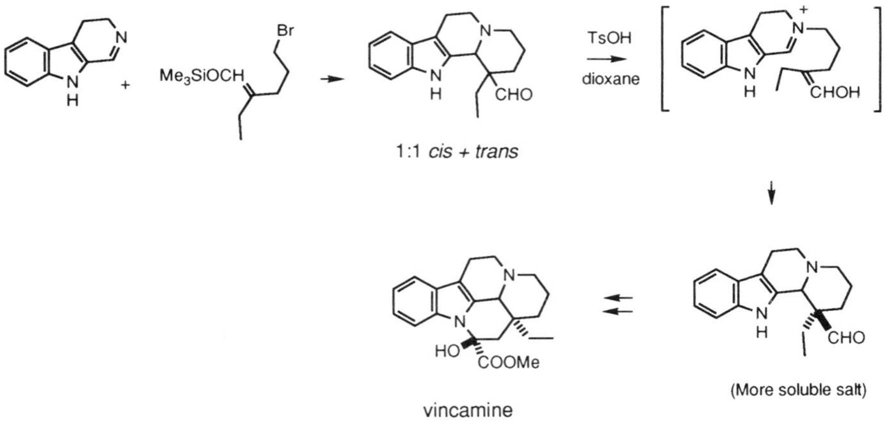

(-)-CSA, EtOAc, Δ

(-)-emetine

TsOH, dioxane

1:1 *cis* + *trans*

(More soluble salt)

vincamine

Carvone is a versatile chirogen because both enantiomers are readily available. In fact the interconversion of them is also achievable via a Wharton rearrangement of the epoxide and oxidation of the resulting carveol. Using the natural carvones two different routes have been developed which converge in (+)-*trans*-chrysanthemic acid [Torii, 1983].

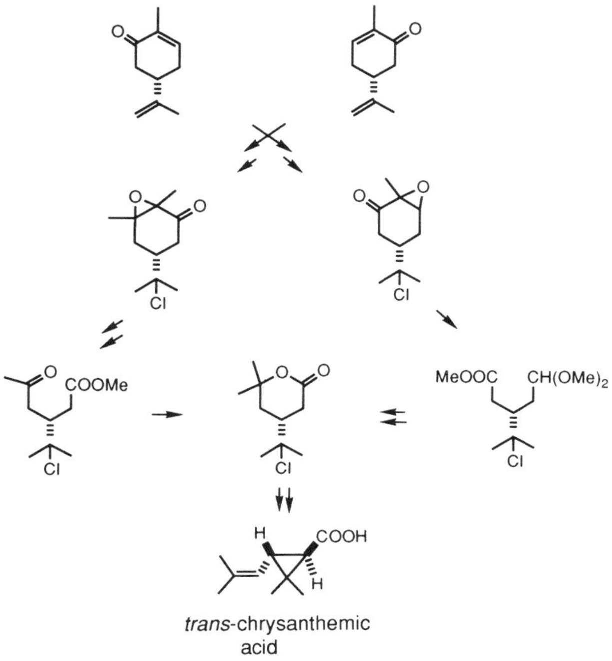

trans-chrysanthemic acid

The intramolecular alkylation of an anisole ring with an enamide under the influence of protic acids, generated the desired intermediate of lycopodine [Stork, 1968]. The cyclization was facilitated by the lower strain of the iminium species in comparison with the alternative ion. The latter featured a boat-like cyclohexane ring.

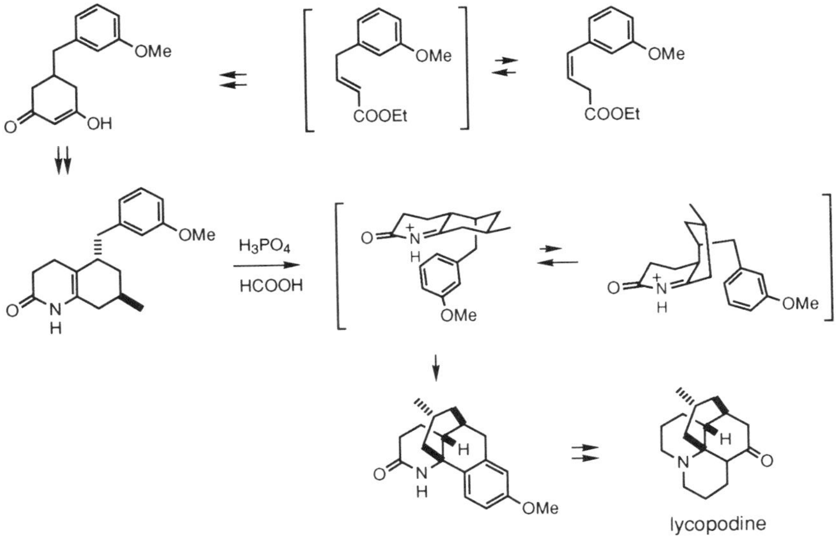

H_3PO_4

HCOOH

lycopodine

Double bond migration during a reaction is a useful feature which permits the employment of a positional isomer as reactant. In the same lycopodine synthesis the preparation of 5-*m*-anisylmethyl-1,3-cyclohexanedione by a Michael–Claisen reaction tandem used a styrene as the Michael acceptor.

β,*γ*-Unsaturated acids undergo cyclodehydration to give cyclopentenones when treated with strong acids. This reaction has been used in a synthesis of muscopyridine [Biemann, 1957]. Note that the starting material was readily available from a Stobbe condensation.

HOOC COOEt PPA COOEt

muscopyridine

1,4-Cyclohexadienes are Birch reduction products. They yield Diels–Alder products on heating with dienophiles in the presence of dichloromaleic anhydride. This *in situ* conjugation expedites syntheses such as that of nootkatone [Dastur, 1974].

OMe Cl Cl Δ OMe COOMe Δ OMe COOMe OMe OH HCOOH OCHO

nootkatone

Skeletal isomerization upon reduction of a 4-formylcoumarin was crucial to the assembly of the furano[2,3-*b*]benzofuran ring system of aflatoxin-B_1 [Büchi, 1967a]. This is an example of *β*-acyllactone rearrangement.

BnO CHO MeO Zn HOAc BnO CHO COOH MeO OH BnO H MeO H MeO H H

aflatoxin-B1

An approach to [2]-[cyclohexatetraoctane][cyclooctacosane]catenane [Schill, 1983] also took advantage of an acid-catalyzed equilibrium between trityl ether and the trityl cation–alcohol system. In the presence of a macrocycloalkane, a long chain diol bistrityl ether after undergoing mono-deetherification may penetrate the ring and be recapped in the rotaxane conformer. With the chain equipped with two arylsulfonyl units to enable alkylation with an ω-bromoalkyne at two ends of the rotaxane, a catenane can be constructed by an intramolecular Glaser coupling of the diyne.

RSO_2-$(CH_2)_{20}$-SO_2R + $(CH_2)_{28}$ ⇌ (TsOH) RSO_2CH_2—$(CH_2)_{18}$—CH_2SO_2R / $(CH_2)_{28}$

R = —C_6H_4—$(CH_2)_{11}OCPh_3$

RSO_2CH—$(CH_2)_{28}$—$CHSO_2R$ / $(CH_2)_{11}$ $(CH_2)_{28}$ $(CH_2)_{11}$ ← ($Cu(OAc)_2$, py) RSO_2CH—$(CH_2)_{28}$—$CHSO_2R$ / $(CH_2)_{11}$ $(CH_2)_{28}$ $(CH_2)_{11}$

Deliberate isomerization of the major *trans*-7,7-dimethylbicyclo[4,2,0]octan-2-one in a mixture obtained from [2 + 2]-photocycloaddition to the *cis*-isomer before proceeding further in a synthesis of the caryophyllenes [Corey, 1964a] was a well-conceived tactical decision. It was considered that upon establishment of the 4:9-fused ring the stability relationship between the isomeric ketones would reverse.

KOH
MeOH
OH
OTs
caryophyllene
OH
OTs
isocaryophyllene

The synthesis of aspidospermine [Stork, 1963] via Fischer indolization of a tricyclic aminoketone could be accomplished without paying attention to the relative stereochemistry at the ring junctures. Because of the equilibration at these carbon sites during the indolization, by way of a reversible Mannich reaction, the most stable arrangement of the three contiguous asymmetric centers which corresponds to that of the alkaloid, would result.

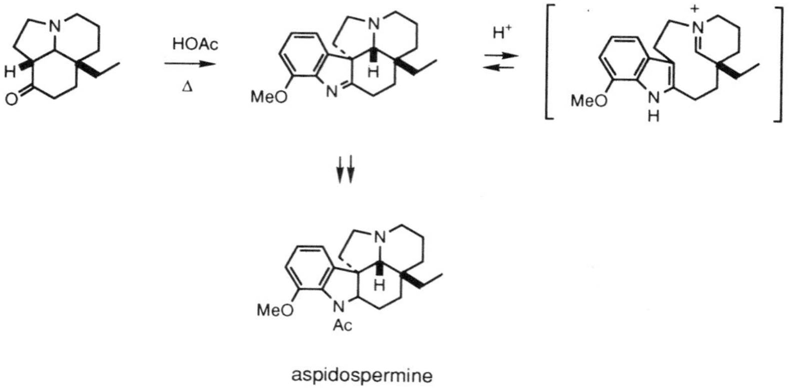

HOAC
Δ
MeO
H^+
aspidospermine

Almost all synthetic efforts towards naturally occurring spiroketals involved equilibration conditions to set up the structural unit, whose stability is the outcome of an interplay of anomeric and steric effects [Boivin, 1987]. An example shown below concerned a synthesis of calcimycin [Evans, 1979].

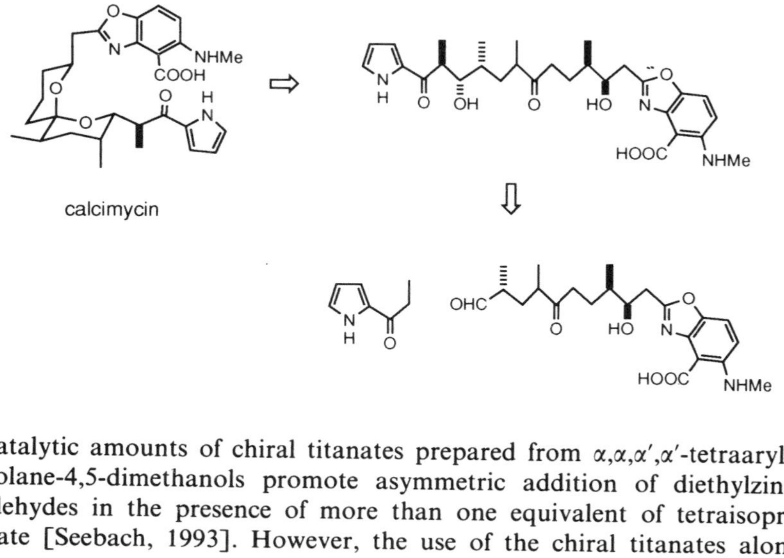

calcimycin

Catalytic amounts of chiral titanates prepared from $\alpha,\alpha,\alpha',\alpha'$-tetraaryl-1,3-dioxolane-4,5-dimethanols promote asymmetric addition of diethylzinc to araldehydes in the presence of more than one equivalent of tetraisopropyl titanate [Seebach, 1993]. However, the use of the chiral titanates alone in equimolar (or excess) quantities gives much inferior results. It has been suggested that the most active and desirable catalysts are those in which the titanium atom is surrounded by one chiral dioxide unit and two isopropoxy groups. Tetraisopropyl titanate serves as a cleansing agent, reconstituting the desired catalysts from the less selective species.

The access to both enantiomers of 1,7-dioxaspiro[5.5]undecane from a chiral sulfinyltetrahydropyran [Iwata, 1985b] exploited kinetic cyclization which led to one enantiomer bearing an axial sulfinyl group. Isomerization of this compound also caused configuration inversion of the spirocyclic center allowing entry into the enantiomeric series.

(R) (S)

1,7-dioxaspiro[5.5]undecane

Diastereomeric 2,3-disubstituted *N,N*-diethyl-4-pentenamides can be acquired as major products from allylic alcohols and *N,N*-diethylamino-1-propyne in the presence or absence of an acid catalyst [Bartlett, 1979]. The addition of the alcohol to the keteniminium species from the side away from the methyl group would lead to the (*E*)-ketene *N,O*-acetal which undergoes Claisen rearrangement to give the *threo*-product. On the other hand, the presence of an acid allows isomerization of the initial adduct to the more stable (*Z*)-isomer before rearrangement, eventually resulting in the *erythro*-amide.

A remarkable method for the synthesis of certain rotanes [Ashton, 1993] by heating a mixture of a 4,4′-bipyridinium salt and a crown ether capitalized on spatial donor-acceptor interactions between the aromatic moieties of the components. When established upon slippage of the rod-like component through the macrocycle, the electronic interactions provide a thermodynamic trap which largely prevents disengagement of the complex.

It is inappropriate to be too enthusiastic in holding a belief that equilibration and isomerization are panaceas to synthetic problems. On the contrary, stereochemistry often rears its ugly head [Ireland, 1969] in a complex synthesis, such as was confronted in an approach to lycopodine [Wiesner, 1968]. The generation of epilycopodine must have been a disappointment.

9.2. ALLOSTERIC CONTROL

Although this aspect of control by distant elements as applied to synthesis has received attention in sporadic places of this book, control which does not fall neatly into those categories is delineated in this section.

The alkyne linkage occupies a very small space and therefore it exerts practically no steric effects on its vicinity. It is no surprise that alkynyl cyclopropyl carbinols give a mixture of (*E*)-and (*Z*)-isomers on being subjected to the Julia cleavage. However, the (*E*)-enynes are obtained when the carbinols are converted into the dicobalt octacarbonyl adducts before treatment with hydrobromic acid (and subsequent demetallation with ferric ion) [Descoins, 1976].

R' R OH HBr ZnBr2 R R' Br + R' R Br

Co2(CO)8 Fe3+

R' R OH (OC)3Co Co(CO)3 HBr ZnBr2 R' R Br (OC)3Co Co(CO)3

While 1,2-asymmetric induction is most effective, it is interesting to note that regiochemistry of a reaction can be determined by a remote substituent. For example, the Dieckmann cyclization of a diester derived from cleavage of the Diels–Alder adduct of butadiene and angelicalactone proceeded unidirectionally (of course the reaction is reversible and thermodynamically controlled) [Taub, 1970]. Apparently the secondary methyl and the carbonyl groups at the subangular positions played a dominant role such that the product has the least nonbonding interactions. This ketoester was used in a synthesis of the prostaglandins.

O O COOMe COOMe O O tAmONa O COOMe O O

O MeOOC O O

In an approach to vancomycin the biaryl subunit was united by oxidative coupling with VOF_3 which required an additional oxygen functionality for activation and regiocontrol. However, the presence of this group caused the formation of the "unnatural" atropoisomer due to its induction of $A^{1,3}$-strain with the benzylic amido substituent. Fortunately, studies have shown [Evans, 1993] that the natural conformer would definitely emerge upon removal of the extra group and establishment of the bridge containing the nearby diaryl ether.

9.3. BIOMIMETIC TACTICS

The wonder of nature's capability of producing the great variety of carbogens has been a paradigm for the chemist to emulate. At the structural determination of a natural product the chemist would ponder its genesis and consider its synthesis in the laboratory. If a crucial intermediate could be persuaded to undergo reaction similar to that which is conjectured to be used by nature, high efficiency and selectivity would be expected [van Tamelen, 1961]. Furthermore, the reaction conditions would be relatively mild.

The pioneering work based on this principle is a synthesis of tropinone [Robinson, 1917] by combining succindialdehyde, methylamine, and acetone (greatly improved by replacing acetone with acetonedicarboxylic acid, which requires the further operation of acidification and thermal decarboxylation).

CHO CHO + $MeNH_2$ + COOH =O COOH → $(-CO_2)$ NMe =O

tropinone

It is apparent that the Daphniphyllum alkaloids are derived in nature from an oxidized squalene and ammonia. Upon preparation of dihydrosqualenedialdehyde, the biogenetically pattern formation of protodaphniphylline was indeed realized when it was treated with ammonia in the presence of triethylamine, and acetic acid in succession [Heathcock, 1992]. Similar results were obtained from either the (*E*)- or the (*Z*)-aldehyde, but the use of methylamine instead of ammonia increased the yield of the tetracyclic base by more than fourfold, presumably due to the more nucleophilic *N*-methylenamine intermediate. Note the product is dihydrodaphniphylline.

X X $\xrightarrow[HOAc]{NH_3}$ [N → N] → HN

X = CHO

protodaphniphylline

The fascinating group of indole alkaloids having the strychnos, iboga, and aspidosperma skeletons are biogentically related. The conversion of stemmadenine into tabersonine and catharanthine [A.I. Scott, 1975], via the hypothetical dihydrosecodine intermediate gives credence to the relationship. The independently synthesized secodine-like structure furnished *in situ* vincadifformine [Kuehne, 1978, 1979a] and the generation of andranginine by thermolysis of precondylocarpine acetate [Kan-Fan, 1974] is obviously the result of a retro-Mannich fission, isomerization to a vinyldihydropyridine, and an intramolecular Diels–Alder reaction. The last process presumably is involved in the biogenesis of the alkaloid.

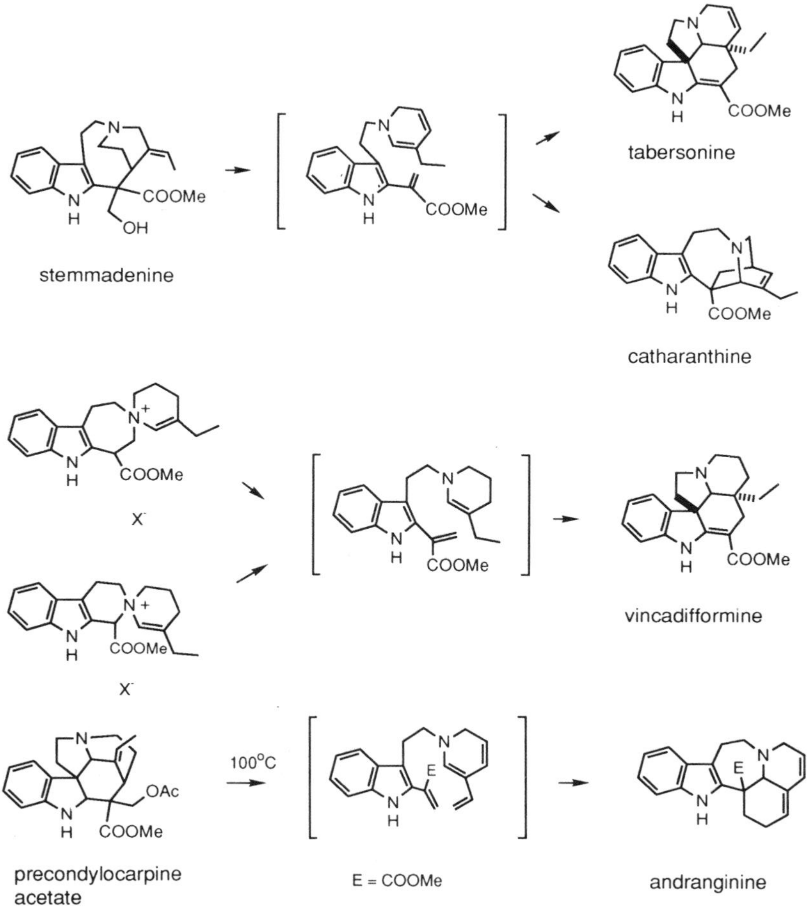

stemmadenine

tabersonine

catharanthine

vincadifformine

100°C

precondylocarpine acetate

E = COOMe

andranginine

Surprisingly numerous natural products are (obviously) derived from the Diels–Alder reaction, the evidence of which is particularly convincing when both the addends and the adducts are isolated from the same source. The coexistence of piperstachine and cyclostachine-A [Joshi, 1975] inspired a synthesis of the latter compound from an alkatrienoic ester. Notwithstanding the lack of diastereocontrol in the Diels–Alder reaction which produced both *exo*- and *endo*-isomers, the *endo* product was transformed into cyclostachine-A.

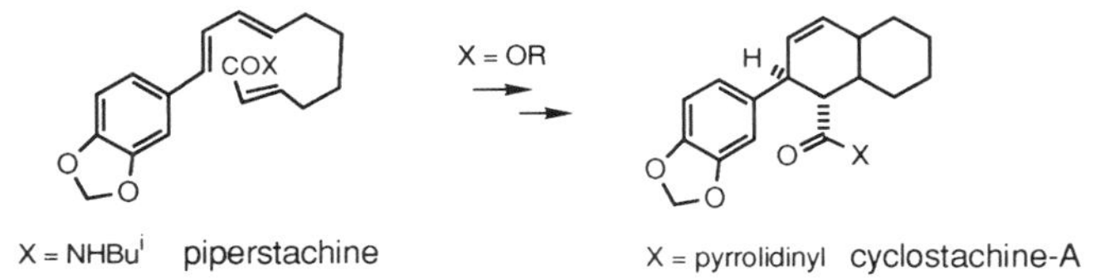

X = OR

X = NHBu^i piperstachine

X = pyrrolidinyl cyclostachine-A

A convenient synthesis of lachnanthocarpone consisted of oxidation of a catechol with sodium periodate at room temperature [Bazan, 1978]. The *o*-quinone underwent intramolecular Diels–Alder reaction immediately.

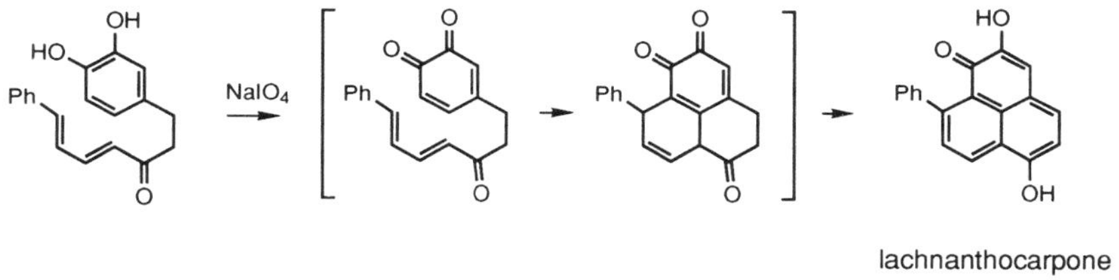

$NaIO_4$

lachnanthocarpone

The fungal pigments badione-A and -B are thought to arise by oxidative dimerization of pulvinic acids. A model study which gave the naphtho[1,8-*bc*] pyrandione core [Steglich, 1985] seemed to bear out the hypothesis.

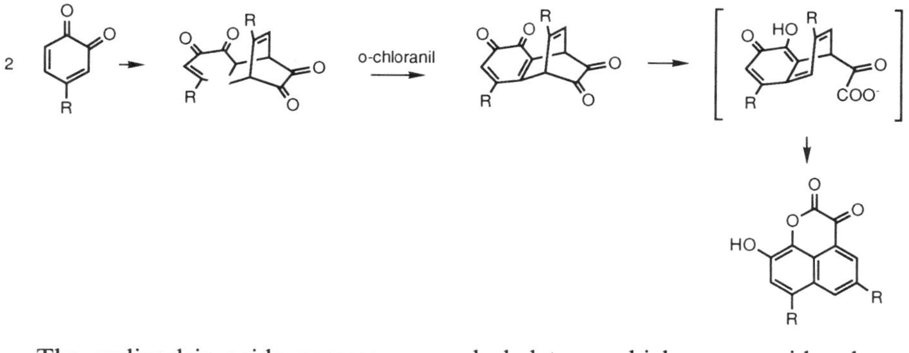

o-chloranil

The endiandric acids possess unusual skeletons which are considered as arising from aliphatic precursors by a series of electrocyclization and Diels–Alder reactions. A biomimetic synthesis of their methyl esters has been accomplished [Nicolaou, 1982b,d]. When the proper polyenediynes were submitted to Lindlar hydrogenation, products containing the bicyclo[4.2.0]octane framework were detected. On heating to 100°C the tetracyclic product in one series proved to be the methyl ester of endiandric acid-A and two compounds derived from the ethenolog correspond to acid-B and acid-C. It is remarkable that the thermal transformation resulted the generation of four rings and eight asymmetric centers.

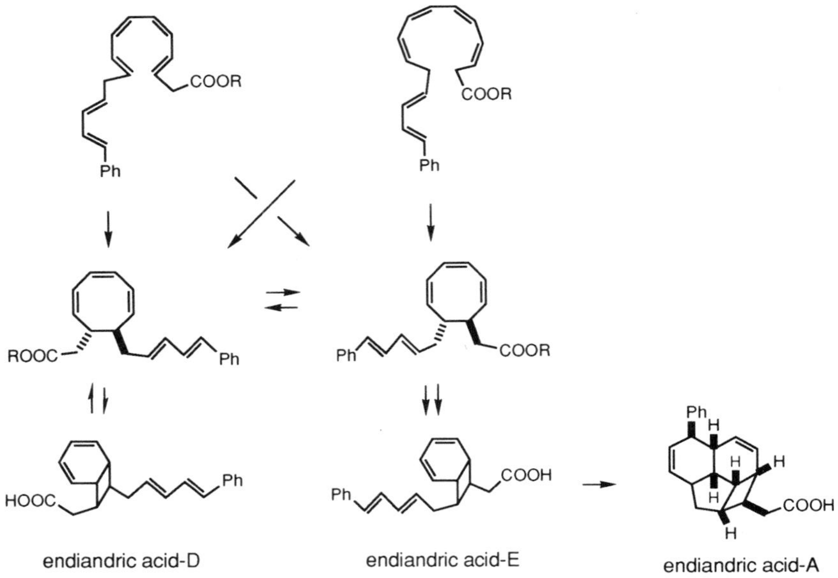

endiandric acid-D

endiandric acid-E

endiandric acid-A

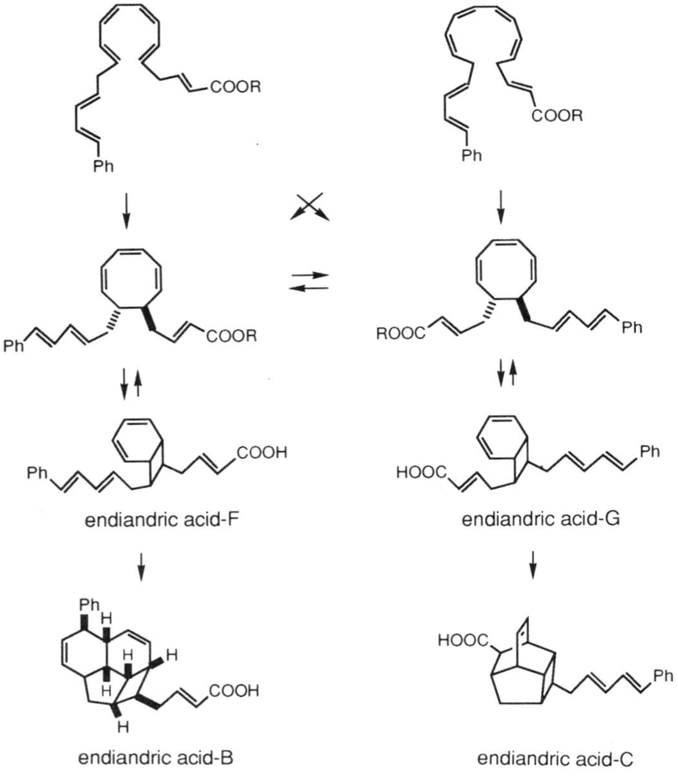

endiandric acid-F

endiandric acid-G

endiandric acid-B

endiandric acid-C

Alkaloids including morphine are biosynthesized by oxidative phenolic coupling. A simper type is the dibenzoazonines which are plausible intermediates on the biosynthetic pathway to erythrina alkaloids. Laboratory synthesis of erysodienone [Gervay, 1966; Mondon, 1966] by such a method has been demonstrated. Note a desymmetrization by an intramolecular Michael addition.

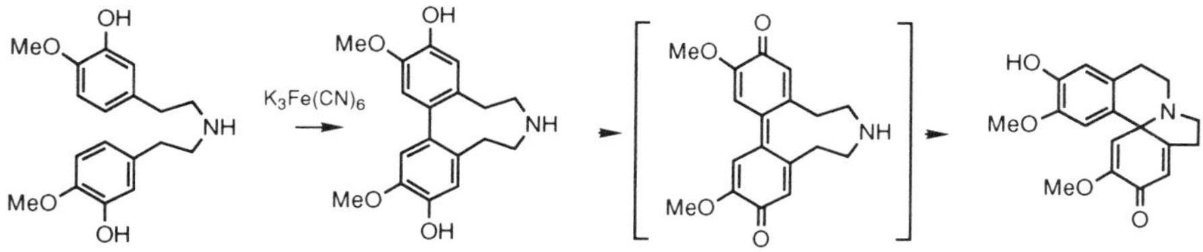

$K_3Fe(CN)_6$

erysodienone

Besides alkaloids the formation of many other families of compounds also involve phenolic coupling, and therefore they are amenable to biomimetic synthesis, an example of which is usnic acid [Barton, 1956] from methylphloracetophenone.

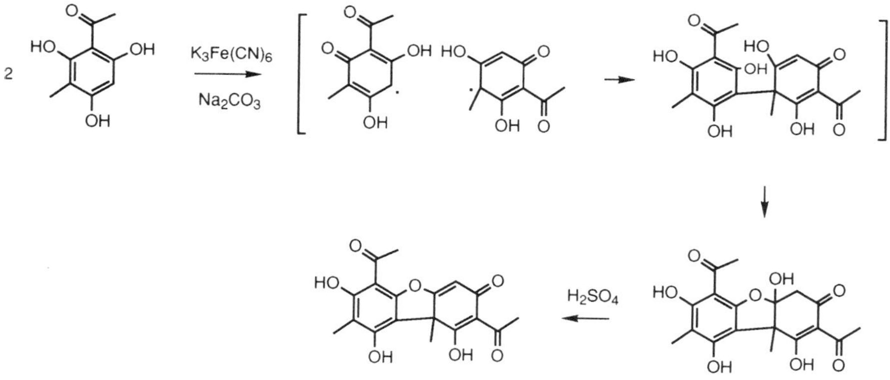

$K_3Fe(CN)_6$, Na_2CO_3

H_2SO_4

usnic acid

Extensive cationic reactions are involved in terpene biogenesis. The heroic efforts of W.S. Johnson in effecting cyclization of polyene substrates have borne abundant fruits. Much simpler but interesting transformations include the hydration with ring opening that converted thujopsene into widdrol [Dauben, 1972], cyclization of a spirocyclic precursor (acorane type) to give cedrene [Corey, 1969b; Crandall, 1969], preparation of isolongiborneol and its rearrangement into longifolene [Kuo, 1988], and the generation of patchouli alcohol by epoxidation of α-patchoulene [Büchi, 1961, 1964]. (For a documentation of biogenetic-type rearrangements of terpenes, see Coates [1976].)

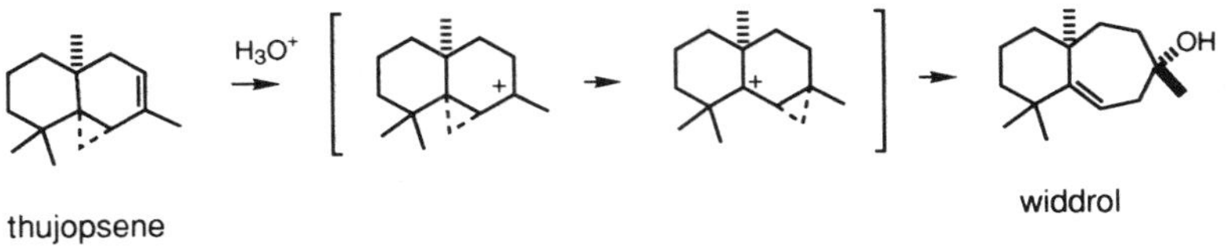
H3O+
thujopsene
widdrol

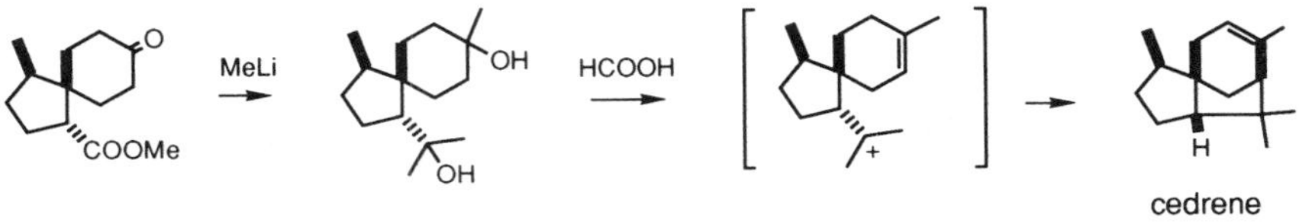
MeLi
HCOOH
COOMe
OH
OH
H
cedrene

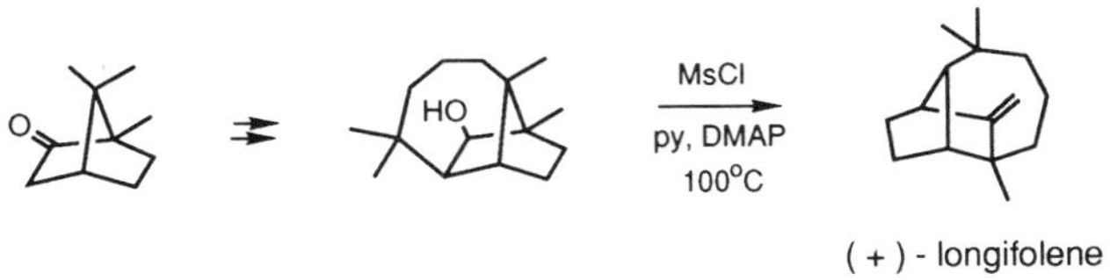
MsCl
py, DMAP
100°C
HO
(+) - longifolene

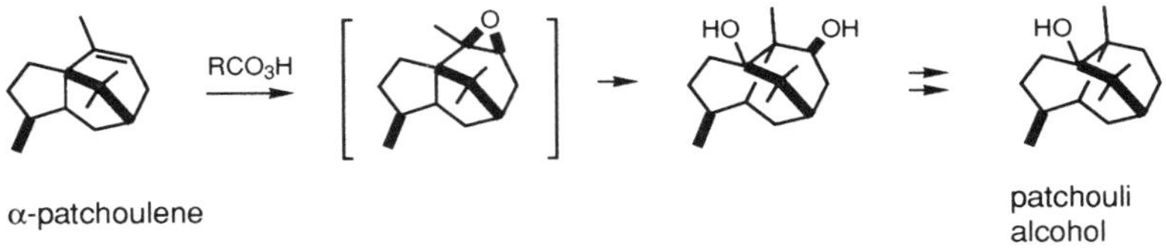
RCO3H
HO
OH
HO
α-patchoulene
patchouli
alcohol

9.4. DIVERGENCY

Convergent strategies of synthesis and enantioconvergency were discussed at the beginning of this book and in this chapter, respectively. It is appropriate to conclude the volume on a note of divergency.

Divergent synthesis is valuable but not always feasible. In this respect the chemist is hopeless in mimicking nature which has shown marvellous economy in biosynthesis. In other words, nature abhors waste and she directs the production of a great variety of substances from some common intermediates. The key to such a success is the availability of enzymes.

Enzymes can differentiate enantiotopic groups and desymmetrize *meso* compounds. Selective reaction of a C_2-symmetric compound so that the identical half of the molecule remains unperturbed represents a powerful technique of desymmetrization. As an illustration the conversion of a diol derived from the acetonide of dialkyl D-tartrate to several natural products [Kotsuki, 1990] is instructive. Sequential protection/activation of the two hydroxyl groups by tosylation and triflation accomplished this goal. A Grignard reagent in the presence of copper(I) bromide effects displacement of the triflate, whereas a diorganocuprate reagent prefers reaction at the tosyloxy carbon.

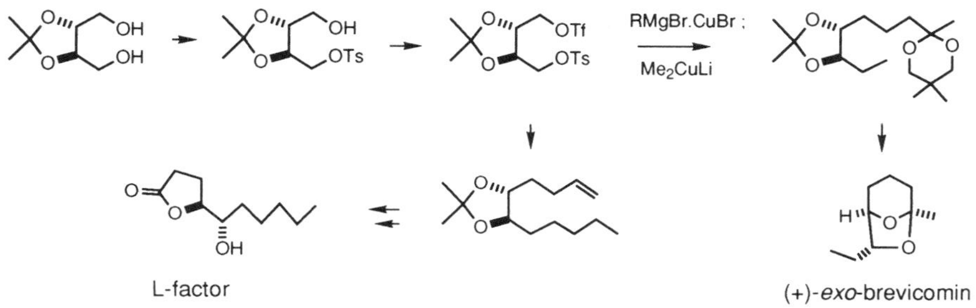

RMgBr.CuBr ; Me_2CuLi

L-factor

(+)-*exo*-brevicomin

A C_2-symmetric dienophile prepared from D-mannitol reacts with cyclopentadiene to give two diastereomeric adducts in a 3:2 ratio, having all (*S*) and (*S*,*S*; *R*,*S*) configurations respectively at the new asymmetric centers [Takano, 1987b]. Both compounds are formed via anti-Felkin–Ahn transition states, so as to avoid the considerable dipole–dipole and nonbonding interactions.

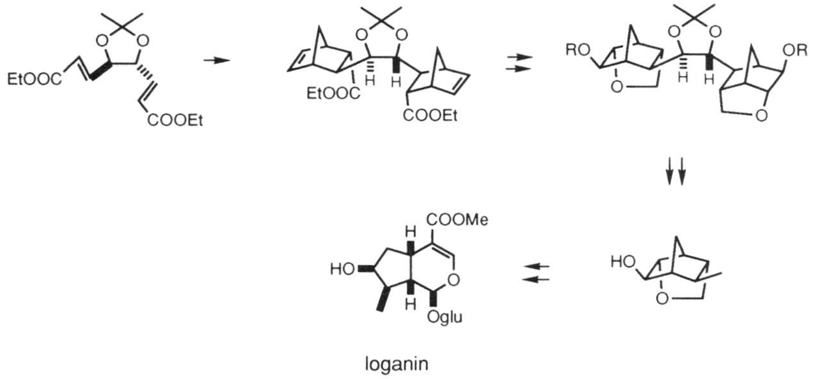

loganin

The diester derived from dimethyl (2*R*,3*R*)-*O*-isopropylidenetartrate also underwent cyclopropanation with Wittig reagent. The product was readily converted into natural *trans*-chrysanthemic acid [Krief, 1988].

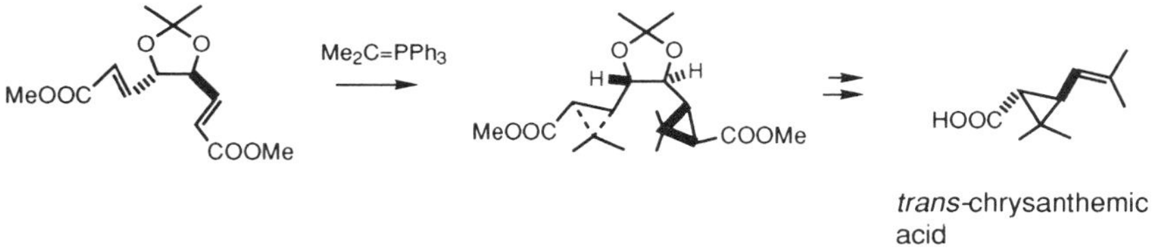

$Me_2C{=}PPh_3$

trans-chrysanthemic acid

An analysis of the tricarbocyclic portion of ikarugamycin [Whitesell, 1987] indicated the existence of pseudosymmetry elements. In synthesis both enantiomers of a bicyclo[2.2.0]octadienecarboxylic ester were used and the two halves were joined and electrocyclized under photochemical conditions. The stereochemistry of the cyclohexene ring was later adjusted.

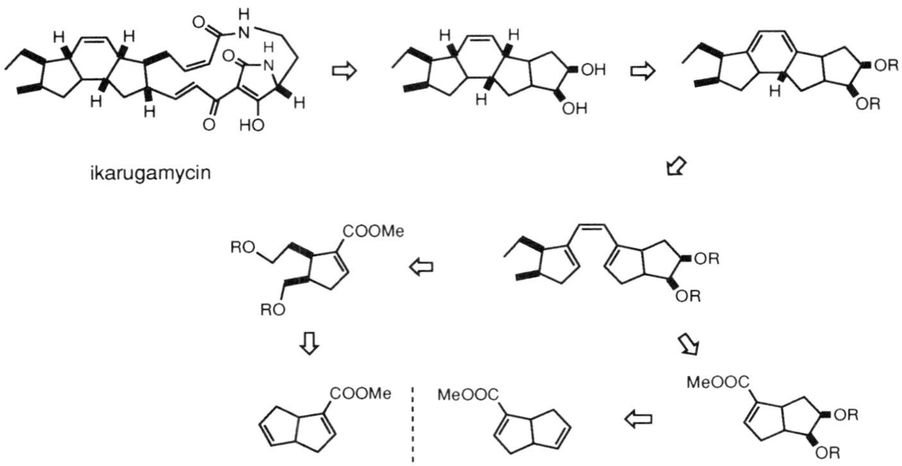

ikarugamycin

A related aspect of great synthetic expediency is what one can call "point divergency". Steric effects often play a role in the segregation of two identical groups, as shown in a synthesis of aldosterone [Heusler, 1959]. Here the elements for the D-ring and the angular aldehyde group were each provided by methallyl group, conveniently introduced by dialkylation of the tricyclic ketone. Differentiation of the two chains was aided by participation of the axial hydroxyl group at C-11 (steroid numbering).

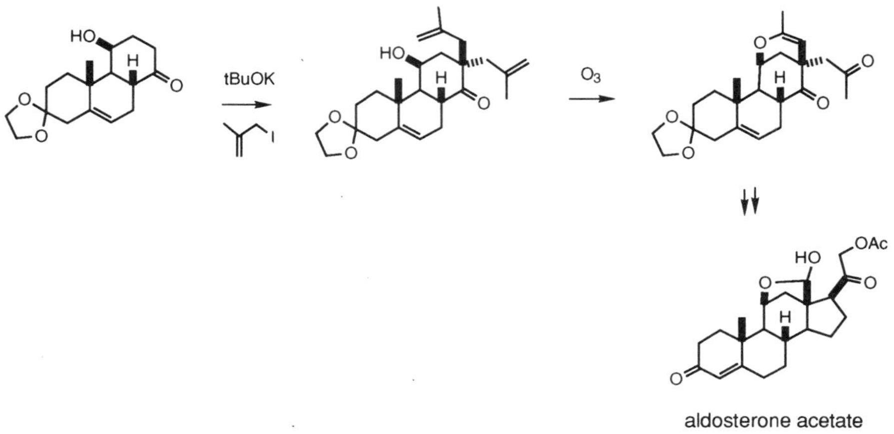

tBuOK

O_3

aldosterone acetate

When a compound containing a single asymmetric center to serve as reference point is converted into two diastereomers, inversion of the reference center in one of them would lead to an enantiomeric substance. Through such operations two optical isomers of a synthetic target molecule can be prepared. In such a manner, the Geissman–Waiss lactone, a useful intermediate in the synthesis of pyrrolizidine alkaloids, can now be obtained [Thaning, 1990].

MeO OR MeOOC N OR MeOOC N OR MeOOC N O O H H R' N (+)- Geissman-Waiss lactone O O H H R' N (-)- R' = H

It is also effective to use an intermediate divergently and converge the individual derivatives at a later stage. This tactic was practiced in a construction of the C-1/C-12 segment of amphotericin-B [Solladie, 1987].

BnO O HO BnO S S RO OH BnO S S O O BnO CHO K_2CO_3 O O BnO CHO O O BnO S S OH O O TfO OR O O S S O O BnO OH OR

Biogenetic considerations of binaphthyl alkaloids have guided the synthesis of both the naphthalene unit and the isoquinoline unit of ancistrocladeine [Bringmann, 1982] from an indane precursor.

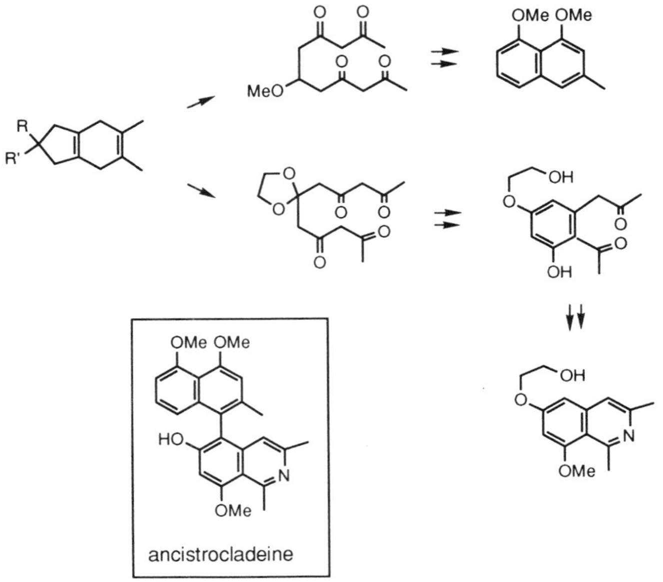

ancistrocladeine

Chiral compounds become useful building blocks (chirons) for synthesis only when they match structural (most importantly configurational) requirements with target molecules. To make such connections a chemist must possess adequate knowledge and ingenuity. The ability to design ways to employ a building block to construct different targets or a pair of enantiomeric targets is a sign of a versatile chemist.

(*S*)-4-Trityloxy-4-butanolide which is readily obtained from (*S*)-glutamic acid or D-mannitol is an extremely useful chiral building block. The large substituent directs the stereochemical course of reactions at both C-2 and C-3. It is of particular interest to note the development of synthetic schemes of (−)-velbanamine [Takano, 1980b], (+)-velbanamine [Takano, 1982b], (+)-quebrachamine [Takano, 1980a], and (−)-quebrachamine [Takano, 1981c] based on this substance.

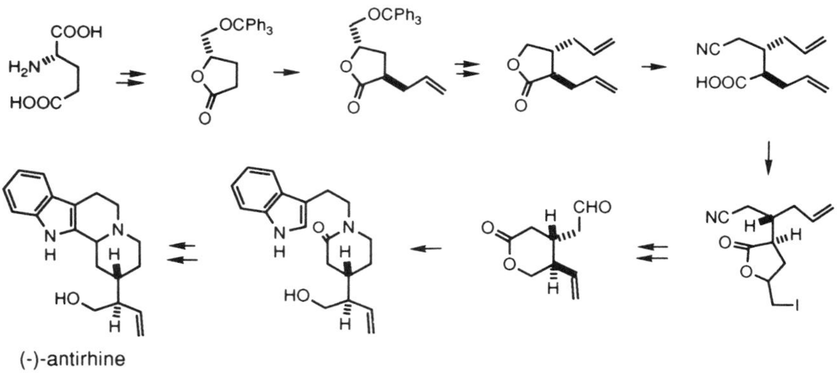

(-)-antirhine

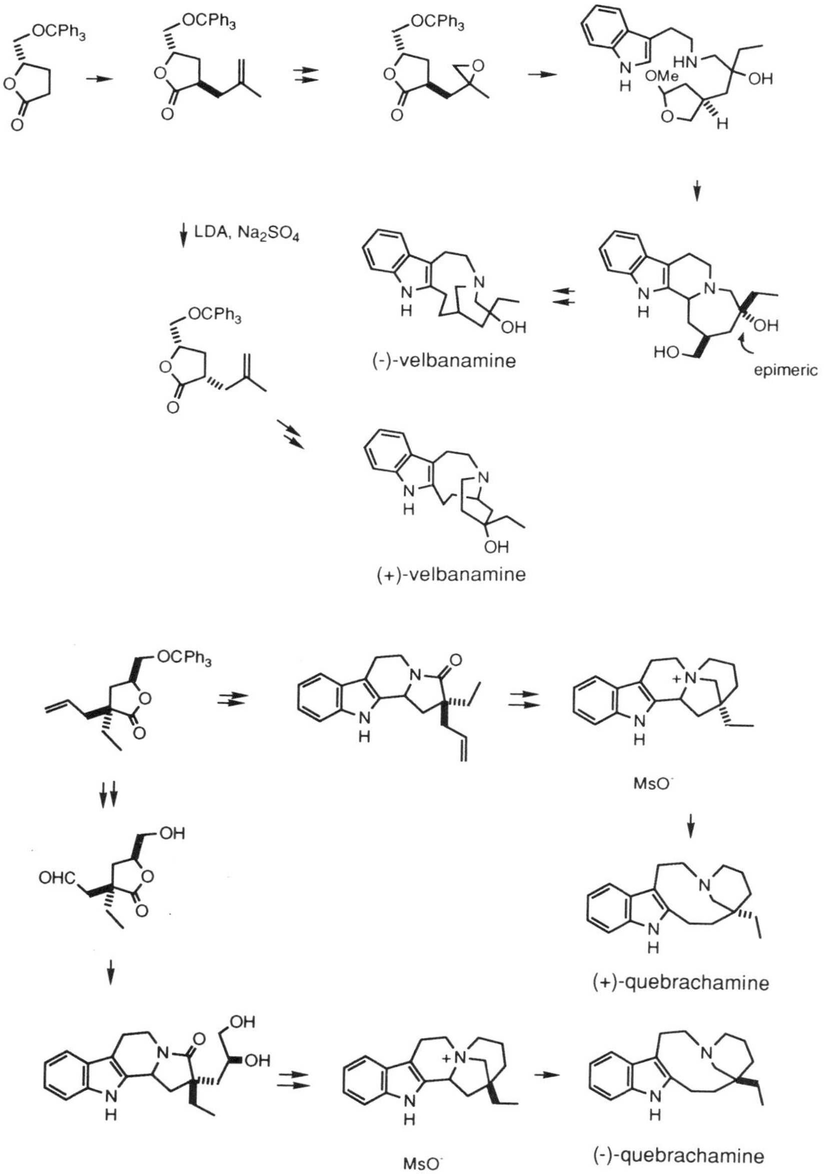

LDA, Na_2SO_4

(-)-velbanamine

epimeric

(+)-velbanamine

(+)-quebrachamine

(-)-quebrachamine

Other utility of the chiral butanolide includes access to both *syn*- and *anti*-2,6-dimethyl-1,7-heptanediol monosilyl ethers [Hanessian, 1985a], differentiated ethers of *syn*- and *anti*-3,5-dimethyl-1,7-heptanediols [Hanessian, 1985b], and monoethers of all-*syn* and *syn,anti*-1,2,4,6,7-heptanepentaols [Hanessian, 1985c].

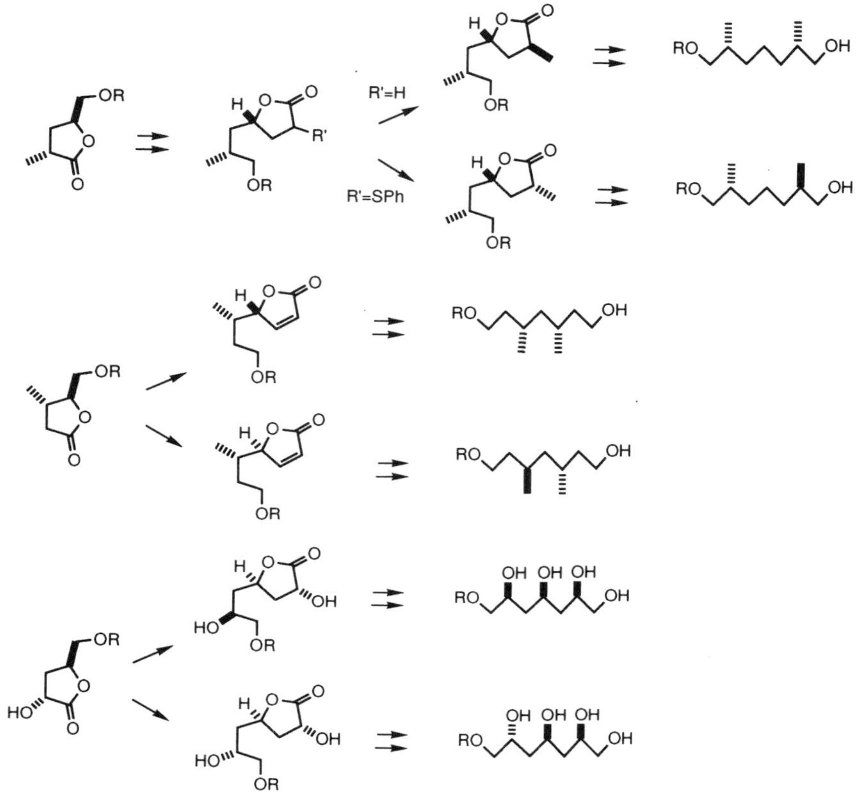

(*S*)-5-Hydroxymethylbuten-2-olide undergoes Diels–Alder reaction with cyclopentadiene. Conversion of the adduct into both enantiomers of β-santalene and epi-β-santalene has been effected [Takano, 1987a].

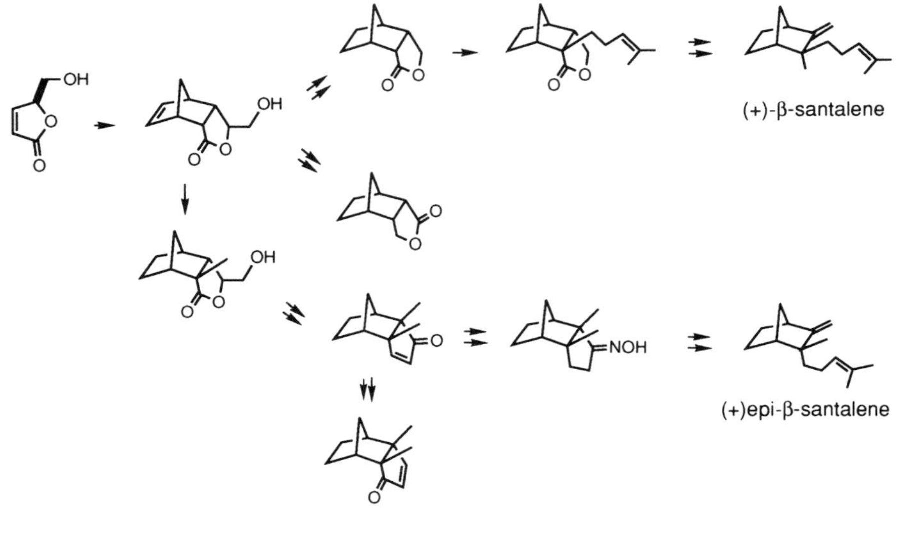

(+)-β-santalene

(+)epi-β-santalene

Sequential allylation and hydroxylation at the α-position of (*S*)-4-trityloxy-4-butanolide, followed by reduction of the lactone carbonyl group to a methyl led to a chiral compound in which the quaternary carbon atom is linked to two different latent acetic acid chains [Takano, 1984]. Accordingly, the synthesis of both (+)- and (−)-mevalonolactone is a matter of functional manipulation. Another enantiodivergent approach made use of an unsaturated δ-lactone [Takano, 1990].

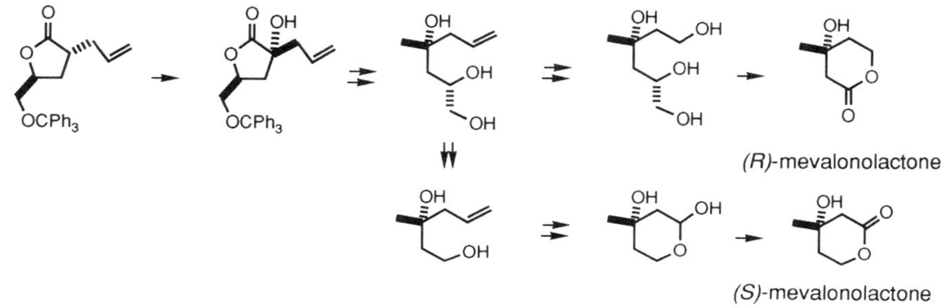

(*R*)-mevalonolactone

(*S*)-mevalonolactone

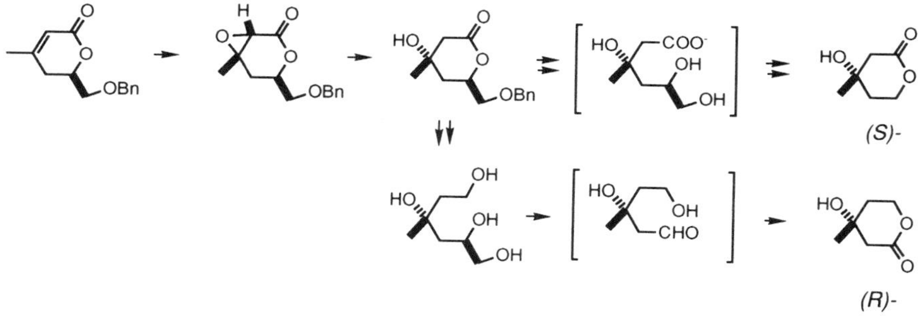

(*S*)-

(*R*)-

Enantiomeric indolizidinediols have been prepared from a common intermediate derived from D-isoascorbic acid [Heitz, 1989]. One enantiomer was acquired via cyclization to a γ-lactam and conversion of the latter into an iminium ion to be intercepted by a vinylsilane, whereas in reaching the other enantiomer the γ-hydroxy-γ-lactam was the source of the (acyl)iminium species for inducing the cyclization.

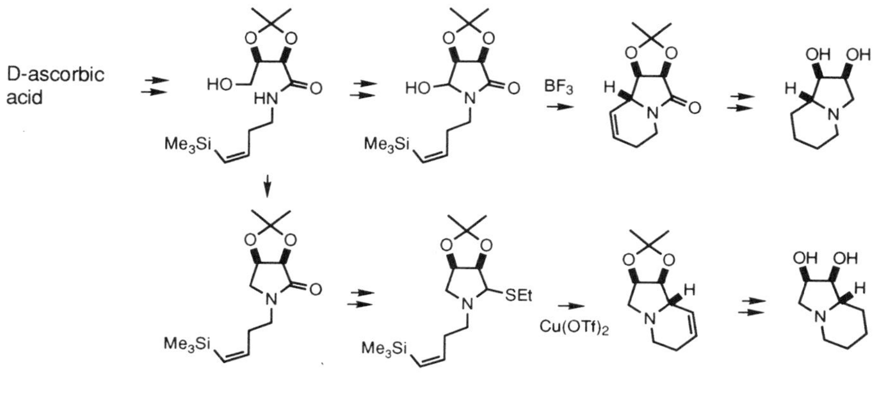

D-ascorbic acid

A bicyclic aminal obtained from (*S*)-proline gives various ketones on reaction with Grignard reagents at its ester group. These ketones can undergo Grignard reaction a second time, and on acid hydrolysis liberate chiral α-hydroxyaldehydes. By variation of the order of the two Grignard reaction steps, enantiomeric products are generated. The synthesis of (*R*)- and (*S*)-frontalin [Mukaiyama, 1979; Sakito, 1979] by this tactic has been realized.

MeOOC — NPh; RMgBr; MeMgBr; NPh; MeMgBr ; HCl; CHO; OH; (*S*)- frontalin; RMgBr ; HCl; CHO; OH; (*R*)- frontalin

The acetonide of (4*S*,5*R*)-dihydroxy-3-methoxy-2-cyclopentenone can be persuaded to undergo chemoselective reactions. Consequently, enantiomeric 3-alkyl-4,5-dihydroxy-2-cyclopentenones are accessible [Bestmann, 1986].

COOH; COOMe; PPh_3; COOMe; Δ; PhMe; PPh_3; COOMe; R_2CuLi; RLi; R; OMe; R; (*R*,*R*)-; (*S*,*S*)-

An enantiodivergent approach to (+)- and (−)-eburnamonine [Hakam, 1987] consisted of cyclopropane opening by different nucleophiles. With cyanoacetic ester anion as reagent the original lactam must be cleaved to release carbon chain that would join the indolic nitrogen atom. On the other hand, the chain created by cyanide ion attack on the cyclopropane was directly involved in the lactamization with the indole moiety (inversion at C-3 accompanied the latter process), and the remaining task concerned with ring expansion.

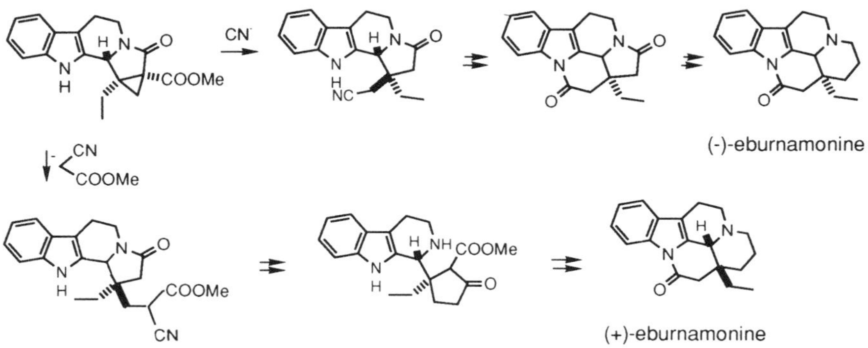

The convenient desymmetrization of *meso*-diesters by means of enzymatic reactions was mentioned in the last chapter. The (*R*)(+)-2-methyl-2-*p*-tolylmalonic acid monomethyl ester, obtained from a selective hydrolysis, possesses two readily differentiable functional groups and is therefore convertible into (+)-α-cuparenone, (−)-*β*-cuparenone, and (+)-*β*-cuparenone [Canet, 1992].

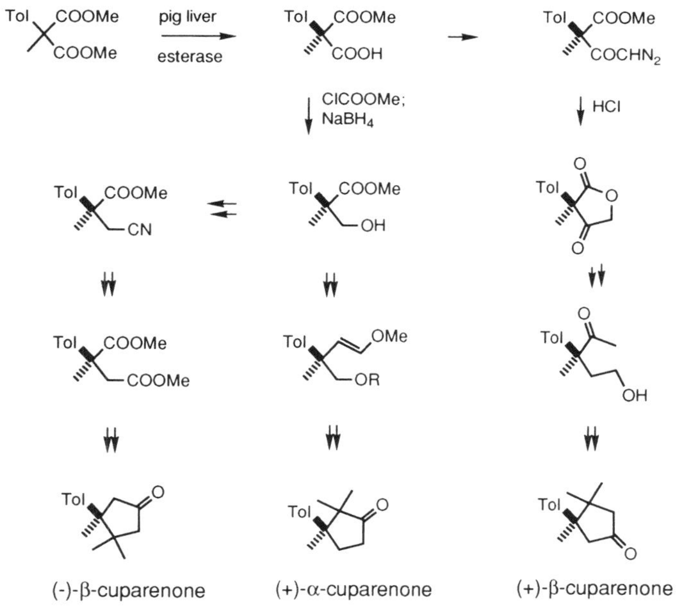

Whenever synthesis of one particular chiral target from an arbitrary enantiomer of a building block becomes feasible, the situation is highly opportune. This possibility has been clearly demonstrated in the elaboration of protoemetine from norcamphor [Takano, 1982a], although the model study (two routes by exchanging reactivity of two sidechains) was performed using racemic material.

protoemetine

Structural units corresponding to (+)- and (−)-nonactic acid series are accessible from an open-chain *syn*-1,3-dihydroxy ketoester [Bartlett, 1984b]. The formation of a crucial chiral center by cyclization, using the ketone carbonyl (inversion) or the secondary hydroxyl group (retention) as the nucleophile, determines the two other chiral centers at C-5 of the tetrahydrofuran ring and the adjacent carbon in the sidechain, as asymmetric induction in such systems is very effective. The two diastereomers differ in the configuration of the remaining hydroxyl group, the one leading to (−)-nonactic acid derivatives necessitates mesylation and S_N2 displacement.

(-)-nonactic acid deriv. (+)-nonactic acid deriv.

Diastereoselection during an intermolecular condensation depends on the configurational integrity of both reactants and a defined transition state. With such restrictions it is of interest to note that the exchange of chiral 1-carbamyloxyallyllithium with a ligand on certain achiral titanium reagents can lead to organotitanium species with retention or inversion of configuration [Krämer, 1987]. Excellent diastereoselectivities have been obtained from condensation of the resulting species with chiral aldehydes.

OH RO iPr$_2$N O O CHO RO Ti(OPr^i)$_4$ O iPr$_2$N O CHO RO OH RO O iPr$_2$N O

(iPrO)$_4$Ti

Li-TMEDA O iPr$_2$N O

(Et$_2$N)$_3$TiCl

OH RO O iPr$_2$N O CHO RO Ti(NEt$_2$)$_3$ O iPr$_2$N O CHO RO OH RO O iPr$_2$N O

REFERENCES

Abdallah, H., Grée, R., Carrie, R. (1982) *Tetrahedron Lett.* **23**: 503.

Abelman, M.M., Funk, R.L., Munger, J.D. (1982) *J. Am. Chem. Soc.* **104**: 4030.

Adams, C., Walker, F.J., Sharpless, K.B. (1985) *J. Org. Chem,* **50**: 420.

Agami, C., Meyner, F., Puchot, C., Guilhem, J., Pascard, C. (1984) *Tetrahedron* **40**: 1031.

Ahlbrecht, H. (1977) *Chimia* **31**: 391.

Ahlbrecht, H., Raab, W., Vonderheid, C. (1979) *Synthesis* 127.

Ahlbrecht, H., Bonnet, G., Enders, D., Zimmermann, G. (1980) *Tetrahedron Lett.* 3175.

Ahlbrecht, H., Sudheendranath, C.S. (1982) *Synthesis* 717.

Aicher, T.D., Buszek, K.R., Fang, F.G., Forsyth, C.J., Jung, S.H., Kishi, Y., Matelich, M.C., Scola, P.M., Spero, D.M., Yoon, S.K. (1992) *J. Am Chem. Soc.* **114**: 3162.

Albrecht, R., Tamm, C. (1957) *Helv. Chim. Acta* **40**: 2216.

Alexakis, A., Mangeney, P. (1990a) *Tetrahedron*: *Asymmetry* **1**: 477.

Alexakis, A., Sedrani, R., Mangeney, P. (1990b) *Tetrahedron Lett.* **31**: 345.

Alexakis, A., Lensen, N., Mangeney, P. (1991) *Tetrahedron Lett.* **32**: 1171.

Alexakis, A., Lensen, N., Transchier, J.-P., Mangeney, P. (1992) *J. Org. Chem.* **57**: 4563.

Ali, S.M., Finch, M.A.W., Roberts, S.M., Newton, R.F. (1980) *Chem. Commun.* 74.

Alonso, R.A., Burgey, C.S., Rao, B.V., Vite, G.D., Vollerthun, R. Zottola, M.A., Fraser-Reid, B. (1993) *J. Am. Chem. Soc.* **115**: 6666.

Altani, P.M., Biellmann, J.F., Dubé, S., Vicens, J.J. (1974) *Tetrahedron Lett.* 2665.

Amupitan, J.A., Huq, E., Mellor, M., Scovell, E.G., Sutherland, J.K. (1983) *J. Chem. Soc. Perkin Trans I* 747.

Anderson, H.L., Sanders, J.K.M. (1990) *Angew. Chem. Intern. Ed. Engl.* **29**: 1400.

Anderson, R.J., Henrick, C.A. (1975) *J. Am. Chem. Soc.* **97**: 4327.

Anderson, S., Anderson, H.L., Sanders, J.K.M. (1992) *Angew, Chem, Intern. Ed. Engl.* **31**: 907.

Anderson, S., Anderson, H.L., Sanders, J.K.M. (1993) *Acc. Chem. Res.* **26**: 469.

Andersson, C.-M., Larsson, J., Hallberg, A. (1990) *J. Org. Chem.* **55**: 5757.

Andres, C., Gonzalez, A., Pedrosa, R., Perez-Encabo, A., Garcia-Granda, S., Salvado, M.A., Gomez-Beltran, F. (1992) *Tetrahedron Lett.* **33**: 4743.

Angel, S.R., Arnaiz, D.O., (1989) *Tetrahedron Lett.* **30**: 515.

Anh, N.T. (1980) *Top. Curr. Chem.* **88**: 145.

Antczak, K., Kingston, J.F., Fallis, A.G. (1985) *Can. J. Chem*, **63**: 993.

Appel, R., Hünerbein, J., Knoch, F. (1983) *Angew. Chem. Intern. Ed Engl.* **22**: 61.

Arai, Y., Kontani, T., Koizumi, T. (1991) *Chem. Lett.* 2135.

Aratani, M., Dunkerton, L.V., Fukuyama, T., Kishi, Y., Kakoi, H., Sugiura, S., Inoue, S. (1975) *J. Org. Chem.* **40**: 2009.

Aratani, T., Yoneyoshi, Y., Nagase, T. (1975) *Tetrahedron Lett.* 1707; (1982) *Tetrahedron Lett.* **23**: 685.

Armer, R., Simkins, N.S. (1993) *Tetrahedron Lett.* **34**: 363.

Asaoka, M., Fujii, N., Takei, H. (1988a) *Chem. Lett.* 805.

Asaoka, M., Fujii, N., Takei, H. (1988b) *Chem. Lett.* 1665.

Asaoka, M., Shima, K., Fujii, N., Takei, H. (1988c) *Tetrahedron* 44: 4757.

Asaoka, M., Takenouchi, K., Takei, H. (1988d) *Tetrahedron Lett.* **29**: 325.

Asaoka, M., Takenouchi, K., Takei, H. (1988e) *Chem. Lett.* 921.

Asaoka, M., Takenouchi, K., Takei, H. (1988f) *Chem. Lett.* 1225.

Asaoka, M., Sonoda, S., Takei, H. (1989a) *Chem. Lett.* 1847.

Asaoka, M., Takei, H. (1989b) *Heterocycles* **29**: 243.

Asoaka, M., Sakurai, M., Takei, H. (1990a) *Tetrahedron Lett.* **31**: 4759.

Asaoka, M., Takei, H. (1990b) *J. Synth. Org. Chem. Jpn.* **48**: 216.

Asaoka, M., Sakurai, M., Takei, H. (1991) *Tetrahedron Lett.* **32**: 7567.

Ashton, P.R., Brown, G.R., Isaacs, N.S., Giuffrida, D., Kohnke, F.H., Mathias, J.P., Slawin, A.M.Z., Smith, D.R., Stoddart, J.F., Williams, D.J. (1992) *J. Am. Chem. Soc.* **114**: 6330.

Ashton, P.R., Belohradsky, M. Philp, D., Stoddart, J.F. (1993) *Chem. Commun.* 1269.

Attenburrow, J., Cameron A.F.B., Chapman, J.H., Evans, R.M., Hems, B.A., Jansen, A.B.A., Walker, T. (1952) *J. Chem. Soc.* 1094.

Auerbach, J., Weinreb, S.M. (1974) *Chem. Commun.* 298.

Auricchio, S., Morrocchi, S., Ricca, A. (1974) *Tetrahedron Lett.* 2793.

Ayer, W.A., Browne, L.M. (1974) *Can. J. Chem.* **52**: 1352.

Babudri, F., Fiandanese, V., Marchese, G. Naso, F. (1991) *Chem. Commun*, 237.

Bac, N.Y., Langlois, Y. (1982) *J. Am. Chem. Soc.* **104**: 7666.

Bac, N.Y., Fall, Y., Langlois, Y. (1986) *Tetrahedron Lett.* **27**: 841.

Baggiolini, E.G., Lee, H.L., Pizzolato, G., Uskokovic, M.R. (1982) *J. Am. Chem. Soc.* **104**: 6460.

Baker, R. (1973) *Chem. Rev.* **73**: 487.

Baker, R., Sims, R.J. (1981) *J. Chem. Soc. Perkin Trans. I* 3087.

Baker, R., Herbert, R., Parton, A.H. (1982) *Chem. Commun.* 601.

Bakuzis, P., Bakuzis, M.L.F., Weingartner, T.F. (1978) *Tetrahedron Lett.* 2371.

Bal, S.A., Marfat, A., Helquist, P. (1982) *J. Org. Chem*, **47**: 5045.

Balanson, R.D., Kobal, V.M., Schumaker, R.R. (1977) *J. Org. Chem.* **42**: 393.

Balci, M., Schalenbach, R., Vogel, E. (1981) *Angew. Chem. Intern. Ed. Engl.* **20**: 809.

Baldwin, J.E., Höfle, G.A., Lever, O.W. (1974) *J. Am. Chem. Soc.* **96**: 7125.

Baldwin, J.E., Bailey, P.D., Gallacher, G., Singleton, K.A., Wallace, P.M. (1983) *Chem. Commun.* 1049.

Baldwin, J.E., Jones, R.H., Najera, C., Yus, M. (1985) *Tetrahedron* **41**: 699.

Baldwin, S.W., Fredericks, J.E. (1982a) *Tetrahedron Lett.* **23**: 1235.

Baldwin, S.W., Landmesser, N.G. (1982b) *Tetrahedron Lett.* **23**: 4443.

Baldwin, S.W., Martin, G.F., Nunn, D.S. (1985) *J. Org. Chem.* **50**: 5720.

Ballini, R., Marcantoni, E., Petrini, M. (1992) *J. Org. Chem.* **57**: 1316.

Ban, Y., Yoshida, K. Goto, J., Oishi, T., Takeda, E. (1983) *Tetrahedron* **39**: 3657.

Barborak, J.C., Watts, L., Pettit, R. (1966) *J. Am. Chem. Soc.* **88**: 1328.

Barclay, L.R.C., Milligan, C.E., Hall, N.D. (1962) *Can. J. Chem.* **40**: 1664.

Barrett, A.G.M., Itoh, T., Wallace, E.M. (1993) *Tetrahedron Lett.* **34**: 2233.

Bartlett, P.A., Hahne, W.F. (1979) *J. Org. Chem.* **44**: 882.

Bartlett, P.A., Pizzo, C.F. (1981) *J. Org. Chem.* **46**: 3896.

Bartlett, P.A. (1984a) in Morrison, J.D. (ed.) *Asymmetric Synthesis* **3B**: 411.

Bartlett, P.A., Meadows, J.D., Ottow, E. (1984b) *J. Am. Chem. Soc.* **106**: 5304.

Barton, D.H.R., Deflorin, A.M., Edwards, O.E. (1956) *J. Am. Chem. Soc.* 530.

Barton, D.H.R., Beaton, J.M. (1961) *J. Am. Chem. Soc.* **83**: 4083.

Bauermeister, H., Riechers, H. Schomberg, D., Washausen, P., Winterfeldt, E. (1991) *Angew, Chem. Intern. Ed. Engl.* **30**: 191.

Bauman, J.G., Hawley, R.C., Rapoport, H. (1984) *J. Org. Chem.* **49**: 3791.

Bazan, A.C., Edwards, J.M., Weiss, U. (1978) *Tetrahedron* **34**: 3005.

Beak, P., Carter, L.G. (1981) *J. Org. Chem.* **46**: 2363.

Beak, P., Snieckus, V. (1982) *Acc. Chem. Res.* **15**: 306.

Beckmann, M., Hildebrandt, H., Winterfeldt, E. (1990) *Tetrahedron: Asymmetry* **1**: 335.

Beddoes, R.L., Davies, M.P.H., Thomas, E.J. (1992) *Chem. Commun.* 538.

Bell, A., Davidson, A.H., Earnshaw, C., Norrish, H.K., Torr, R.S., Warren, S. (1978) *Chem. Commun.* 988.

Bell, R.A., Ireland, R.E., Partyka, R.A. (1966) *J. Org. Chem.* **31**: 2530.

Belokon, Yu. N., Zel'tzer, L.E., Ryzhov, M.G., Saporovskaya, M.B., Bakhmutov, V.I., Belikov, V.M. (1982) *Chem. Commun.* 180.

Ben, I., Castedo, L., Saa, J.M., Seijas, J.A., Suau, R., Tojo, G. (1985) *J. Org. Chem.* **50**: 2236.

Benson, W., Winterfeldt, E. (1979) *Angew, Chem. Intern. Ed. Engl.* **18**: 862.

Beresford, K.J.M., Howe, G.P., Procter, G. (1992) *Tetrahedron Lett.* **33**: 3355.

Bergman, J., Pelcman, B. (1989) *J. Org. Chem.* **54**: 824.

Berlage, U., Schmid, J., Peter, U., Welzel, P. (1987) *Tetrahedron Lett.* **28**: 3091.

Berson, J.A., Olin, S.S. (1969) *J. Am. Chem. Soc.* **91**: 777.

Bertrand, M., Gras, J.L. (1974) *Tetrahedron* **30**: 793.

Bessière, Y., Vogel, P. (1980) *Helv. Chem. Acta* **63**: 232.

Bestmann, H.J., Schobert, R. (1985) *Angew. Chem. Intern. Ed. Engl.* **24**: 790.

Bestmann, H.J., Moenius, T. (1986) *Angew, Chem. Intern. Ed. Engl.* **25**: 994.

Biellmann, J.F., Ducep, J.B. (1968) *Tetrahedron Lett.* 5629.

Biemann, K., Büchi, G., Walker, B.H. (1957) *J. Am. Chem. Soc.* **79**: 5558.

Bindra, J.S., Grodski, A., Schaaf, T.K., Corey, E.J. (1973) *J. Am. Chem. Soc.* **95**: 7522.

Birch, A.J., Macdonald, P.L., Powell, V.H. (1970) *J. Chem. Soc. C* 1469.

Bloch, R., Gilbert, L. (1987) *Tetrahedron Lett.* **28**: 423.

Bloch, E., Putman, D. (1990) *J. Am. Chem. Soc.* **112**: 4072.

Boeckman, R.K., Bruza, K.J. (1974) *Tetrahedron Lett.* 3365.

Boeckman, R.K., Blum, D.M., Arthur, S.D. (1979) *J. Am. Chem. Soc.* **101**: 5060.

Boeckman, R.K., Naegely, P.C., Arthur, S.D. (1980) *J. Org. Chem.* **45**: 752.

Boeckman, R.K., Cheon, S.H. (1983) *J. Chem. Soc.* **105**: 4112.

Boeckman, R.K., Arvanitis, A., Voss, M.E. (1989a) *J. Am. Chem. Soc.* **111**: 2737.

Boeckman, R.K., Springer, D.M., Alessi, T.R. (1989b) *J. Am. Chem. Soc.* **111**: 8284.

Boger, D.L., Weinreb, S.M. (1987) *Hetero, Diels–Alder Methodology in Organic Synthesis*, Academic Press, San Diego.

Boger, D.L., Zhang, M. (1991) *J. Am. Chem. Soc.* **113**: 4230.

Boivin, T.L.B. (1987) *Tetrahedron* **43**: 3309.

Bottaro, J.C., Berchtold, G.A. (1980) *J. Org. Chem.* **45**: 1176.

Bozzato, G., Bachmann, J.-P., Pesaro, M. (1974) *Chem. Commun.* 1005.

Brennan, J. (1981) *Chem. Commun.* 880.

Breslow, R., Wife, R.L., Prezant, D. (1976) *Tetrahedron Lett.* 1925.

Breslow, R., Corcoran, R.J., Snider, B.B., Doll, R.J., Khanna, P.L., Kaleya, R. (1977a) *J. Am. Chem. Soc.* **99**: 905.

Breslow, R., Maresca, L.M. (1977b) *Tetrahedron Lett.* 623.

Breslow, R. (1980) *Acc. Chem. Res.* **13**: 170.

Bringmann, G. (1982) *Angew. Chem. Intern. Ed. Engl.* **21**: 200.

Bringmann, G., Jansen, J.R., Rink, H.-P. (1986) *Angew. Chem. Intern. Ed. Engl.* **25**: 913.

Bringmann, G., Künkel, G., Geuder, T. (1990) *Synlett* 253.

Brocard, J., Mahmoudi, M., Pelinski, L., Maciejewski, L. (1989) *Tetrahedron Lett.* **30**: 2549.

Broussard, M.E., Juma, B., Train, S.G., Peng, W.J., Laneman, S.A., Stangley, G.G. (1993) *Science* **260**: 1784.

Brown, C.A., Yamaichi, A. (1979) *Chem. Commun.* 100.

Brown, P.J., Jones, D.N., Khan, M.A., Meanwell, N.A., Richards, P.J. (1984) *J. Chem. Soc. Perkin Trans. I* 2049.

Brown-Wensley, K.A., Buchwald, S.L., Cannizzo, L., Clawson, L., Ho, S., Meinhardt, D., Stille, J.R., Straus, D., Grubbs, R.H. (1983) *Pure Appl. Chem.* **55**: 1733.

Bruhn, J., Heimgartner, H., Schmid, H. (1979) *Helv. Chim. Acta* **62**: 2630.

Brunet, E., Batra, M.S., Aguilar, F.J., Garcia-Ruano, J.L. (1991) *Tetrahedron Lett.* **32**: 5423.

Brünjes, R., Tilstam, U., Winterfeldt, E. (1991) *Chem. Ber.* **124**: 1677.

Buchanan, G.L., Young, G.A.R. (1973) *J. Chem. Soc. Perkins Trans. I* 2404.

Büchi, G., Erickson, R.E., Wakabayashi, N. (1961) *J. Am. Chem. Soc.* **83**: 927.

Büchi, G., MacLeod, W.D., Padilla, J. (1964) *J. Am. Chem. Soc.* **86**: 4438.

Büchi,G., Coffen, D.L., Kocsis, K., Sonnet, P.E., Ziegler, F.E. (1966a) *J. Am. Chem. Soc.* **88**: 3099.

Büchi, G., Wüest, H. (1966b) *J. Org. Chem.* **31**: 977.

Büchi, G., Foulkes, D.M., Kurono, M., Mitchell, G.F., Schneider, R.S. (1967a) *J. Am. Chem. Soc.* **89**: 6745.

Büchi, G., Schneider, R.S., Wild, J. (1967b) *J. Am. Chem. Soc.* **89**: 2776.

Büchi, G., Wüest, H. (1969) *J. Org. Chem.* **34**: 1122.

Büchi, G., Kulsa, P., Ogasawara, K., Rosati, R. (1970a) *J. Am. Chem.* **95**: 799.

Büchi, G., Powell, J.E. (1970b) *J. Am. Chem. Soc.* **92**: 3126.

Büchi, G., Carlson, J.A., Powell, J.E., Tietze, L.-F. (1973) *J. Am. Chem. Soc.* **95**: 540.

Büchi, G., Fliri, H., Shapiro, R. (1978) *J. Org. Chem.* **43**: 4765.

Büchi, G., Chu, P.S. (1979) *J. Am. Chem. Soc.* **101**: 6767.

Büchi, G., Rodriguez, A.D., Yakushijin, K. (1989) *J. Org. Chem.* **54**: 4494.

Buchta, F., Merk, W. (1966) *Liebigs Ann. Chem.* **694**: 1.

Bunce, R.A., Schlecht, M.F., Dauben, W.G., Heathcock, C.H. (1983) *Tetrahedron Lett.* **24**: 4943.

Burdi, D., Hoyt, S., Begley, T.P. (1992) *Tetrahedron Lett.* **33**: 2133.

Burgstahler, A.W., Bithos, Z.J. (1959) *J. Am. Chem. Soc.* **81**: 503.

Burke, S.D., Armistead, D.M., Fevig, J.M. (1985) *Tetrahedron Lett.* **26**: 1163.

Burke, S.D., Cobb, J.E. (1986a) *Tetrahedron Lett.* **27**: 4237.

Burke, S.D., Schoenen, F.J., Murtiashaw, C.W. (1986b) *Tetrahedron Lett.* **27**: 449.

Burke, S.D., Chandler, A.C., Nair, M.S., Campopiano, O. (1987a) *Tetrahedron Lett.* **28**: 4147.

Burke, S.D., Schoenen, F.J., Nair, M.S. (1987b) *Tetrahedron Lett.* **28**: 4143.

Burke, S.D., Silks III, L.A., Strickland, S.M.S. (1988) *Tetrahedron Lett.* **29**: 2761.

Burkholder, T.P., Fuchs, P.L. (1990) *J. Am. Chem. Soc.* **112**: 9601.

Burton, A., Hevesi, L., Dumont, W., Cravador, A., Krief, A. (1979) *Synthesis* 877.

Bycroft, B.W., Lee, G.R. (1975) *Chem. Commun.* 988.

Cabal, M.P., Coleman, R.S., Danishefsky, S. (1990) *J. Am. Chem. Soc.* **112**: 3253.

Cacciapaglia, R., Casnati, A., Mandolini, L., Ungaro, R. (1992) *Chem. Commun.* 1291.

Gagniant, P., Cagniant, D. (1953) *Bull. Soc. Chim. Fr.* 62; (1955) *Bull. Soc. Chim. Fr.* 680.

Caille, J.C., Bellamy, F., Guilard, R. (1984) *Tetrahedron Lett.* **25**: 2345.

Caine, D., Gupton, III, J.T. (1974) *J. Org. Chem.* **39**: 2654.

Cairns, N., Harwood, L.M., Astles, D.P. (1987) *Chem. Commun.* 400.

Callant, P., Ongena, R., Vandewalle, M. (1981) *Tetrahedron* **37**: 2085.

Canet, J.-L., Fadel, A., Salaün, J. (1992) *J. Org. Chem.* **57**: 3463.

Capon, B., McManus, S.P. (1976) *Neighboring Group Participation*, "Plenum Press, New York.

Cardillo, G., Orena, M., Sandri, S. (1976) *Tetrahedron Lett.* 3985.

Carey, F.A., Court, A.S. (1972) *J. Org. Chem.* **37**: 1926.

Carey, J.S., Thomas, E.J. (1992) *Synlett* 585.

Carey, J.T., Knors, C., Helquist, P. (1986) *J. Am. Chem. Soc.* **108**: 8313.
Cargill, R.L., Wright, B.W. (1975) *J. Org. Chem.* **40**: 120.
Carroll, W.A., Grieco, P.A. (1993) *J. Am. Chem. Soc.* **115**: 1164.
Carruthers, W. (1990) *Cycloaddition Reactions in Organic Synthesis*, Pergamon, Oxford.
Cartier, D., Ouahrani, M., Levy, J. (1989) *Tetrahedron Lett.* **30**: 1951.
Castedo, L., Harley-Mason, J., Kaplan, M. (1969) *Chem. Commun.* 1444.
Castro, J. Sörensen, H. Riera, A., Morin, C., Moyano, A. Pericas, M.A., Greene, A.E. (1990) *J. Am. Chem. Soc.* **112**: 9388.
Catoni, G., Galli, C., Mandolini, L. (1978) *J. Org. Chem.* **45**: 1906.
Cauwberghs, S.G., De Clercq, P.J. (1988) *Tetrahedron Lett.* **29**: 6501.
Cervinka, O., Hub, L. (1968) *Coll. Czech. Chem. Commun.* **33**: 2927.
Challand, B.P., Kornis, G., Lange, G.L., deMayo, P. (1967) *Chem. Commun.* 704.
Chamberlain, T., Fu, X., Pechacek, J.T., Peng, X., Wheeler, D.M.S., Wheeler, M.M. (1991) *Tetrahedron Lett.* **32**: 1707.
Chan, J.H.H., Rickborn, B. (1968) *J. Am. Chem. Soc.* **90**: 6406.
Chan, T.H., Pellon, P. (1989) *J. Am. Chem. Soc.* **111**: 8737.
Chapman, O.L., Engel, M.R., Springer, J.P., Clardy, J.C. (1971) *J. Am. Chem. Soc.* **93**: 6696.
Charette, A.B., Coté, B., Marcous, J.-F. (1991) *J. Am. Chem. Soc.* **113**: 8166.
Chass, D.A., Buddhasukh, D., Magnus, P. (1978) *J. Org. Chem.* **43**: 1750.
Chemla, P. (1993) *Tetrahedron Lett.* **34**: 7391.
Chen, X., Hortelano, E.R., Eliel, E.L., Frye, S.V. (1992) *J. Am. Chem. Soc.* **114**: 1778.
Cherest, M., Felkin, H., Prudent, N. (1968) *Tetrahedron Lett.* 2199.
Chida, N., Ohtsuka, M. Nakazawa, K., Ogawa, S. (1991) *J. Org. Chem.* **56**: 2976.
Chin, J., Kim, J.H. (1990) *Angew. Chem. Intern. Ed. Engl.* **29**: 523.
Chou, T.-s., Tso, H.-H. (1989) *Org. Prep. Proc. Intern.* **21**: 259.
Choy, W., Reed, L.A., Masamune, S. (1983) *J. Org. Chem.* **48**: 1137.
Chu, K.S., Negrete, G.R., Konopelski, J.P., Lakner, F.J., Woo, N.-T., Olmstead, M.M. (1992) *J. Am. Chem. Soc.* **114**: 1800.
Chu-Moyer, M.Y., Danishefsky, S.J. (1992) *J. Am. Chem. Soc.* **114**: 8333.
Ciufolini, M.A., Byrne, N.E. (1991) *J. Am. Chem. Soc.* **113**: 8016.
Clark, D.E., Meredith, P.R.K. Ritchie, A.C., Walker, T. (1962) *J. Chem Soc.* 2490.
Clark, R.D., Jahangir (1988) *J. Org. Chem.* **53**: 2378.
Clive, D.L.J., Khodabocus, A., Cantin, M., Tao, Y. (1991) *Chem. Commun.* 1755.
Clive, D.L.J., Manning, H.W. (1993) *Chem. Commun.* 666.
Coates, R.M. (1976) *Fortschr. Chem. Org. Naturstoffe* **33**: 73.
Cochrane, J.S., Hanson, J.R. (1972) *J. Chem. Soc. Perkin Trans. I* 361.
Collum, D.B., Still, W.C., Mohamadi, F. (1986) *J. Am. Chem. Soc.* **108**: 2094.
Comins, D.L., Brown, J.D. (1984) *J. Org. Chem.* **49**: 1078.
Compernolle, F., Saleh, M.A., Van den Branden, S., Toppet, S., Hoornaert, G. (1991) *J. Org. Chem.* **56**: 2386.
Confalone, P.N., Pizzolato, G. (1981) *J. Am. Chem. Soc.* **103**: 4251.
Cookson, R.C., Parsons, P.J. (1976) *Chem. Commun.* 990.

Cooper, J., Knight, D.W., Gallagher, P.T. (1987) *Chem. Commun.* 1220.

Corey, E.J., Hess, H.-J., Proskow, S. (1963) *J. Am. Chem. Soc.* **85**: 3979.

Corey, E.J., Mitra, R.B., Uda, H. (1964a) *J. Am. Chem. Soc.* **86**: 485.

Corey, E.J., Nozoe, S. (1964b) *J. Am. Chem. Soc.* **86**: 1652.

Corey, E.J., Ohno, M., Mitra, R.B., Vatachencherry, P.A. (1964c) *J. Am. Chem. Soc.* **86**: 478.

Corey, E.J., Nozoe, S. (1965a) *J. Am. Chem. Soc.* **87**: 5728.

Corey, E.J., Nozoe, S. (1965b) *J. Am. Chem. Soc.* **87**: 5733.

Corey, E.J., Seebach, D. (1965c) *Angew. Chem Intern. Ed. Engl.* **4**: 1075.

Corey, E.J., Hamanaka, E. (1967) *J. Am. Chem. Soc.* **89**: 2758.

Corey, E.J., Katzenellenbogen, J.A., Gilman, N.W., Roman, S.A., Erickson, B.W. (1968a) *J. Am. Chem. Soc.* **90**: 5618.

Corey, E.J., Vlattas, I., Andersen, N.H., Harding, K. (1968b) *J. Am. Chem. Soc.* **90**: 3247.

Corey, E.J., Broger, E.A. (1969a) *Tetrahedron Lett.* 1779.

Corey, E.J., Girotra, N.N., Mathew, C.T. (1969b) *J. Am. Chem. Soc.* **91**: 1557.

Corey, E.J., Vlattas, I., Harding, K. (1969c) *J. Am. Chem. Soc.* **91**: 535.

Corey, E.J., Weinshenker, N.M., Schaaf, T.K. Huber, W. (1969d) *J. Am. Chem. Soc.* **91**: 5675.

Corey, E.J., Erickson, B.W., Noyori, R. (1971) *J. Am. Chem. Soc.* **93**: 1724.

Corey, E.J., Katzenellenbogen, J.A., Roman, S.A., Gilman, N.W. (1971b) *Tetrahedron Lett.* 1821.

Corey, E.J., Becker, K.B., Varma, R.K. (1972a) *J. Am. Chem. Soc.* **94**: 8616.

Corey, E.J., Fuchs, P.L. (1972b) *J. Am. Chem. Soc.* **94**: 4014.

Corey, E.J., Balanson, R.D. (1973) *Tetrahedron Lett.* 3153.

Corey, E.J., Balanson, R.D. (1974a) *J. Am. Chem. Soc.* **96**: 6516.

Corey, E.J., Nicolaou, K.C. (1974b) *J. Am. Chem. Soc.* **96**: 5614.

Corey, E.J., Ensley, H.E. (1975a) *J. Am. Chem. Soc.* **97**: 6908.

Corey, E.J., Ulrich, P. (1975b) *Tetrahedron Lett.* 3685.

Corey, E.J., Wollenberg, R.H., Williams, D.R. (1977) *Tetrahedron Lett.* 2243.

Corey, E.J., Danheiser, R.L., Chandrasekaran, S., Siret, P., Keck, G.E., Gras, G.-L. (1978a) *J. Am. Chem. Soc.* **100**: 8031.

Corey, E.J., Danheiser, R.L., Chandrasekaran, S., Keck, G.E., Gopalan, B., Larsen, S.D., Siret, P., Gras, G.-L. (1978b) *J. Am. Chem. Soc.* **100**: 8034.

Corey, E.J., Kim, S., Yoo, S.e., Nicolaou, K.C., Melvin, L.S., Brunelle, D.J., Falck, J.R., Trybulski, E.J., Lett, R., Sheldrake, P.W. (1978c) *J. Am. Chem. Soc.* **100**: 4620.

Corey, E.J., Trybulski, E.J., Melvin, L.S., Nicolaou, K.C., Secrist, J.A., Lett, R. Sheldrake, P.W., Falck, J.R., Brunelle, D.J., Haslanger, M.F., Kim, S., Yoo, S.e. (1978d) *J. Am. Chem. Soc.* **100**: 4618.

Corey, E.J., Niwa, H., Falck, J.R. (1979a) *J. Am. Chem. Soc.* **101**: 1586.

Corey, E.J., Pearce, H.L. (1979b) *J. Am. Chem. Soc.* **101**: 5847.

Corey, E.J., Tius, M.A., Das, J. (1980a) *J. Am. Chem. Soc.* **102**: 1742; (1980b) *J. Am. Chem. Soc.* **102**: 7612.

Corey, E.J., De, B. (1984) *J. Am. Chem. Soc.* **106**: 2735.

Corey, E.J., Niimura, K., Konishi, Y. Hashimoto, S. Hamada, Y. (1986) *Tetrahedron Lett.* **27**: 2199, 3556.

Corey, E.J., Bakshi, R.K., Shibata, S. (1987a) *J. Am. Chem. Soc.* **109**: 5551.

Corey, E.J., Hannon, F.J. (1987b) *Tetrahedron Lett.* **28**: 5233, 5237.

Corey, E.J., Su, W.-G. (1988) *Tetrahedron Lett.* **29**: 3423.

Corey, E.J., Cheng, X.-M. (1989a) *The Logic of Chemical Synthesis.* Wiley, New York.

Corey, E.J., Da Silva Jardine, P. (1989b) *Tetrahedron Lett.* **30**: 7297.

Corey, E.J., Hahl, R.W. (1989c) *Tetrahedron Lett.* **30**: 3023.

Corey, E.J., (1990) *Pure Appl. Chem.* **63**: 1209.

Corey, E.J., Cheng, X.-M., Cimprich, K.A. (1991a) *Tetrahedron Lett.* **32**: 6839.

Corey, E.J., Imai, N., Zhang, H.-Y. (1991b) *J. Am. Chem. Soc.* **113**: 728.

Corey, E.J., Loh, T.-P. (1991c) *J. Am. Chem. Soc.* **113**: 8967.

Corey, E.J., Link, J.O. (1992a) *J. Am. Chem. Soc.* **114**: 1906.

Corey, E.J., Sarshar, S., Bordner, J. (1992b) *J. Am. Chem. Soc.* **114**: 7938.

Corey, E.J., Noe, M.C., Shieh, W.-C. (1993a) *Tetrahedron Lett.* **38**: 5995.

Corey, E.J., Wu, L.I. (1993b) *J. Am. Chem. Soc.* **115**: 9327.

Cornforth, J.W., Cornforth, R.H., Mathew, K.K. (1959) *J. Chem. Soc.* 112, 2539.

Cox, P.J., Simpkins, N.S. (1991) *Tetrahedron: Asymmetry* **2**: 1.

Coxon, A.C., Laidler, D.A., Pettman, R.B., Stoddart, J.F. (1978) *J. Am. Chem. Soc.* **100**: 8260.

Cram, D.J., Abd Elhafez, F.A. (1952) *J. Am. Chem. Soc.* **74**: 5828, 5951.

Cram, D.J., Kopecky, K.R. (1959) *J. Am. Chem. Soc.* **81**: 2748.

Crandall, T.G., Lawton, R.G. (1969) *J. Am. Chem. Soc.* **91**: 2127.

Crimmins, M.T., O'Mahony, R (1989a) *J. Org. Chem.* **54**: 1157.

Crimmins, M.T., Thomas, J.B. (1989b) *Tetrahedron Lett.* **30**: 5997.

Crimmins, M.T., Jung, D.K., Gray, J.L. (1992) *J. Am. Chem. Soc.* **114**: 5445.

Cupas, C., Schleyer, P.v.R., Trecker, D.J. (1965) *J. Am. Chem. Soc.* **87**: 917.

Curphey, T.J., Kim, H.L. (1968) *Tetrahedron Lett.* 1441.

Curren, D.P., Rakiewicz, D.M. (1985) *J. Am. Chem. Soc.* **107**: 1448.

Curran, D.P., Kuo, S.-C. (1987) *Tetrahedron* **43**: 5653.

Curran, D.P., Choi, S.-M., Gothe, S.A., Lin, F.-T. (1990) *J. Org. Chem.* **55**: 3710.

Curran, D.P., Liu, H. (1992a) *J. Am. Chem. Soc.* **114**: 5863.

Curran, D.P., Yu, H. (1992b) *Synthesis* 123.

Dale, J. Daasvatn, K. (1976) *Chem. Commun.* 295.

Danheiser, R.L., Morin, J.M., Salaski, E.J. (1985) *J. Am. Chem. Soc.* **107**: 8066.

Danheiser, R.L., Cha, D.D. (1990) *Tetrahedron Lett.* **31**: 1527.

Danieli, B., Lesma, G., Palmisano, G., Riva, R. (1984) *Chem. Commun.* 909.

Danishefsky, S., Crawley, L.S., Solomon, D.M., Heggs, P. (1971) *J. Am. Chem. Soc.* **93**: 2356.

Danishefsky, S., Nagel, A. (1972) *Chem. Commun.* 373.

Danishefsky, S., Cain, P., Nagel, A. (1975) *J. Am. Chem. Soc.* **97**: 380.

Danishefsky, S., McKee, R. Singh, R.K. (1977a) *J. Am. Chem. Soc.* **99**: 4783.

Danishefsky, S., Schuda, P.F., Kitahara, T., Etheredge, S.J. (1977b) *J. Am. Chem. Soc.* **99**: 6066.

Danishefsky, S., Hirama, M., Gombatz, K., Harayama, T., Berman, E. Schuda, P.F. (1979) *J. Am. Chem. Soc.* **101**: 7020.

Danishefsky, S., Zamboni, R., Kahn, M., Etheredge, S.J. (1981) *J. Am. Chem. Soc.* **102**: 3460.

Danishefsky, S., Kobayashi, S., Kerwin, J.F. (1982) *J. Org. Chem.* **47**: 1981.

Danishefsky, S., De Ninno, S., Lartey, P. (1987) *J. Am. Chem. Soc.* **109**: 2082.

Danishefsky, S., Cabal, M.P., Chow, K. (1989a) *J. Am. Chem. Soc.* **111**: 3456.

Danishefsky, S., Lee, J.Y. (1989b) *J. Am. Chem. Soc.* **111**: 4829.

Dastur, K.P. (1974) *J. Am. Chem. Soc.* **96**: 2605.

Dauben, W.G., Ashcraft, A.C. (1963) *J. Am. Chem. Soc.* **85**: 3673.

Dauben, W.G., Freidrich. L.E., Oberhänsli, P., Aoyagi, E.I. (1972) *J. Org. Chem.* **37**: 9.

Dauben, W.G., Kessel, C.R., Takemura, K.H. (1980) *J. Am. Chem. Soc.* **102**: 6894.

Dauben, W.G., Brookhart, T. (1981) *J. Am. Soc.* **103**: 237.

Davidson, A.H., Floyd, C.D., Jones, A.J., Myers, P.L. (1985) *Chem. Commun.* 1662.

Davies, J., Roberts, S.M., Reynolds, D.P., Newton, R.F. (1981) *J. Chem. Soc. Perkin Trans. I* 1729.

Davies, S.G., Dordor, I.M., Warner, P. (1984) *Chem. Commun.* 956.

Davies, S.G., Newton, R.F., Williams, J.M.J. (1989) *Tetrahedron Lett.* **30**: 2967.

Davis, F.A., Vishwakarma, L.C. (1985) *Tetrahedron Lett.* **26**: 3539.

Debal, A., Cuvigny, T., Larcheveque, M. (1977) *Tetrahedron Lett.* 3187.

DeClercq, P., Vandewalle, M. (1977) *J. Org. Chem.* **42**: 3447.

Demole, E., Enggist, P. (1969) *Chem. Commun.* 264.

Demuth, M., Ritterskamp, P., Schaffner, K. (1984) *Helv. Chem. Acta* **67**: 2023.

Descoins, C., Samain, D. (1976) *Tetrahedron Lett.* 745.

Deslongchamps, P., Moreau, C., Frehel, D., Chênevert, R. (1975) *Can. J. Chem.* **53**: 1204.

Desmaële, D., Ficini, J., Guingant, A., Kahn, P. (1983) *Tetrahedron Lett.* **24**: 3079.

Dewanckele, J.M., Zutterman, F., Vandewalle, M. (1983) *Tetrahedron* **39**: 3235.

Deyrup, J.A., Clough, S.C. (1974) *J. Org. Chem.* **34**: 902.

Dietrich-Buchecker, C., Sauvage, J.P. (1992) *Bull. Soc. Chim. Fr.* **129**: 113.

"Dinsburg, G." (alias of Ginsburg, D.) (1982) *Nouv. J. Chem.* **6**: 175.

Disanayaka, B.W., Weedon, A.C. (1985) *Chem. Commun.* 1282.

Dolle, R.E., Osifo, K.I., Li, C.-S. (1991) *Tetrahedron Lett.* **32**: 5029.

Dominguez, X.A., Delgado, J.G., Reeves, W.P., Gardner, P.D. (1967) *Tetrahedron Lett.* 2493.

Donaldson, R.E., McKenzie, A. Byrn, S., Fuchs, P. (1983) *J. Org. Chem.* **48**: 2167.

Donaldson, W.A., Craig, R., Spanton, S. (1992) *Tetrahedron Lett.* **33**: 3967.

Dondoni, A., Fantin, G., Fogagnolo, M. Medici, A. Pedrini, P. (1989) *J. Org. Chem.* **54**: 693.

Dorsch, M. Jäger, V., Spönlein, W. (1984) *Angew, Chem. Intern. Ed. Engl.* **23**: 798.

Dötz, K.H. (1984) *Angew. Chem. Intern. Ed. Engl.* **23**: 587.

Dötz, K.H., Popall, M. (1985) *Tetrahedron* **41**: 5797.

Drouin, J., Leyendecker, F., Conia, J.M. (1975) *Tetrahedron Lett.* 4053.

Dubois, J.E., Felimann, P. (1975) *Tetrahedron Lett.* 1225.

Decoux, J.-P., Le Menez, P., Kunesch, N., Kunesch, G., Wenkert, E. (1990) *Tetrahedron Lett.* **31**: 2595.

Duerr, B.F., Czarnik, A.W. (1989) *Tetrahedron Lett.* **30**: 6951.

Dussault, P.H., Hayden, M.R. (1992) *Tetrahedron Lett.* **33**: 443.

Earl, R.A., Vollhardt, K.P.C. (1984) *J. Org. Chem.* **49**: 4786.

Eaton, P.E., Cole, T.W. (1964) *J. Am. Chem. Soc.* **86**: 962, 3157.

Edwards, M.P., Ley, S.V., Lister, S.G., Palmer, B.D. (1983) *Chem. Commun.* 630.

Edwards, M.P., Ley, S.V., Lister, S.G., Palmer, B.D., Williams, D.J. (1984) *J. Org. Chem.* **49**: 3503.

Ehlinger, E., Magnus, P. (1980) *J. Am. Chem. Soc.* **102**: 5004.

Eisch, J.J., Merkley, J.H., Galle, J.E. (1979) *J. Org. Chem.* **44**: 587.

Eliel, E.L. (1974) *Tetrahedron Lett.* **30**: 1503.

Eliel, E.L., Koskimies, J.K., Lohri, B., Frazee, W.J., Morris-Natschke, S., Lynch, J.E., Soai, K. (1982) in (Eliel, E.L., Otsuke, S., eds.) *Asymmetric Reactions and Processes in Chemistry* ACS, Washington, DC.

Enders, D., Papadopoulos, K. (1983) *Tetrahedron Lett.* 4967.

Enders, D. (1984) in *Asymmetric Synthesis* (Morrison, J.D., ed.) **3**: 275, Academic Press, N.Y.

Enders, D., Mannes, D., Raabe, G. (1992) *Synlett* 837.

Ensley, H.E., Parnell, C.A., Corey, E.J. (1978) *J. Org. Chem.* **43**: 1610.

Eschenmoser, A., Felix, D., Ohloff, G. (1967) *Helv. Chim. Acta* **50**: 708.

Eschenmoser, A. (1970) *Quart. Rev.* **24**: 366.

Eschenmoser, A. (1976) *Chem. Soc. Rev.* **5**: 377.

Eschenmoser, A. (1988) *Angew. Chem. Intern. Ed. Engl.* **27**: 6.

Essiz, M., Coudert, G. Guillaumet, G., Caubere, P. (1976) *Tetrahedron Lett.* 3185.

Evans, D.A., Sims, C.L. (1973) *Tetrahedron Lett.* 4691.

Evans, D.A., Andrews, G.C., Buckwalter, B. (1974a) *J. Am. Chem. Soc.* **96**: 5560.

Evans, D.A., Carroll, G.L., Truesdale, L.K. (1974b) *J. Org. Chem.* **39**: 914.

Evans, D.A., Billargeon, D.J., Nelson, T.V. (1978) *J. Am. Chem. Soc.* **100**: 2242.

Evans, D.A., Sacks, C.E., Kleschick, W.A., Taber, T.R. (1979) *J. Am. Chem. Soc.* **101**: 6789.

Evans, D.A., Takacs, J.M. (1980) *Tetrahedron Lett.* **21**: 4233.

Evans, D.A., McGee, L.R. (1981a) *J. Am. Chem. Soc.* **103**: 2876.

Evans, D.A., Tanis, S.P., Hart, D.J. (1981b) *J. Am. Chem. Soc.* **103**: 5813.

Evans, D.A., Ennis, M.D., Mathre, D.J. (1982a) *J. Am. Chem. Soc.* **104**: 1737.

Evans, D.A., Thomas, E.W., Cherpeck, R.E. (1982b) *J. Am. Chem. Soc.* **104**: 3695.

Evans, D.A. (1984a) in Morrison, J.D. (Ed.) *Asymmetric Synthesis*, Vol. **3B**: Chap. 1 Academic Press. Orlando.

Evans, D.A., Chapman, K.T., Bisaha, J. (1984b) *Tetrahedron Lett.* **25**: 4071.

Evans, D.A., Britton, T.C., Dorow, R.L., Dellaria, J.F. (1986) *J. Am. Chem. Soc.* **108**: 6395.

Evans, D.A., Chapman, K.T., Bisaha, J. (1988a) *J. Am. Chem. Soc.* **100**: 1238.

Evans, D.A., Chapman, K.T., Carreira, E.M. (1988b) *J. Am. Chem. Soc.* **110**: 3560.

Evans, D.A., Carreira, E.M. (1990a) *Tetrahedron Lett.* **31**: 4703.

Evans, D.A., Clark, J.S., Metternic, R., Novack, V.J., Sheppard, G.S. (1990b) *J. Am. Chem. Soc.* **112**: 866.

Evans, D.A., Dow, R.L., Shih, T.L., Takacs, J.M., Zahler, R. (1990c) *J. Am. Chem. Soc. J. Am. Chem. Soc.* **112**: 5290.

Evans, D.A., Gage, J.R. (1990d) *Tetrahedron Lett.* **31**: 6129.

Evans, D.A., Hoveyda, A.H. (1990e) *J. Am. Chem. Soc.* **112**: 6447.

Evans, D.A., Gauchet-Prunet, J.A., Carreira, E.M., Charette, A.B. (1991) *J. Org. Chem.* **56**: 741.

Evans, D.A., Black, W.C. (1992a)*J. Am. Chem. Soc.* **114**: 2260.

Evans, D.A., Fu, G.C., Anderson, B.A. (1992b) *J. Am. Chem. Soc.* **114**: 6679.

Evans, D.A., Fu, G.C., Hoveyda, A.H. (1992c) *J. Am. Chem. Soc.* **114**: 6671.

Evans, D.A., Dinsmore, C.J. (1993) *Tetrahedron Lett.* **34**: 6029.

Fahrenholtz, K.E., Lurie, M., Kierstead, R.W. (1967) *J. Am. Chem. Soc.* **89**: 5934.

Faller, J.W., John, J.A., Mazzieri, M.R. (1989) *Tetrahedron Lett.* **30**: 1769.

Felber, H., Kresze, G., Prewo, R., Vasella, A. (1986) *Helv. Chim. Acta* **69**: 1137.

Feldman, K.S., Lee, Y.B. (1987) *J. Am. Chem. Soc.* **109**: 5850.

Feldman, K.S., Wu, M.-J. Rotella, D.P. (1990) *J. Am. Chem. Soc.* **112**: 8490.

Feldman, K.S., Ensel, S.M., Weinreb, P.H. (1992) *J. Org. Chem.* **57**: 2199.

Fessner, W.-D., Prinzbach, H., Rihs, G. (1983) *Tetrahedron Lett.* **24**: 5857.

Ficini, J., d'Angelo, J., Noire, J. (1974) *J. Am. Chem. Soc.* **96**: 1213.

Finan, J.M., Kishi, Y. (1982) *Tetrahedron Lett.* **23**: 2719.

Finn, M.G., Sharpless, K.B. (1985) in Morrison, J.D. (ed) *Asymmetric Synthesis* **5**: 247. Academic Press, Orlando.

Fitjer, L., Kanschik, A., Majewski, M. (1988) *Tetrahedron Lett.* **29**: 5525.

Fleming, I. (1981) *Chem. Soc. Rev.* **10**: 83.

Fleming, I., Lawrence, N.J. (1988) *Tetrahedron Lett.* **29**: 2077.

Fouque, E., Rousseau, G., Seyden-Penne, J. (1990) *J. Org. Chem.* **55**: 4807.

Franck-Neumann, M., Martina, D., Heitz, M.P. (1966) *J. Organomet. Chem.* **301**: 61.

Frater, G., Müller, U., Günther, W. (1984) *Tetrahedron Lett.* **40**: 1269.

Fried, J., Lin, C.H., Sih, J.C., Dalven, P., Cooper, G.F. (1972) *J. Am. Chem. Soc.* **94**: 4343.

Fried, J., Sih, J.C. (1973) *Tetrahedron Lett.* 3899.

Fronz, G., Fuganti, G., Grasselli, P., Petrocchi-Fanton, G., Zirotti, C. (1982) *Tetrahedron Lett.* **23**: 4143.

Frydman, B., Reil, S., Despuy, M.E., Rapoport, H. (1969) *J. Am. Chem. Soc.* **91**: 2368.

Fu, G.C., Grubbs, R.H. (1992) *J. Am. Chem. Soc.* **114**: 5426, 7324.

Fuhrer, W., Gschwend, H.W. (1979) *J. Org. Chem.* **44**: 1133.

Fujimoto, R., Kishi, Y., Blount, J.F. (1980) *J. Am. Chem. Soc.* **102**: 7154.

Fujimura, O., Takai, K., Utimoto, K. (1990) *J. Org. Chem.* **55**: 1705.

Fujisawa, T., Funabora, M., Ukaji, Y., Sato, T. (1980a) *Chem Lett.* 59.

Fujisawa, T., Mobele, B.A., Shimizu, M. (1991) *Tetrahedron Lett.* **32**: 7055.

Fukuyama, T., Yang, L.-H. (1987) *J. Am. Chem. Soc.* **109**: 7881.

Funk, R.L., Munger, J.D. (1984) *J. Org. Chem.* **49**: 4319.

Funk, R.L., Munger, J.D. (1985) *J. Org. Chem.* **50**: 707.

Funk, R.L., Abelman, M.M. (1986) *J. Org. Chem.* **51**: 3247.

Funk, R.L., Daily, W.J., Parvez, M. (1988a) *J. Org. Chem.* **53**: 4141.

Funk, R.L., Olmstead, T.A., Parvez, M. (1988b) *J. Am. Chem. Soc.* **110**: 3298.

Furber, M., Mander, L.N. (1987) *J. Am. Chem. Soc.* **109**: 6389.

Fürstner, A., Weidmann, H. (1990) *J. Org. Chem.* **55**: 1363.

Furuichi, K., Miwa, T. (1974) *Tetrahedron Lett.* 3689.

Furuta, K., Shimizu, S., Miwa, Y., Yamamoto, H. (1989) *J. Org. Chem.* **54**: 1481.

Furuta, K., Maruyama, T., Yamamoto, H. (1991) *J. Am. Chem. Soc.* **113**: 1041.

Gadwood, R.C., Lett, R.M., Wissinger, J.E. (1986) *J. Am. Chem. Soc.* **108**: 6343.

Galina, C., Ciattini, P.G. (1979) *J. Am. Chem. Soc.* **101**: 1035.

Gallagher, T., Magnus, P., Huffman, J.C. (1982) *J. Am. Chem. Soc.* **104**: 1140.

Galli, C. (1992) *Org. Prep. Proc. Intern.* **24**: 287.

Gange, D., Magnus. P. (1978) *J. Am. Chem. Soc.* **100**: 7746.

Garner, P., Sunitha, K., Ho, W.B., Youngs, W.J., Kennedy, V.O., Djebli, A. (1989) *J. Org. Chem.* **54**: 2041.

Gates, M. (1950) *J. Am. Chem. Soc.* **72**: 228.

Gates, M., Tschudi, G. (1956) *J. Am. Chem. Soc.* **78**: 1380.

Gawley, R.E. (1976) *Synthesis* 777.

Geiss, K., Seuring, R., Pieter, R., Seebach, D. (1974) *Angew. Chem. Intern. Ed. Engl.* **13**: 479.

Gennari, C., Colombo, L., Bertolini, G. (1986) *J. Am. Chem. Soc.* **108**: 6394.

Gensler, W.J., Chan, S., Ball, D.B. (1975) *J. Am. Chem. Soc.* **97**: 436.

Gerlach, H., Thalmann, A. (1974) *Helv. Chim. Acta* **57**: 2661.

Germanas, J., Aubert, C., Vollhardt, K.P.C. (1991) *J. Am. Chem. Soc.* **113**: 4006.

Gervay, J.E., McCapra, F., Money, T., Sharma, G.M. (1966) *Chem Commun*, 142.

Giarrusso, F., Ireland, R.E. (1968) *J. Org. Chem.* **33**: 3560.

Gibbons, E.G. (1982) *J. Am. Chem. Soc.* **104**: 1767.

Gillard, J.W., Fortin, R., Grimm, E.L., Maillard, M. Tjepkema, M. Bernstein, M.A., Glaser, R. (1991) *Tetrahedron Lett.* **32**: 1145.

Gilman, H., Breuer, F. (1934) *J. Am. Chem. Soc.* **56**: 1123.

Glas, R.S., Deardorff, D.R., Gains, L.H. (1978) *Tetrahedron Lett.* 2965.

Godleski, S.A., Villhauer, E.B. (1986) *J. Org. Chem.* **51**: 486.

Goering, H.L., Kantner, S.S., Tseng, C.C. (1983) *J. Org. Chem.* **48**: 715.

Goldfarb, Ya. L., Fabrichnyi, B.P., Shalavina, I.F. (1962) *Tetrahedron* **18**: 21.

Goldsmith, D.J., Thottathil, J.K. (1981) *Tetrahedron Lett.* **22**: 2447.

Gooding, O.W. (1990) *J. Org. Chem.* **55**: 4209.

Grass, J.-L., Poncet, A., Nouguier, R. (1992) *Tetrahedron Lett.* **33**: 3323.

Gratten, T.J., Whitehurst, J.S. (1988) *Chem. Commun.* 43.

Gray, M., Parsons, P.J., Neary, A.P. (1992) *Synlett* 597.

Green, B.S., Hagler, A.T., Rabinson, Y., Rejto, M. (1976/1977) *Isr. J. Chem.* **15**: 124.

Greene, A.E. (1980) *Tetrahedron Lett.* 3059.

Greene, A.E., Luche, M.-J., Serra, A.A. (1985) *J. Org. Chem.* **50**: 3957.

Greene, A.E., Charbonnier, F., Luche, M.-J., Moyano, A. (1987) *J. Am. Chem. Soc.* **109**: 4752.

Greene, T.W., Wuts, P.G.M. (1991) *Protective Groups in Organic Synthesis*, 2nd. ed., Wiley, New York.

Greenlee, W.J., Woodward, R.B. (1976) *J. Am. Chem. Soc.* **98**: 6075.

Grieco, P.A., Ohfune, Y., Majetich, G. (1977) *J. Am. Chem. Soc.* **99**: 7393.

Grieco, P.A., Ohfune, Y., Yokoyama, Y., Owens, W. (1979a) *J. Am. Chem. Soc.* **101**: 4749.

Grieco, P.A., Takigawa, T., Moore, D.R. (1979b) *J. Am. Chem. Soc.* **101**: 4380.

Grieco, P.A., Ohfune, Y. (1980) *J. Org. Chem.* **45**: 2251.

Grieco, P.A., Lis, B., Zeller, R.E., Finn, J. (1986) *J. Am. Chem. Soc.* **108**: 5908.

Grieco, P.A., Hon, Y.S., Perez-Medrano, A. (1988) *J. Am. Chem. Soc.* **110**: 1630.

Grieco, P.A., Stuk, T.L. (1990) *J. Am. Chem. Soc.* **112**: 7799.

Gröbel, B.-T., Seebach, D. (1974) *Angew. Chem. Intern. Ed. Engl.* **13**: 83.

Grotjahn, D.B., Anderson, N.H. (1981) *Chem. Commun.* 306.

Gschwend, H.W., Rodriguez, H.R. (1979) *Org. React.* **26**: 1.

Guidon, Y., Simoneau, B., Yoakim, C., Gorys, V., Lemieux, R., Ogilvie, W. (1991) *Tetrahedron Lett.* **32**: 5453.

Gupta, A.K., Fu, X., Snyder, J.P., Cook, J.M. (1991) *Tetrahedron* **47**: 3665.

Guthrie, R.W., Valenta, Z., Wiesner, K. (1966) *Tetrahedron Lett.* 4645.

Gutte, B., Merrifield, R.B. (1969) *J. Am. Chem. Soc.* **91**: 501.

Hafner, K., Golz, M. (1982) *Angew. Chem. Intern. Ed. Engl.* **21**: 695.

Hagiwara, H., Uda, H., Kodama, T. (1980) *J. Chem. Soc. Perkin Trans. I*963.

Hakam, K., Thielmann, M., Thielmann, T., Winterfeldt, E. (1987) *Tetrahedron* 43: 2035.

Halcomb, R.L., Danishefsky, S.J. (1989) *J. Am. Chem. Soc.* **111**: 6661.

Hale, M.R., Hoveyda, A.H. (1992) *J. Org. Chem.* **57**: 1643.

Hall, E.S., McCapra, F., Scott, A.I. (1967) *Tetrahedron* **23**: 4131.

Hamon, D.P.G., Spurr, P.R. (1982) *Chem. Commun.* 372.

Hanamoto, T., Baba, Y., Inanaga, J. (1993) *J. Org. Chem.* **58**: 299.

Handa, Y., Inanaga, J. (1987) *Tetrahedron Lett.* **28**: 5717.

Hanessian, S. (1966) *Carbohydr. Res.* **2**: 86.

Hanessian, S. (1983) *Total Synthesis of Natural Products: The 'Chiron' Approach*, Pergamon, Oxford.

Hanessian, S., Murray, P.J., Sahoo, S.P. (1985a) *Tetrahedron Lett.* **26**: 5623.

Hanessian, S., Murray, P.J., Sahoo, S.P. (1985b) *Tetrahedron Lett.* **26**: 5627.

Hanessian, S., Sahoo, S.P., Murray, P.J. (1985c) *Tetrahedron Lett.* **26**: 5631.

Hanson, R.M. (1991) *Chem. Rev.* **91**: 437.

Hansson, S., Miller, J.F., Liebeskind, L.S. (1990) *J. Am. Chem. Soc.* **112**: 9660.

Harada, T., Hayashiya, T., Wada, I., Iwa-ake, N. Oku, A. (1987) *J. Am. Chem. Soc.* **109**: 527.

Harada, T., Matsuda, Y., Wada, I., Uchimura, J., Oku, A. (1990) *Chem. Commun.* 21.

Harada, T., Takahashi, T., Takahashi, S. (1992) *Tetrahedron Lett.* **33**: 369.

Harding, K.E., Clement, K.S., Tseng, C.-Y. (1990) *J. Org. Chem.* **55**: 4403.

Harirchian, B., Magnus, P. (1977) *Chem. Commun.* 522.

Harris, T.M., Harris, C.M. (1969) *Org. React.* **17**: 155.

Harris, T.M., Webb, A.D., Harris, C.M., Wittek, P.J., Murray, T.P. (1976) *J. Am. Chem. Soc.* **98**: 6065.

Hart, D.J. (1981) *J. Org. Chem.* **46**: 3576.

Hartmann, J., Stähle, M., Schlosser, M. (1974) *Synthesis* 888.

Hase, T.A. (ed). (1987) *Umpoled Synthons*, Wiley, New York.

Hassel, T., Seebach, D. (1979) *Angew, Chem. Intern. Ed. Engl.* **18**: 399.

Hatakeyama, S., Sakurai, K., Numata, H., Ochi, N., Takano, S. (1988) *J. Am. Chem. Soc.* **110**: 5201.

Hattori, K., Yamamoto, H. (1993) *J. Org. Chem.* **58**: 5301.

Hauser, F.M., Ellenberger, S.R., Clardy, J.C., Bass, L.S. (1984) *J. Am. Chem. Soc.* **106**: 2458.

Hawker, C.J., Frechet, J.M.J. (1990) *J. Am. Chem. Soc.* **112**: 7638.

Hayashi, K., Hamada, Y., Shioiri, T. (1991) *Tetrahedron Lett.* **32**: 7287.

Hayashi, T., Konishi, M., Fukushima, M., Kanehira, K., Hioki, T., Kumada, M. (1983) *J. Org. Chem.* **48**: 2195.

Hayashi, T. (1991) *Tetrahedron Lett.* **32**: 5369.

Heathcock, C.H., Kleinman, E.F., Binkley, E.S. (1982) *J. Am. Chem. Soc.* **104**: 1054.

Heathcock, C.H., Davidsen, S.K., Mills, S. Sanner, M.A., (1986) *J. Am. Chem. Soc.* **108**: 5650.

Heathcock, C.H. (1992) *Angew. Chem. Intern. Ed. Engl.* **31**: 665.

Hedstrand, D.M., Byrn, S.R., McKenzie, A.T., Fuchs, P.L. (1987) *J. Org. Chem.* **52**: 592.

Heitz, M.-P., Overman, L.E. (1989) *J. Org. Chem.* **54**: 2591.

Helmchen, G., Schmierer, R. (1981) *Angew. Chem. Intern. Ed. Engl.* **20**: 205.

Henbest, H.B., Lovell, B.J. (1957a) *J. Chem. Soc.* 1965.

Henbest, H.B., Wilson, R.A. (1957b) *J. Chem. Soc.* 1958.

Hendrickson, J.B. (1976) *Top. Curr. Chem.* **62**: 49.

Hendrickson, J.B. (1977) *J. Am. Chem. Soc.* **99**: 5439.

Hendrickson, J.B., Palumbo, P.S. (1985) *J. Org. Chem.* **50**: 2110.

Heusler, K., Wieland, P., Wettstein, A. (1959) *Helv. Chim. Acta* **42**: 1586.

Hewson, A.T., MacPherson, D.T. (1985) *J. Chem. Soc. Perkin Trans. I* 2625.

Hikino, H., Suzuki, N., Takemoto, T. (1966) *Chem. Pharm. Bull.* **14**: 1441.

Hilpert, H., Siegfried, M.A., Dreiding, A.S. (1985) *Helv. Chim. Acta* **68**: 1670.

Hirama, M., Uei, M. (1982) *J. Am. Chem. Soc.* **104**: 4251.

Hirst, G.C., Howard, P.N., Overman, L.E. (1989) *J. Am. Chem. Soc.* **111**: 1514.

Hirst, S.C. Hamilton. A.D. (1991) *J. Am. Chem. Soc.* **113**: 382.

Ho, P.-T. Davies, N. (1983) *Synthesis* 462.

Ho, T.-L. (1974a) *Synth. Commun.* **4**: 189.

Ho, T.-L. (1974b) *Synthesis* 715.

Ho, T.-L., Wong, C.M. (1974c) *Synth. Commun.* **4**: 307.

Ho, T.-L., Din, Z.U. (1982a) *Synth. Commun.* **12**: 257.

Ho, T.-L., Liu S.H. (1982b) *Synth. Commun.* **12**: 501.

Ho, T.-L. (1988) *Carbocycle Construction in Terpene Synthesis*, VCH, New York.

Ho, T.-L. (1991) *Polarity Control for Synthesis*, Wiley, New York.

Ho, T.-L. (1992a) *Enantioselective Synthesis: Natural Products from Chiral Terpenes*, Wiley, New York.

Ho, T.-L. (1992b) *Tandem Organic Reactions*, Wiley, New York.

Ho, T.-L., Yeh, W.-L., Yule, J., Liu, H.-J. (1992c) *Can. J. Chem.* **70**: 1375.

Ho, T.-L., (manuscript in preparation) *Symmetry: A Basis for Synthesis Design.*

Hodgson, S.T., Hollinshead, D.M., Ley, S.V. (1985) *Tetrahedron* **41**: 5871.

Holoboski, M.A., Koft, E. (1992) *J. Am. Chem. Soc.* **57**: 965.

Hoffman, R.V., Shechter, H. (1971) *J. Am. Chem. Soc.* **93**: 5940.

Hoffmann, R.W. (1982a) *Angew. Chem. Intern. Ed Engl.* **21**: 555.

Hoffmann, R.W., Kemper, B. (1982b) *Tetrahedron Lett.* **23**: 845.

Hoffman, R.W., Helbig, W., Ladner, W. (1982c) *Tetrahedron Lett.* **23**: 3479.

Hoffman, R.W. (1988) *Pure Appl. Chem.* **60**: 123.

Holmes, T.L., Stevenson, R. (1970) *Tetrahedron Lett.* 199.

Holton, R.A. (1977) *J. Am. Chem. Soc.* **99**: 8083.

Holton, R.A., Sibi, M.P., Murphy, W.S. (1988) *J. Am. Chem. Soc.* **110**: 314.

Hoppe, D., Krämer, T., Schwark, J.-R., Zschage, O. (1990) *Pure Appl. Chem.* **62**: 1999.

Horiguchi, Y., Komatsu, M., Kuwajima, I. (1989) *Tetrahedron Lett.* **30**: 7087.

Horikawa, M., Hashimoto, K., Shirahama, H. (1993) *Tetrahedron Lett.* **34**: 331.

Hortmann, A.G., Daniel, D.S., Martinelli, J.E. (1973) *J. Org. Chem.* **38**: 728.

Horton, M., Pattenden, G. (1984) *J. Chem. Soc. Perkin Trans. I* 811.

Hosokawa, T., Uno, S., Inui, S., Murahashi, S.I. (1981) *J. Am. Chem. Soc.* **103**: 2318.

Hosokawa, T., Aoki, S., Takano, M., Nakahira, T., Yoshida, Y., Murahashi, S.-I. (1991) *Chem. Commun.* 1559.

Hosomi, A., Hashimoto, H., Sakurai, H. (1978) *J. Org. Chem.* **43**: 2551.

Hoye, T.R., North, J.T., Yao, L.J. (1993) *206th ACS Nat. Meet*, ORGN 170.

Hudlicky, T., Fleming, A., Radesca, L. (1989) *J. Am. Chem. Soc.* **111**: 6691.

Hudlicky, T., Fan, R., Luna, H., Olivo, H., Price, J. (1992) *Pure Appl. Chem.* **64**: 1109.

Huguet, J., Karpf, M., Dreiding, A.S. (1982) *Helv. Chim. Acta* **65**: 2413.

Hünig, S., Wehner, G. (1975) *Synthesis* 180.

Hunt, E., Lythgoe, B. (1972) *Chem. Commun.* 757.

Hutchins, R.O., Maryanoff, B.E. (1972) *J. Org. Chem.* **37**: 1829.

Ibuka, T., Minakata, H., Mitsui, Y., Tabushi, E., Taga, T., Inubushi, Y. (1982) *Chem. Pharm. Bull.* **30**: 2840.

Ichihara, A., Kobayashi, M., Oda, K., Sakamura, S., Sakai, R. (1979) *Tetrahedron* **35**: 2861.

Ichihara, A., Kimura, R., Yamada, S., Sakamura, S. (1980) *J. Am. Chem. Soc.* **102**: 6353.

Ihara, M., Toyota, M., Fukumoto, K. Kametani, T. (1986) *J. Chem. Soc. Perkin Trans. I* 2151.

Ihara, M., Suzuki, S., Fukumoto, K., Kabuto, C. (1990) *J. Am. Chem. Soc.* **112**: 1164.

Ihara, M., Taniguchi, N., Suzuki, S., Fukumoto, K. (1992) *Chem. Commun.* 976.

Ikeda, N., Omori, K., Yamamoto, H. (1986) *Tetrahedron Lett.* **27**: 1175.

Ikemoto, N., Schreiber, S.L. (1992) *J. Am. Chem. Soc.* **114**: 2524.

Ikota, N., Hanaki, A. (1987) *Chem. Pharm. Bull.* **35**: 2140.

Imai, T., Mineta, H., Nishida, A. (1990) *J. Org. Chem.* **55**: 4986.

Imanishi, T., Kurumada, T., Maezaki, N., Sugiyama, K., Iwata, C. (1991) *Chem. Commun.* 1409.

Inanaga, J., Baba, Y., Hanamoto, T. (1993) *Chem. Lett.* 241.

Inomata, K., Igarashi, S., Mohri, M., Yamamoto, T., Kinoshita, H., Kotake, H. (1987) *Chem. Lett.* 707.

Inouye, Y., Shirai, M., Michino, T., Kakisawa, H. (1993) *Bull. Chem. Soc. Jpn.* **66**: 324.

Ireland, R.E., Kierstead, R.C. (1962) *J. Org. Chem.* **27**: 703.

Ireland, R.E., Kierstead, R.C. (1966) *J. Org. Chem.* **31**: 2543.

Ireland, R.E., (1969) *Organic Synthesis*, Prentice-Hall, Englewood Cliffs, N.J.

Ireland, R.E., Dawson, M.I., Welsh., S.C., Hagenbach, A. Bordner, J. Trus, B. (1973) *J. Am. Chem. Soc.* **95**: 7829.

Ireland, R.E., Mueller, R.H., Willard, A.K. (1976) *J. Am. Chem. Soc.* **98**: 2868.

Ireland, R.E., Thaisrivongs, S., Wilcox, C.S. (1980) *J. Am. Chem. Soc.* **102**: 1155.

Ireland, R.E., Varney, M.D. (1984) *J. Am. Chem. Soc.* **106**: 3668.

Ireland, R.E., Norbeck, D.W., Mandel, G.S., Mandel, N.S. (1985) *J. Am. Chem. Soc.* **107**: 3285.

Ireland, R.E., Armstrong, J.D., Lebreton, J., Meissner, R.S., Rizzacasa, M.A. (1993) *J. Am. Chem. Soc.* **115**: 7152.

Irie, H., Katakawa, J., Mizuno, Y., Udaka, S., Taga, T., Osaki, K. (1978) *Chem. Commun.* 717.

Ishibashi, Ho., So, T.S., Nakatani, H., Minami, K., Ikeda, M. (1988) *Chem. Commun.* 827.

Isler, O., Lindlar, H., Montavon, M., Ruegg, R., Zeller, P. (1956) *Helv. Chim. Acta* **39**: 249.

Isler, O., Schudel, P. (1963) *Adv. Org. Chem.* **4**: 115.

Isobe, M., Iio, H., Kawai, T., Goto, T. (1978) *J. Am. Chem. Soc.* **100**: 1940.

Isobe, M., Kitamura, M., Goto, T. (1980) *Tetrahedron Lett.* **21**: 4727; (1981) *Tetrahedron Lett.* **22**: 239.

Isobe, M., Funabashi, Y., Ichikawa, Y., Mio, S., Goto, T. (1984) *Tetrahedron Lett.* **25**: 2021.

Isobe, M. (1986a) in Steyn, P.S., Vleggaar, R. (eds.) *Mycotoxins & Phycotoxins*, p.231, Elsevier, Amsterdam.

Isobe, M., Ichikawa, Y., Funabashi, Y., Mio, S., Goto, T. (1986b) *Tetrahedron* **42**: 2863.

Ito, S., Hirata, Y. (1977) *Bull. Chem. Soc. Jpn.* **50**: 1813.

Ito, T., Tomiyoshi, K., Nakamura, K., Azuma, S., Izawa, M., Muruyama, F., Yanagiya, M., Shirahama, H., Matsumoto, T. (1984) *Tetrahedron* **40**: 241.

Ito, Y., Saegusa, T. (1977) *J. Org. Chem.* **42**: 2326.

Ito, Y., Nakajo, E., Nakatsuka, M., Saegusa, T. (1983) *Tetrahedron Lett.* **24**: 2881.

Ito, Y., Katsuki, T., Yamaguchi, M. (1984) *Tetrahedron Lett.* **25**: 6015.

Itsuno, S., Frechet, J.M.J. (1987) *J. Or. Chem.* **52**: 4140.

Iwasawa, N., Mukaiyama, T., (1982) *Chem, Lett.* 1441.

Iwata, C., Yamada, M., Shinoo, Y. (1979) *Chem. Pharm. Bull.* **27**: 274.

Iwata, C., Fusaka, T., Fujiwara, T., Tomita, K., Yamada, M. (1981a) *Chem. Commun.* 463.

Iwata, C., Miyashita, K., Ida, Y., Yamada, M. (1981b) *Chem. Commun.* 461.

Iwata, C., Morie, T., Maezaki, N., Shimamura, H., Tanaka, T., Imanishi, T. (1984) *Chem. Commun.* 930.

Iwata, C., Nakamura, S., Shinoo, Y., Fusaka, T., Okada, H., Kishimoto, M., Uetsuji, H., Maezaki, N., Yamada, M., Tanaka, T. (1985a) *Chem. Pharm. Bull.* **33**: 1961.

Iwata, C., Fujita, M., Hattori, K., Uchida, S., Imanishi, T. (1985b) *Tetrahedron Lett.* **26**: 2221.

Iyoda, M., Kushida, T., Kitami, S., Oda, M. (1986) *Chem. Commun.* 1049.

Jackson, A.H., Shannon, P.V.R., Wilkins, D.J. (1987) *Tetrahedron Lett.* **28**: 4901.

Jackson, W.R., Moffat, M.R., Perlmutter, P., Tasdelen, E.E., (1992) *Aus. J. Chem.* **45**: 823.

Jacobi, P.A., Craig, T.A., Walker, D.G., Arrick, B.A., Frechette, R.F. (1984) *J. Am. Chem. Soc.* **106**: 5585.

Jansen, J.F.G.A., Feringa, B.L. (1991) *Tetrahedron Lett.* **32**: 3239.

Jaouen, G., Meyer, A. (1975) *J. Am. Chem. Soc.* **97**: 4667.

Jenkinson, J.J., Parsons, P.J., Eyley, S.C. (1992) *Synlett* 679.

Jeong, K.-S., Parris, K., Ballester, P., Rebek, J. (1990) *Angew, Chem. Intern. Ed. Engl.* **29**: 555.

Johnson, C.R., Barbachyn, M.R. (1984) *J. Am. Chem. Soc.* **106**: 2459.

Johnson, C.R., Chen, Y.-F. (1991) *J. Am. Org. Chem.* **56**: 3344, 3352.

Johnson, C.R., Adams, J.P., Bis, S.J., De Jong, R.L., Golebiowski, A., Medich, J.R., Penning, T.D., Senanayake, C.H., Steensma, D.H., Van Zandt, M.C. (1992a) *Pure Appl. Chem.* **64**: 1115.

Johnson, C.R., Ple, P.A., Su, L., Heeg, M.J., Adams, J.P. (1992b) *Synlett* 338.

Johnson, E.P., Vollhardt, K.P.C. (1991) *J. Am. Chem. Soc.* **113**: 381.

Johnson, R.A., Sharpless, K.B. (1991) in Trost, B.M., Fleming, I. (eds.) *Comprehensive Organic Synthesis* **7**: 389.

Johnson, W.S., Collins, J.C., Pappo, R., Rubin, M.B., Kropp, P.J., Johns, W.F., Pike, J.E., Bartmann, W. (1963) *J. Am. Chem. Soc.* **85**: 1409.

Johnson, W.S., Semmelhack, M.F., Sultanbawa, M.U.S., Dolak, L.A. (1968) *J. Am. Chem. Soc.* **90**: 2994.

Johnson, W.S., Li, T., Harbert, C.A., Bartlett, W.R., Herrin, T.R., Staskun, B., Rich, D.H. (1970a) *J. Am. Chem. Soc.* **92**: 4461.

Johnson, W.S., Werthemann, L., Bartlett, W.R., Brocksom, T.J., Li, T., Faulkner, D.J., Peterson, M.R. (1970b) *J. Am. Chem. Soc.* **92**: 741.

Johnson, W.S., Brocksom, T.J., Loew, P., Rich, D.W., Werthemann, L., Arnold, R.A., T.t. Li., Faulkner, D.J. (1970c) *J. Am. Chem. Soc.* **92**: 4463.

Johnson, W.S., Gravestock, M.B., McCarry, B.E., (1971) *J. Am. Chem. Soc.* **93**: 4332.

Johnson, W.S. (1991) *Tetrahedron* **47**(41): xi.

Jommi, G., Orsini, F., Rosmini, M., Sisti, M. (1991) *Tetrahedron Lett.* **32**: 6969.

Jones, J.B., Finch, M.A.W., Jakovac, I.J. (1982) *Can. J. Chem.* **60**: 2007.

Jones, J.B. (1986) *Tetrahedron* **42**: 3351.

Jones, T.K., Denmark, S.E. (1983) *Helv. Chim. Acta* **66**: 2377.

Joshi, B.A., Viswanathan, N., Gawad, D.H., Balakrishnan, V., von Philipsborn, W., Quick, A. (1975) *Experientia* **31**: 880.

Julia, M., Julia, S., Guegan, R. (1960) *Bull. Soc. Chim. Fr.* 7072.

Julia, M., LeGoffic, F., Igolen, J., Baillarge, M. (1969) *Tetrahedron Lett.* 1569.

Julia, M., Bagot, J., Siffert, O. (1973) *Bull. Soc. Chim. Fr.* (II): 1424.

Julia, M., Schouteenten, A., Baillarge, M. (1974) *Tetrahedron Lett.* 3433.

Jung, M.E., Hudspeth, J.P., (1980) *J. Am. Chem. Soc.* **102**: 2463.

Jung, M.E., Halweg, K.M. (1981) *Tetrahedron Lett.* **22**: 2735.

Jung, M.E., Castro, C. (1993) *J. Org. Chem.* **58**: 807.

Just, G., Liak, T.J., Lim, M.I., Potvin, P., Tsantrizos, Y.S. (1980) *Can. J. Chem.* **58**: 2024.

Kadota, S., Takeshita, M., Makino, K., Kikuchi, T. (1989) *Chem. Pharm. Bull.* **37**: 843.

Kalivretenos, A., Stille, J.K., Hegedus, L.S. (1991) *J. Org. Chem.* **56**: 2883.

Kalvoda, J. (1968) *Helv. Chim. Acta* **51**: 267.

Kametani, T., Matsumoto, H., Nemoto, H., Fukumoto, K. (1978a) *J. Am. Chem. Soc.* **100**: 6218; (1978b) *Tetrahedron Lett.* 2425.

Kametani, T., Suzuki, K., Nemoto, H. (1981) *J. Am. Chem. Soc.* **103**: 2890.

Kametani, T., Toya, T., Ueda, K., Tsubuki, M., Honda, T. (1988) *J. Chem. Soc. Perkin Trans. I.* 2433.

Kametani, T., Tsubuki, M., Tatsuzaki, Y., Honda, T. (1990) *J. Chem. Soc. Perkin Trans. I.* 639.

Kanazawa, R., Kotsuki, H., Tokoroyama, T. (1975) *Tetrahedron Lett.* 3651.

Kanemasa, S., Kobayashi, S., Nishiuchi, M., Yamamoto, H., Wada, E. (1991) *Tetrahedron Lett.* **32**: 6367.

Kan-Fan, C., Massiot, G., Ahond, A., Das, B.C., Husson, H.-P., Potier, P., Scott, A.I., Wei, C.C. (1974) *Chem. Commun.* 164.

Kashima, C., Mukai, N., Yamamoto, Y., Tsuda, Y., Omote, (1977) *Heterocycles* **7**: 241.

Katagiri, N., Akatsuka, H., Haneda, T., Kaneko, C., Sera, A. (1988) *J. Org. Chem.* **53**: 5464.

Kato, N., Kusakabe, S., Wu, X., Kamitamari, M., Takeshita, H. (1993) *Chem. Commun.* 1002.

Katritzky, A.R., Cozens, A.J. (1983) *J. Chem. Soc. Perkin Trans. I.* 2611.

Kauffmann, T., Mitschker, A. (1973) *Tetrahedron Lett.* 4039.

Kauffmann, T., Abel, T., Schreer, M (1988) *Angew. Chem. Intern. Ed. Engl.* **27**: 944.

Kauffman, T. (1989) in Werner, H., Erker, G. (eds.) *Organometallics in Organic Synthesis*, **2**: 161, Springer Verlag, Berlin.

Kauffmann, T., Stach, D. (1991) *Angew, Chem. Intern. Ed. Engl.* **30**: 1684.

Kawanami, Y., Ito, Y., Kitagawa, T., Taniguchi, Y., Katsuki, T., Yamaguchi, M. (1984) *Tetrahedron Lett.* **25**: 857.

Keay, B.A., Rodrigo, R. (1982) *J. Am. Chem. Soc.* **104**: 4725.

Keck, G.E., Webb, R.R., II (1982) *J. Org. Chem.* **47**: 1302.

Kelly, T.R., Gillard, J.W., Goerner, R.N., Lyding, J.M. (1977) *J. Am. Chem. Soc.* **99**: 5513.

Kelly, T.R., Montury, M. (1978) *Tetrahedron Lett.* 4309.

Kelly, T.R., Whiting, A., Chandrakumar, N.S. (1986) *J. Am. Chem Soc.* **108**: 3510.

Kelly, T.R., Kim, M.H. (1992) *J. Org. Chem.* **57**: 1593.

Kende, A.S., Rizzi, J.P. (1981) *J. Am. Chem. Soc.* **103**: 4247.

Kende, A.S., Koch, K., Dorey, G., Kaldor, I., Lio, K. (1993) **115**: 9842.

Kim, S., Kim, Y.G., Kim, D. (1992) *Tetrahedron Lett.* **33**: 2565.

Kitagawa, I., Tsujii, S., Nishikawa, F., Shibuya, H. (1983) *Chem. Phasm. Bull.* **31**: 2639.

Kitahara, T., Mori, K., Matsui, M. (1984) *Tetrahedron* **40**: 2935.

Kitamura, M., Suga, S., Kawai, K., Noyori, R. (1986) *J. Am. Chem. Soc.* **108**: 6071.

Kito, M., Sakai, T., Yamada, K., Matsuda, F., Shirahama, H. (1993) *Synlett.* 158.

Klein, L.L. (1985) *J. Am. Chem. Soc.* **107**: 2573.

Klumpp, G.W., Veefkind, A.H., deGraaf, W.L., Bickelhaupt, F. (1967) *Liebigs Ann. Chem.* **706**: 47.

Knapp, S., Kukkola, P.J., (1990) *J. Org. Chem.* **55**: 1632.

Knox, L.H., Blossey, E., Carpino, H., Cervantes, L., Crabbe, P., Velarde, E., Edwards, J.A. (1965) *J. Org. Chem.* **30**: 2198.

Ko, S.Y., Lee, A.W.M., Masamune, S., Reed, L.A., Sharpless, K.B., Walker, F.J. (1983) *Science* **220**: 949.

Kobayashi, S., Koyayashi, S., Mukaiyama, T. (1974) *Chem. Lett.* 1425.

Koch, H., Runsink, J., Scharf, H.-D. (1983) *Tetrahedron Lett.* **24**: 3217.

Kocienski, P., Yeates, C. (1983) *Tetrahedron Lett.* **24**: 3905.

Kocienski, P., Wadman, S., Cooper, K. (1988) *Tetrahedron Lett.* **29**: 2357.

Kodama, M., Takahashi, T., Kurihara, T., Ito, S. (1980) *Tetrahedron Lett.* 2811.

Kogen, H., Tomioka, K., Hashimoto, S., Koga, K. (1981) *Tetrahedron* **37**: 3951.

Kon, K., Isoe, S. (1980) *Tetrahedron Lett.* **21**: 3399.

Kondo, K., Negishi, A., Matsui, K., Tunemoto, D., Masamune, S. (1972) *Chem. Commun.* 1311.

Kondo, K., Tunemoto, D. (1975) *Tetrahedron Lett.* 1397.

Kondo, T., Mukai, T., Watanabe, T. (1991) *J. Org. Chem.* **56**: 487.

Konopelski, J.P., Chu, K.S., Negrete, G.R. (1991) *J. Org. Chem.* **56**: 1355.

Koot, W.-J., van Ginkel, R., Kranenburg, M., Hiemstra, H., Louwrier, S., Moolenaar, M.J., Speckamp, W.N. (1991) *Tetrahedron Lett.* **32**: 401.

Koreeda, M., George, I.A. (1986) *J. Am. Chem. Soc.* **108**: 8098.

Kotani, E., Miyazaki, F., Tobinaga, S. (1974) *Chem. Commun.* 300.

Kotsuki, H., Ushio, Y., Kadota, I., Ochi, M. (1989) *J. Org. Chem.* **54**: 5153.

Kotsuki, H., Kadota, I., Ochi, M. (1990) *J. Org. Chem.* **55**: 4417.

Kozikowski, A.P., Scripko, J.G., (1984) *J. Am. Chem. Soc.* **106**: 353.

Krafft, M.E., Juliano, C.A., Scott, I.L., Wright, C., McEachin, M.D. (1991) *J. Am. Chem. Soc.* **113**: 1693.

Kramer, U., Guggisberg, A., Hesse, M., Schmid, H. (1978) *Angew, Chem. Intern. Ed. Engl.* **17**: 200.

Krämer, T., Hoppe, D. (1987) *Tetrahedron Lett.* **28**: 5149.

Krapcho, A., Vivelo, J.A. (1985) *Chem. Commun.* 233.

Kraus, G.A., Roth, B., Frazier, K., Shimagaki, M. (1982) *J. Am. Chem. Soc.* **104**: 1114.

Kraus, G.A., Hagen, M.D. (1983) *J. Org. Chem.* **48**: 3265.

Kraus, G.A., Nagy, J.O. (1985) *Tetrahedron* **41**: 3537.

Kraus, G.A., Shi, J. (1991) *J. Org. Chem.* **56**: 4147.

Krief, A., Dumont, W., Pasau, P. (1988) *Tetrahedron Lett.* **29**: 1079.

Krohn, K., Tolkiehn, K. (1979) *Chem. Ber.* **112**: 3453.

Kuehne, M.E., Roland, D.M., Hafter, R. (1978) *J. Org. Chem.* **43**: 3705.

Kuehne, M.E., Huebner, J.A., Matsko, T.H. (1979a) *J. Org. Chem.* **44**: 2477.

Kuehne, M.E., Matsko, T.H., Bohnert, J.C., Kirkemo, C.L. (1979b) *J. Org. Chem.* **44**: 1063.

Kumada, M. (1980) *Pure Appl. Chem.* **52**: 669.

Kündig, E.P., Ripa, A., Bernardinelli, G. (1992) *Angew. Chem. Intern. Ed. Engl.* **31**: 1071.

Kunz, H., Müller, B., Schanzenbach, D. (1987) *Angew. Chem. Intern. Ed. Engl.* **26**: 267.

Kunz, H., Mohr, J. (1988) *Chem. Commun.* 1315.

Kunz, H., Sager, W., Schanzenbach, D., Decker, M. (1991) *Liebigs Ann. Chem.* 649.

Kuo, D.L., Money, T. (1988) *Can. J. Chem.* **66**: 1794.

Kurth, M.J., Brown, E.G., Hendra, E., Hope, H. (1985) *J. Org. Chem.* **50**: 1115.

Kutney, J.P., Abdurahaman, N., Le Quesne, P., Piers, E., Vlattas, I. (1966) *J. Am. Chem. Soc.* **88**: 3656.

Kwantes, P.M., Klumpp, G.W. (1976) *Tetrahedron Lett.* 707.

Landry, D.W. (1983) *Tetrahedron* **39**: 2761.

Lansbury, P.T., Vacca, J.P. (1982) *Tetrahedron Lett.* **23**: 2623.

Larock, R.C., Lee, N.H. (1991) *J. Am. Chem. Soc.* **113**: 7815.

Laronze, J.-Y., Laronze- Fontaine, J., Levy, J., Le Men, J. (1974) *Tetrahedron Lett.* 491.

Larsen, S.D., Monti, S.A. (1977) *J. Am. Chem. Soc.* **99**: 8015.

Larsen, S.D., Monti, S.A. (1979) *Synth. Commun.* **9**: 141.

Larson, E.R., Raphael, R.A. (1982) *J. Chem. Soc. Perkin Trans. I* 521.

Lau, P.W.K., Chan, T.H. (1978) *Tetrahedron Lett.* 2383.

Laumen, K., Schneider, M. (1984) *Tetrahedron Lett.* 25: 5875.

Lautens, M., Belter, R.K. (1992) *Tetrahedron Lett.* **33**: 2617.

Lavielle, S., Bory, S., Moreau, B., Luche, M.J., Marquet, A. (1978) *J. Am. Chem. Soc.* **100**: 1558.

Le Dreau, M.-A., Desmaële, D., Dumas, F., d'Angelo, J. (1993) *J. Am. Chem. Soc.* **58**: 2933.

Lee, E., Paik, Y.H., Park, S.K. (1982) *Tetrahedron Lett.* **23**: 2671.

Lee, M.D., Dunne, T.S., Chang, C.C., Siegel, M.M., Morton, G.O., Ellestad, G.A., McGahren, W.J., Borders, D.B. (1992) *J. Am. Chem. Soc.* **114**: 985.

Lee, S.Y., Niwa, M., Snider, B.B. (1988) *J. Org. Chem.* **53**: 2356.

Lee, T.-J., Hoffman, W.F., Holtz, W.J., Smith, R.L. (1992) *J. Org. Chem.* **57**: 1966.

Lehr, F., Gonnermann, J., Seebach, D. (1979) *Helv. Chim. Acta* **62**: 2258.

Lei, B., Fallis, A.G. (1990) *J. Am. Chem. Soc.* **112**: 4609.

Le Merrer, Y., Dureault, A. Gravier, C., Languin, D., Depezay, J.C. (1985) *Tetrahedron Lett.* **26**: 319.

Leonard, N.J., Blum, S.W. (1960) *J. Am. Chem. Soc.* **82**: 503.

Leonard, N.J., Sato, T. (1969) *J. Org. Chem.* **34**: 1066.

Lewis, K.G. (1961) *J. Am. Chem. Soc.* 4690.

Ley, S.V., Armstrong, A., Diez-Martin, D., Ford, M.J., Grice, P., Knight, J.G., Kolb, H.C., Madin, A., Marby, C.A., Mukherjee, S., Shaw, A.N., Slawin, A.M.Z., Vile, S., White, A.D., Williams, D.J., Woods, M. (1989) *J. Chem. Soc. Perkin Trans. I* 667.

Ley, S.V., Leslie, R., Tiffin, P.D., Woods, M. (1992a) *Tetrahedron Lett.* **33**: 4767.

Ley, S.V., Woods, M., Zanotti-Gerosa, A. (1992b) *Synthesis* 52.

Liebeskind, L.S., Welker, M.E. (1984) *Tetrahedron Lett.* **25**: 4341.

Linstead, R.P., Lunt, J.C., Weedon, B.C.L. (1950) *J. Chem. Soc.* 3331; (1951) *J. Chem. Soc.* 1130.

Lipshutz, B.H., Kozlowski, J.A. (1984) *J. Org. Chem.* **49**: 1147.

Liu, H.-J., Sato, Y., Valenta, Z., Wilson, J.S., Yu, T.T.-J. (1976) *Can. J. Chem.* **54**: 97.

Lombardo, L., Mander, L.N., Turner, J.V. (1980) *J. Am. Chem. Soc.* **102**: 6626.

Lu, L.D.-L., Johnson, R.A., Finn, M.G., Sharpless, K.B. (1984) *J. Org. Chem.* **49**: 728.

Lu, T., Yoo, H.K., Zhang, H., Bott, S., Atwood, J.L., Echegoyen, L., Gokel, G.W. (1990) *J. Org. Chem.* **55**: 2269.

Lukes, R., Srogl, J. (1961) *Coll. Cech. Chem. Commun.* **26**: 2238.

Machinaga, N., Kibayashi, C. (1991) *J. Org. Chem.* **56**: 1386.

Machinaga, N., Kibayashi, C. (1993) *Tetrahedron Lett.* **34**: 841.

Madin, A., Overman, L.E. (1992) *Tetrahedron Lett.* **33**: 4859.

Magar, S.S., Desai, R.C., Fuchs, P.L. (1992) *J. Org. Chem.* **57**: 5360.

Magnus, P., Schultz, J., Gallagher, T. (1985) *J. Am. Chem. Soc.* **107**: 4984.

Magnus, P., Cairns, P.M., Moursounidis, J. (1987) *J. Am. Chem. Soc.* **109**: 2469.

Magnus, P., Carter, P.A. (1988) *J. Am. Chem. Soc.* **110**: 1626.

Magnus, P., Annoura, H., Harling, J. (1990) *J. Org. Chem.* **55**: 1709.

Magnus, P., Giles, M., Bonnert, R., Kim, C.S., McQuire, L., Merritt, A., Vicker, N. (1992a) *J. Am. Chem. Soc.* **114**: 4403.

Maitra, U., Bag, B.G. (1992) *J. Org. Chem.* **57**: 6979.

Mandai, T., Matsumoto, T., Kawada, M., Tsuji, J. (1992) *J. Org. Chem.* **57**: 1326.

Mandolini, L. (1986) *Pure Appl. Chem.* **58**: 1485.

Manh, D.D.K., Ecoto, J., Fetizon, M., Colin, H., Diez-Masa, J.-C. (1981) *Chem. Commun.* 953.

Marfat, A., McGuirk, P.R., Helquist, P. (1979) *J. Org. Chem.* **44**: 1345.

Marinier, A., Deslongchamps, P. (1988) *Tetrahedron Lett.* **29**: 6215.

Marino, J.P., Long, J.K. (1988) *J. Am. Chem. Soc.* **110**: 7916.

Marshall, J.A., Rudin, R.A. (1970) *Tetrahedron Lett.* 1239.

Marshall, J.A., Ruth, J.A., (1974) *J. Org. Chem.* **39**: 1971.

Marshall, J.A., Ellison, R.H. (1976) *J. Am. Chem. Soc.* **98**: 4312.

Marshall, J.A., Conrow, R.E. (1983) *J. Am. Chem. Soc.* **105**: 5679.

Marshall, J.A., Wang, X. (1992) *J. Org. Chem.* **57**: 1242.

Martin, S.F., DuPriest, M.T. (1977a) *Tetrahedron Lett.* 3925.

Martin, S.F., Garrison, P.J. (1977b) *Tetrahedron Lett.* 3875.

Martin, S.F., Dappen, M.S., Dupre, B., Murphy, C.J. (1987a) *J. Org. Chem.* **52**: 3706.

Martin, S.F., Grzejszczak, S., Rüegar, H., Williamson, S.A. (1987b) *J. Am. Chem. Soc.* **109**: 6124.

Martin, S.F., Pacofsky, G.J., Gist, R.P., Lee, W.-C. (1989) *J. Am. Chem. Soc.* **111**: 7634.

Martin, V.S., Woodward, S.S., Katsuki, T., Yamada, Y., Ikeda, M. Sharpless, K.B. (1981) *J. Am. Chem. Soc.* **103**: 6237.

Maruoka, K., Itoh, T., Yamamoto, H. (1985) *J. Am. Chem. Soc.* **107**: 4573.

Maruoka, K., Ito, T., Sakurai, M., Nonoshita, K., Yamamoto, H. (1988) *J. Am. Chem. Soc.* **110**: 3588.

Maruoka, K., Sato, J., Banno, H., Yamamoto, H. (1990) *Tetrahedron Lett.* **31**: 377.

Maruoka, K., Sato, Yamamoto, H. (1992) *J. Am. Chem. Soc.* **114**: 1089.

Masaki, Y., Nagata, K., Serizawa, Y., Kaji, K. (1982) *Tetrahedron Lett.* **23**: 5553.

Masamune, S., Ang, S.K., Egli, C., Nakatasuka, N., Sarkar, S.K., Yasunari, Y. (1967) *J. Am. Chem. Soc.* **89**: 2506.

Masamune, S., Kim, C.U., Wilson, K.E., Spessard, G.O., Georghiou, P.E., Bates, G.S. (1975) *J. Am. Chem. Soc.* **97**: 3512.

Masamune, S., Reed, L.A., Davis, J.T., Choy, W. (1983) *J. Org. Chem.* **48**: 4441.

Masamune, S., Choy, W., Petersen, J.S., Sita, L.R. (1985) *Angew. Chem. Intern. Ed. Engl.* **24**: 1.

Massanet, G.M., Pando, E., Rodriguez-Luis, F., Salva, J. (1987) *Heterocyles* **26**: 1541.

Masuyama, Y., Hayashi, R., Otake, K., Kurusu, Y. (1988) *Chem. Commun.* 44.

Mathews, R.S., Whitesell, J.K. (1975) *J. Org. Chem.* **40**: 3312.

Matsutomo, T., Ichihara, A., Yanagiya, M., Yuzawa, T., Sannai, A., Oikawa, H., Sakamura, S. Eugster, C.H. (1985) *Helv. Chim. Acta* **68**: 2324.

McCrombie, S.W., Cox, B., Lin, S.-I., Ganguly, A.K., McPhail, A.T. (1991) *Tetrahedron Lett.* **32**: 2083.

McCurry, P.M., Abe, K. (1973) *J. Am. Chem. Soc.* **95**: 5824.

McGarvey, G.J., Williams, J.M., Hiner, R.N., Matsubara, T., Oh, T. (1986) *J. Am. Chem. Soc.* **108**: 4943.

McKew, J.C., Kurth, M.J. (1993) *J. Org. Chem.* **58**: 4589.

McKillop, A., McLaren, L., Watson, R.J., Taylor, R.J.K., Lewis, N. (1993) *Tetrahedron Lett.* **34**: 5519.

McMurry, J.E. (1968) *J. Am. Chem. Soc.* **90**: 6821.

McMurry, J.E., Glass, T.E. (1971) *Tetrahedron Lett.* 2575.

McMurry, J.E., Isser, S.J. (1972) *J. Am. Chem. Soc.* **94**: 7132.

McMurry, J.E., Fleming, M.P. (1974) *J. Am. Chem. Soc.* **96**: 4708.

McMurry, J.E., Blaszczak, L.C., Johnson, M.A. (1978) *Tetrahedron Lett.* 1633.

McMurry, J.E. (1989) *Chem. Rev.* **89**: 1513.

McNeill, A.H., Thomas (1990) *Tetrahedron Lett.* **31**: 6239; (1992) *Tetrahedron Lett.* **33**: 1369.

Mehta, G., Murthy, A.N., Reddy, D.S., Reddy, A.V. (1986) *J. Am. Chem. Soc.* **108**: 3443.

Mehta, G., Viswanath, M.B., Sastry, G.N., Jemmis, E.D., Reddy, D.S.K. Kunwar, A.C. (1992) *Angew. Chem. Intern. Ed. Engl.* **31**: 1488.

Mekelburger, H.-B., Jaworek, W., Vögtle, F. (1992) *Angew. Chem. Intern. Ed. Engl.* **31**: 1571.

Melder, J.-P., Pinkos, R., Fritz, H., Prinzbach, H. (1989) *Angew. Chem. Intern. Ed. Engl.* **28**: 305.

Meric, R., Vigneron, J. (1973) *Bull. Soc. Chim. Fr.* 327.

Merlini, L., Nasini, G. (1967) *Gazz. Chim. Ital.* **97**: 1915.

Merrifield, R.B. (1964) *J. Am. Chem. Soc.* **86**: 304; *Biochemistry* **3**: 1385.

Meyers, A.I., Whitten, C.E. (1975) *J. Am. Chem. Soc.* **97**: 6266.

Meyers, A.I., Knaus, G., Kamata, K., Ford., M.E. (1976) *J. Am. Chem. Soc.* **98**: 567.

Meyers, A.I., Williams, D.R., Erickson, G.W., White, S., Druelinger, M. (1981) *J. Am. Chem. Soc.* **103**: 3081.

Meyers, A.I., Babiak, K.A., Campbell, A.L., Comins, D.L., Fleming, M.P., Henning, R., Heuschmann, M., Hudspeth, J.P., Kane, J.M., Reider, P.J., Roland, D.M., Shimizu, K., Tomioka, K., Walkup, R.D. (1983) *J. Am. Chem. Soc.* **105**: 5015.

Meyers, A.I., Yamamoto, Y. (1984) *Tetrahedron* **40**: 2309.

Meyers, A.I., Brown, J.D., Laucher. D. (1987a) *Tetrahedron Lett.* **28**: 5283.

Meyers, A.I., Dickman, D.A. (1987b) *J. Am. Chem. Soc.* **109**: 1263.

Micovic, V.M., Stojcic, S., Bralovic, M., Mladenovic, S., Jeremic, D., Stefanovic, M. (1969) *Tetrahedron* **25**: 985.

Midland, M.M., McDowell, D.C., Hatch, R.L., Tramontano, A. (1980) *J. Am. Chem. Soc.* **102**: 867.

Midland, M.M., Graham, R.S. (1984a) *J. Am. Chem. Soc.* **106**: 4294.

Midland, M.M., McLoughlin, J.L. (1984b) *J. Org. Chem.* **49**: 1316.

Mikami, K., Loh, T.-P., Nakai, T. (1988) *Tetrahedron Lett.* **29**: 6305.

Mikami, K., Takahashi, K., Nakai, T. (1990) *J. Am. Chem. Soc.* **112**: 4035.

Mitchell, D., Koenig, T.M., (1972) *Tetrahedron Lett.* **33**: 3281.

Miyano, S., Handa, S., Tobita, M., Hashimoto, H. (1986) *Bull. Chem. Soc. Jpn.* **59**: 235.

Miyano, S., Fukushima, H., Handa, S., Ito, Hashimoto, H., (1988) *Bull. Chem. Soc. Jpn.* **61**: 3249.

Miyashita, M., Yoshikoshi, A. (1974) *J. Am. Chem. Soc.* **96**: 1917.

Mizuno, Y., Tomita, M., Irie, H. (1980) *Chem. Lett.* 107.

Molander, G.A., Etter, J.B., Harring. L.S., Thorel, P.-J. (1991a) *J. Am. Chem. Soc.* **113**: 8036.

Molander, G.A., Haar, J.P. (1991b) *J. Am. Chem. Soc.* **113**: 3608.

Molander, G.A., Bobbitt, K.L., Murray, C.K. (1992) *J. Am. Chem. Soc.* **114**: 2759.

Momose, T., Toyooka, N., Seki, S., Hirai, Y. (1990) *Chem. Pharm. Bull.* **38**: 2072.

Mondon, A., Ehrhardt, M. (1966) *Tetrahedron Lett.* 2557.

Montpert, A., Martelli, J., Gree, R., Carrie, R. (1981) *Tetrahedron Lett.* **22**: 1961.

Moody, C.J., Roberts, S.M., Toczek, J. (1988) *J. Chem. Soc. Perkin Trans. I* 1401.

Moody, C.J., Rees, C.W., Thomas, R. (1990) *Tetrahedron Lett.* **31**: 4375.

Moody, C.J., Rahimtoola, K.F., Porter, B., Ross, B.C. (1992) *J. Org. Chem.* **57**: 2105.

Morgans, D.J. (1981) *Tetrahedron Lett.* **22**: 3721.

Mori, A., Ishihara, K., Arai, I., Yamamoto, H. (1987) *Tetrahedron* **43**: 755.

Mori, K., Watanabe, H. (1986) *Tetrahedron* **42**: 273.

Mori, M., Chuman, T., Kato, K., Mori, K. (1982) *Tetrahedron Lett.* **23**: 4593.

Mori, Y., Suzuki, M. (1989a) *Tetrahedron Lett.* **30**: 4383; (1989b) *Tetrahedron Lett.* **30**: 4387.

Mori, Y., Kohchi, Y., Ota, T., Suzuki, M. (1990) *Tetrahedron Lett.* **31**: 2915.

Motherwell, W.B., Shipman, M. (1991) *Tetrahedron Lett.* **32**: 1103.

Mpango, G.B., Snieckus, V. (1980) *J. Am. Chem. Soc.* **111**: 34827.

Mukaiyama, T. (1976) *Angew. Chem. Intern. Ed. Engl.* **15**: 94.

Mukaiyama, T., Takeda, T., Osaki, M. (1977) *Chem. Lett.* 1165.

Mukaiyama, T., Sakito, Y., Asami, M. (1978a) *Chem. Lett.* 1253.

Mukaiyama, T., Takeda, T., Fujimoto, K. (1978b) *Bull. Chem. Soc. Jpn.* **51**: 3368.

Mukaiyama, T., Sakito, Y., Asami, M. (1979) *Chem. Lett.* 705.

Mukaiyama, T., Isawa, N. (1981) *Chem. Lett.* 29.

Mukaiyama, T., Yamada, T., Suzuki, K. (1983) *Chem. Lett.* 705.

Mukaiyama, T., Yamada, T., Suzuki, K. (1983) *Chem. Lett.* 5.

Mukaiyama, T., Iwasawa, N. (1984) *Chem. Lett.* 753.

Müller, I., Jäger, V. (1982) *Tetrahedron Lett.* **23**: 4777.

Mulzer, J., Schöllhorn, B. (1990) *Angew. Chem. Intern. Ed. Engl.* **29**: 1476.

Murad, G., Cagniant, D., Cagniant, P. (1973) *Bull. Chem. Chim. Fr.* (*II*) 343.

Murai, A., Sato, S., Masamune, T. (1981) *Tetrahedron Lett.* **22**: 1033.

Murakami, M., Hayashi, M., Ito, Y. (1992) *J. Org. Chem.* **57**: 793.

Murakami, M., Suginome, M., Fujimoto, K., Nakamura, H., Andersson, P.G., Ito, Y. (1993) *J. Am. Chem. Soc.* **115**: 6487.

Muratake, H., Mikawa, A., Natsume, M. (1992) *Tetrahedron Lett.* **32**: 4595.

Muthukrishnan, R., Schlosser, M. (1976) *Helv. Chim. Acta* **59**: 13.

Muxfeldt, H., Hardtmann, G., Kathawala, F., Vedejs, E., Mooberry, J.B. (1968) *J. Am. Chem. Soc.* **90**: 6534.

Myers, A.G., Widdowson, K.L. (1990) *J. Am. Chem. Soc.* **112**: 9672.

Myers, A.G., Kephart, S.E., Chen, H. (1992a) *J. Am. Chem. Soc.* **114**: 7922.

Myers, A.G., Widdowson, K.L., Kukkola, P.J. (1992b) *J. Am. Chem. Soc.* **114**: 2765.

Myles, D.C., Danishefsky, S.J., Schulte, G. (1990) *J. Am. Chem. Soc.* **55**: 1636.

Näf, F., Decorzant, R., Giersch, W., Ohloff, G. (1981) *Helv. Chim. Acta* **64**: 1387.

Nagai, M., Gaudino, J.J., Wilcox, C.S. (1992) *Synthesis* 163.

Nagaoka, H., Ohsawa, K., Takata, T., Yamada, Y. (1984) *Tetrahedron Lett.* **25**: 5389.

Nagaoka, H., Kobayashi, K., Matsui, T., Yamada, Y. (1987) *Tetrahedron Lett.* **28**: 2021.

Nagaoka, H., Kobayashi, K., Yamada, Y. (1988) *Tetrahedron Lett.* **29**: 5945.

Nagumo, S., Suemune, H., Sakai, K. (1990) *Chem. Commun.* 1778.

Naito, T., Iida, N., Ninomiya, I. (1986) *J. Chem. Soc. Perkin Trans. I* 99.

Naito, T., Habu, Y., Miyata, O., Ninomiya, I., Ohishi, H. (1992) *Chem. Phar. Bull.* **40**: 602.

Nakamura, E., Kuwajima, I. (1977) *J. Am. Chem. Soc.* **99**: 7360.

Nakata, T., Suenaga, T., Oishi, T. (1989) *Tetrahedron Lett.* **30**: 6525.

Nakatsuka, M., Ragan, J.A., Sammakia, T., Smith, D.B., Uehling, D.E., Schreiber, S.L. (1990) *J. Am. Chem. Soc.* **112**: 5583.

Nakayama, J., Yamaoka, S., Nakanishi, T., Hoshino, M. (1988) *J. Am. Chem. Soc.* **110**: 6598.

Nakayama, J., Hasemi, R. (1990) *J. Am. Chem. Soc.* **112**: 5654.

Naota, T., Taki, H., Mizuno, M., Murahashi, S.-I. (1989) *J. Am. Chem. Soc.* **111**: 5954.

Narasaka, K., Miwa, T., Hayashi, H., Ohta, M. (1984a) *Chem. Lett.* 1399.

Narasaka, K., Pai, F.-C. (1984b) *Tetrahedron* **40**: 2233.

Narasaka, K., Shimada, S., Osoda, K., Iwasawa, N. (1991) *Synthesis* 1171.

Nazarov, I.N., Gusev, B.P., Gunar, V.I. (1958) *Zhur. Obshch. Khim. SSSR* **28**: 1444.

Neeson, S.J., Stevenson, P.J. (1988) *Tetrahedron Lett.* **29**: 813.

Negishi, E., Luo, F.-T., Pearson, A.J., Silveira, A. (1983) *J. Org. Chem.* **48**: 2427.

Negishi, E., Iyer, S., Rousset, C.J. (1989) *Tetrahedron Lett.* **30**: 291.

Negishi, E. (1991) in Trost, B.M., Fleming, I. (ed) *Comprehensive Organic Synthesis* **5**: 1163.

Negishi, E. (1992) *Pure Appl. Chem.* **64**: 323.

Nerz-Stormes, M., Thornton, E.R. (1991) *J. Org. Chem.* **56**: 2489.

Newman-Evans, R.H., Carpenter, B.K. (1985) *Tetrahedron Lett.* **26**: 1141.

Newton, R.F., Roberts, S.M. (1980) *Tetrahedron* **36**: 2163.

Nicholas, K.M. (1987) *Acc. Chem. Res.* **20**: 207.

Nicolaou, K.C., Petasis, N.A., Uenishi, J., Zipkin, R.E. (1982a) *J. Am. Chem. Soc.* **104**: 5557.

Nicolaou, K.C., Petasis, N.A., Zipkin, R.E. (1982b) *J. Am. Chem. Soc.* **104**: 5560.

Nicolaou, K.C., Uenishi, J. (1982c) *Chem. Commun.* 1292.

Nicolaou, K.C., Zipkin, R.E., Petasis, N.A. (1982d) *J. Am. Chem. Soc.* **104**: 5558.

Nicolaou, K.C., Veale, C.A., Webber, S.E., Katerinopoulos, H. (1985) *J. Am. Chem. Soc.* **107**: 7515.

Nicolaou, K.C., Ahn, K.H. (1989) *Tetrahedron Lett.* **30**: 1217.

Nicolaou, K.C. (1991a) *Chemtracts-Org. Chem.* **4**: 181.

Nicolaou, K.C., Veale, C.A., Hwang, C.-K., Hutchinson, J., Prasad, C.V.C., Ogilvie, W.W. (1991b) *Angew. Chem. Intern. Ed. Engl.* **30**: 299.

Ninomiya, I., Naito, T. (1983) in Brossi, A. (ed.) *The Alkaloids* **22**: 189, Academic press, N.Y.

Node, M., Hao, X.-j., Nagasawa, H., Fuji, K. (1989) *Tetrahedron Lett.* **30**: 4141.

Nomoto, T., Takayama, H. (1985) *Heterocycles* **23**: 2913.

Nonoshita, K., Banno, H., Maruoka, K., Yamamoto, H. (1990) *J. Am. Chem. Soc.* **112**: 316.

Noyori, R., Baba, Y., Hayakawa, Y. (1974) *J. Am. Chem. Soc.* **96**: 3336.

Noyori, R., Baba, Y., Makino, S., Takaya, T. (1978) *J. Am. Chem. Soc.* **100**: 1786.

Noyori, R., Suzuki, M., (1990) *Chemtracts-Org. Chem.* **3**: 173.

Nunn, K., Mosset, P., Gree, R., Saalfrank, R.W. (1988) *Angew. Chem. Intern. Ed. Engl.* **27**: 1188.

Nuss, J.M., Rennels, R.A., Levine, B.H. (1993) *J. Am. Chem. Soc.* **115**: 6991.

Obaza-Nutaitis, J.A., Gribble, G.W. (1986) *J. Nat. Prod.* **49**: 449.

O'Brien, M.K., Pearson, A.J., Pinkerton, A.A., Schmidt, W., Willman, K. (1989) *J. Am. Chem. Soc.* **111**: 1499.

Ochiai, M., Iwaki, S., Ukita, T., Matsuura, Y., Shiro, M., Nagao, Y. (1988) *J. Am. Chem. Soc.* **110**: 4606.

Oda, H., Sato, M., Morizawa, Y., Oshima, K., Nozaki, H. (1983) *Tetrahedron Lett.* **24**: 2877.

Odinokov, V.N., Kukovinets, O.S., Sakharova, N.I., Tolstikov, G.A. (1985) *Zh. Org. Khim.* **21**: 1180.

Ogawa, T., Fang, C.-L., Suemune, H., Sakai, K. (1991) *Chem. Commun.* 1438.

Ogura, K., Tsuchihashi, G. (1971) *Tetrahedron Lett.* 3151.

Ogura, K., Yamashita, M., Suzuki, M., Tsuchihashi, G. (1974) *Tetrahedron Lett.* 3653.

Ogura, K., Yamashita, M., Furukawa, S., Suzuki, M., Tsuchihashi, G. (1975) *Tetrahedron Lett.* 2767.

Ohashi, M., Maruishi, T., Kakisawa, H. (1968) *Tetrahedron Lett.* 719.

Ohfune, Y., Tomita, (1982) *J. Am. Chem. Soc.* **104**: 3511.

Ohkita, M., Tsuji, T., Nishida, S. (1991) *Chem. Commun.* 37.

Ohno, A., Ikeguchi, M., Kimura, T., Oka, S. (1979) *J. Am. Chem. Soc.* **101**: 7036.

Ohno, M., Kobayashi, S., Kurihara, M. (1986) *Org. Synth. Chem. Jpn.* **44**: 38.

Ohnuma, T., Tabe, M., Shiiya, K., Ban, Y., Date, T. (1983) *Tetrahedron Lett.* **24**: 4249.

Ohtsuka, Y., Nittsuma, S., Tadokoro, H., Hayashi, T., Oishi, T. (1984) *J. Am. Chem. Soc.* **49**: 2326.

Ohwa, M., Eliel, E.L. (1987) *Chem. Lett.* 41.

Oinuma, H., Dan, S., Kakisawa, H. (1983) *Chem. Commun.* 654.

Oishi, T., Nakata, T. (1984) *Acc. Chem. Res.* **17**: 338.

Ojima, I., Inaba, S. (1980) *Tetrahedron Lett.* **31**: 2077, 2081.

Okada, K., Mizuno, Y., Tanino, H., Kakoi, H., Inoue, S. (1992) *Chem. Pharm. Bull.* **40**: 1110.

Okamura, H., Mitsuhira, Y., Miura, M., Takei, H. (1978) *Chem. Lett.* 517.

Ono, N., Miyake, H., Kaji, A. (1982) *Chem. Commun.* 33.

Openshaw, H.T., Whittaker, N. (1963) *J. Am. Chem. Soc.* 1461.

Oppolzer, W., Hauth, H., Pfäffli, P. Wenger, R. (1977) *Helv. Chim. Acta* **60**: 1801.

Oppolzer, W., Godel, T. (1978) *J. Am. Chem. Soc.* **100**: 2583.

Oppolzer, W., Roberts, D.A. (1980) *Helv. Chim. Acta* **63**: 1703.

Oppolzer, W., Francotte, E., Bättig, K. (1981a) *Helv. Chim. Acta* **64**: 478.

Oppolzer, W., Löher, H.J. (1981b) *Helv. Chim. Acta* **64**: 2808.

Oppolzer, W., Bättig, K. (1982a) *Tetrahedron Lett.* **23**: 4669.

Oppolzer, W., Pitteloud, R. (1982b) *J. Am. Chem. Soc.* **104**: 6478.

Oppolzer, W. (1983a) in Nozaki, H. (ed.) *Current Trends in Organic Chemistry*, Pergamon, Oxford; p. 131.

Oppolzer, W., Moretti, R., Godel, A., Meunier, A., Löher, H. (1983b) *Tetrahedron Lett.* **24**: 4971.

Oppolzer, W., Robbiani, C. (1983c) *Helv. Chim. Acta* **66**: 1119.

Oppolzer, W., Dupuis, D. (1985) *Tetrahedron Lett.* **26**: 5437.

Oppolzer, W., Radinov, R.N. (1988) *Tetrahedron Lett.* **29**: 5645.

Oppolzer, W. (1989) *Angew. Chem. Intern. Ed. Engl.* **28**: 38.

Oppolzer, W., Tamura, O., Sundarababu, G., Signer, M. (1992) *J. Am. Chem. Soc.* **114**: 5900.

Orban, J., Turner, J.V. (1983) *Tetrahedron Lett.* **24**: 2697.

Orito, K., Yorota, K., Suginome, H. (1991) *Tetrahedron Lett.* **32**: 5999.

Oshima, K., Yamamoto, H., Nozaki, H. (1973) *J. Am. Chem. Soc.* **95**: 7926.

Ostrowicki, A., Koepp, E., Vögtle, F. (1991) *Top. Curr. Chem.* **161**: 37.

Otterbach, A., Musso, H. (1987) *Angew. Chem. Intern. Ed. Engl.* **26**: 554.

Overman, L.E., Campbell, C.B. (1974) *J. Org. Chem.* **39**: 1474.

Overman, L.E., Malone, T.C. (1982) *J. Org. Chem.* **47**: 5297.

Overman, L.E., Sworin, M., Burk, R.M. (1983) *J. Org. Chem.* **48**: 2685.

Overman, L.E., Wild, H. (1989) *Tetrahedron Lett.* **30**: 647.

Paddon-Row, M.N., Rondan, N.G., Houk, K.N. (1982) *J. Am. Chem. Soc.* **104**: 7162.

Page, P.C.B., van Niel, M.B., Prodger, J.C. (1989) *Tetrahedron* **45**: 7643.

Palmisano, G., Danieli, B., Lesma, G., Passarella, D., Toma, L. (1991) *J. Org. Chem.* **56**: 2380.

Palomo, C., Aizpurua, J.M., Urchegui, R., Garcia, J.M. (1993) *J. Org. Chem.* **58**: 1646.

Pandey, G., Reddy, P.Y., Bhalero, U.T. (1991) *Tetrahedron Lett.* **32**: 5147.

Pansegrau, P.D., Rieker, W.F., Meyers, A.I. (1988) *J. Am. Chem. Soc.* **110**: 7178.

Paquette, L.A., Wyvratt, M.J., Berk, H.C., Moerck, R.E. (1978) *J. Am. Chem. Soc.* **100**: 5845.

Paquette, L.A. (1984) in Lindberg, T. (ed.), *Strategies and tactics in organic synthesis*, Academic Press, Orlando, FL.

Paquette, L.A. (1990a) *Angew. Chem. Intern. Ed. Engl.* **29**: 609.

Paquette, L.A., Macdonald, D., Anderson, L.G. (1990b) *J. Am. Chem. Soc.* **112**: 9292.

Paquette, L.A., Maleczka, R.E. (1992) *J. Org. Chem.* **57**: 7118.

Parker, K.A., Fokas, D. (1992a) *J. Am. Chem. Soc.* **114**: 9688.

Parker, K.A., Kim, H.-J. (1992b) *J. Org. Chem.* **57**: 752.

Parker, W., Raphael, R.A., Wilkinson, D.I. (1959) *J. Chem. Soc.* 2433.

Parker, W.L., Johnson, F. (1969) *J. Am. Chem. Soc.* **91**: 7208.

Parnell, C.A., Vollhardt, K.P.C. (1985) *Tetrahedron* **41**: 5791.

Partridge, J.J., Chadha, N.K., Uskokovic, M.R. (1973) *J. Am. Chem. Soc.* **95**: 532.

Paterson, I. Tillyer, R.D. (1992) *Tetrahedron Lett.* **33**: 4233.

Pattenden, G., Teague, S.J. (1984) *Tetrahedron Lett.* **25**: 3021.

Pauls, H.W., Fraser-Reid, B. (1983) *J. Org. Chem.* **48**: 1392.

Pauson, P.L. (1985) *Tetrahedron Lett.* **41**: 5855.

Pearlman, B.A. (1979) *J. Am. Chem. Soc.* **101**: 6404.

Pearson, A.J., Lai, Y.S. (1988) *Chem. Commun.* 442.

Pearson, A.J., Mortezrei, R. (1989) *Tetrahedron Lett.* **30**: 5049.

Pearson, A.J. (1990) *Synlett* 10.

Pearson, A.J., Lai, Y.-S., Srinivasan, K. (1992) *Aust. J. Chem.* **45**: 109.

Pedersen, C.J., Frensdorff, H.K. (1972) *Angew. Chem. Intern. Ed. Engl.* **11**: 241.

Peel, R., Sutherland, J.K. (1974) *Chem. Commun.* 151.

Pelter, A., Singaram, B., Wilson, J. (1983) *Tetrahedron Lett.* **24**: 631.

Pierre, J.-L., Gagnaire, G., Chautemps, P. (1992) *Tetrahedron Lett.* **33**: 217.

Piers, E., Abeysekera, B.F., Herbert, D.J., Sucking, I.D. (1985) *Can. J. Chem.* **63**: 3418.

Piers, E., Roberge, J.Y. (1991) *Tetrahedron Lett.* **32**: 5219.

Pinkos, R., Rihs, G., Prinzbach, H. (1989) *Angew. Chem. Intern. Ed. Engl.* **28**: 303.

Posner, G.H., Lentz, C.M. (1978) *Tetrahedron Lett.* 3769.

Posner, G.H., Mallamo, J.F., Black, A.Y. (1981) *Tetrahedron Lett.* **37**: 3921.

Posner, G.H., Mallamo, J.P., Hulce, M., Frye, L.L. (1982) *J. Am. Chem. Soc.* **104**: 4180.

Posner, G.H., Miura, K., Mallamo, J.P., Hulce, M., Kogan, T.P. (1983) in Nozaki, H. (ed.), "*Current Trends in Organic Synthesis*", Pergamon Press, Oxford, p. 177.

Posner, G.H., Hulce, M. (1984) *Tetrahedron Lett.* **25**: 379.

Posner, G.H., Weitzberg, M., Hamill, T.G., Asirvatham, E., Cun-heng, H., Clardy, J. (1986) *Tetrahedron Lett.* **42**: 2919.

Poss, C.S., Rychnovsky, S.D., Schreiber, S.L. (1993) *J. Am. Chem. Soc.* **115**: 3360.

Poulter, C.D., Friedrich, E.C., Winstein, S. (1969) *J. Am. Chem. Soc.* **91**: 6892.

Prelog, V. (1953) *Helv. Chim. Acta* **36**: 308.

Quinkert, G., Schmieder, K.R., Dürner, G., Hache, K., Stegk, A., Barton, D.H.R. (1977) *Chem. Ber.* **110**: 3582.

Quinkert, G., Adam, F., Dürner, G. (1982) *Angew. Chem. Intern. Ed. Engl.* **21**: 856.

Rama Rao, A.V., Chakraborty, T.K., Joshi, S.P. (1992) *Tetrahedron Lett.* **33**: 4045.

Ramig, K., Kuzemko, M.A., McNamara, K., Cohen, T. (1992) *J. Org. Chem.* **57**: 1968.

Ranganathan, D., Rathi, R., Sharma, S. (1990) *J. Org. Chem.* **55**: 4006.

Rawal, V.H., Michoud, C. (1993) *J. Org. Chem.* **58**: 5583.

Rebek, J., Costello, T., Marshall, L., Watthey, R., Gadwood, R.C., Onan, K. (1985a) *J. Am. Chem. Soc.* **107**: 7481.

Rebek, J., Costello, T., Watthey, R. (1985b) *J. Am. Chem. Soc.* **107**: 7487.

Reetz, M.T. (1982a) *Top, Curr. Chem.* **106**: 1.

Reetz, M.T., Steinbach, R., Kessler, K. (1982b) *Angew. Chem. Intern. Ed. Engl.* **21**: 864.

Reetz, M.T., Wenderoth, B. (1982c) *Tetrahedron Lett.* **23**: 5259.

Reetz, M.T., Jung, A. (1983a) *J. Am. Chem. Soc.* **105**: 4833.

Reetz, M.T., Kesseler, K., Schmidtberger, S., Wenderoth, B., Steinbach, R. (1983b) *Angew. Chem. Intern. Ed. Engl.* **22**: 989.

Reetz, M.T., Wenderoth, B., Peter, R. (1983c) *Chem. Commun.* 406.

Reetz, M.T. (1984a) *Angew. Chem. Intern. Ed. Engl.* **23**: 556.

Reetz, M.T., Kesseler, K., Jung, A. (1984b) *Tetrahedron Lett.* **25**: 729.

Reetz, M.T., Westermann, J., Kyung, S.H. (1985) *Chem. Ber.* **118**: 1050.

Reetz, M.T., Jung, A., Bolm, C. (1988) *Tetrahedron Lett.* **44**: 3889.

Reetz, M.T. (1991) *Angew. Chem. Intern. Ed. Engl.* **30**: 1531.

Reetz, M.T. (1993) *Acc. Chem. Res.* **26**: 462.

Regen, S.L. (1975) *J. Am. Chem. Soc.* **97**: 5956.

Reginato, G., Ricci, A., Roelens, S., Scapecchi, S. (1990) *J. Org. Chem.* **55**: 5132.

Renger, B., Seebach, D. (1977) *Cherm. Ber.* **110**: 2334.

Rigby, J.H., Senanayake, C. (1987) *J. Am. Chem. Soc.* **109**: 3147.

Rigby, J.H., Ateeq, H.S., Charles, N.R., Cuisiat, S.V., Ferguson, M.D., Henshilwood, J.A., Krueger, A.C., Ogbu, C.O., Short, K.M., Heeg, M.J. (1993) *J. Am. Chem. Soc.* **115**: 1382.

Robin, J.P., Gringore, O., Brown, E. (1980) *Tetrahedron Lett.* 2709.

Robinson, R. (1917) *J. Chem. Soc.* 762.

Rosenmund, P., Hosseini-Merescht, M. (1992) *Liebigs Ann. Chem.* 1321.

Rosini, G., Ballini, R., Sorrenti, P. (1983) *Tetrahedron* **39**: 4127.

Rossiter, B.E. (1985) in Morrison, J.D. (ed.) *Asymmetric Synthesis* **5**: 193.

Roush, W.R., Gillis, H.R., Ko, A.I. (1982) *J. Am. Chem. Soc.* **104**: 2269.

Roush, W.R., Park, J.C. (1991) *Tetrahedron Lett.* **32**: 6285.

Rowley, M., Tsukamoto, M., Kishi, Y. (1989) *J. Am. Chem. Soc.* **111**: 2735.

Rychnovsky, S.D., Griesgraber, G. (1992a) *J. Org. Chem.* **57**: 1559.

Rychnovsky, S.D., Rodriguez, C. (1992b) *J. Org. Chem.* **57**: 4793.

Sabbioni, G., Shea, M.L., Jones, J.B. (1984) *Chem. Commun.* 236.

Sachdev, K., Sachdev, H.S. (1976) *Tetrahedron Lett.* 4223.

Saito, S., Yamamoto, T., Matsuoka, M., Moriwake, T. (1992) *Synlett* 239.

Sakan, T., Fujino, A., Murai, F., Suzui, F., Butsugan, Y. (1960) *Bull.Chem. Soc. Jpn.* **33**: 1737.

Sakan, T., Mori, Y. (1973) *Chem. Lett.* 713.

Sakito, Y., Mukaiyama, T. (1979) *Chem Lett.* 1207.

Salomon, R.G., Sachinvala, N.D., Raychaudhuri, S.R., Miller, D.B. (1984) *J. Am. Chem. Soc.* **106**: 2211.

Sardina, F.J., Howard, M.H., Morningstar, M., Rapoport, H. (1990) *J. Org. Chem.* **55**: 5025.

Sato, M., Abe, Y., Takayama, K., Sekiguchi, K., Kaneko, C., Inoue, N., Furuya, T., Inukai, N. (1991) *J. Heterocycl. Chem.* **28**: 241.

Sato, T., Ito, R., Hayakawa, Y., Noyori, R. (1978) *Tetrahedron Lett.* 1829.

Sato, T., Goto, Y., Fujisawa, T. (1982) *Tetrahedron Lett.* 4111.

Sato, T., Okazaki, H., Otera, J., Nozaki, H. (1988) *J. Am. Chem. Soc.* **110**: 5209.

Saulnier, M.G., Gribble, G.W. (1982) *J. Org. Chem.* **47**: 757.

Schiehser, G.A., White, J.D. (1980) *J. Org. Chem.* **45**: 1864.

Schill, G. (1967) *Chem. Ber.* **100**: 2021.

Schill, G., Logemann, E., Vetter, W. (1972) *Angew. Chem. Intern. Ed. Engl.* **11**: 1089.

Schill, G., Schweickert, N., Fritz, H., Vetter, W. (1983) *Angew. Chem. Intern. Ed. Engl.* **22**: 889.

Schlessinger, R.H., Wood, J.L., Poss, A.J., Nugent, R.A., Parsons, W.H. (1983) *J. Org. Chem.* **48**: 1147.

Schleyer, P.v.R., Donaldson, M.M. (1960) *J. Am. Chem. Soc.* **82**: 4645.

Schmittenhenner, H.F., Weinreb, S.M. (1980) *J. Org. Chem.* **45**: 3373.

Schöllkopf, U. (1983) *Pure Appl. Chem.* **55**: 1799.

Scholz, G., Tochtermann, W. (1971) *Tetrahedron Lett.* **32**: 5535.

Schöpf, C., Lehmann, G., Arnold, W. (1937) *Angew. Chem.* **50**: 783.

Schore, N.E. (1988) *Chem. Rev.* **88**: 1081.

Schreiber, J., Leimgruber, W., Pesaro, M., Schudel, P., Threlfall, T., Eschenmoser, A. (1961) *Helv. Chim. Acta* **44**: 540.

Schreiber, J., Felix, D., Eschenmoser, A., Winter, M., Gautschi, F., Schulte-Elte, K.H., Sundt, E., Ohloff, G., Kalvoda, J., Kaufmann, H., Wieland, P., Anner, G. (1967) *Helv. Chim. Acta* **50**: 2101.

Schreiber, S.L., Sommer, T.J. (1983) *Tetrahedron Lett.* **24**: 4781.

Schreiber, S.L., Hoveyda, A.H. (1984a) *J. Am. Chem. Soc.* **106**: 7200.

Schreiber, S.L., Santini, C. (1984b) *J. Am. Chem. Soc.* **106**: 4038.

Schreiber, S.L., Satake, K. (1984c) *J. Am. Chem. Soc.* **106**: 4186.

Schreiber, S.L., Wang, Z. (1985) *J. Am. Chem. Soc.* **107**: 5303.

Schreiber, S.L. (1987a) *Chem. Scripta* **27**: 563.

Schreiber, S.L., Goulet, M.T. (1987b) *Tetrahedron Lett.* **28**: 1043.

Schreiber, S.L., Goulet, M.T., Schulte, G. (1987c) *J. Am. Chem. Soc.* **109**: 4718.

Schreiber, S.L., Meyers, H.V. (1988) *J. Am. Chem. Soc.* **110**: 5198.

Schregenberger, C., Seebach, D. (1984) *Tetrahedron Lett.* **25**: 5881.

Schultz, A.G., Dittami, J.P. (1984) *J. Org. Chem.* **49**: 2615.

Schultz, A.G., Puig, S. (1985) *J. Org. Chem.* **50**: 915.

Schultz, A.G., Macielag, M., Sundararaman, P., Tavaras, A.G., Welch, M. (1988) *J. Am. Chem. Soc.* **100**: 7828.

Schwartz, M.A., Marni, I.S. (1975) *J. Am. Chem. Soc.* **97**: 1239.

Scott, A.I., McCapra, F., Buchanan, R.L., Day, A.C., Young, D.W. (1965) *Tetrahedron Lett.* **21**: 3605.

Scott, A.I. (1975) *Recent Adv. Phytochem.* **9**: 189.

Scott, J.W., Borer, R., Saucy, G. (1972a) *J. Org. Chem.* **37**: 1659.

Scott, J.W., Saucy, G. (1972b) *J. Org. Chem.* **37**: 1652.

Scott, L.T., Hashemi, M.M., Meyer, D.T., Warren, H.B. (1991) *J. Am. Chem. Soc.* **113**: 7082.

Scott, W.L., Evans, D.A. (1972) *J. Am. Chem. Soc.* **94**: 4779.

Seebach, D., Jones, N.R., Corey, E.J. (1968) *J. Org. Chem.* **33**: 300.

Seebach, D., Enders, D. (1975) *Angew. Chem. Intern. Ed. Engl.* **14**: 15.

Seebach, D., Seuring, B., Kalinowski, H.-O., Lubosch, W., Renger, B. (1977) *Angew. Chem. Intern. Ed. Engl.* **89**: 270.

Seebach, D. (1979) *Angew. Chem. Intern. Ed. Engl.* **18**: 239.

Seebach, D., Hungerbühler, E. (1980) in Scheffold, R., ed. *Modern Synthetic Methods* **2**: 91.

Seebach, D., Beck, A.K., Mukhopadhyay, T., Thomas, E. (1982) *Helv. Chim. Acta* **65**: 1101.

Seebach, D., Knochel, P. (1984) *Helv. Chim. Acta* **67**: 261.

Seebach, D., Plattner, D.A., Beck, A.K., Wang, Y.M., Hunziker, D., Petter, W. (1992) *Helv. Chim. Acta* **75**: 2171.

Seki, K., Ohnuma, T., Oishi, T., Ban, Y. (1975) *Tetrahedron Lett.* 723.

Semmelhack, M.F. (1972) *Org. React.* **19**: 115.

Semmelhack, M.F., Chong, B.P., Stauffer, R.D., Rogerson, T.D., Chong, A., Jones, L.D. (1975) *J. Am. Chem. Soc.* **97**: 2507.

Semmelhack, M.F., Tomoda, S., Hurst, K.M. (1980) *J. Am. Chem. Soc.* **102**: 7567.

Semmelhack, M.F., Bozell, J.J., Keller, L., Sato, T., Spiess, E.J., Wulff, W., Zask, A. (1985a) *Tetrahedron* **41**: 5803.

Semmelhack, M.F., Keller, L., Sato, T., Spiess, E.J., Wulff, W. (1985b) *J. Org. Chem.* **50**: 5566.

Semmelhack, M.F., Jeong, N., Lee, G.R. (1990) *Tetrahedron Lett.* **31**: 609.

Seto, H., Fujimoto, Y., Tatsuno, T., Yoshioka, H. (1985) *Synth. Commun.* **15**: 1217.

Shanzer, A. (1980) *Tetrahedron Lett.* **21**: 221.

Sharpless, K.B., Verhoeven, T.R. (1979) *Aldrichimica Acta* **12**: 63.

Shave, T.T., Meyers, A.I. (1991) *J. Org. Chem.* **56**: 2751.

Shea, K.J., Staab, A.J., Zandi, K.S. (1991) *Tetrahedron Lett.* **32**: 2715.

Shibasaki, M., Iseki, K., Ikegami, S. (1980) *Tetrahedron Lett.* 3587.

Shigeno, K., Sasai, H., Shibasaki, M. (1992) *Tetrahedron Lett.* **33**: 4937.

Shimizu, I., Nakagawa, H. (1992) *Tetrahedron Lett.* **33**: 4957.

Shimizu, M., Ukaji, Y., Tanizaki, J., Fujisawa, T. (1992) *Chem Lett.* 1349.

Shishido, K., Shitara, E., Komatsu, H., Hiroya, K., Fukumoto, K., Kametani, T. (1986) *J. Org. Chem.* **51**: 3007.

Shono, T. (1988) *Top. Curr. Chem.* **148**: 131.

Siddall, J.B., Marshall, J.P., Bowers, A., Cross, A.D., Edwards, J.A., Fried, J.H. (1966) *J. Am. Chem. Soc.* **88**: 379.

Sigrist, R., Rey, M., Dreiding, A.S. (1986) *Chem. Commun.* 944.

Simmons, H.E., Smith, R.D. (1958) *J. Am. Chem. Soc.* **80**: 5323.

Simmons, H.E., Cairns, T.L., Vladuchick, S.A., Hoiness, C.M., (1973) *Org. React.* **20**: 1.

Sisko, J., Balog, A., Curran, D.P. (1992) *J. Org. Chem.* **57**: 4341.

Skattebol, L., Stenstrom, Y., (1989) *Acta Chem. Scand.* **43**: 93.

Slocum, D.W., Jennings, C.A. (1976) *J. Org. Chem.* **41**: 3653.

Smith, A.B., Liverton, N.J., Hrib, N.J., Sivaramakrishnan, H., Wizenberg, K. (1985) *J. Org. Chem.* **50**: 3239.

Smith, A.B., Dorsey, B.D., Visnick, M., Maeda, T., Malamus, M.S. (1986) *J. Am. Chem. Soc.* **108**: 3110.

Smith, A.B., Keenan, T.P., Holcomb, R.C., Sprengeler, P.A., Guzman, M.C., Wood, J.L., Carroll, P.J., Hirschmann, R. (1992) *J. Am. Chem. Soc.* **114**: 10672.

Smith, A.L., Hwang, C.-K., Pitsinos, E., Scarlato, G.R., Nicolaou, K.C. (1992) *J. Am. Chem. Soc.* **114**: 3134.

Smith, R., Livinghouse, T. (1985) *Tetrahedron* **41**: 3559.

Snider, B.B., Phillips, G.B., Cordova, R. (1983) *J. Org. Chem.* **48**: 3003.

Snider, B.B., Buckman, B.O. (1992a) *J. Org. Chem.* **57**: 4883.

Snider, B.B., Shi, Z. (1992b) *J. Am. Chem. Soc.* 114: 1790.

Snieckus, V. (1990) *Chem. Rev.* **90**: 879.

Soai, K., Ookawa, A., Kaba, T., Ogawa, K. (1987) *J. Am. Chem. Soc.* **109**: 7111.

Solladie, G., Hutt, J. (1987) *Tetrahedron Lett.* **28**: 797.

Solladie, G., Maestro, M.C., Rubio, A., Pedregal, C., Carrene, M.C., Garcia Ruano, J.L. (1991) *J. Org. Chem.* **56**: 2317.

Solladie, G., Almario, A. (1992) *Tetrahedron Lett.* **33**: 2477.

Solladie, G., Lohse, O. (1993) *J. Org. Chem.* **58**: 4555.

Sonnet, P.E., Heath, R.R. (1980) *J. Org. Chem.* **45**: 3137.

South, M.S., Liebeskind, L.S. (1984) *J. Am. Chem. Soc.* **106**: 4181.

Speckamp, W.N., Hiemstra, H. (1985) *Tetrahedron* **41**: 4367.

Stack, J.G., Curran, D.P., Geib, S.V., Rebek, J., Ballester, P. (1992) *J. Am. Chem. Soc.* **114**: 7007.

Steglich, W., Huppertz, H.-T., Steffan, B. (1985) *Angew. Chem. Intern. Ed. Engl.* **24**: 711.

Steliou, K. Szczygielska-Nowosielska, A., Favre, A., Poupart, M.A., Hanessian, S. (1980) *J. Am. Chem. Soc.* **102**: 7578.

Stephenson, G.R., Finch, H., Owen, D.A., Swanson, S. (1993) *Tetrahedron* **49**: 5649.

Sternbach, D.D. (1989) in Lindberg, T. (ed.) *Strategies and Tactics in Organic Synthesis* Vol. 2, Chap. 12, Academic Press, San Diego, CA.

Sternberg, E.D., Vollhardt, K.P.C. (1984) *J. Org. Chem.* **49**: 1574.

Stetter, H., Thomas, H.G. (1968) *Chem. Ber.* **101**: 1115.

Stetter, H., Elfert, K. (1974) *Synthesis* 36.

Stetter, H. (1976) *Angew. Chem. Intern. Ed. Engl.* **15**: 639.

Stevens, R.V., Wentland, M.P., (1968) *J. Am. Chem. Soc.* **90**: 5580.

Stevens, R.V., Lee, A.W.M. (1979) *J. Am. Chem. Soc.* **101**: 7032.

Stevens, R.V., Pruitt, J. (1983) *Chem. Commun.* 1425.

Stevens, R.V., Beaulieu, N., Chan, W.H., Daniewski, A.R., Takeda, T., Waldner, A., Willard, P.G., Zutter, U. (1986) *J. Am. Chem. Soc.* **108**: 1039.

Still, W.C., Macdonald, T.L. (1976) *Tetrahedron Lett.* 2659.

Still, W.C., Schneider, M.J. (1977) *J. Am. Chem. Soc.* **99**: 948.

Still, W.C. (1979) *J. Am. Chem. Soc.* **101**: 2493.

Still, W.C., Darst, K.P. (1980a) *J. Am. Chem. Soc.* **102**: 7385.

Still, W.C., McDonald, J.H. (1980b) *Tetrahedron Lett.* 1031.

Still, W.C., Schneider, J.A., (1980c) *Tetrahedron Lett.* 1035.

Still, W.C., Schneider, J. (1980d) *J. Org. Chem.* **45**: 3375.

Still, W.C., Sreekumar, C. (1980e) *J. Am. Chem. Soc.* **102**: 1201.

Still, W.C., Tsai, M.Y. (1980f) *J. Am. Chem. Soc.* **102**: 3654.

Still, W.C., Shaw, K.R. (1981) *Tetrahedron Lett.* **22**: 3725.

Still, W.C., Murata, S., Revial, G., Yoshihara, Y. (1983) *J. Am. Chem. Soc.* **105**: 625.

Still, W.C. (1984a) in Bartmann, W., Trost, B.M. (eds.) *Selectivity—a Goal for Synthetic Efficiency*, p. 263, Verlag Chemie, Weinheim.

Still, W.C., MacPherson, L.J., Harada, T., Callahan, J.F., Rheingold, A.L. (1984b) *Tetrahedron* **40**: 2275.

Stille, J.R., Grubbs, R.H. (1986) *J. Am. Chem. Soc.* **108**: 855.

Stork, G., van Tamelen, E.E., Friedman, L.J., Burgstahler, A.W. (1953) *J. Am. Chem. Soc.* **75**: 384.

Stork, G., Darling, S.D., Harrison, I.T., Wharton, P.S. (1962) *J. Am. Chem. Soc.* **84**: 2018.

Stork, G., Dolfini, J.E., (1963) *J. Am. Chem. Soc.* **85**: 2872.

Stork, G., Borch, R. (1964a) *J. Am. Chem. Soc.* **86**: 935.

Stork, G., Tomasz, M. (1964b) *J. Am. Chem. Soc.* **86**: 471.

Stork, G., McMurry, J.E. (1967) *J. Am. Chem. Soc.* **89**: 5464.

Stork, G. (1968) *Pure Appl. Chem.* **17**: 383.

Stork, G., Stotter, P.L. (1969) *J. Am. Chem. Soc.* **91**: 7780.

Stork, G., Nelson, G.L., Rouessac, F., Gringore, O. (1971) *J. Am. Chem. Soc.* **93**: 3091.

Stork, G., Guthikonda, R.M. (1972a) *J. Am. Chem. Soc.* **94**: 5109.

Stork, G., Tabak, J.M., Blount, J.F. (1972b) *J. Am. Chem. Soc.* **94**: 4735.

Stork, G., Isobe, M. (1975) *J. Am. Chem. Soc.* **97**: 6260.

Stork, G., Macdonald, T.L. (1976) *J. Am. Chem. Soc.* **98**: 2370.

Stork, G., Ozorio, A.A., Leong, A.Y.W. (1978) *Tetrahedron Lett.* 5175.

Stork, G., Clark, G., Shiner, C.S. (1981) *J. Am. Chem. Soc.* **103**: 4948.

Stork, G., Paterson, I., Lee, F.K.C. (1982) *J. Am. Chem. Soc.* **104**: 4686.

Stork, G., Kahne, D.E. (1983) *J. Am. Chem. Soc.* **105**: 1072.

Stork, G., Kahn, M. (1985) *J. Am. Chem. Soc.* **107**: 500.

Stork, G., Saccomano, N.A. (1986a) *Nouv. J. Chim.* **10**: 677.

Stork, G., Sher, P.M., Chen, H.L. (1986b) *J. Am. Chem. Soc.* **108**: 6384

Stork, G., Rychnovsky, S.D. (1987) *Pure Appl. Chem.***59**: 345.

Stork, G. (1989a) *Pure App. Chem.* **61**: 439.

Stork, G., Mah, R. (1989b) *Tetrahedron Lett.* **30**: 3609.

Stork, G., Chan, T.Y., Breault, G.A. (1992a) *J. Am. Chem. Soc.* **114**: 7578.

Stork, G., Kim., G. (1992b) *J. Am. Chem. Soc.* **114**: 1087.

Stotter, P.L., Hornish, R.E. (1973) *J. Am. Chem. Soc.* **95**: 4444.

Streitwieser, A., Ewing, S.P. (1975) *J. Am. Chem. Soc.* **97**: 190.

Sugimura, T., Koguro, K., Tai, A (1993) *Tetrahedron Lett.* **34**: 509.

Sundberg, R.J., Russell, H.R. (1973) *J. Org. Chem.* **38**: 3324.

Sundberg, R.J., Parton, R.L. (1976) *J. Org. Chem.* **41**: 163.

Suzuki, M., Kimura, Y., Terashima, S. (1986) *Bull. Chem. Soc. Jpn.* **59**: 3559.

Suzuki, T., Sato, E., Unno, K., Kametani, T. (1986) *Chem. Pharm. Bull.* **34**: 3135.

Swiss, K.A., Liotta, D.C., Maryanoff, C.A. (1990) *J. Am. Chem. Soc.* **112**: 9393.

Swiss, K.A., Maryanoff, C.A., Liotta, D.C., (1992) *Synthesis* 127.

Takahashi, H., Yoshioka, M., Ohno, M., Kobayashi, S. (1992) *Tetrahedron Lett.* **33**: 2575.

Takahashi, M., Moritani, Y., Ogiku, T., Ohmizu, H., Kondo, K., Iwasaki, T. (1992) *Tetrahedron Lett.* **33**: 5103.

Takahashi, O., Mikami, K., Nakai, T. (1987) *Chem. Lett.* 69.

Takahashi, T., Ikeda, H., Tsuji, J. (1981a) *Tetrahedron Lett.* **22**: 1359.

Takahashi, T., Nagashima, T., Tsuji, J. (1981b) *Tetrahedron Lett.* **22**: 1359.

Takahashi, T., Nemoto, H., Kanda, Y., Tsuji, J., Fukazawa, Y., Okajima, T. Fujise, Y. (1987) *Tetrahedron* **43**: 5499.

Takahashi, T., Shimizu, K., Doi, T., Tsuji, J., Fukazawa, T. (1988) *J. Am. Chem. Soc.* **110**: 2674.

Takano, S., Tanigawa, K., Ogasawara, K. (1976) *Chem. Commun.* 189.

Takano, S., Chiba, K., Yonaga, M., Ogasawara, K. (1980a) *Chem. Commun.* 616.

Takano, S., Yonaga, M., Chiba, K., Ogasawara, K. (1980b) *Tetrahedron Lett.* 3697.

Takano, S., Ogawa, N., Ogasawara, K. (1981a) *Heterocycles* **16**: 915.

Takano, S., Tamura, N., Ogasawara, K. (1981b) *Chem. Commun.* 1155.

Takano, S., Yonaga, M., Ogasawara, K. (1981c) *Chem. Commun.* 1153.

Takano, S., Hatakeyama, S., Takahashi, Y., Ogasawara, K. (1982a) *Heterocycles* **17**: 263.

Takano, S., Uchida, W., Hatakeyama, S., Ogasawara, K. (1982b) *Chem. Lett.* 733.

Takano, S., Morimoto, M., Ogasawara, K. (1984) *Chem. Commun.* 82.

Takano, S., Yonaga, M., Morimoto, M., Ogasawara, K. (1985) *J. Am. Chem. Soc. Perkin Trans. I* 305.

Takano, S., Inomata, K., Kurotake, A., Ohkawa, T., Ogasawara, K. (1987a) *Chem. Commun.* 1720.

Takano, S., Kurotake, A., Ogasawara, K. (1987b) *Tetrahedron Lett.* **28**: 3991.

Takano, S., Iwabuchi, Y., Ogasawara, K. (1988a) *Chem. Commun.* 1204.

Takano, S., Shimazaki, Y., Takahashi, M., Ogasawara, K. (1988b) *Chem. Commun.* 1004.

Takano, S., Shimazaki, Y., Iwabuchi, Y., Ogasawara, K. (1980) *Tetrahedron Lett.* **31**: 3619.

Takano, S., Moriya, M., Ogasawara, K. (1992a) *Tetrahedron Lett.* **33**: 329; (1992b) *Tetrahedron Lett.* **33**: 1909.

Takaya, H., Hayakawa, Y., Makino, S., Noyori, R. (1978) *J. Am. Chem. Soc.* 100: 1778.

Takeshita, H., Iwabuchi, H., Kouno, I., Iino, M., Nomura, D. (1979) *Chem. Lett.* 649.

Tamai, Y., Akiyama, M., Okamura, A., Miyano, S. (1992) *Chem. Commun.* 687.

Tamao, K., Ishida, N., Kumada, M. (1983) *J. Org. Chem.* **48**: 2120.

Tamao, K., Tanaka, T., Nakajima, T., Sumiya, R., Arai, H., Ito, Y. (1986) *Tetrahedron Lett.* **27**: 3377.

Tamao, K., Maeda, K., Tanaka, T., Ito, Y. (1988a) *Tetrahedron Lett.* **29**: 6955.

Tamao, K., Nakagawa, Y., Arai, H., Higuchi, N., Ito, Y. (1988b) *J. Am. Chem. Soc.* **110**: 3712.

Tamao, K., Nakagawa, Y., Ito. (1990) *J. Org. Chem.* **55**: 3438.

Tamao, K., Kobayashi, K., Ito, Y. (1992) *Synlett* 539.

Tamura, R., Watabe, K. One, N., Yamamoto, Y. (1992) *J. Org. Chem.* **57**: 4895.

Tamura, O., Yamaguchi, T., Noe, K., Sakamoro, M. (1993) *Tetrahedron Lett.* **34**: 4009.

Tanabe, M., Crowe, D.F., Dehn, R.L., Detre, G. *Tetrahedron Lett.* 3739.

Tanaka, H., Torii, S. (1975) *J. Org. Chem.* **40**: 462.

Tanaka, M., Tomioka, K. Koga, K. (1985) *Tetrahedron Lett.* **26**: 3035.

Tanaka, M., Suemune, H., Sakai, K. (1988) *Tetrahedron Lett.* **29**: 1733.

Tanaka, M., Sakai, K. (1991) *Tetrahedron Lett.* **32**: 5581.

Tanaka, T., Toru, T., Okamura, N., Hazato, A., Sugiura, S., Manabe, K., Kurozumi, S. (1983) *Tetrahedron Lett.* **24**: 4103.

Tanner, D., Birgersson, C. (1991) *Tetrahedron Lett.* **32**: 2533.

Taub, D., Zelawski, Wendler, N.L. (1970) *Tetrahedron Lett.* 3667.

Taylor, S.L., Lee, D.Y., Martin, J.C. (1983) *J. Org. Chem.* **48**: 4156.

Terashima, S., Yamada, S. (1977) *Tetrahedron Lett.* 1001.

Terashima, S., Jew, S.-s., Koga, K. (1978) *Tetrahedron Lett.* 4937.

Ternansky, R.J., Balogh, D.W., Paquette, L.A., (1982) *J. Am. Chem. Soc.* **104**: 4503.

Thaning, M., Wistrand, L.-G. (1990) *J. Org. Chem.* **55**: 1406.

Thomas, A.F. (1969) *J. Am. Chem. Soc.* **91**: 3281.

Thompson, H.W., McPherson, E. (1977) *J. Org. Chem.* **42**: 3350.

Tiecco, M., Testaferri, L., Tingoli, M., Chianelli, D., Wenkert, E. (1982) *Tetrahedron Lett.* **23**: 4629.

Tietze, L.F., Kiedrowski, G., Berger, B. (1982) *Synthesis* 683.

Tietze, L.F., Brand, S., Pfeiffer, T., Antel, J., Harms, K., Sheldrick, G.M. (1987) *J. Am. Chem. Soc.* **109**: 921.

Tobe, Y., Yamashita, S., Kakiuchi, K., Odaira, Y. (1984) *Chem. Commun.* 1259.

Tobe, Y., Kishida, T., Yamashita, T., Kakiuchi, K., Odaira, Y. (1985a) *Chem. Lett.* 1437.

Tobe, Y., Yamashita, T., Kakiuchi, K., Odaira, Y. (1985b) *Chem. Commun.* 898.

Tokoroyama, T., Matsuo, K., Kanazawa, R., Kotsuki, H., Kubota, T. (1974) *Tetrahedron Lett.* 3093.

Tolbert, L.M., Ali, M.M. (1982) *J. Am. Chem. Soc.* **104**: 1742.

Tomioka, K., Shimizu, K., Yamada, S., Koga, K. (1977) *Heterocycles* **6**: 1752.

Tomioka, K., Koga, (1979) *Tetrahedron Lett.* 3315.

Tomioka, K., Ishiguro, T., Koga, K. (1980) *Tetrahedron Lett.* 2973.

Tomioka, K., Kawasaki, H., Iitaka, Y., Koga, K. (1985) *Tetrahedron Lett.* **26**: 903.

Tomioka, K., Sugimori, M., Koga, K. (1987) *Chem. Pharm. Bull.* **35**: 906.

Torii, S., Tanaka, H., Tomotaki, Y. (1977) *Bull. Chem. Soc. Jpn.* **50**: 537.

Torii, S., Inokuchi, T., Oi, R. (1983) *J. Org. Chem.* **48**: 1944.

Toshima, K., Yoshida, T., Mukaiyama, S., Tatsuta, K. (1991a) *Tetrahedron Lett.* **32**: 4139; (1991b) *Carbohydr. Res.* **222**: 173.

Toth, J.E., Hamann, P.R., Fuchs, P.L. (1988) *J. Org. Chem.* **53**: 4694.

Tou, J.S., Reusch, W. (1980) *J. Org. Chem.* **45**: 5012.

Trimble, L.A., Vederas, J.C. (1986) *J. Am. Chem. Soc.* **108**: 6397.

Trost, B.M., Bogdanowicz, M.J., (1973) *J. Am. Chem. Soc.* **95**: 5321.

Trost, B.M., Keeley, D.E. (1975a) *J. Org. Chem.* **40**: 2013.

Trost, B.M., Preckel, M., Leichter, L.M. (1975b) *J. Am. Chem. Soc.* **97**: 2224.

Trost, B.M., Keeley, D.E., Arndt, H.C., Bogdanowicz, M.J. (1977) *J. Am. Chem. Soc.* **99**: 3088.

Trost, B.M., Timko, J.M., Stanton, J.L. (1978) *Chem. Commun.* 436.

Trost, B.M., Bernstein, P.R., Fünfschilling, P.C. (1979a) *J. Am. Chem. Soc.* **101**: 4378.

Trost, B.M., Shuey, C.D., DiNinno, F. (1979b) *J. Am. Chem. Soc.* **101**: 1284.

Trost, B.M., O'Krongly, D., Belletire, J.L. (1980a) *J. Am. Chem. Soc.* **102**: 7595.

Trost, B.M., Verhoeven, T.R. (1980b) *J. Am. Chem. Soc.* **102**: 4730.

Trost, B.M., Verhoeven, T.R. (1980c) *J. Am. Chem. Soc.* **102**: 4743.

Trost, B.M., Renaut, P. (1982) *J. Am. Chem. Soc.* **104**: 6668.

Trost, B.M. (1986) *Angew. Chem. Intern. Ed. Engl.* **25**: 1.

Trost, B.M., Kuo, G.H., Benneche, T. (1988a) *J. Am. Chem. Soc.* **110**: 621.

Trost, B.M., Sudhakar, A.R. (1988b) *J. Am. Chem. Soc.* **110**: 7933.
Trost, B.M., Lee, D.C. (1989a) *J. Org. Chem.* **54**: 2271.
Trost, B.M., Holcomb, R.C. (1989b) *Tetrahedron Lett.* **30**: 7157.
Trost, B.M. (1990a) *Acc. Chem. Res.* **23**: 34.
Trost, B.M., Greenspan, P.D., Yang, B.V., Saulnier, M.G. (1990b) *J. Am. Chem. Soc.* **112**: 9022.
Trost, B.M., Urabe, H. (1990c) *J. Org. Chem.* **55**: 3982.
Trost, B.M., Tasker, A.S., Rüther, G., Brandes, A. (1991) *J. Am. Chem. Soc.* **113**: 670.
Trost, B.M. (1992a) *Pure Appl. Chem.* **64**: 315.
Trost, B.M., Kulawiec, R.J., (1992b) *J. Am. Chem. Soc.* **114**: 5579.
Tsuchihashi, G., Mitamura, S., Ogura, K. (1976) *Tetrahedron Lett.* 855.
Tsuji, J., Shimizu, I., Suzuki, H., Naito, Y. (1979) *J. Am. Chem. Soc.* **101**: 5070.
Tufariello, J.J., Meckler, H., Pushpananda, K., Senaratne, K.P.A. (1985) *Tetrahedron* **41**: 3447.
Uemura, M., Kobayashi, T., Isobe, K., Minami, T., Hayashi, Y. (1986) *J. Am. Chem. Soc.* **51**: 2859.
Uemura, M., Minami, T., Hayashi, Y. (1987) *J. Am. Chem. Soc.* **109**: 5277.
Uemura, M., Minami, T., Hayashi, Y. (1988) *Tetrahedron Lett.* **29**: 6271.
Uemura, M., Nishimura, H., Hayashi, Y. (1989) *J. Organomet. Chem.*, **376**: C3.
Ukaki, Y., Watai, T., Sumi, T., Fujisawa, T. (1991) *Chem. Lett.* 1555.
Utimoto, K., Kato, S., Tanaka, M., Hoshino, Y., Fujikura, S., Nozaki, H. (1982) *Heterocycles* **18**: 149.
Uyehara, T., Chiba, N., Suzuki, I., Yamamoto, Y. (1991) *Tetrahedron Lett.* **32**: 4371.
van Draanen, N.A., Arseniyadis, S., Crimmins, M.T., Heathcock, C.H. (1991) *J. Org. Chem.* **56**: 2499.
van Hijfte, L., Vandewalle, M. (1984) *Tetrahedron* **40**: 4371.
van Tamelen, E.E., Shamma, M., Burgstahler, A.W., Wolinsky, J., Tamm, R., Aldrich, P.E. (1958) *J. Am. Chem. Soc.* **80**: 5006.
van Tamelen, E.E., Foltz, R.L. (1960) *J. Am. Chem. Soc.* **82**: 2400.
van Tamelen, E.E. (1961) *Fortschr, Chem. Org. Naturstoffe* **19**: 242.
Vedejs, E. (1984a) *Acc. Chem. Res.* **17**: 358.
Vedejs, E., Reid, J.G., (1984b) *J. Am. Chem. Soc.* **106**: 4617.
Vedejs, E., Buchanan, R.A., Conrad, P., Meier, G.P., Mullins, M.J., Watanabe, Y. (1987) *J. Am. Chem. Soc.* **109**: 5878.
Vedejs, E., Rodgers, J.D., Wittenberger, S.J., (1988) *J. Am. Chem. Soc.* **110**: 4822.
Vedejs, E., Haight, A.R., Moss, W.O. (1992) *J. Am. Chem. Soc.* **114**: 6556.
Velluz, L., Valls, J., Mathieu, J. (1967) *Angew. Chem. Intern. Ed. Engl.* **6**: 778.
Vigneron, J.P., Kagan, H., Horeau, A. (1968) *Tetrahedron Lett.* 5681.
Vlattas, I., DellaVecchia, L., Lee, A.O., (1976) *J. Am. Chem. Soc.* **98**: 2008.
Vogel, E., Böll, W.A. (1964a) *Angew. Chem. Intern. Ed. Engl.* **3**: 642.
Vogel, E., Roth, H.D. (1964b) *Angew. Chem. Intern. Ed. Engl.* **3**: 228.
Vogel, E., Deger, H.M., Sombroek, J., Palm, J., Wagner, A., Lex, J. (1980) *Angew. Chem. Intern. Ed. Engl.* **19**: 41.

Vogel, E., Wieland, H., Schmalstieg, L., Lex, J. (1984) *Angew. Chem. Intern. Ed. Engl.* **23**: 717.

Vögtle, F., Rossa, L. (1979) *Angew. Chem. Intern. Ed. Engl.* **18**: 515.

Volkmann, R.A., Andrews, G.C., Johnson, W.S. (1975) *J. Am. Chem. Soc.* **97**: 4777.

Vollhardt, K.P.C. (1980) *Ann. N.Y. Acad. Sci.* **333**: 241.

Vollhardt, K.P.C. (1984) *Angew. Chem. Intern. Ed. Engl.* **23**: 539.

Vorländer, E., Erig, J. (1897) *Liebigs Ann. Chem.* **294**: 314.

Wagemann, W., Iyoda, M., Deger, H.M.. Sombroek, J., Vogel, E. (1987) *Angew. Chem. Intern. Ed. Engl.* **17**: 956.

Walborsky, H.M., Barash, L., Davis, T.C., (1963) *Tetrahedron Lett.* 2333.

Walborsky, H.M., Ronman, P. (1978) *J. Org. Chem.* **43**: 731.

Waldmann, H. (1988) *J. Org. Chem.* **53**: 6133.

Walter, C.J., Anderson, H.L., Sanders, J.K.M. (1993) *Chem. Commun.* 458.

Wang, C.-L.J., Calabrese, J.C. (1991) *J. Org. Chem.* **56**: 4341.

Wasserman, H.H., Berger, G.D. (1983a) *Tetrahedron* **39**: 2459.

Wasserman, H.H., Robinson, R.P., Carter, C.G. (1983b) *J. Am. Chem. Soc.* **105**: 1697.

Watanabe, M., Snieckus, V. (1980) *J. Am. Chem. Soc.* **102**: 1457.

Watt, D.S. (1976) *J. Am. Chem. Soc.* **98**: 271.

Webb, I.D., Borcherdt, G.T. (1951) *J. Am. Chem. Soc.* **73**: 2654.

Wehle, D., Fitjer, L. (1987) *Angew. Chem. Intern. Ed. Engl.* **26**: 130.

Weidmann, B., Seebach, D. (1983) *Angew. Chem. Intern. Ed. Engl.* **22**: 31.

Weinreb, S.M., Garigipati, R.S. (1989) *Pure Chem.* **61**: 435.

Welch, S.C., Rao, A.S.C.P., Gibbs, C.G., Wong, R.Y. (1980) *J. Am. Chem. Soc.* **45**: 4077.

Wender, P.A., Hubbs, J.C. (1980) *J. Org. Chem.* **45**: 365.

Wender, P.A., Howbert, J.J. (1981) *J. Am. Chem. Soc.* **103**: 688.

Wender, P.A., Holt, D.A., Sieburth, S.M. (1983) *J. Am. Chem. Soc.* **105**: 3348.

Wender, P.A., Ternansky, R.J. (1985) *Tetrahedron Lett.* **26**: 2625.

Wender, P.A., Schaus, J.M., White, A.W. (1987) *Heterocycles* **25**: 263.

Wenkert, E., Mueller, R.A., Reardon, E.J., Sathe, S.S., Scharf, D.J., Tosi, G. (1970) *J. Am. Chem. Soc.* **92**: 7428.

Wenkert, E., Berges, D.A., Golob, N.F. (1978a) *J. Am. Chem. Soc.* **100**: 1263.

Wenkert, E., Buckwalter, B.L., Craveiro, A.A., Sanchez, E.L., Sathe, S.S. (1978b) *J. Am. Chem. Soc.* **100**: 1267.

Wenkert, E., Leftin, M., Michelotti, E.L. (1984a) *Chem. Commun.* 617.

Wenkert, E., Michelotti, E.L., Swindell, C.S., Tingoli, M. (1984b) *J. Org. Chem.* **49**: 4894.

Wenkert, E., Kim, H.-S. (1989) in Atta-ur-Rahman (ed.), *Studies in Natural Product Chemistry*, Vol. 3B, Elsevier, Amsterdam, p. 287.

Wharton, P.S., Sundin, C.E., Johnson, D.W., Kluender, H.C. (1972) *J. Org. Chem.* **37**: 34.

White, J.D., Fukuyama, Y. (1979) *J. Am. Chem. Soc.* **101**: 226.

White, J.D., Somers, T.C., (1987) *J. Am. Chem. Soc.* **109**: 4424.

White, J.D., Butlin, R.J., Hahn, H.-G., Johnson, A.T. (1990) *J. Am. Chem. Soc.* **112**: 8595.

White, J.D., Dillon, M.P., Butlin, R.J. (1992) *J. Am. Chem. Soc.* **114**: 9673.

Whitesell, J.K., Felman, S.W. (1977a) *J. Org. Chem.* **42**: 1663.

Whitesell, J.K., Whitesell, M.A. (1977b) *J. Org. Chem.* **42**: 377.

Whitesell, J.K., Bhattacharya, A., Henke, K. (1982) *Chem. Commun.* 988.

Whitesell, J.K., Minton, M.A. (1987) *J. Am. Chem. Soc.* **109**: 6403.

Wiesner, K., Musil, V., Wiesner, K.J. (1968) *Tetrahedron Lett.* 5643.

Wiesner, K., Poon, L., Jirkovsky, I., Fishman, M. (1969) *Can. J. Chem.* **47**: 433.

Wilke, G. (1978) *Pure Appl. Chem.* **50**: 677.

Williams, D.R., Osterhout, M.H. (1992) *J. Am. Chem. Soc.* **114**: 8750.

Williams, R.M., Glinka, T., Kwast, E. (1988) *J. Am. Chem. Soc.* **110**: 5927.

Williams, V.Z., Schleyer, P.v.R., Gleicher, G.J., Rodewald, C.B. (1966) *J. Am. Chem. Soc.* **88**: 3862.

Willoughby, C.A., Buchwald, S.L. (1992) *J. Am. Chem. Soc.* **114**: 7562.

Willstätter, R., Bommer, M. (1920) *Annalen* **422**: 15.

Wilson, J.W. (1980) *J. Organomet. Chem.* **186**: 297.

Wilson, S.R., Mao, D.T. (1978a) *J. Am. Chem. Soc.* **100**: 6289.

Wilson, S.R., Sawicki, R.A. (1978b) *Tetrahedron Lett.* 2969.

Winkler, J.D., Hey, J.P., Williard, P.G. (1986) *J. Am. Chem. Soc.* **108**: 6425.

Winkler, J.D., Henegar, K.E., Williard, P.G. (1987) *J. Am. Chem. Soc.* **109**: 2850.

Winkler, J.D., Hey, J.P., Williard, P.G. (1988) *Tetrahedron Lett.* **29**: 4691.

Winkler, J.D., Hershberger, P.M. (1989) *J. Am. Chem. Soc.* **111**: 4852.

Winkler, J.D., Scott, R.D., Williard, P.G. (1990) *J. Am. Chem. Soc.* **112**: 8971.

Winstein, S., Sonnenberg, J. (1961) *J. Am. Chem. Soc.* **83**: 3235.

Winterfeldt, E. (1993) *Chem. Rev.* **93**: 827.

Wiseman, J.R., Lee, S.Y. (1986) *J. Org. Chem.* **51**: 2485.

Wittig, G., Otlen, J. (1963) *Tetrahedron Lett.* 601.

Wittman, M.D., Kallmerten, J. (1988) *J. Org. Chem.* **53**: 4631.

Wolf, J.F., Murray, T.P. (1970) *Chem. Commun.* 336.

Wong, K.-T., Luh, T.-Y. (1992)*J. Am. Chem. Soc.* **114**: 7308.

Wood, J.L., Porco, J.A., Taunton, J., Lee, A.Y., Clardy, J., Schreiber, S.L. (1992) *J. Am. Chem. Soc.* **114**: 5898.

Woodward, R.B., Doering, W.v.E. (1945) *J. Am. Chem. Soc.* **67**: 860.

Woodward, R.B., Sondheimer, F., Taub, D., Heusler, K., McLamore, W.M. (1952) *J. Am. Chem. Soc.* **74**: 4223.

Woodward, R.B., Bader, F.E., Bickel, H., Frey, A.J., Kierstead, R.W. (1958) *Tetrahedron Lett.* **2**: 1.

Woodward, R.B. (1960) *Angew. Chem.* **72**: 651.

Woodward, R.B. (1961) *Pure Appl. Chem.* **2**: 383.

Woodward, R.B., Cava, M.P., Ollis, W.D., Hunger, A., Daeniker, H.U., Schenker, K. (1963) *Tetrahedron* **19**: 247.

Woodward, R.B. (1963–4) *The Harvey Lectures* **59**: 31.

Woodward, R.B., Fukunaga, T., Kelly, R.C. (1964) *J. Am. Chem. Soc.* **86**: 3162.

Woodward, R.B. (1966) *Science* **153**: 487.

Woodward, R.B. (1968) *Pure Appl. Chem.* **17**: 519.

Woodward, R.B. (1971) *Pure Appl. Chem.* **25**: 283.

Woodward, R.B. (1973a) *Pure Appl. Chem.* **33**: 145.

Woodward, R.B., Gosteli, J., Ernest, I., Friary, R.J., Nestler, G., Raman, H., Sitrin, R., Suter, C., Whitesell, J.K. (1973b) *J. Am. Chem. Soc.* **95**: 6853.

Woodward, R.B., et al. (1981) *J. Am. Chem. Soc.* **103**: 3210, 3213, 3215.

Wright, J., Drtina, G.J., Roberts, R.A., Paquette, L.A. (1988) *J. Am. Chem. Soc.* **110**: 5806.

Wrobel, J., Takahashi, K., Honkan, V., Lannoye, G., Cook, J.M., Bertz, S.H. (1983) *J. Org. Chem.* **48**: 139.

Wulff, W.D., Tang, P.C. (1984) *J. Am. Chem. Soc.* **106**: 434.

Wulff, W.D. (1989) in Liebeskind, L.S. (ed.) *Advances in Metal-Organic Chemistry* **1**: 209.

Wynberg, H., Logothetis, A. (1956) *J. Am. Chem. Soc.* **78**: 1958.

Wynberg, H., Staring, E.G.J. (1982) *J. Am. Chem. Soc.* **104**: 166; (1985) *J. Org. Chem.* **50**: 1977.

Wynberg, H. (1986) *Top. Stereochem.* **16**: 87.

Yamada, K., Kyotani, Y., Manabe, S., Suzuki, M. (1979) *Tetrahedron* **35**: 293.

Yamada, S. (1992) *J. Org. Chem.* **57**: 1591.

Yamada, T., Manabe, Y., Miyazawa, T., Kuwata, S., Sera, A. (1984) *Chem. Commun.* 1500.

Yamaguchi, M., Mukaiyama, T. (1982) *Chem. Lett.* 237.

Yamaguchi, M. (1989) *Pure App. Chem.* **61**: 413.

Yamakawa, K., Sakaguchi, R., Nakamura, T., Watanabe, K. (1976) *Chem. Lett.* 991.

Yamamoto, H., Maruoka, K. (1981a) *J. Am. Chem. Soc.* **103**: 4186; (1981b) *J. Am. Chem. Soc.* **103**: 6133.

Yamamoto, H., Mori, A. (1987) *J. Synth. Org. Chem. Jpn.* **45**: 944.

Yamamoto, Y., Yatagai, H., Maruyama, K. (1980) *J. Org. Chem.* **45**: 195.

Yamamoto, Y., Nishii, S., Yamada, J. (1986) *J. Am. Chem. Soc.* **108**: 7116.

Yamamoto, Y., Furuta, T. (1990) *J. Org. Chem.* **55**: 3971.

Yamamoto, Y., Yamada, J., Kadota, (1991) *Tetrahedron Lett.* **32**: 7069.

Yan, T.-H., Chu, V.-V., Lin, T.-C., Tseng, W.-H., Cheng, T.-W. (1991) *Tetrahedron Lett.* **32**: 5563.

Yasuda, A., Tanaka, S., Yamamoto, H., Nozaki, H. (1979) *Bull. Chem. Soc. Jpn.* **52**: 1701.

Yasukouchi, T., Kanematsu, K. (1989) *Tetrahedron Lett.* **30**: 6559.

Yokoyama, Y., Kawashima, H., Kohno, M., Ogawa, Y., Uchida, S. (1991) *Tetrahedron Lett.* **32**: 1479.

Yoshida, T., Saito, S. (1982) *Bull. Chem. Soc. Jpn.* **55**: 3931.

Yoshida, J., Maekawa, T., Morita, Y., Isoe, S. (1992) *J. Org. Chem.* **57**: 1321.

Yoshikoshi, A., Miyashita, M. (1985) *Acc. Chem. Res.* **18**: 284.

Yoshino, T., Okamoto, S., Sato, F. (1991) *J. Org. Chem.* **56**: 3205.

Zhao, B., Snieckus, V. (1991) *Tetrahedron Lett.* **32**: 5277.

Ziegler, F.E., Kloek, J.A., Zoretic, P.A. (1969) *J. Am. Chem. Soc.* **91**: 2342.

Ziegler, F.E., Reid, G.R., Studt, W.L., Wender, P.A. (1977) *J. Org. Chem.* **42**: 1991.

Ziegler, F.E., Piwinski, J.J. (1982) *J. Am. Chem. Soc.* **104**: 7181.

Ziegler, F.E., Jaynes, B.H., Saindane, M.T. (1985) *Tetrahedron Lett.* **26**: 3307;(1987) *J. Am. Chem. Soc.* **109**: 8115.

Ziegler, K., Eberle, H., Ohlinger, H. (1933) *Liebigs Ann. Chem.* **504**: 94.

Zimmerman, H.E., Grunewald, G.L., Paufler, R.M., Sherwin, M.A. (1969) *J. Am. Chem. Soc.* **91**: 2330.

Zurflüh, R., Wall, E.N., Siddall, J.B., Edwards, J.A. (1968) *J. Am. Chem. Soc.* **90**: 6224.

Zurflüh, R., Dunham, L.L., Spain, V.L., Siddall, J.B. (1970) *J. Am. Chem. Soc.* **92**: 425.

INDEX